THE NEXT STEP

From Nazi Ideology and Scientific Authority to the New Biopolitics of Health Security

THE NEXT STEP

From Nazi Ideology and Scientific Authority to the New Biopolitics of Health Security

David Horsfall

The Next Step: From Nazi Ideology and Scientific Authority to the New Biopolitics of Health Security

ISBN: 9798996741502 (hardback)
ISBN: 9798996741526 (paperback)
ISBN: 9798996741519 (ebook)

Published by David Horsfall
Printed in the United States of America
First Edition

O Lord, the hope of Israel, LLC 2017-2026
https://www.olordthehopeofisrael.com

Acknowledgements

All glory and thanks to God the Most High and His beloved Son, the Lord Jesus Christ, Yeshua HaMashiach, the Jewish Messiah for the Jewish people. His guidance, inspiration, and wisdom have sustained me throughout this project. Their insight shaped both the content and structure of this report, providing the principal source of inspiration and direction. Deep gratitude is also owed to my wife, Gaylene Horsfall, for her unwavering support, counsel, and encouragement along the way.

Book One

BOOK ONE

PART I. PAST HISTORICAL DEVELOPMENTS AND REVIEW

We need a better knowledge of man,
Health through natural living,
Harmony between Blood, Soil and Cosmos, Life
reform as a national aim,
Knowledge and Life,
The Rule of the Living.

—Bramwell, A. C. (1985). Blood and soil, 176–77, quoting Demeter, 1939, 155. The poem was entitled, 'The Primacy of the Living.'

BOOK ONE

THE NEXT STEP: FROM NAZI IDEOLOGY AND SCIENTIFIC AUTHORITY TO THE NEW BIOPOLITICS OF HEALTH SECURITY

David Horsfall

Abstract

Was Covid-19 a global pilot program and dress rehearsal for a future contagion outbreak designed to enforce a strict segregation regime? The Nazi regime was founded on the belief that they were descendants of a superior Aryan master race—considered god-men on earth. Francis Bacon's New Atlantis is the bedrock for the themes within this paper. Their primary target was the Jewish population, but they also persecuted other ethnic groups, social misfits, non-conformists, the elderly, individuals with cognitive, congenital disorders, and chronic illnesses. This pseudoscientific worldview provided a pretext for segregation and led to numerous racial purity laws aimed at restoring the Aryan genetic status. These measures escalated into a eugenics and racial hygiene program to purge German society of perceived genetic threats. The Nazis extended their ideology by preventing sexual relations among "inferior" groups and by labelling Jews and others as contagious carriers of disease, segregating them into ghettos. This narrative of disease served as a justification for the creation of death camps, ostensibly to eliminate the "infection." This paper examines parallels between Nazi-era policies and the Covid-19 lockdowns, highlighting emerging patterns of control and segregation that suggest a new form of global totalitarian governance. Understanding this historical continuum is critical to unveiling a potentially sinister future.

Keywords: Anti-Semitism, Anthropology, Atlantis, Bacteriology, Bioethics, Biosecurity, Classical Greek Philosophy, COVID Lockdowns, Democracy, Enlightenment, Epidemics, Eugenics, Genocide, Global Health Governance, Governmentality, Holocaust, Huıman Rights, Infectious Diseases, Modern Physics, New Atlantis, Nazi Germany, Occultism, Quarantine Camps, Race Science, Racial Hygiene, Scientific Biopolitics, Segregation, Social Control, Social Darwinism, Surveillance, Technocracy, Vaccine Policy

Disclaimer Introduction

The present author wishes to emphasise that this paper is written with the sole intention of drawing critical lessons from history. The content herein in no way diminishes, undermines, or detracts from the horror of the Holocaust, nor does it seek to appropriate or relativise the suffering of its victims. Rather, this study seeks to understand the mechanisms and patterns by which ideologies of hatred and exclusion gain traction, so that we may remain vigilant against similar developments in our own time.

Gisela Bock (1983) for example, expressed that it was necessary for her to undertake a comparative historical analysis "*. . . to conceptualize the connection between racism* *and sexism . . . .*"[1] In that spirit, this paper seeks to trace the conditions under which societies have succumbed to similarly destructive ideologies. While any comparison to the Nazi regime may be met with discomfort or opposition, we have a duty to understand what happened in the past in order to recognise the emergence of comparable societal patterns and trends in the present and future. Appleby et al. (1994) and Bergan (1998) eloquently expressed that "*[w]hat historians do best is to make connections with the past in order to illuminate the problems of the present and the potential of the future.*"[2]

Gavin Langmuir (1990), renowned for his scholarship on medieval antisemitism, poignantly compared the defenselessness of medieval Jewish communities against their persecutors' delusions to the analogous dynamics underpinning Hitler's genocidal ideology. Langmuir's parallel is not intended to diminish the singular horror of the Holocaust, but rather to illuminate recurring mechanisms of hatred and violence throughout Jewish history.[3] Furthermore, as Katz (1993), a Holocaust survivor, reflected in his introduction to his book, "*Ordinary People and Extraordinary Evil,*" he asserted the sentiment that "*[w]e do have a duty to remember what happened when evil prevailed. Yet the greatest challenge is to develop effective knowledge, and this goes far beyond remembering and documenting what happened.*"[4]

More importantly, a cadre of recent publications and orators persist in denying that the Holocaust ever happened. Accordingly, in the words of Kressel (1996), it continues to be necessary to affirm the incontrovertible facts: millions of Jews were forcibly removed from their homes, segregated, stripped of their rights, and ultimately murdered in concentration camps.[5]

1 Bock, G., (1983). Racism and Sexism, 403–04.
2 Appleby, J. et al. (1994). Telling The Truth. 8, *10; Bergen, B. J. (1998). The Banality of Evil, 1.
3 Langmuir, G. I. (1990). History, Religion, and Antisemitism.
4 Katz, F. E. (1993). Ordinary People and Extraordinary Evil, 3.
5 Kressel, N. J. (1996). Mass Hate, 8.

Introduction

A SUBSTANTIAL BODY OF scholarship for the past century has focused extensively on the genocide aspects of the Holocaust. Other studies have examined related fields of Nazi political infrastructure, the electoral system dynamics, and the development of racist and genetic discourse. Likewise, the literature on Nazi Racial Hygiene policy, (the focus of this paper) are exhaustive just as much as they extensive. Scholars, Hannah Arendt, Naomi Baumslag, Omer Bartov, Wolfgang Bialas, Michael Burleigh, Nicholas Goodrick-Clarke, Henry Friedlander, Raul Hilberg, Eric Kurlander, George L. Mosse, Robert Proctor, Sheila Faith Weiss, Paul Weindling, Peter Weingart, Gretchen Schafft, et al., have written monumental works that move well beyond the foundational records, exposing the ideology underpinning medical crimes in the Nazi era.

As such, this study does not seek to replicate existing contributions. Rather, its first aim is to provide a chronological analysis of the history of science and medicine, from Plato and Aristotle through Francis Bacon, tracing the intellectual and historical developments that contributed to the conditions leading to the Holocaust. Its second aim is to extend this analysis into modernity, with particular attention to German history and the influence of Bacon's *New Atlantis*. The third aim is to provide a comparative analysis of the Nazi scientific framework and modern developments in human genetics and their trajectory. This includes an examination of the political and scientific structures of Nazi Germany, with particular emphasis on the construction of disease-based pretexts for quarantine and the resulting systems of segregation.

This research argues that emerging global genetic technologies presage a new dystopian system yet to be unveiled. This analysis will explore scholarship discourses from the field of genocide studies, the history of science and of medicine, the historical analysis of antisemitism. In particular, it traces how anthropology, biology, among others, informed and contributed to the rise of authoritarian medicine, the precursor to the Holocaust.

When presented with a significant virus outbreak, segregation is an anticipated and acceptable outcome in the eyes of many. However, history informs us that there is a correlation between a virus alleged to be an epidemic/pandemic, with the creation of the narrative to form a collaborative authoritarian rule as a result. This includes the extreme form of segregation, and the subsequent establishment of genocide camps. The Nazi government's intended goal for racial hygiene had a layered systemic strategy.[6] The bigger picture was to create the conditions—the framework with which to justify segregation—the lie that Jewish souls were diseases carriers or infectious pathogen

6 See Weindling, P. J. (2000). Epidemics and Genocide in Eastern Europe.

carriers.[7] Scientific racism has been a well-researched subject over the centuries, but there is another angle that requires further examination.

The relationship and continuity between European Medieval era anti-Semitism with the formation of Scientific racism, *The Protocols of the Elders of Zion* and their amalgamation with Social Darwinism. The ideology of social seclusion to prevent procreation with the genetically inferior leading to degenerate prodigy, was a soft approach. Darwinian evolutionary theory later appropriated and adapted by Ernst Hacekel within emerging Social Darwinist and eugenic ideologies. This represented the early development of the biologically mediocre elements of the less genetic fit would swamp, outbreed, and multiply faster than the socially elite strata of society.[8] Proctor (1988) notes that Social Darwinists who frequented in social circles with the genetically weak or racily contaminated, viewed this as an obstacle to the betterment of the race and society, framing it as "*race suicide.*"[9] Weiss's (1987a) research argues that natural selection provided tools to enable elimination of the "*biologically unfit*" from the population through racial hygiene programmes.[10] This study builds on this literature to examine broader intellectual and historical trajectories, and a contemporary analysis of biotechnological developments and their future implications.

7 Ibid.
8 Ibid.; Hutton, C., (2005). Race and the Third Reich, 10.
9 Goldstein, J. A., (2011). On Racism, 62; Proctor, R. N., (1988). Racial Hygiene, 14-15, 19–20; Seidelman, W. E., (1992). Racial Hygiene Review, 148.
10 Weiss, S. F., (1987a). The Race Hygiene Movement, 199.

CHAPTER 1. GERMAN ARYANISM AND SCIENTIFIC RACIAL POLICY DEVELOPMENT FOUNDATIONS

A. Genesis of Nazi Eugenics and Racial Hygiene (1450s–1800s)

i. Francis Bacon's New World of Atlantis and Aldous Huxley's Brave New World

FRANCIS BACON'S (1561–1626) sole work of fiction, New Atlantis, published posthumously in 1627, has long captivated literary scholars seeking to unravel its imagery. Bacon played a pivotal role in (i) establishing the empirical sciences de facto as a modality of "*learning*" rather than dogma, (ii) articulating the vision and goals of racial philosophy, and (iii) advancing the formal development of scientific racism, culminating in the classification of racial distinctions and their practical application.[1] Bierman (2020) captures the essence of New Atlantis, not merely as a paradigm, but forecasts "*. . . an image of science in operation and of the relationship between scientists and society . . . .*"[2]

Bacon's version loosely bears conceptual ties to Plato's History of Atlantis to which he had read.[3] While eugenics, selective breeding, and anthropological ideas trace back to Plato's *Republic*, Herodotus (c.484–420 BCE), and Hippocrates (c.469-399 BCE), the modern formulation of eugenics emerged in the 1800s.[4] Interestingly, despite Hippocrates' association with the ethical principles enshrined in the Hippocratic Oath,[5] Plato himself expressed fervent enthusiasm for eugenic ideals, advocating for state-controlled reproduction in the pursuit of societal perfection—

> [T]he offspring of inferior parents, and any deformed child born . . . they will conceal in some secret and obscure place as they ought. 'Yes,' he said, 'that must be done if the breed of the guardians is to be preserved in its purity.'[6]

In Plato's *Republic* it is argued that to breed and maintain noble humans, the best individuals should reproduce with each other at the prime of their lives "*. . . to improve*

---

1 Matthews, S. (2008). Theology 104; Agassi, J. (2012). The Very Idea, 11–21; Bierman, J. (2020). Science, 492.

2 Bierman, 492.

3 Ibid., 494; Ellis, R. (1998). Imagining Atlantis, 30–34.

4 Galton, D. J., (1998). Greek Theories; Ludmerer, K. M. & Lappé, M. (1978). Eugenics, 457; Hannaford, I. (1996). Race, 18, 25–30.

5 Cf. Bruns, F. (2014). Turning away, 228–30, on how the Hippocratic Oath was received generally in Nazi Germany; Lifton, R. J. (1986). The Nazi Doctors, 32, commenting on Himmler's embrace of Hippocrates as a model for SS doctors.

6 Plato, (1918). The Republic of Plato, Book V, 274. See also Galton, D. J. (1998). Greek Theories.

the quality of their progeny."[7] Hannaford (1996) attributes the philosophy of the "*biological theory of social dynamics*" or in plain terms, social equilibrium, to Plato's *Timaeus.*[8] Plato's political vision is revered as a utopian programme for the future, though in reality it is a rigidly stratified order rather than a classless society. According to Popper (1963), this vision resembles a socialist or communist regime[9]—dominated by a ruling elite, the guardian class, which he describes as the "*. . . premier class of Athenian citizens,*" a kind of "*master race*" entrusted with governance.[10]

Popper makes the vital distinction that Plato was concerned only with the rulers. The auxiliaries (the military) however, could be promoted to guardians, so together they formed a single ruling caste, while everyone else belonged to the working class.[11] Plato advocated eugenic stratification both for breeding a future superior race and for maintaining purity—ideas that, in part, resonated with later utopian thought, including Bacon's vision in *New Atlantis.* The old Atlantis legend, framed as a quasi-historical narrative, first appears in *Timaeus* and *Critias* (or Kritias), where it is alleged that Egyptian priests imparted the tale to Solon, an Athenian statesman. It was then passed down to Greek philosophical circles, and subsequently recorded by Plato, blending the fabled continent with intellectual intrigue.

The Atlantis myth was situated west of the Pillars of Herakles (e.g. the Strait of Gibraltar), ruled by the god Poseidon, father of the Atlanteans, a war-like people who attempted to conquer the Mediterranean world, but were ultimately defeated by the Athenians. Following their military downfall, a cataclysmic earthquake caused the continent to sink beneath the sea.[12] Donnelly (1882) considers Plato's account as a probable narrative but does exhibit shortcomings. Nevertheless, De Camp (1970) explains that Donnelly effectively transformed Plato's theory into a cult-like framework, beginning with his declaration of thirteen "*theses.*"[13] Zhirov (1970), a Soviet chemist, conducted a study on Plato's depiction of Atlantis, despite controversial points and surmises, Godwin (2010) asserts that Zhirov's work carries "*great conviction*" as to its existence, supported by tables, maps and charts, that it is located on "*. . . the northeastern slopes of the mid-Atlantic ridge . . . .*"[14] Furthermore, Zhirov notes that occultists developed a fascination with esoteric legends about Atlantis, drawn to the glamour of lost knowledge, the birth place of magic, and for its geometrical significance.[15]

Aristotle was the pupil and protégé of Plato during his time at Plato's Academy in Athens, where he resided for the next twenty years. It is understood that Aristotle loosely reworked Plato's views to develop his own 'biological theory of reproduction', and then he

7 Galton (1998), 264; Leunissen, M., (2017). Eugenics, 86–87, citing Rep. 459a1–561e9.
8 Hannaford, 35.
9 Popper, K. R. (1963). The Open Society, 45–46, 48, generally, 45–54.
10 Galton (1998), 264; Plato. (1955). The Republic of Plato, 62.
11 Popper, K. R. (1963). The Open Society, 46-47.
12 Craig, T. L. (2010). On the Significance, 232–33; Ellis, 7–27; Godwin, J. (2010). Atlantis, ix, 25, 29–30, 65–66; James, P. (1996). The Sunken Kingdom, 12–18, 59–60; Hammer, O. (2003). Claiming Knowledge, 99–100: Plato. (1804). Introduction to Critias, 227–28; Weinberger, J. (1976). Science, 878.
13 De Camp, L. S. (1970). Lost continents, 1–20; Ellis, 39–41.
14 Donnelly, I. (1882). Atlantis, 5–21, 22–30, respectively; De Camp, 37–*39, 40, 41–44; Flowers, S. E. (2022). The Occult in National Socialism, 64; Godwin (2010) ix–4; James, 37. See generally James (1996) for an analytical overview.

Other postulations for the Atlantis location include the Canary and Madiera Islands, the Aegean Sea (James, 1996), the Arctic Circle, Scandinavian regions within the North Sea (King, 2005), (Spanuth, 1956), (Spanuth, 1979). Additionally, some theories suggest Atlantis may have been located in the Irish Sea, under Cornish control, extending from Scotland to Libya (Dunbavin, 2003). For further discussion on alternative locations, see Godwin (2010), 18–23.
15 Godwin (2010), 6–7, 30, 34, 58; James, 5; Michell, J. F. (2001). The Dimensions; Zhirov, N. (1970). Atlantis, 99–108.

prescribed his own eugenic recommendations based on Plato's foundation.[16] Aristotle's discourse on genetics was *avant la lettre*, establishing the foundational architectures of hereditary knowledge that would shape the future of scientific thought.[17] Nordenskiöld (1936) highlights that after Aristotle laid the foundations for "*. . . a specialized form of biological research . . .* ," primarily established in Alexandria, Egypt, under the Ptolemaic dynasty, subsequent centuries witnessed an "*abundant harvest*" of biological inquiry.[18]

Kohn (1960) traces this earliest form of superior and inferior races, citing Aristotle's notation on the hostility and warfare between the Greeks and the Barbarians. Aristotle expanded upon Plato's acceptance of warfare relations to develop a hierarchical theory of humanity. Aristotle attributed egalitarian value distinctions, as did Plato, to the terms of 'Greeks' and 'freemen' as the superior race, contrasting them with all non-Hellenes as intrinsically inferior 'Barbarians' and 'slaves.' Aristotle's perspective on slavery was rooted in his hierarchical conception of nature. He believed that slavery was expedient, since savages and savage cultures merited such treatment because they were undisciplined, and lacked sufficient rationality to govern themselves.[19] Accordingly, Poliakov (1974) and Kohn (1960), citing Aristotle, both affirm, with Kohn in particular writing—

> Hellenes rule Barbarians, not that alien yoke rest on Hellenes, Mother. They be bondmen, we be freeborn folk. [. . .] "[F]rom the hour of their birth some are marked out for subjection, others for rule. . . . The lower sort are by nature slaves and it is better for them as for all inferior that they should be under the rule of the master. . . . It is clear then, that some men are by nature free and others slaves, and for this latter slavery is both expedient and right."[20] (Omission in original)

Aristotle's belief was that these less cultured groups were naturally predisposed to subservience, serving as tools for the more rational and disciplined members of society. In terms of blood testing to ascertain racial purity, a primitive form dates back to the medieval epoch after the Reconquista of Spain (722–1492 AD), following Muslim rule and the subsequent expulsion of Jews from Spain in 1492. This practice "*. . . became one of major importance, necessarily accompanied by extensive hereditary investigation.*"[21] In Bacon's time, we can identify a further primitive form of scientific racial theory already circulating in Anglo-Saxon (410-1066 AD) literature, frequently referred to as 'race hygiene.'[22] Interestingly, this topic was a particular focus for Bacon. Serjeantson (2018) states that Bacon prized medicine above all, and it was in his work *New Atlantis* that he developed it most extensively—

> Medicine in the Renaissance was often divided into two broad parts: the cure of disease – therapy – and the prevention of disease – hygiene. Unlike the various forms of therapy, hygiene dealt with ways of pre-serving health and prolonging

16 Nordenskiöld, E. (1936). The History of Biology, 34, 36–38.
17 Sharpley, C. F. et al. (2022). Aristotle's Concept.
18 Nordenskiöld, 50.
19 Davies, A. (1980). Racism, 22; Kohn, H. (1960). The Idea, 52; Wokler, R. (1995). Anthropology, 31.
20 Poliakov, L. (1974). The Aryan Myth, 131, citing Kohn, H. (1960). The Idea, 52n.19, quoting Aristotle. (1885). The Politics of Aristotle, Bk. VII [sic], vi, 1-1327b. The correct citation to Aristotle's work should be Bk. I: *The Politics*, 6:1254a et seq.
21 Savage, R. (2007). "Disease Incarnate," 407.
22 Bock (1983), 401.

> life, rather than with dealing with diseases after they had appeared.[23]

The early development of anthropological literature has its roots in the philosophical writings of the Renaissance (c.1450–1600) Notably, two influential texts that emerged during this period were Thomas More's Utopia (1516) and Francis Bacon's *New Atlantis* (1627). More's Utopia pays little attention to the applied sciences per se; instead, it emphasises the social rationality of a 'natural social science' or 'applied natural sciences', highlighting the methodical use of reason. Knowledge and science are existential in the organisation of social relations, with science broadly permeating all aspects of society.[24] While Bacon's work was essentially a derivative of Utopia; it introduced new content, in particular, experimental science and the systematic organisation of knowledge.[25] Bacon envisioned a utopian society (Marxist) called Bensalem, situated on an island in the South Pacific.[26] This was a philosophical fictional, idealised authoritarian utopia, which combined elements of biology, communism, socialism, and mythology. It was a reformulation and refinement of Plato's ancient "*Atlantean dialogues*" found in Plato's works *Timaeus* and *Critias*. The intent was to supplant the old methodology with the *New Atlantis*—a vision of an advanced, scientifically based society.[27]

Indeed, while *New Atlantis* is a myth framed on the original Atlantis of antiquity, the new iteration is more than a manual or prophecy; it presents a schema, methodology, and plan "*. . . for making the world calculable.*"[28] The consensus view was that the ultimate goal was scientific conquest. Bacon conceived an audacious plan for the "*restoration of knowledge,*" a work he entitled *Great Instauration* (1620). Clarke (1972) labelled Bacon's plan the "*Baconian scheme*" and then went on to say—

> [T]he New Atlantis became the bell-ringer of the new sciences. It offered a most attractive model of what could be achieved by the still undeveloped powers of science[.][29]

Bacon presented his model to realise his aspirations, consisted of six parts[30]—

- **Part one**: division of the sciences . . . .
- **Part two**: the new inductive method or the "*art itself of interpreting nature*" and was provided in the *Novum Organum* . . . .
- **Part three**: foundational natural history, or the actual "*first food*" or factual material on which the new method was to work . . . .

---

[23] Serjeantson, R., (2018). Bacon's New Atlantis, 92, citing Mikkeli, H. (1999), 32–40n.38. See also Clarke, I. F., (1972). 4. Bacon's New Atlantis, 274; Serjeantson (2018)n.34, cites Bacon, De augmentis, in Works, I, 598 (4.2)(trans. IV, 290); Bacon, Historia vitae, in Works, II, 103, 155 (trans. V, 215, 263). Bacon was in fact repeating a common accusation that learned physicians made against 'empirics' at the time: see Cook, H. J., (1990). The New Philosophy, 409–10n.38.

[24] Adams, R. P. (1949). The Social, 377–81; Bierman, 493.

[25] Agassi, 11,19.

[26] Gasman, D. (1971). The Scientific Origins, xii; Lucas, P. (2018). Bacon's New Atlantis; Smith, J. E. H., (2015). Nature, 13, 21.

[27] Bury, J. B. (1920). The Idea of Progress, 59–63; Craig, 228–29; Gutmann, J., (1978). III. Western Philosophical Thought, 240. See Adams (1949), 374–98, Craig (2010), 228–39, and Weinberger (1976), 872–80, for a comparative analysis of the ancient Atlantis and Bacon's New Atlantis. Lucas (2018) offers a brief examination of both the old Atlantis, New Atlantis, and More's utopia. For a comparison between the ancient Atlantis and Huxley's Brave New World, see also Franck, M. J. (2013), Brave 'New World.'

[28] Horkheimer & Adorno, 4.

[29] Clarke, 277.

[30] See Bacon, F. (1986). The Great Instauration, viii–x.

- **Part four:** "*set forth examples of inquiry and invention*" . . . .
- **Part five:** Bacon's own provisional discoveries not "*according to the true rules and methods of interpretation, but by the ordinary use of the understanding in inquiring and discovering . . . .*"
- **Part six:** New Philosophy; or Active Science. This described the end of the new science, and although no Baconian work is identified . . . the *New Atlantis* fits the description of this part as described in the *Great Instauration*.[31]

Bacon predicated Bensalem by its pursuit of science, and wholeness: "*. . . a nation to be compounded of all goodness,*" the epitome of "*. . . a well-polished state of utopian perfection . . . ,*" as Smith (2008) described it.[32] It was a hub for international exchange, where all European languages were spoken, and visitors from the Middle Eastern and Far Eastern countries interacted. The Atlanteans residing there possessed a comprehensive knowledge, including mystical and supernatural insights. Some have even speculated that this place was an international, mystical, supernatural city, even the "*. . . source of divine, possibly extra-terrestrial, racial and spiritual perfection.*"[33] This is why Kurlander (2017a) posits that the combination of science and the supernatural, among other elements, proved especially attractive to many Nazis.[34] This intersection of literature, philosophy, and utopian ideals reflects the rich history of anthropological thought and its engagement with imaginative narratives. As anthropologists continue to explore diverse writing styles, the boundaries between fiction and truth blur, inviting new perspectives on human cultures and societies.

To this end, Kohn's (1960) comparison with Utopia notes that Bacon regarded *New Atlantis* as "*. . . the vehicle for bringing in the millennium. All the last part of The New Atlantis was a glorification of scientific inventions and inventors,*" a work which Brooke (1991) places in the locus of experimental science.[35] While the Renaissance (c.1450–1699) laid the groundwork for scientific inquiry and methodology, Bacon's version was a contributory milestone to the formalisation of empirical observation and experimentation. Building on this, the Enlightenment championed reason, societal change, and the sciences, which were crucial for systematic experimentation, rigorous data collection and testing, and the pursuit of knowledge based on evidence. Culver & Ianna (1984) elaborates on its monumental significance—

> [T]he modern scientific method consists of four basic components including (1) experimentation, (2) recognition of laws or behaviour patterns in the experimental results, (3) describing the entire collections of laws with an overall theory, and (4) predictive testing of the theory . . . [T]he most elegant of these components is experimentation. It is the linchpin of the scientific method. It permits the development of scientific laws and the testing of theories. It is the bridge between the "real world" and our scientific representations of that world . . . it is the final arbitrator of any and all scientific disputes.[36]

31 Ibid., x–xi—n.4, 19.
32 Ibid., xi, 59; Smith, S, 117. See also Feder, K. L. (2010). Encyclopedia of Dubious Archaeology.
33 Bacon, 37, 49–52; Kurlander, E. (2017b). One Foot in Atlantis, 82. Kurlander, E. (2017a). Hitler's Monsters, 16.
34 Kurlander, E. (2017a). Hitler's Monsters, 17. See also Goodrick-Clarke, N. (2004). The Occult Roots, 24-26, 58–61.
35 Brooke, J. H. (1991). Science and Religion, 22; Kohn, 162.
36 Culver & Ianna, 84, 168. See also Barbour, I. G. (1997). Religion and Science, 106–09; and Jones, R. H. (1986). Science and Mysticism, 19–33, for applied commentary.

In essence, the quote suggests that any emerging hypothesis is constantly probed, prodded, and tested—subjected to rigorous and continuous examination to replicate results, generate predictions, and enable practical applications across diverse contexts. Hammer (2003) summarised it further as a process that is "*. . . repeatable and error correcting.*"[37] For this reason, the scientific community considered Bacon's work a milestone and a prophetic key text: "*. . . the chief 'projector' of the scientific revolution . . . ,*"—especially within the late Renaissance culture, and in the industrial revolution (1760–1840).[38] The Enlightenment advanced this approach, using reason, empirical methods and the standard of calculability to dismantle incorrect scientific doctrines. Bacon's notion of science as "*knowledge is power*" became a cornerstone of Enlightenment thought, and technology is the distilled essence of this knowledge.[39] This laid the foundation for the Enlightenment (c.1700–1820), the Scientific Revolution, which built upon the advancements of the Renaissance. While a discussion on occultism is not appropriate here, it is important to briefly acknowledge that German occultism, which emerged in the 1890s, annexed Bacon's scientific method. According to Staudenmaier (2015), they integrated it within what he described as an esoteric framework and viewed, "*. . . their activities as an innovative type of science.*"[40]

Contemporaries of Bacon perceived that *New Atlantis* was a pipe dream, but in retrospect it was a philosophical treatise for the advancement of scientific knowledge, and a manual for learning, even though the modern term 'science' did not exist at the time of Bacon's writing. Was it a pipe dream though? In the *New Atlantis* remodelled version, it is governed and controlled by modern technological advances in science, which in turn is conquered and controlled by scientific authority, technological innovation, and rational thought.[41] Essentially, *New Atlantis* is a synonym for the prospective new sciences. In short, a "*. . . new society should be founded on the basis of scientific socialism in the free world.*"[42] Further, in view of Nazi racial science under the guise of eugenics/genetics, and today's present-day advanced technologies, all-round, "*. . . many of the scientific possibilities [Bacon] sketched have since become [stark] realities.*"[43] Bacon envisaged science as the foundation of power and a revolution on the horizon for new uses of science: "*Finis scientiarum a nemine adhuc bene positus est.*"[44] It is not so unsurprising then that Dr. Alfred Ploetz (1860-1940), the founding father of the eugenics, racial hygiene movement in Germany, engineered a plan to make the *New Atlantis* a reality. He called the Nazi regime the new elite utopian society, a "*. . . Constitution of the Freeland Society' which placed socialism on racial principles.*"[45]

The expansion of Bacon's (1627) and More's (1516) texts was furthered by Aldous Huxley's (1894-1963) *Brave New World* (1932) and George Orwell's (1903–1950) literary work, '*1984*.' Huxley's vision echoes Bacon's thematic authoritarian society, just on a

---

37 Hammer, 204.
38 Serjeantson, 84–85. See also Albanese, D. (1990). The New Atlantis, 506; Summers, C. J. et al., (2002). Fault Lines and Controversies, 23; Wallace, A. F. C. (2003). The Social Context of Innovation.
39 Adams (1949), 395; Horkheimer, M. & Adorno, T. W. (2002). Dialectic of Enlightenment, 1-3; Jones, 19; Marchand, S. L. (2009). German Orientalism, 19.
40 Staudenmaier, P. (2015). Esoteric Alternatives, 30.
41 Bacon, F. (1986). The Great Instauration, xiii; Craig, T. L. (2010). Literary Character of Francis Bacon's, 228–29; Feder, K. L. (2010). Encyclopedia, 29–30; Friend, J. M. (2014). The Biopolitics of Policy-Making, 35; Serjeantson, R. (2018). Francis Bacon's New Atlantis; Smith (2015), 13–14; Smith, S. (2008). The New Atlantis, 97.
42 Weindling, P. J. (1993). Health, 67.
43 Bierman, J. (1963). Science and Society, 492; Clarke, I. F. (1972). '4. Bacon's New Atlantis,' 274; Lucas, 114; Serjeantson, 83–84.
44 Kohn, 161.
45 Weindling (1993), 69.

narrower scale. His focus was on a world without disease—not merely curing all diseases, but preventing them. He expanded upon the schema with the inclusion of genetics and evolutionary biology, which fused Darwinist principles with science (discussed later). Indeed, Woiak (2007) elaborates that Huxley's remarkably "*. . . accurate in its predictions about science and technology, economics and politics, and arts and leisure,*" then proceeds to mention the various present-day contemporary future scientific applications in the fields of IVF, cloning and endocrinology, with a view, specifically, to further the science of eugenics for "*race betterment.*"[46] See Portrait of the New Atlantis, 302.

In Huxley's world, the World State had completely severed sexual intercourse from procreation. He envisioned an institution called the "*Hatchery and Conditioning Centre*" to assume all such duties. All babies were conceived on assembly lines, selected *in vitro*, and gestation—following the doctrine of *survival of the fittest* or *superior individuals*—only breeding the best stock of genetically fit individuals.[47] Hadomi (1991) distinguished their defining themes: More's Utopia comprising categories of social equality and organisation, whereas Bacon's, a century later, focused on scientific progress and governance.

Orwell's '1984' depicted the dystopian manifestation of these umbrella categories, while Huxley's text appeared to be a softer version, with "*. . . more effective totalitarianism than Orwell's 'boot smashing down on the face,*" Huxley expressed.[48] The present writer contends that either of these totalitarianisms will mobilise considerable oppression and persecution. Smith (2008) recognised that Bacon's "*. . . scientific visions of the 'relief of the human estate' . . . in New Atlantis can only be accomplished at the cost of a potentially tyrannical politics.*"[49] The most recent correlation of Bacon and Huxley's vision for society, bears out in the lyrics of John Lennon's song, "*Imagine*" (1971). See also discussion in Class-Based Stratification, 283.

Recall Bacon and Huxley's scientific vision, Lennon's core desire, encompasses two key aspects: (i) a world that is a socialist society where everyone is equal and are one, and (ii) a world free from hunger. Bacon's vision suggests that scientific innovations will perpetually satisfy human culinary desires. Additionally, envisioning a world without disease or death presupposes an underlying genetic framework, achieved through germ line engineering, discussed further under heading Genetic Technologies, 326. Lastly, global peace implies hierarchical rulers maintaining law and order, ensuring a war-free existence.

Consider modern film dramatisations of Bacon and Huxley's oeuvres applied to eugenics. Indeed, Allen (2002) posits that eugenicists mobilised mass media to promote social reform sympathetic with eugenic objectives.[50] The film, *The Giver* (1993), is a dystopian adaptation of both ideologies. It represents the *New Atlantis*, a perfect world where scientists and highly advanced technological innovations govern an isolated island situated between the Troposphere and Stratosphere. It fulfils the Brave New World axioms; human beings are conceived using eugenics methodologies in controlled laboratory environments. However, the grim reality is that babies deemed genetically unfit are acceptably and rationally put to death. Another eugenics film is *Gattaca* (1997), a first-rate sci-fi genre. It portrays a world where only perfect, genetically fit, elite humans

---

46 Woiak, J. (2007). Designing a brave new world, 107, 109. See also Turney, J. (1998). Frankenstein's Footsteps, 91–120.

47 Franck, M. J. (2013). Brave "New World", 75.

48 Ibid., 73; Hadomi, L. (1991). From Technological Dystopia, 110; Wartburg, W. v. & Liew, J. (1999). Gene Technology, 22, 92–93.

49 Smith, S. (2008). The New Atlantis, 100.

50 Allen, G. E. (2002). The Ideology of Elimination, 22.

are permitted entrance into white-collar professions, the sciences, and cultured society. These individuals are designated as super-humans, possessing perfect 20/20 eyesight and eugenically defined 'perfect' bodies that never incur illness or disease. Notably, this mirrors aspects of Nazi Germany, where the inferior were kept separate to prevent interaction. Hence, a daily randomised testing regimen ensured the establishment of biogenetic superiority fitness, preventing security breaches and potential hybrid breeding. This process involved electronic pinpricks for blood sampling. Interestingly, Weindling (1993) expressed an interesting passing comment on the prophetic quality of these dramatisations, although his perspective was not definitive—

> Novels, dramas and philosophies publicized the reforming implications of medicine and biology. They prophesied degeneration if industrial and urban development continued to disregard the laws of heredity and health. They emphasized the physical and psychological limits to human perfectibility.[51]

ii. The Founders of Anthropology and the Emergence of Racial Hierarchies

THE EXOTERIC ALLURE of Bacon's New Atlantis resembled a sublime, picturesque paradise. Just like the French Riviera, the aesthetics of the location and surroundings, New Atlantis held a similar charm for both the learned and the layperson. However, beneath this surface beauty lay esoteric content of far greater significance. Bacon's work not only envisioned scientific and technological advances but also laid the groundwork for future generations, including present-day scientific technologies. Unfortunately, this blueprint extended beyond scientific progress; it influenced tyrannical governance and contributed to the emergence of modern racism and the scientific method (beginning with Nazi Germany). In fact, Bacon set in motion the formal founding of anthropological ideology and, more broadly, the sciences, which eventually became the cradle of modern racism and scientific methodology.[52], [53]

While *New Atlantis* is considered incomplete due to its lack of guidance on government or political rule, it played a cardinal role in Bacon's overarching plan—the *Great Instauration.* For a detailed exploration of what "*completes*" this vision, see The Philosophy of Progress, 291. Weinberger (1976) stipulated, this work provided a primitive framework comprehensive, "*. . . reformation of human science and knowledge, which will 'lay the foundation, not of any sect or doctrine, but of human power and utility' . . . .*"[54] Scientific research, in turn, represented the most accurate approach to understanding, leading to the eventual adoption of scientific methods to address societal problems and maintenance.[55] The solution, in process of time, devolved into

51 Weindling (1993), 80.
52 Banton, M. (1978). The Idea of Race, 14; Barzun, J. (1938, 1965). Race, 51–52; Hannaford, 148-49; Mosse (2020), xxiv, xxvii, xxxix, 3, respectively; Phillips, D. (2015), 388; Smith (2015), 7.
53 For Anti-Semitic racism grounded in the Medieval ages to the Enlightenment, but not all references, see I. Bancel et al., (Eds.). (2014). The Invention of Race, 1; Berger, D. (Ed.). (1997). History and Hate; Endelman, T. M. (1997). Comparative perspectives, 97; Katz (1980). 13; Mosse (2020), 3; Popkin, R. H. (1974a). The Philosophical Basis of Modern Racism;
II. On the sixteenth-century origins of racism, see Puzzo, D. A. (1964). 'Racism and the Western Tradition; Popkin, R. H. (1974a); Scammell, G. V. (1991). Essay and Reflection;
III. The conception that Scientific Racism emerged during the Enlightenment, see Augstein, H. F. (1996). Race, ix–xxxiii; Hudson, N. (1996). From "Nation to "Race"; Popkin, R. H. (1974b). The Philosophical Basis of Eighteenth-Century Racism; Stuurman (2000), 1n.1.
54 Weinberger, J. (1976). Science, 867. See also Adams (1949), 385.
55 Banton (1978), 89–90; Ludmerer & Lappé, 458.

Social Darwinism, which later became official policy during the Nazi era. See heading Pioneering the Next Phase, 35. Hence, the spectre of "*Social*" Darwinism' was focused on society. Richard Hofstadter, American historian, critically examined how the social milieu intersected with Darwinism—

> An ideology using a competitive view of nature and Darwin's concept of the struggle for existence as a basis for social theory . . . in its early phase Social Darwinism served primarily as a justification for ideas of laissez faire, since it stressed individualist competition. . . . This second phase of Social Darwinism emphasized a collectivist struggle and arose in conjunction with progressivism.[56]

The Renaissance and the Enlightenment were transformative periods that marked the emergence of several pivotal figures who played a foundational role in the establishment of anthropology and the construction of racial hierarchies. This transition facilitated a reimagining of how history, culture, and society were perceived and analysed, culminating in the early development of modern biology and anthropology.[57] Browning (2020) elaborates upon this notion that the emerging sciences such as anthropology, physiognomy, and polygenesis (the study of diverse origins of the human face) became prominent fields of inquiry. Similarly, Davies (1980) accentuates the Enlightenment's vigorous dedication to data classification propelled a more fervent anthropological focus on race thinking, and positioned polygenesis as a central tenet of this paradigm.[58] These intellectual explorations laid the groundwork for many of the critical thinkers[59] who became seminal contributors to these disciplines, including—

i. **John Locke (1632–1704)**: A philosopher known for his ideas on individual rights and empiricism.
ii. **George Berkeley (1685–1753)**: An Irish philosopher who explored the nature of perception and reality.
iii. **Georges-Louis Leclerc, Comte de Buffon (1701–1788)**: A French naturalist who made significant contributions to natural history.
iv. **David Hume (1711–1776)**: A Scottish philosopher and historian.
v. **Voltaire (François-Marie Arouet) (1694–1778)**: A French writer, philosopher, and advocate for freedom of thought.
vi. **Baron Charles Montesquieu (1689–1755)**: A French jurist and political philosopher.
vii. **Johann Blumenbach (1752–1840)**: A German anatomist and early anthropologist.
viii. **Immanuel Kant (1724–1804)**: A prominent German philosopher and Enlightenment thinker.

The roots of anthropology can be traced back to ancient intellectuals such as Hippocrates, Herodotus, Aristotle, Pliny, and Galen. Wokler (1995) threads this invention of anthropology from ancient Greece and its redefinition during the Enlightenment—

56 Hofstadter, R. (1955). Social Darwinism, 5-6 passim; Weikart, R. (1993). The Origins 469.
57 Wolff, L. & Cipolloni, M. (2007). The anthropology, xi, xiii, 8–11; Nordenskiöld, 82–83.
58 Browning's Introduction in Mosse (2020), xiii; Davies, 23–24; Wokler, 31–47; See also Mosse (2020), xxv.
59 Lucas, 115; Hannaford, 59; Phillips, D. (2015), 383; Poliakov (1974), 169, 327, 330.

> [T]he genesis of sexuality, the treatment of the insane, or the comparative anatomy of apes or races—were all thought essential to a proper development of the science of man, itself an expression, together with the term anthropology . . . invented or substantially redefined in the course of the Enlightenment.[60]

The establishment of a rigorous scientific framework was pivotal for the progression of anthropology, as well as contemporary biology and genetics. These disciplines drew legitimacy from contemporary advancements in biology, genetics, and ethnology, epitomising what Davies (1980) poetically articulated that they donned the "*. . . imposing [fashionable] robes of science . . .*" with which to draw these proofs.[61] In Germany, this scholarly structure was encapsulated by the concept of "*Wissenschaft,*" broadly translated as science.[62]

In summary, the concept of race was intertwined with broader intellectual developments of the time, and was essentially and respectably scientific.[63] The evolution of anthropology into a modern discipline was deeply rooted in the rigorous scientific frameworks of the Enlightenment. This synthesis of interdisciplinary approaches not only reinforced anthropology's academic standing but also accorded with broader intellectual movements, embedding the study of race and human diversity within the scientific discourse of its time. By integrating these elements, anthropology emerged as a field capable of addressing the intricate intersections of biology, culture, and society.

In the context of the Enlightenment and Romanticism, significant developments occurred in the understanding of human nature and differences in the "*. . . seventeenth and the beginning of the eighteenth century.*"[64] On this point, in Hudson's (1996) discourse on race, the scientific community had not yet embarked on the intellectual endeavour of categorising humans into 'varieties,' 'tribes,' or 'races' until the Renaissance, with François Bernier (1620–1688), French traveller, physician, and Gassendist philosopher, making key contributions. In Europe, the focus on physical characteristics led to the emergence of ethnology, which w the precursor to modern anthropology.[65] Nevertheless, the Enlightenment and German Romanticism were crucial epochs in the foundation of modern racism. The key differences between the two, illustrating the features that are shared by both are as follows—

Table I-1. Enlightenment and Romanticism: Divergence and Convergence

Enlightenment	Romanticism
Reason	Sentiment
Objective norms	Subjectivity
Rigid forms	Lack of form, fragmentarism
Universality	Nationalism
Man	Co-national
Secularism	Religion
Protestantism	Catholicism
Freedom	Authority

60 Wokler, 32, 48n.3.
61 Davies, 26.
62 Phillips, D. (2015). Francis Bacon and the Germans.
63 Mosse (1964), 88; Müller-Hill, 82.
64 Smith (2015), 7.
65 Barzun (1938, 1965), 511; Schafft, G. E. (2004). From Racism to Genocide, 38, 40, *52.

Responsibility	Obedience
Modernism	Medievalism
Clarity	Obscurity
Manliness	Femininity

Source data: Reichmann, E. G. (1951). Hostages, 84, 138.

Bernier published his paper, *New Division of the Earth* (1684) in the prestigious, *Journal des Scavan*, marking the first attempt to formally classify the world's population. This intellectual milestone may be perceived by some that this was the opening of Pandora's box, initiating a complex and somewhat problematic discourse on human classification. Johann Friedrich Blumenbach (1752–1840) further expanded upon Bernier's five racial classifications during the Enlightenment, deepening the scientific exploration of human diversity.[66]

The Enlightenment era promoted scientific inquiry, while Romanticism contributed by solidifying Aryanism in a nationalist context. See also Romanticism Dynamics, 250. They shaped opinion and beliefs about human nature and human difference in the "*. . . seventeenth and the beginning of the eighteenth century.*"[67] The sciences track, it is generally credited to Buffon and Blumenbach, who played pivotal roles in the development of scientific racism, but is hotly contested.[68]

While Buffon's seminal work, *Histoire Naturelle* (1749–1789), published in 36 volumes, primarily focused on the study of animals, he was the first since Aristotle to apply his animal hypotheses, either by humanising animals or animalising humans. Buffon organised his theory along primitive Darwinian evolutionary lines by questioning the "*. . . fixity of species and groping toward a proto-evolutionary theory of the earth and its life forms,*" as Stuurman (2017) expressed it. Accordingly, "*[Buffon] classified humans with apes, monkeys, lemurs, and bats based on teeth and mammae.*"[69] However, Buffon's teleology of human classification, articulated through his definition of 'biological species,' was fundamentally flawed. It relied on broad, inconsistent general impressions rather than precise scientific measurement.[70]

Buffon's approach to categorizing humans lacked the rigorous methodology that later became standard. Furthermore, his assessment of the Amazonian climate, in a generic sense, associated "*nonwhite races with degeneration,*" which was the forerunner to Joseph-Arthur de Gobineau's humankind degeneration treatise.[71] In contrast, Blumenbach, while he adopted Buffon's method, he elevated science above racial prejudice. Blumenbach is considered to be the founder of physical anthropology, who pioneered craniometrics, and laid the foundation for objective scientific inquiry published c.1765.[72] Comparatively, Blumenbach's work focused more on racial categorisation, whereas Buffon's work emphasises the hybridisation of humans and animals.[73]

---

[66] Bernier, F. (2001). A New Division of the Earth; Hoquet, 19–21; Hudson, N. (1996). From "Nation to "Race", 252; Stuurman, S. (2000). François Bernier; Stuurman, S. (2017). The Invention, 301–02; Snyder, L. L. (1962). The Idea, 11–12, *103.

[67] Smith (2015), 7.

[68] Hoquet, T. (2014). Biologization, 17; Blumenbach, J. F. ([1865] (1978)). The Anthropological Treatises; Lovejoy, 178; Stuurman (2000), 1–2; Michael, J. S. (2017). Nuance Lost in Translation; Nordenskiöld,306–09; Poliakov (1974), 168.

[69] Hoquet, 34; Stuurman (2017), 304. See also Wokler, 43–44.

[70] Hudson, 254–55; Popkin (1974a), 142.

[71] Stuurman (2017), 308.

[72] See Hudson (1996), 255; Hoquet, 22; Meijer, M. C. (2014). Cranial Varieties, 34; Michael, J. S. (2017). Nuance Lost in Translation; Poliakov (1974), 173.

[73] Stuurman (2017), 312. For a comparative analysis of the differences between Buffon and Blumenbach, see 309–16.

Immanuel Kant, although drawing inspiration from Buffon, also made significant contributions to the discourse on race. His race principles had a lasting impact on late 18th century life science in Germany and contributed to the inclusion of 'race' as a topic in scientific literature. Kant's views on race were complex, as he both defined human 'races' and ranked them hierarchically, positioning 'whites' at the top, based on alleged deficiencies in the capacity for thought and action. His racial views were closely connected to his endorsement of European colonial rule and race-based slavery during that period.[74]

As early as 1775, in Kant's essay, '*Of the Different Races of Human Beings*' he theoretically examines the question of race, in which he connects race with ancestry, by defining racial features. In '*Determination of the Concept of a Human Race*' (1785), Kant delineates race theory, addressing physical differences between humans. Then in '*On the Use of Teleological Principles in Philosophy*' (1788), Kant endorses slavery and introduces his theory of racial hierarchy, placing the superior races at the top tier and the inferior, weak races at the bottom.[75] Müller-Wille & Rheinberger (2007) argue that this theory was the "*... prototypical for the emergence of heredity.*"[76] Although Kant was the first to theorise race, his hierarchy theory was predicated on, and derivative of Aristotle's treatise on hierarchy classifications, who introduced the concept of slavery, reproduction and genetic transmission *avant la lettre*. Hannaford (1996) remarked that Romanticism and the modern nationalist movements, "*... had their proper beginning in the age of the French Revolution.*"[77] The pioneers of anthropology were German figureheads, who "*... contributed to every phase of the Third Reich. They provided theory, policy formation, enforcement, and proactive engagement.*" Although they were not involved in the commission of any crimes or genocidal acts, they were the anthropology theorists that laid the foundation for their Nazi cohorts, who did. These were the pen pushers that paved the way for the killings.[78]

What began as a scientific endeavour in the seventeenth and eighteenth centuries, laying the groundwork for racial policy, included elements of anthropology and philology. This framework changed form in the nineteenth century with the further development of philology, synthesised with Aryan distinctions.[79] By 1860, physical anthropology had emerged as an independent scientific discipline, led by prominent French figures such as Geoffroy Saint-Hilaire, Quatrefages, and Paul Broca. For further analysis of Broca, see heading The Confluence of German Aryanism, 44. In 1869, Germany followed suit by founding its first anthropological society, established by the physician and anthropologist Rudolf Virchow (1821–1902), who maintained a pivotal presence in German physical anthropology until his death in 1902.[80]

Mosse (1964, 1975) and Zmarzlik (1972, 2011) stress that during this crucial period, the scientific constructs of physical anthropology became corrupted through the reengagement of Gobineau's taxonomy of racial hierarchy, grounded in notions

---

74 Kleingeld, P. (2024). Anti-Racism and Kant Scholarship; Kleingeld, P. (2007). Kant's Second Thoughts; Lu-Adler, H. (2023). Kant, Race, and Racism: Views from Somewhere; Poliakov (1974), 171 et seq.

75 Barzun, 51; Kant, I. (2007). On the use of teleological principles; Kleingeld, P. (2007). Kant's Second Thoughts; Lovejoy, A. O. (1959). Kant, 179; Montagu, A. (1974). Man's Most Dangerous Myth, 33; Mosse (1964), 88, 99; Sandford, S. (2018). Kant, Race, and Natural History.

Kant's view racial on differences included not only physical but also intellectual, moral and psychological differences.

76 Müller-Wille, S. & Rheinberger, H.-J. (2007). Heredity 18.

77 Mosse (2020), xxvi, 3; Mosse (1964), 88; Cf. Cohn, N. (1981). Warrant, 25; Hannaford, 14.

78 Schafft, 222, 224, respectively.

79 Augstein, ix–xxxiii; Hannaford, 306; Mosse (1964), 88–89; Snyder, L. L., 32, 40–41.

80 Massin, 82–83; Wegner, G. (2002). Anti-Semitism, 12.

of superiority and inferiority. Dilettantish race anthropologists carried out this work; Zmarzlik specifically notes, "*[w]ith the help of skull and body measurements as well as pigment specifications . . .*"—methods infused with romanticised precepts and values that deviated from rigorous scientific criteria. Consequently, the scientific framework of 'race' was adulterated, being transposed from a scientific to a metaphysical domain around 1799, during the era of Romanticism and Social Darwinism[81] On this point, Mosse (1978, 1985), cited by Wegner, explains why Virchow's survey could do nothing to alter German perceptions—

> Race had been infused with myths, stereotypes, and subjectives long ago, and a scientific survey could change little. The ideal of pure, superior races and the concept of a racial enemy solved too many pressing problems to be easily discarded.[82]

iii. Origins of Racist Aryan belief and the Atlantis Connection (1788–1859)

In Feder's (2010) critique of Plato's Atlantis, he asserted that it represented a 'perfection' that was a conceited and corrupted culture—

> Atlantis is the embodiment of a materially wealthy, technologically advanced, and militarily powerful nation that has become corrupted by its wealth, sophistication, and might.[83]

The Nazis also believed themselves to be the offspring (sub-race) of the Atlantean race. They adopted the brutal behavioural characteristics of the Atlanteans for the betterment of the German race.[84] Helena Blavatsky inspired German belief that some Atlanteans escaped Atlantis either before or after a great flood, but prior to its final sinking. See heading The Superman Doctrine, 72. These survivors fled to and settled in Tibet and Sanskrit, India. It is said they imparted their esoteric knowledge and wisdom to their descendants, the Aryans, who went on to establish an advanced civilisation. In doing so, it was believed that the Aryans inherited superhuman mental and physical capabilities. This belief later informed the aims of the famous Nazi Tibetan expedition conducted between April 1938 and August 1939—the quest to uncover remnants of lost Aryan knowledge.✪ [85]

In Poliakov's *Aryan Myth* (1974), he establishes the Tibetan narrative as a hoax. The Nazis claimed that their Aryan ancestors and their superiority originated from

---

81 Augstein, ix–xxxiii; Hannaford, 306; Mosse (1964), 88–89; Mosse, G. L. (1975). The Nationalization, 5; Zmarzlik, H.-G. (1972). Social Darwinism in Germany, 444–45; Zmarzlik (2011), *12–13. See also Snyder, L. L., 20, 41. Snyder recorded Charles White (1728–1813) an English surgeon, provided a romantic, elegant and flowery delineation of the superior race traits of Aryan beauty.

82 Mosse, 91–92; Wegner, 12.

83 Feder, 31.

84 Spielvogel, J. & Redles, D. (2011). Hitler's Racial Ideology, 86.

85 See Hale, C. (2003). Himmler's Crusade; Junginger, H. (2007). From Buddha, 143; Kurlander (2017a), 184–85, 187–88, 352n.229, citing Kaufmann, W. (2009). Das Dritte, 347–40; Kurlander (2017b), 94; Kurlander (2019), 16; Longerich, P. (2012). Heinrich Himmler, 281–82.

✪ See The Superman Doctrine, 72 for how Blavatsky's writings inspired many, including the formation of occult societies in Germany, which were rooted in a belief system that fused Atlantis (or Thule)–Aryan narratives with emerging strains of antisemitism.

ex oriente lux—a mysterious semi-Oriental land in Sanskrit, India.[86] In this context, Thapar (1996) asserts that the claim to Aryan origins was not only a marker of racial purity but also conferred an exclusive social status, one that was tightly guarded by the upper-caste Hindus. Thapar argues, "*. . . [this] excluded not only the lower castes but also the non-Hindus . . . Aryanism, therefore, became an exclusive status.*"[87]

With this understanding, one might further contend that their ruthlessness represented an effort to emulate their ancient motherland, Atlantis, to construct a "*New Atlantis*": an idealised cultural order achieved through the systematic purification of the nation from genetic, racial, and social "*undesirables.*" This envisaged utopia was predicated on the reestablishment of governance by the purportedly scientifically and biologically superior Nazi Aryan elite, destined, in their view, as rulers of the world.[88] In terms of eugenics, Socrates said "*. . . the children of the good parents were to be reared, but those of the bad were to be sent secretly to various other parts of the city.*"[89] The writing was on the wall as the next predictable scientific application of *New Atlantis*, which was, of course, cultivated to formulate the scientific racial, eugenics policy of Nazi Germany. Bacon's *New Atlantis* is a refinement and perfection of the old version. Implicit in his text were the instructions and scientific schema to formulate an anthropological treatise. Lucas (2018) exhorts *New Atlantis* proponents that it should be not read because of one's "*. . . fascination with the Atlantis myth . . . as the history of a storehouse of ancient wisdom.*"[90]

The Nazis claimed that their ancestors could be traced back to the lost civilisation of Atlantis. Herein lies the paradox, the old Atlantis myth looks back, to examine antiquity, the supernatural, architecture, and cultural legends. See comparison between the old Atlantis, Nazi Atlantis and future Atlantis in Figure IV-2, 310. The purpose was to establish their ancestry, the greatness and glory of their dynasty, to unveil their master race heritage. The refurbished version, *New Atlantis*, contains the new methodology that is forward-looking. It seeks to engineer a future utopia—a sanitary bio-utopia.[91] Thus, *New Atlantis* holds a twofold application. First, it cast Nazi Germany as a shadow of the impending utopia. Second, it hinted at a yet-to-be-realised eugenics-based regime. Unlike traditional Nazi racial hygiene, but with a twist, in that the future regime caste would likely be stratified based on genetics "*Newgenics,*" rather than race or ethnicity.[92] See heading Class-Based Stratification, 283. Intriguingly, defrocked Nazi Hermann Rauschning (2005) attributed Hitler's success to the idea that "*. . . every German has one foot in Atlantis and one in Tibet, where he seeks a better fatherland.*"[93] Kurlander (2017a) elaborates further that Rauschning was referring to the—

> [E]thno-religious connections between the lost Ario-Germanic civilization of the Thule (Atlantis) and an Indo-Aryan civilization centred in northern India. Many Nazis believed that Tibet was a refuge, in which important elements of an "Aryan-Nordic-Atlantean Ur-culture" had survived.[94]

86 See also Godwin's (2010), 125–27, documented migration narrative from Atlantis.
87 Thapar, R. (1996). The Theory of Aryan Race, 9–10.
88 Ogilvie, J. & Annandale, C. (1883). 'Aryan,' I:159; Poliakov (1974), 213.
89 Plato (2000). Timaeus, 3, cited in Isaac, B. (2006). The Invention of Racism, 125; Popkin (1974a), 126; Todorov, 106–112.
90 Lucas, 120.
91 Hutton, 113; Gutmann, 240.
92 Black, E. (2012). War Against the Weak, 427–44.
93 Kurlander (2017a). Hitler's Monsters, xi; Kurlander (2017b). 'One Foot in Atlantis, 81, citing Rauschning, H. (2005), 208.
94 Kurlander (2017a), 184. See also Goodrick-Clarke, N. (2002). Black Sun, 4.

It was for this reason that Heinrich Himmler (1900–1945) led a SS expedition of Nazi scientists to Tibet in 1938–39. They believed that the "*. . . Aryans . . . had descended directly and fully formed 'from heaven,'*" and sought to substantiate this postulate by locating the Atlantis civilisation, Himmler aimed to find evidence supporting these claims.[95] Accordingly, Friedrich Schlegel, and numerous German patriotic thinkers of the 18th and 19th centuries, considered Germany to be the Orient of Europe.[96] In Hutton's (2005) and Poliakov's (1974) exhaustive discussion on the matter, they deciphered the web of complexity distinctions between linguistic and anthropological classifications, identifying the Japheth line as pure Aryan stock while characterizing the Semitic line as impure. The table of nations' ancestry is derived from Noah's sons: Shem, Ham, and Japheth. The Semitic line encompasses Jewish and Asian lineages, Ham's descendants represent the African branch, and Japheth is the ancestor of Europeans,[97] who migrated to Europe following the Tower of Babel incident when God confused the languages.

In linguistic terms, they claimed that the first Aryan was Japheth, and that the language of the Garden of Eden was German. This thesis, established in '*The Book of a Hundred Chapters*' written in Alsace between 1490 and 1510, asserts that "*Adam was a German.*"[98] The actual origin of the Aryan question, however, dates back to Sir William Jones (1746–1794) a pioneering scholar of Sanskrit, who addressed the Royal Asiatic Society in 1786. A decade later, c.1788–1799, concepts of superior groups began to emerge.[99] Jones identified the linguistic connection and historical affinities between Aryan with Sanskrit, India, and European languages.[100] The development of Aryan theory was first attempted by the leading Sanskrit scholar Friedrich Max Müller (1823–1900), an Anglo-German philologist and Orientalist.[101] Between 1869 and 1871, Müller advocated in England his version of the Indian Aryan discovery. He defined Aryan-Europeans as the ancestors of a division of the human race that included Hindus, Persians, Celts, Greeks, Italians, Slavs, and Teutons. According to Müller, the parent stock was—

> [A] small clan settled on the highest elevation of Central Asia speaking a language, not yet Sanskrit, or Greek, or German, but containing the dialectic germs of all. The [first Aryan-Europeans] . . . . were living together [under] the same roof . . . .[102]

95 Hale, 60. See also Goodrick-Clarke (2002), 123–24; Kurlander, E. (2012). Hitler's Monsters; Kurlander, E. (2013). The Orientalist Roots; Kurlander, E. (2017a). Hitler's Monsters; Kurlander, E. (2017b). 'One Foot in Atlantis; Kurlander, E. (2019). Nazism, 15–18; Mosse, G. L. (1964). The Crisis; Poewe, 6–7, 10.

96 Field, G. G. (1981). Evangelist of Race, 201-04, 210; Germana, N. A. (2010). Self-Othering, 80–90; Manjapra, K. (2014). Age of Entanglement, 56–108; Mosse (1964), 89-90; Poliakov (1974), 193n.29, quoting Gérard, R. (1963). L'Orient, 132; See generally Marchand, S. L. (2009). German Orientalism.

97 Hutton, 81–83; Hale, 60–61; Montriou, J. A. L. (1788). Elements, 4; Poliakov (1974), 7, 22; Snyder, L. L., 40.

98 Hutton, 80, citing Webb, J. (1669); Poliakov (1974), 7, 16, 22, 35, *76–77.

99 Poliakov (1974), 190; Snyder, L. L., 20–23; Tenenbaum, J. (1956). Race and Reich, 9. See also Augstein, H. F. (1996). Race.

100 Barzun, 135; Hale, 61–62; Dunlap, K. (1944). The Great Aryan Myth, 296; Goodrick-Clarke, N. (1998). Hitler's Priestess, 31; Hutton, 83–84; Kurlander (2017b), 96; Marchand, 19–20; Poewe, K. O. (2006). New Religions; Poliakov (1974), 190; Snyder, L. L., 40–41; Stuurman (2017), 352–53. See Thapar, R. (1965). A History of India (Vol. I) and Thapar, R. (1996). The Theory of Aryan Race, for a full analytical discussion on the linguistic elements.

101 Barzun, 140–41; Poliakov (1974), 199; Snyder, L. L., 41–42.

102 Ogilvie, J., & Annandale, C. (Eds.). (1883). The Imperial Dictionary, entry for "Aryan" 1:159, col. B; See also Dunlap, K. (1944). The Great Aryan Myth, 297; Poliakov (1974), 213.

Consequently, from this period onwards, a group of Philologists started to identify language with race, namely, that the Indo-European group of languages were the same as the Indo-European race, which was then later renamed to *Indo-Germanic*, coined by the Orientalist Julius von Klaproth in 1823.[103] Mallory (1989) refutes Aryan supremacy of the Nazis arguing it was—

> [I]nconsistent with Aryan as a linguistic term . . . while Romany-speaking Gypsies had a far better claim to the title of Aryans than any North European. Thus, the myth of Aryan supremacy . . . was the misappropriation of a linguistic concept and its subsequent grafting onto an already existing framework of prejudices, speculations and political aspirations.[104]

German ancestral heritage is understood to be the miscegenation between Hinduism and Buddhism, with the inclusion of a pre-Christian Germanic Faith. This formed the basis for their Indo-Germanic tradition. Then, the renowned racial ideologist Joseph-Arthur de Gobineau's (1816–1882) legitimised it with playfulness of genius. In Gobineau's book, *Essai Sur l'inégalité des Races Humaines,* 4 vols., (c.1853), he lent credibility and form to the Aryan myth, earning him the title of the "*Father of Racism.*"[105] Synder (1962) sums up Gobineau's thesis in one sentence: "*. . . all human races are anatomically, physically, and psychologically unequal.*"[106] Gobineau's social stratification revolves around the core concept of 'beauty.' He believed this was an absolute and necessary principle, which was not subjective or open to optional application. Thus, Gobineau employed romanticised language to describe the native German race as "*. . . the most moral, the most intelligent, energetic, beautiful, and the strongest.*" This perspective found a welcome audience among the Nazis, in line with the views of his ideological successors such as Ploetz, Lenz, and Rosenberg.[107]

Gobineau's emphasis was not so much about the ascendancy of the Aryan race, rather, he lamented the complete loss of this superior race due to miscegenation and degeneration, rendering them irredeemable. According to Gobineau, the Aryans had become a mixed race, corrupted by inferior genetics as a result of Mongol invasions and ancient migratory patterns, which led to racial miscegenation and subsequent cultural decline, and extinction.[108] For this reason, Fischer (1913) employed the term in 1870 'Bastardvolk' translated as "*racial bastards*" or "*nation of bastards,*" to describe the German mixed-race population, but yet maintained the assumption that the "*. . . Nordic race predominated.*"[109] Interestingly, Snyder (1962) noted that Gobineau rejected Blumenbach's five racial categories and sub-categories, arguing that they too were corrupted as "*notorious hybrids.*"[110]

---

[103] Hutton, 83–84 (see also 94–100, for discussion on the Aryan myth debate); Kurlander (2017b), 96; Poliakov (1974), 193; Poewe, K. O. (2006). New Religions; Snyder, L. L., 40–41; generally, 39–53.
[104] Mallory, J. P. (1989). In Search of the Indo-Europeans, 269–70. See also Hutton, 96.
[105] Arendt, H. (1944). Race, 54–55; Baumslag, 36; Barzun, 73, 98; Field, 207; Gasman, xxiv–xxv; Günther, H. F. K. (1927). The Racial Elements, 254–56; Mosse (1964), 90–91; Mundy, R. (2006). Musical Evolution, 16; Schaff, 39; Stein, G. J. (1988). Biological Science, 54.
[106] Snyder, L. L., 46.
[107] Reynaud-Paligot, C. (2014). Construction, 93; Todorov, 128, 130; generally, 123–40.
[108] Arendt (1944), 57–58; Banton (1978), 42–45, 54, 80; Butler, 135–36; Hutton, 81, 85; Kurlander (2017b), 85, 92; Marten, H. G. (1999). Racism, 27–29; Reynaud-Paligot, 95; Voegelin, E. (1940). The Growth, 299; Todorov, 137–38.
[109] Fischer, E. (1913). Die Rehobother Bastards, 29, 45, 190, cited in Hutton, 70–71.
[110] Snyder, L. L., 47–48.

Gobineau's Aryans, Fortier (1967) wrote, "*. . . were blond, blue-eyed giants of great muscular force and physical beauty. Utilitarian, hard-headed, and warlike, they were born conquerors.*"[111] Gobineau went further to argue that there was a hierarchy language that corresponded to the hierarchy of races.[112] However, with that said, the national mood for that period, Poliakov (1974) observed that anthropological literature was already deeply rooted in his time, to draw support for his own race supremacy hierarchy thesis.[113] This provided a platform with which to entrench the idea further in the eyes of German proponents by which, the subsequent extraction of antisemitism in 1880 operated nicely within racial concepts.

Therefore, in an effort to undo and clarify the confusion surrounding Aryan ideology, the discourse became entangled in a "*labyrinth of contradictions,' false trails and surprise traps,*" regarding the origin of race and its supposed link to language differentiation. Race theorists ultimately objected to the conflation of racial and linguistic origins.[114] They argued there was a distinction between the linguistic and racial identity. The general consensus was that the term 'Aryan' was primarily a linguistic term, not to be applied to make racial identity distinctions.[115] Yet, it continued to be a fundamental affinity and area of study by the Third Reich. They viewed it as the outer sign of the unity of the *Volk*—the mother-tongue ideology was a precious possession, and still compatible with racial ideology—"*[t]he 'mother tongue' was the most vital expression of the Volk [and collectivity, which] could be used as a criterion to differentiate one people from another.*"[116]

Obviously, as the time differential increased between the original Japheth dynasty and their descendants, the existential prospect of race destruction would have increased, due to migrations or more invasions. Effectively, the closer the language of their ancestors, the more original it was.[117] Consequently, it was argued that before their perceived racial decline and linguistic pollution, they were the true Aryans. Another body of academic and scientific opinion flatly refuted the congruence as complete Aryan myth and an absurdity. In Snyder's (1962) own words, he asserts—

> This identity of language and race was assumed even though there was not the slightest proof for the identification.
>
> . . . .
>
> Then the term Indo-European was replaced by Indo-Germanic, a designation made general in the works of James Cowles Prichard (1786-1848), a British philologist and ethnologist, and Franz Bopp (1791-1867), a German linguist.[118]

Jacques Barzun (1938) concurs and appears to make light of racial classifications, concluding that the term 'Aryan' is effectively defunct and analytically unworkable—

111 Fortier, P. A. (1967). Gobineau, 342.
112 Weindling (1993), 51.
113 Poliakov (1974), 233–34.
114 Benedict, R. (2000). Race : What it is not; Poliakov (1974), 2, *188–89, 205. See also Barzun, 13.
115 Hutton, 23, citing Wirth, A. (1914). Rasse und Volk, 33–34.
116 Hutton, 19–20, and 142–43, citing Schmidit-Rohr, G. (1934), 203, Schmidit-Rohr (1939) and Vossler, K. (1936), 362.
117 Baur, E. et al., (1931). Human Heredity, 647; Hutton, 81, 86.
118 Snyder, L. L., 40, 42, 45. See also Poliakov (1974), 189.

> No one today would deny that [race is a] great catchword[] about which ink and blood are everywhere spilled in reckless quantities. Yet no agreement seems to exist about what Race means.[119]

Herein lies the dichotomy. The Nazi Aryan theorists had plunged themselves into somewhat of a dilemma. On one hand, there was the linguistic affinity of Sanskrit, India, the original Aryans, but, on the other, they belonged to the same linguistic family as the Europeans languages. This was indicative of the belief that linguistic and racial identity were congruent—the dark-skinned Sanskrit Hindu and the white-skinned European belonged to the same race. India or the Himalayas was the "*. . . imagined birthplace of white humanity.*"[120] In short, the Semite, which is the Jewish and Asian line, and Japheth, which is the European line, both were cut from the same cloth. This negates the Nazi claim that Aryans were the superior race. In other words, the half-Hindu, half-European, white–black-skinned Aryan is *contradictio in adjecto* to the Aryan myth. Their origin was rooted in an Indian geographical location, and their progeny descended from a dark-skinned race; yet, they deemed black and Jewish people inferior, despite allegedly sharing the same dynasty.[121]

Hitler ([1925] (1971)) intimated that miscegenation was a selfless act, whereby Aryans gave up the purity of their blood, "*. . . became submerged in the racial mixture, . . . and lost his cultural capacity.*"[122] Gruber (1927) claimed positively that breeding a high standing race with a low race "*. . . was not good for the higher race as the offspring would lie in the middle,*" adding that only science has proved it.[123] However, Ehrenreich (2007) maintains that overt claims that 'race-mixing' caused hereditary damage persisted, were rare; by contrast, "*[t]he idea that there were distinct human lineages . . . was clearly grist for the racist mill.*"[124] The Nazis' principal contention was that non-Aryans, such as blacks, Orientals, Jews *inter alia* were inferior; yet the Nazis themselves were cut from the same cloth. Put simply, the Nazis were hybrids too![125] Although, Eugen Fischer (1874–1967), co-author of *Human Heredity* (1931), rejected this view on the grounds of 'as long both parents were of equal racial value, this was desirable and a positive phenomenon.' On the other hand, he also contended that a racial trait beneficial to one group could be destructive to another. Therefore, Kohn-Behrens (1934) argued that the harmful forms of mixing had to be rooted out and destroyed.[126] Poliakov (1974) adds some fresh perspective, rather than endlessly trying to resolve and decipher the correct Aryan distinctions, one might instead question the Aryan premise itself as an inherently romanticised fable? Poliakov quotes archaeologist Salomon Reinach (1852–1932) for this insight—

> To speak of an Aryan race existing three thousand years ago, [Reinach] wrote, is to put forward a gratuitous hypothesis; to speak about it as though it still existed

---

[119] Barzun, 3, 5; Hutton, 65, 73; Poliakov (1974), 72–77; See also Günther (1927), 257–58, and Massin's (1996) brief comments, 114. Cf. Schafft, 203–04, discussion about no consensus on the definition of race.
[120] Hutton, 85; Sayce, A. H. (1888). Address, 167, 171. See also Poliakov (1974), 101, *207.
[121] See Hutton's excellent thematic discussion at 83–86. Cf. Ettinger, 231.
[122] Hitler, A. ([1925] (1971)). Mein Kampf, 296.
[123] Ehrenreich, E. (2007). Nazi Ancestral Proof, 16, 28; Hutton, 125, citing Gruber, M. v. (1927), 247. See also Bartov, O. (2000). Mirrors of Destruction, 105.
[124] Ehrenreich, 28.
[125] Hutton, 113–14, 117–20, 132.
[126] Hutton, 135–36, citing Kratz, R. G. (2009). Eyes and Spectacles, 57–60; See also Hutton, 145, for critique of Fischer's descent.

> today is quite simply to talk nonsense.[127]

The general Aryan rebuttal to the claim was to assert theory of Aryan ascendancy, dismissing it as imprecise linguistic usage or "*terminological confusion.*"[128] In particular, Georges Vacher de Lapouge's (1854–1936) seminal work, *L'Aryan* (1899) reformulated Gobineau's Aryan theories in biological terms. However, Lapouge's rebuttal also stated that the term 'Aryan' was far from satisfactory, as it indicated that Indo-European peoples descended from an Aryan stock, "*. . . which is supposed to have swarmed all over Europe and a part of Asia.*" Lapouge went on to argue that many individuals who were not of the original type, namely, "*. . . long-skulled blond-haired type he referred to as Homo Europeans, the dominant element of the original Aryans.*"[129] Virchow, for example, studied race in 1871 to test the proposition of the existence of pure races. He surveyed 6,760,000 schoolchildren, within which were both Jews, and Aryan Germans, who were examined separately. The result contradicted the beliefs of those who subscribed to Aryan supremacy and the purity of Aryan bloodlines. Accordingly, Wegner (2002) to put it mildly, wrote that the "*. . . acclaimed blue-eyed, blond-haired Germans as a strong numerical presence in the country took a severe hit.*"[130] The data is summarised as follows—

Table I-2. Rudolf Virchow's Survey Results indicating flawed theory of Aryan Supremacy

Sample Size 6,760,000 schoolchildren Based on Hair Colour[131]

German Subset (n=6,684,623)			Jewish Subset (n=75,377)		
Blonde	Brunettes	Mixed	Blonde	Black-Hair	Mixed
31.8%	14.05%	54.15%	11%	42%	47%

Moreover, through the lens of physiognomy and pseudoscientific racialisation, Gilman (1991, 1992, 2000) explained that the Jew was characterised and viewed as black, because—

> [. . .] Jews look different, they have a different appearance, and this appearance has pathognomonic significance. Skin colour [stamped] the Jew as both different and diseased.[132] (First set of brackets in source)

In this context, racialisation was used to differentiate Jews from the dominant population, reinforcing stereotypes and justifying discrimination. The quotation emphasises that Jews were perceived as having a distinct appearance with "*pathognomonic significance,*" meaning their physical traits were seen as indicative of disease or abnormality. Here, it is metaphorically applied to suggest that Jewish physical traits

[127] Reinach, S. (1892). L'origine, 90, quoted in Poliakov (1974), 267 et seq.
[128] Hutton, 97.
[129] Hutton, 85–86, citing Lapouge, G. V. d. (1899). L'Aryen, 330, 341, respectively. Weindling (1993), 52. See also Günther (1927), 257.
[130] Wegner, 12.
[131] Ackerknecht, E. H. (1953). Rudolf Virchow, 210, 214; Gilman, S. L. (1991). The Jew's Body, 177–78; Gilman, S. L. (1992). The Jewish Body, 234; Gilman, S. L. (2000). Are Jews White, 233–34; Mosse, G. L. (1978). Toward the Final Solution, 91–92; Wegner, G. (2002). Anti-Semitism, 12; Weindling (1993), 49.
[132] Gilman (1991), 174–75; Gilman (1992), 232–33; Gilman (2000) 231, 233.

were viewed as inherently pathological. The idea that "*skin colour [stamped] the Jew*" underlines the extent to which racial and medical discourses were intertwined. This perception of Jews as racially and physically distinct contributed to their marginalisation was employed to justify social hierarchies and exclusion. Thus, German dominance and Jewish differentiation were framed as necessary to protect the German nation from a perceived Jewish population explosion, believed to surpass that of the Germans. Consequently, one rationale for Germany's resort to genocide was to maintain its power and dominance.[133] See headings Racial Hygiene at its Zenith, 160 and Hatred transmission, 181. Despite the racial criticism, German theorists maintained their Aryan claims, whilst acknowledging the problems with the "*Aryan*" term, they continued with it. First, because they viewed themselves as "*racially heterogeneous*" and, second, they were superior over all races.[134]

The racist viewpoints expressed by Gobineau in *Essai*, published before Darwin's *Origin* were well-received by German composer Richard Wagner (1813-1883). Wagner forged a deep friendship with Gobineau in 1880.[135] He then began a study on Gobineau's *Inegalité des races humaines*. His study helped to define his ideas by providing a philosophical and "*scientific*" basis that confirmed and substantiated his tentative views on race and blood. They further fortified and encouraged him in the production of his vitriolic and influential racist writings. The following year, 1881, Wagner wrote *Herodom and Christendom*, published in the *Bayreuther Blatter*.[136] As a consequence, he too developed a "*fanatical hatred*" for the Jews. Wagner and Chamberlain, two peas in a pod, were the most famous antisemitic writers. Any wonder then that Hitler shared the same deep-seated hatred.[137] Mosse (1964, 1975) stated that Wagner was held in high esteem as a prophet of Germanism, exerting a formative and malleable influence on Hitler's mental development, while Poilakov (1974) expressed that Wagner was held as the chief prophet of Aryanism.[138] Hall (2017) noted that "*Hitler was a Wagnerian from an early age*"—he was 12 years old when he saw Wagner's opera *Lohengrin* c.1901–1902).[139] Thus, as Hall insightfully explains, these reverences shed light on why Hitler worshipped him from an early age.

Wagner's dream was German regeneration, national rebirth and unification, founded upon German culture underpinned by a strategy of production and reproduction of myths, shaped predominately by German art and music, particularly Wagner's music—with an elimination of all things Jewish.[140] Wagner was pivotal. Moreover, Hitler and August Kubizek (1888–1956) were close teenage friends. Between 1906 and 1907, when Hitler was approximately 17 years old, he is believed to have watched Wagner's opera *Rienzi*, which later biographical accounts purport had a significant, life-changing impact on young Hitler. So much so, it moved him with complete ecstasy and rapture. Kubizek recounted that after the performance, they climbed Freinberg mountain, whereupon Hitler had a political awakening. Hitler draw an analogy from Wagner's opera about

---

133 Fein, H. (1979). Accounting, 7–9; Fein, H. (1993). Accounting, 89; Staub, E. (1999). The Roots of Evil, 182, 184.
134 Hutton, 86, 96; Poliakov (1974), 72, citing Nietzsche, F. W. (1967). Beyond Good and Evil, 197.
135 Flowers, 31–33; Weindling (1993), 107; Weiss (1987b), 93.
136 Stein, L. (1950). The Racial Thinking, 36.
137 Hall, D. I. (2017). Wagner, 168; Shirer, 101. See also Weikart's (2004, 124) discussion on Chamberlain.
138 Mosse (1964), 93; Mosse (1975), 100–01, 109; Poliakov (1974), 313.
139 Hall, 168–69.
140 Arvidsson, S. (1999). Aryan Mythology, 328, 331; Hall, 164–65, 167; Shirer, W. L. (1960). The Rise, 97, 101; Stein, L. 31, 36, 44; Montagu, 47; Mosse (1964), 93–94, 97.

the rise and fall of the Roman tribune *Rienzi.* Young Hitler prophesied that he would become political leader, a notion he could not yet fully grasp at that time.[141]

However, Kershaw (1998) and Hall (2017) express scepticism and caution regarding this account, arguing that the Freinberg mystical experience is highly fanciful and lacks substantial evidence.[142] Webb (1976) reiterates this sentiment, noting that Kubizek had been "*. . . caught out in inaccuracy and distortion.*"[143] Despite these reservations, what we do know is that according to Synder (1962), Wagner did play a pivotal role in moulding Hitler's racial ideologies. Wagner's music and writings were imbued with the concept of the German Hero-Spirit, a notion he believed was exclusive to individuals of Nordic descent. This worldview, interwoven with the tenets of Social Darwinism, exerted a profound influence on Hitler during his formative years, contributing significantly to the development of his race theories. See also The Superman Doctrine, 72. The impact of Wagner's works on Hitler is underscored by Hitler's own words: "*Whoever wants to understand National Socialist Germany must know Wagner. . . . What joy each of Wagner's works has given me!*"[144]

This connection extends to Houston Stewart Chamberlain (1855–1927), Wagner's son-in-law and racial ideologist,[145] the bridge between earlier theories of racism and the emerging Nazi ideology. In fact, Chamberlain's works were, in turn, relied upon by National Socialism for intellectual legitimacy.[146] Gobineau's works were taken over by Chamberlain. He re-worked the architecture of Gobineau's classification caste theory.[147] This was a concoction and fusion of Kantian philosophy, racist antisemitism notions, selected conclusions of anthropology, biology, other sciences, and his own historical mythology insights, to which Chamberlain applied a nationalistic interpretation. This culminated in the completed work, re-housed in *Die Grundlagen des neunzehnten Jahrhunderts* (trans. by John Lees as *Foundations of the Nineteenth Century*) (1899)). It was distributed widely, especially among the educated middle classes.[148] It became the new Bible for thousands of German racists, and their antisemitic entourage, including German emperor, Kaiser Wilhelm II, who expressed that it was his favourite book, then, it was conveyed to Hitler.[149]

Like his father-in-law, Chamberlain was hailed as an apostle of race, and as a result, his book enjoyed enormous popularity and widespread vogue.[150] Alfred Rosenberg (1893-1946) the Nazi party's ideologist, Reich Minister, and philosopher of the Nazi movement, characterised Chamberlain as '*The Prophet and Founder of a German Future*.' Rosenberg based his entire ideology on Chamberlain's racial theory, revising it in conjunction with Romanticism principles—stripping it of reason and social morality, while granting the right to indulge in primitive desires without restraint.[151] See Table I-1, and heading Romanticism Dynamics, 250, for full discussion. Astonishingly, despite the acclaim surrounding Chamberlain's book, Oakesmith (1919) asserted that

[141] Kubizek, A. (2011). The Young Hitler, 116–18. See also Goodrick-Clarke (2004), 193, 194.
[142] Hall, 169; Kershaw, I. (1998). Hitler, 20–21, 610n.128.
[143] Webb, J. (1976). The Occult Establishment, 300.
[144] Krausnick, H. (1970). The Persecution of the Jews, 26–27; Snyder, L. L., 82–83.
[145] Montagu, 47; Mosse (1964), 93.
[146] Beyerchen, A. (1977). Scientists under Hitler, 124; Field, 1–2; Mosse (1964), 94-97. See also Hutton, 86, citing Chamberlain, H. S. (1905). Arische Weltanschauung 3, and Chamberlain, H. S. (1899). Die Grundlagen. 314–15;
[147] Neumann, 107–08.
[148] Biddiss, 81; Krausnick, 26; Montagu, 47; Shirer, 106; Gutteridge, R. (1976). The German, 22 et seq, Neumann, F. L. (1944). Behemoth; Weindling (1993), 52, 109, 492.
[149] Montagu, 47; Poliakov (1974), 318–19; Shirer, 106–07; Tenenbaum,10.
[150] Oakesmith, J. (1919). Race, 56, 59.
[151] Biddiss, M. (1997). History, 87; Reichmann, 200.

Die Grundlagen was riddled with self-contradictions and employed a "*childish a priori submission,*" privileging asserted truth than over scientific method. Oakesmith described the work as one of the most foolish books ever written, adding that—

> It is false in its theories; ludicrously inaccurate in its assertions; pompous and extravagant in its style; . . . It is so dominated by a spirit of stormy rhetoric that it contradicts itself with passion at every turn.
>
> . . . .
>
> We say, and say it deliberately, that he is the only author we have read to whose work Sidney Smith's phrase, "the crapulous eructations of a drunken cobbler," could appropriately be applied.[152]

Further, Snyder (1962) confirms Oakesmsith's observations, arguing that they are not those of a dilettante but rather demonstrate Chamberlain's proclivity for Aryan ideology, over and above science—

> Here dogmatism and rationalization were carried to extremes. The logic was . . . . imperfect; the author was entangled in his own mystic and romantic intuition; his arguments were sophistic, his conclusions impulsive. Chamberlain boasted that he was "incapable of lying." Yet, he was so overcome by the urge to prove the superiority of the Teutonic race that he lapsed into what he called "prophetic vision" and "historical intuition" to bolster his thesis.
>
> . . . .
>
> Among the numerous contradictions was Chamberlain's treatment of pure races. He regarded "the sacredness of pure race" as an underlying theme of his work, yet at the same time he held that noble races evolve through inter-mixture and inbreeding.[153]

Chamberlain preferred the ideal of superiority, so he preferred to reserve the Aryan worldview for the *Volk*, and always equated Germany with Aryanism. Chamberlain asserted that only individuals, rather than whole nations, could be considered Aryan. Thus, the inference is that he divorces the Aryan view from science because it undermined what he wanted to be true. In other words, he just discarded the science because it interfered with his intuitive insights.[154] In short, Chamberlain's reasoning was *contradictio in adjecto.* In an effort to extract some sense from an already confused idea, the result was precisely the same as what 'Aryan' Physics had become—muddled and confused, *tout court.* Indeed, not to labour the point, Beyerchen (1977) asserted that its name, *Aryan Physics*, implied "*. . . a set of fundamentally incompatible concepts.*"[155] Günther (1927) expressed that Chamberlain "*recklessly*" displaced "*. . . well-established [scientific] conceptions . . .*" of racial differences and anthropological study of human

[152] Oakesmith, 58. See also Montagu, 58.
[153] Snyder, L. L., 53.
[154] Field, 216-20; Hecht, J. M. (2000). Vacher de Lapouge, 296; Hutton, 86, citing Chamberlain, H. S. (1905), 9, and [1899] (1932), 314-15.
[155] Beyerchen, 140.

beings to serve his purpose.[156] Yet, Hitler was influenced by Chamberlain's book, which, as a Wagnerite devotee, contained Wagner's doctrines—central to Hitler, who regarded Wagner as both his master and hero.[157] Hitler demonstrated his preference for Chamberlian's neurotic reasoning by, "*express[ing] regret that Chamberlain's observations were not more heeded during the second Reich.*"[158] There is a further reason why Hitler was influenced by these literary works, in which Wegner (2002) makes the point that Hitler—

> [D]raw from a rich literature of Volkish thinkers whose writings had increased in popularity . . . during the late nineteenth and early twentieth centuries. . . . Among the most strident voices was that of Houston Stewart Chamberlain . . . .[159]

The significance is that it went beyond mere 'influence' from books upon Hitler. Rather, it was indicative of an agenda on Hitler's part—one that sought to exploit the popularity of the above ideologues to garner support for racial science, antisemitism, and expand these ideologies on a much grander scale. One such character was prolific writer, Hans F. K. Günther (1891-1968). He did much to fuse race hygiene with extreme racist nationalism and virulent antisemitism, with a writing style that a lay person could understand.[160] See Heading Education of Society Matrix, 55. Hitler met with Chamberlain in October 1923 and later incorporated Chamberlain's ideas into Mein Kampf during his imprisonment in July 1924. For Chamberlain, he "*. . . admired Hitler as the savior of the race , and was greatly admired by him.*"[161] Furthermore, according to Wegner (2002), Hitler additionally adopted Gobineau's ideas and the Aryan myth for inclusion in Mein kampf.[162] Weiss (1987b) poignantly expressed that the net of Gobineau's doctrine was cast far beyond Hitler's. Gobineau's ideas were adapted and modified to meet the needs of National Socialist ideology.[163] See headings National Awakening, 121 and From Mystical Origins, 88, for discussion of how Gobineau's ideology was suffused with occultism.

[156] Günther (1927), 256. According to Montagu, he suggests that Gobineau and Chamberlain's works were the spiritual progenitors to Hitler's *Mein Kampf.* See also Montagu, 47–48.
[157] Biddiss, 79–80 et seq; Hall, 164; Krausnick, 37.
[158] Hall, 168; Montagu, 48; Shirer, 109. See also Mosse, G. L. (1966). Nazi Culture, 1–2.
[159] Wegner, 8.
[160] Ibid., 14; Weindling (1993), 312.
[161] Fortier, 344.
[162] Wegner, 9, citing Williams, R. (1989). Hegemony and Selective Tradition, 58. Concerning Wegner's statement that "Gobineau's ideas" are included in Mein Kampf. What he meant was they are, but you will not find identify them by searching for his name 'Gobineau,' since it is not expressly specified.
[163] Weiss (1987b), 93.

Figure I–1. The Progression of Aryan Ideology: From Racist Beliefs to Scientific Racism. The scientific succession of anthropology of Darwin and Haeckel's modifications are not included.

Thus, Figure I–1 diagram demonstrates a clear line of succession and development of racist ideology emerges within this circle, each referencing the other, and reflects the continuity of Aryan racist ideology. These conceptual debates may be interpreted as 'teething' problems, intended to sift out the dross of untenable views and create a 'pseudoscience' that, along with its sub-divisions, possessed the appearance of empirical plausibility[164] (e.g. polygenism, phrenology, ethnology), then still in their infancy. The foregoing seminal chronology is presented to reveal the pivotal continuity of Chamberlain's documented virulent antisemitism, incoherent reasoning, and its transmission.

Applying the principles of *New Atlantis*, the racial scientific revolution had already begun, but it was not yet ready for its debut: (i) due to insufficient progress in resolving the linguistic and anthropological confusion, and (ii) because of the absence of scientific technologies, the *sine qua non* for the Nazis to identify, control, and supervise the population, "*. . . which [only] reached its fully fledged form in the Third Reich.*"[165] This statement has a modicum of truth within the context of the Nazi regime. However, a comparative review of the antecedents to the "*New Atlantis*" utopia, reveals that while totalitarian control elements and basic technological forms[166] were present in the Nazi regime, they were not fully realised at that time. Only now, in the present generation of 2026, do the full array of emerging advanced technologies best mirror the anticipated *New Atlantis* utopia. See Part IV. for full discussion. In regard to the Nazi regime, two key *dramatis personae* in Aryan development were Paul de Lagarde (1827–1891) and Julius Langbehn (1851-1907), "*. . . the intellectual vanguard of Völkisch thinkers . . . ,*" Wegner (2002) described them.[167]

iv. The Nordic Race and Aryan distinctions

THE NORDIC AND Teuton race supremacy ideology[168] comes from the branch as Aryanism. The Aryan and Nordic races were used synonymously; they were not always easily distinguishable, the Indo-Germanic, Frankish Aryan people superiority were identical with the Nordic race.[169] Günther [1922] (1934), an advocate of Aryan supremacy and the Third Reich's most prominent race expert, documented his belief that the Nordic race had assumed its modern form by c.7000 BCE.[170] Similar to Aryanism, as you would expect, the Nordic race is believed to have strong blood-ties with Hinduism and Buddhism.[171] Hutton (2005) distinguishes between the two by asserting that evidence for the original Aryan people was primarily linguistic, centred on the study of Aryan languages, whereas the Nordic ideal focused on the proportions of the human body— an aesthetic sign and indicator of human identity.[172]

Although the term 'Nordic' typically connotes north-western Europe, Günther (1934, 1966) reconciled the concept of an Indian element within the Nordic race by arguing that

[164] See Link, F. & Hare, L. J. (2015). Pseudoscience Reconsidered, 105–31, for a comprehensive analysis and opposing arguments on pseudoscience.
[165] Bartov, 106; Snyder, L. L., 28.
[166] Beyerchen, 135-39.
[167] Wegner, 9. See also Flowers, S. E. (2022). The Occult in National Socialism, 57.
[168] See generally Barzun, Ch. II, The Nordic Myth, 27–51. For an elaborate discussion on the physical Nordic anthropological attributes, with illustrations, see Günther (1927), 11–22.
[169] Field, 210; Hutton, 106; Schafft, 122–23; Weikart (2013), 538, and 546, citing Heberer, G. (1943), 1–4n.43.
[170] Hutton, 104, citing Günther ([1922] (1934). Kleine Rassenkunde, 339; Günther (1927), 122, 134. See also Friedlander, H. (1995). Origins, 10. For a robust commentary of Günther, see Steinweis, 25–46.
[171] Günther (1927), 124–37.
[172] Hutton, 106.

it was linked to the occupation of central and southern Europe, resulting from migration and settlement processes. Yet, paradoxically, he condemns the Jews, stating they are not a race but a nation of mixed races; while at the same time, he maintains that the Nordics themselves are a mixed race, and justified the view.[173] Another popular body of opinion submits that the Nordic race was the original race.

Professor Fritz Lenz (1887-1976) German race theorist, geneticist and Nazi pioneer of racial hygiene, believed the Nordic race was responsible for creating the Indo-Germanic Aryan languages and civilisations. The main line of support for this was that Nordics, as the master race, were formed through 'selection' during the ice-age. Günther (1934) wrote that the Nordics emerged from a small group, which had, "*. . . undergone a rigorous process of selection and in-breeding during the ice-age.*"[174] As a result of the harsh climatic conditions, theorists posit that it "*. . . sharpened the struggle for existence, causing the weak to perish and leaving only the most vigorous.*" This created the Nordic superior race. Weinart (1934) argued that the "*. . . Nordic race had emerged with freeing of Germany from the ice-age . . . .*"[175] The twist is, however, Ploetz adopted this same construct of struggle for existence against the other races, not nostalgically looking back to the ice-age, but rather referring to the future Nazi Germany. The rationale behind "*freeing of Germany,*" evidently, had a double-barrelled meaning.

Krausnick (1970) emphasises Lenz's point that it was the Nordics that were the superior race of antiquity, emerging from the Aryan race. The Nordic race, he argues, is capable of "*. . . transforming Europe into a single powerful community,*" and is the "*. . . unifying element for the rest of humanity.*"[176] Fischer (1931) said that the Nordics, and by implication, the Aryans, predominated. Both possess hereditary gifts, talents, intellectual genius, athletic prowess, and superior character traits—qualities selected for mastery over nature, intellectual endowments, and clarity of mind. Lenz (1931) then adds, "*[p]ersons endowed with these gifts are likely to become leaders, inventors, artists, thinkers, organisers.*"[177] He applies a physical anthropological definition of the Nordic man to possess superior, genius traits—

> A person with good hereditary equipment can be recognised from the size and shape of his forehead, chest, and nose. The size of the head gives some measure of intelligence; the size of the chest gives some idea of vigour. 'Genius' is out of the question in persons the circumference of whose head is less 56 cm.[178]

Völkisch theorists believed that Nordic individuals had stronger constitutions than other racial types, which was supposedly evident in their lower incidence of mental and physical disorders. Accordingly, Lenz cited Fischer's observation that the mentality of the Nordic race was noted for its "*. . . industry, vigorous imagination, intelligence, foresight, organising ability, artistic capacity, individualism, a willingness to obey orders . . . and devotion to plan or idea.*"[179] Proctor (1988) cites Lenz's observations of the Nordic racial traits to be based upon empirical science, arguing that the Nordics were selected

[173] Günther (1927), 123. Günther (1966). The Nordic Race, 63; Günther (1933). Kleine Rassenkunde, 9–13, 21–25, 59; Hutton, 105, citing Günther (1934). Die nordische, 227ff. (trans. *The Nordic race among the Indo-Europeans of Asia*); Snyder, L. L., 50.
[174] Hutton, 104, 114, respectively, citing Günther (1934), 339.
[175] Hutton, 105, citing Weinert, H. (1934). Unsere Eiszeit-Ahnen, 31; Weikart (2013), 538.
[176] Krausnick (1970), 34.
[177] Baur (1931) et al., 655–62; Hutton, 71, citing Fischer (1913). Die Rehobother, 45.
[178] Baur et al., 602; Proctor (1988), 52. See generally Proctor (1988), 46–63.
[179] Proctor (1988), 56.

for intelligence and mastery of techniques, and the mentality of the Nordic include "*. . . self-control, self-respect, respect for life and property, desire to know the unknown, a certain wander instinct, and fondness for the sea.*"[180]

Whether Aryan or Nordic, the ideology and concept are fundamentally identical. The practicalities of maintaining 'purity' to ensure superior eugenic fitness, free from contamination by non-Aryan/Nordic races, were centred on the notion of "*blood.*" Massing (1949) and Snyder (1962) articulated that blood had risen to such prominence that it became the "*Holy Grail.*"[181] Massing wrote that Aryan blood was considered "*. . . the secret of the highest creation . . . ,*" and the ultimate source of authority in all spheres of the nation's life; a cult. Similarly, Beyerchen (1977) described it as the "*. . . romantic mystification of blood . . . .*"[182]

Hitler appeared to found his opinion on Aryan blood from the *fin-de-siècle*. Hitler wrote a prediction in his book, Mein kampf (1936), based upon what Ahlwardt (1890) wrote: the nation of "*purest blood*" is predestined to eventually rule the universe, and to become master of the world.[183] These rulers were identified as, "*. . . leaders in politics and culture are for the most part predominantly Nordic men.*"[184] Moreover, Mosse (1964) revealed that the Nazis "*. . . emphasis on blood was directly derived from the ancient German cult of the Druids . . . .*"[185] Further, in the more the extreme occult rituals, the belief is held that the life force is in the flowing of blood.[186] In fact, tracking the continuity and thematic development of the racial elements from Bacon's *New Atlantis* to the Renaissance, Günther (1927) wrote, "*. . . Italian Renaissance is seen clearly as a renewed flow of Nordic blood into the life of a people and its soul.*"[187] For the Nazis, blood transcended all conceptions of race definitions, suppositions and ideologies of what constituted being Aryan or Nordic. According to the '*German Ten Commandments*' it stipulated "*Thou shalt keep thy blood pure.*"[188] Schleunes (1970) captures the importance of racial purity to the Nazis, stating—

> Racial purity was for the Nazis a state of paradise; the mixing of races—a Nazi equivalent of original sin—defiled this paradise and led to the confusion of good with evil, superior with inferior . . . .[189]

Willibald Hentschel (1858-1947) political activist, proposed that a Germanic colony or caste system be established, so that the racial stock can be isolated, shielded and insulated for preservation of the noble blood, and for breeding purposes, would champion the survival of the race. Whereas the non-fit, Hentschel asserted—

180 Ibid.
181 Massing, 82; Snyder, L. L., 19.
182 Barzun (1938), 7, 13–14; Beyerchen, 96; Massing, W. (1949). Rehearsal For Destruction, 82–83, 103; Schleunes, K. A. (1970). The Twisted Road, 52; Tenenbaum, xiii.
183 Massing, 83–84, quoting Ahlwardt, H. (1890). Der Verzweiflungskampf, 230 and Hitler, A. (1936), Mein Kampf, 782. See also Hitler, A. [1925] (1971), 286;
184 Günther (1927), 218; Massing, 82–83.
185 Mosse (1964), 75.
186 Ibid., 76–77.
187 Günther (1927), 214–15, 218, citing Woltmann, L. (1905). Die Germanen, 53.
188 Massing, 82–83, 306–07. The 'German Ten Commandments' was the creature of Theodor Fritsch in *Antisemiten-Katechismus* (1893), 358. See also Günther (1927), 194.
189 Schleunes, 51.

> To ensure the creation of such a racial and Volkish nobility, Hentschel proposed the establishment of a truly Germanic colony. . . . The settlement was to be called Mittgart after the legendary place to which the Aryan race traces its origins . . . Hentschel claimed that the racial stock must be isolated, or rather insulated, in this way, so that it would be protected from the corrupting effects of modernity. Hentschel asserted that the non-fit should be shipped to the city [or a sanctuary], where the environment would destroy them . . . .
>
> . . . .
>
> Hentschel . . . declared his intent to erect a model community to breed the fittest Aryan specimens, who would champion the survival of the race"[190]

This idea never shaped up, but as we shall see, the idea for separation from mainstream society into segregated Jewish ghettos, did. Therefore, the purpose for the term (i) 'Aryan'/'Nordic' was to establish a clear, defined distinction between "*Aryan*" and "*non-Aryans,*" (ii) the development of Aryanism gaining control of the mass media become a pivotal criterion for antisemitic propaganda, and (iii) Aryanism was a technical and a "*fundamental precondition*" to prepare the Nazi state-in-waiting to eliminate Jewish influence, "*. . . for which Hitler had always striven.*"[191]

However, due to its many ambiguous meanings, it led to so many spurious misleading classifications, made it politically and legally problematic.[192] The testing regimen based upon race was a flawed framework with which to base Aryan and non-Aryans distinctions.[193] Therefore, the race framework failed to identify Jews in any precise or legally defined way. For this reason, Rosenberg abandoned 'Aryan' for the term Nordic, because 'Aryan' was redundant as a acceptable term of racial significance.[194] Yet, the enigma is that Ploetz referred to Jews as Aryans, yet they were not part of the superior race![195] This red-flagged the elementary truth behind their intentions in pursuing a nonsensical ideology, which Tenenbaum (1956) sums up in his operating premise quite succinctly—

> German racists needed both an 'Aryan' and a 'Semitic' race to balance each other off. . . . The superiority of the Aryan is predicated on the inferiority of the Semite.[196]

In short, Aryanism superiority was *a point d'appui* to project new paths of development, scope for unlimited expansion, and to justify committing genocide of the inferior.[197] Nevertheless, in 1935, the term 'Aryan' was abandoned. This is why some chose to favour the Nordic element above other racial elements, and others were calling

[190] Mosse (1964), 113–14; See also Weikart (2003). Progress, 281.
[191] Bartov, 104, 108; Fest, J. C. (1970). The Face, 40; Eltringham, N. (2017). "Nordics" and "Hamites", 252 citing Bartov, 91–142.
[192] Barzun, 143–44; Günther attempts a well reasoned answer, in *The Racial Elements* (1927), 4–9; Hutton, 90, citing Schulle, D. (2001), 91.
[193] See Ehrenreich's analysis, 45–49;
[194] Eltringham, 252, 265; Ehrenreich, 9–11; Field, 210; Hutton, 90; Snyder, L. L. 49–50.
[195] Kelly, A. (1981). The Descent, 116, citing Mosse, G. L. (1978). Toward the Final Solution, 80–81; See also Mosse, G. L. (2020). Toward the Final Solution, 73–74.
[196] Tenenbaum, xiii.
[197] Ibid., 4–5.

for the absolute supremacy of the Nordic race.[198] Nazi Germany, therefore, employed a stricter and clearer definition that became law: the concept of German blood ties or purity of blood, with blood became the predominant factor in racial identification. Accordingly, the law restricted citizenship to those of German or cognate blood, serving as the means of exclusion and, ultimately, in the Third Reich, of elimination.[199] This involved replacing somatic anthropology with a racial diagnostic based on blood group testing, consolidating the practical scientific elements and reinforcing German folkdom, which "*. . . affords a means of investigating racial origins and relationships by correlating the frequencies of the hereditary factors concerned.*"[200] Buchanan and Higley (1921) reported on Landsteiner's (1901) classification of people into four blood groups: Group I (A), Group II (B), Group III (AB), and Group IV (O).[201] These groups were categorised by blood type, each linked to varying predispositions to certain diseases. A second classification system, focusing on blood type variation, facilitated the differentiation of ethnic groups. However, the Nazi regime later insidiously applied these findings to: (i) explore blood group disease susceptibility, and (ii) assess blood type distribution across ethnic groups.

The significance of the scientific discovery of blood types had a crucial bearing on the later German obsession with blood, which extended far beyond the realm of science. The biological role of blood in health coincided with a growing racial fixation on blood purity. This obsession drew deeply from medieval folklore, with blood libel being one example, in which Jews were falsely accused of using Christian blood in rituals. See heading Hatred transmission, 181. Under the Nazis, this myth coalesced with emerging scientific theories about race, legitimising the notion of racial superiority.

As a result, the Nazis adopted a covert system of primary serology tests to determine who was considered Aryan or Nordic, with the Nazis themselves serving as the control group. This approach mirrored that of 15th-century Spain, where blood samples were collected to establish racial purity, particularly targeting Jews. According to the writings of Diego de Simancas (1513–1583), the Spanish Inquisition similarly used blood testing to identify those with "*impure*" bloodlines. The underlying rationale was that Jews had a distinct blood group, that validated the idea of racial division and purity.

The pretext of the 'findings of science,' constituted an exigency to formulate a hard and fast blood sampling policy, in order to identify and detect these disease carriers by chemical means with 90 percent accuracy.[202] One cannot change one's blood, Proctor (1988) cited. Thus, genealogical and racial-hygiene frameworks circulated within genealogical circles during the Weimar Republic (1918–1933), supporting ancestry-based identification of Jewish people.[203] See heading Legislation and Genetic Health Courts, 153, how the blood group testing schema was later applied within the Nazi regime, demonstrated both its partial success and failure. However, Ehrenreich (2007) argued that it was flawed criterion to determine from an examination who was Jewish, given that a Jewish person could be physically identical to an "*Aryan/Nordic*"—tall,

---

[198] Hutton, 114, citing Schallmayer, W. (1918), 385 and Essner, C. (2002), 43.
[199] Hutton, 89–97, *92, 208; Schafft, 217; Snyder, L. H. (1926). Human Blood Groups, 234.
[200] Hutton, 208, citing Geisenhainer, K. (2002), 127ff., 169ff; Mazumdar, M. H. (1990). Blood and Soil; Snyder, L. H. (1926). Human blood groups, 234; Weinreich, M. (1999). Hitler's Professors, 170–77.
[201] Buchanan, J. A. & Higley, E. T. (1921). The relationship of blood-groups, 247–55. See also Mazumdar, P. M. (1990). Blood and soil, 188, generally, 187–218; Schneider, W. H. (1996). The history, 277–303.
[202] Proctor (1988), 79n.47, citing DA, 63(1933), 570; Nazi Manifesto Starts Eugenics Propaganda, New York Tribune, September 6, 1933.
[203] Ehrenreich, 49–50, 57; Proctor (1992). The Nazi Doctors, 20.

blond, blue-eyed, "*Roman-nosed,*" with blood type A and vice versa.[204] Dr Ernst Brandis (1936) a senior legal bureaucrat, defined German blood 'racial mix' as follows—

> The German people . . . is composed of . . . the Nordic, Phalian, Dinaric, Alpine, Mediterranean, East Elbian races . . . and mixtures between these. The blood of all these races and their mixtures, which thus is found in the German people, represents 'German blood.'[205]

This was chosen because as Feldscher (1943) explained, the racial mix of the German *Volk* has existed for millennia, inferring there was a large enough community of shared biological inheritance to test against.[206] The definition of race conformed with the German *Volk*, rendered a narrower meaning which consists of history of the German race: one's physical anthropological make-up, and mind and spirit.[207] While a conceptual definition helped to diffuse the confusion, it was not definitive. By the *fin-de-siècle* period, racial science and hereditarian thought had developed sufficiently to furnish later Nazi ideology with a comprehensive framework for understanding human difference expressed in quasi-biological terms.[208]

[204] Barzun, 147; Ehrenreich, 4; Schafft, 73.
[205] Hutton, 92, citing Brandis, E. (1936), 33.
[206] Hutton, 93, citing Feldscher, W. (1943), 20–21.
[207] Hutton, 156, 209, quoting an unpublished Karl Saller handout.
[208] Ehrenreich, 17–32; Hutton, 208.

CHAPTER 2. CULTURAL FOUNDATIONS OF RACIAL IDEOLOGY AND THE MYTH OF THE NEW GERMAN MAN

A. Pioneering the Next Phase of Racial Policy (1859–1880s)

i. Adoption of Darwinism into Germany Society Continuity, generally

AS AN INTRODUCTORY note, beginning with Bacon's New Atlantis (1627) the transmission of scholarly work on racial science and genetic developments evolved with each epoch, ultimately establishing the sciences.[1] The emergence of the New Atlantis advanced civilisation paradigm, highlighted the mass amount of "*inferiors,*" to which Social Darwinism was offered as the solution. Accordingly, Darwin's hypothesis was the foundation for the Third Reich's scientific basis of antisemitism, Aryanism, and eugenics—three separate branches or tracks, each one distinct. They progressively converged between the 1860s and 1890s, expanded and consolidated, and then converged again during the Third Reich, culminating in the Holocaust. It is also worth noting that, concurrent with the publication of Darwin's *On the Origin of Species* ("Origin") (1859), Aryanism was internationally accepted as general usage by the 1860's. Aryan supremacy ideology had formed an intricate link with Darwinist concepts and merged shortly thereafter in 1863.[2]

It influenced key figures to write theories of human variation, which were subsequently incorporated into scientific racism. Buffon played a pivotal role in shaping early conceptions of human variation and natural hierarchy. Although similar to Darwin's work, in title only, with an emphasis on race, Buffon provided a model for degeneracy theories. He theorised 'race' as evolution of variation into race.[3] In addition, Buffon was considered to be on par with Darwin due to the significant impact of his ideas on man and animals, as well as the lasting influence of his works on later literature. His theories contributed to the perception of European intellectual superiority over Africans—a century before Darwin's ground-breaking hypothesis.

Alexander (2019) affirms while Darwin may have formalised evolutionary theory, the core ideas were already in circulation long before his time. Indeed, Alexander distinguishes these earlier formulations as "*. . . lack[ing] the mechanism of natural selection to explain how new species were produced. . . .*"[4] These were the sources that Darwin and others consulted to formulate his hypothesis, *Origin*, as did Darwin's anthropology predecessors had done. For example, Jean-Baptiste Lamarck's (1744-1829)

---

1 Ehrenreich, 16.

2 Banton, M. (2015). What We Now Know, 18; Endelman, T. M. (1997). Comparative Perspectives, 97; Poliakov, 255–56.

3 Fredrickson, G. M. & Camarillo, A. (2002, 2015). Racism, 58. Poliakov (1974), 165–168; Popkin (1974a), 136; Todorov, T. (1993). On Human Diversity, 96–97, generally, 96–106.

4 Alexander, N. (2019). Race, 54, 56.

hypothesis of evolution shaped early ideas from inheritance of acquired characteristics.*[5] Most notably, Chase (1980) held that the founding father of scientific racism was Thomas Malthus (1766–1834), whose '*Essay on Population*' (1798) introduced the theory of the struggle for existence. In accordance with Merle's (1936) medical dictionary definition of Darwinism, it avers—

> [A] doctrine that a struggle for survival takes place among organisms produced in excessive numbers (struggle for life'), which, by means of natural selection, has the consequence of the survival of the fittest, and finally, on grounds of inheritance and adaptation, gives rise to the emergence of new species.' 'Excessive numbers' points to the influence of the population theorist Thomas Malthus on Darwin's work.[6] (Internal citations omitted)

Other influential figures include Thomas Henry Huxley (1825–1895) and Robert Knox (1791–1862), who, nine years before Darwin in 1850, declared that "*. . . with me, race, or hereditary descent, is everything: it stamps the man.*"[7] Knox further asserted that "*. . . race is everything in human history.*"[8] As already discussed, Blumenbach was held to be one of the founder's of anthropology, but it was—

> [Pierre-Louis Moreau de Maupertuis] (1744, 1746) and Kant (1775, 1785, 1790) [that] had been the first to point out the importance of selection for living beings . . . the influence of the conception of selection only really begins to show itself after the foundations of modem biology were laid by Darwin's [On the Origins of Species by Means of Natural Selection or The Preservation of the Improved Races in the Struggle for Life] in 1859.[9] (Emphasis added)

It is no secret that *Origin* (1859) was primarily concerned with plants and animals competing for space, and speciation. With that, birthed the biological science branch of Darwin's hypothesis. There was no application to humans. In due time, Darwin's

[5] Burkhardt, R. W., Jr. (2013). Lamarck, evolution.

* Many negate this Lemarckism view. One school of thought held a hybrid view implicit of both inherent and environmental factors. They can be selected and altered. See Freeden, M. (1979). Eugenics and Progressive Thought, 646–47. Another theory of German zoologist August Weismann, submits an "independent, immutable germ plasm" theory. He suggests that "single genes" that built "even the most complex" body parts, argued that the social environment "was impotent to alter the human condition." (See Black, E. (2012). War Against the Weak, 17; Friedlander (1995), 2; Graham, L. R. (1977). Science and Values, 1135. Weismann, A. et al., (1891). Essays Upon Heredity and Kindred Biological Problems, 193–94; Weismann, A. et al., (1893). The Germ-plasm: A Theory of Heredity, 282–83). Yet, racial policy assumes that certain characteristics are inheritable, such as mental capacity, energy, or leadership, which modified by environment influences. See Glass, B. (1981). A Hidden Chapter of German Eugenics, 361. Recent studies has shown that particular tobacco and alcohol traits are heritable behaviours, and cancer development is related to genetic and environmental factors. See Feng, X. et al., (2022). Association Between Genetic Risk, Adherence to Healthy Lifestyle Behaviour and Saunders, G. R. B. et al., (2022). Genetic diversity fuels gene discovery.

[6] Merle, W. (1936). Guttmanns Medizinische Terminologie, 226, quoted in Hutton, 173–74.

[7] Alexander (2019), 63; Chase, A. (1980). The Legacy, 6; Claeys, G. (2000). The Survival of the Fittest, 224; Himmelfarb, G. (2014). Evolution and Ethics. See also Nordenskiöld, 285.

[8] Knox, R. (1850). The Races, 6, 8, respectively. See also Hudson, N. (1996). From Nation, 248; Stuurman (2017), 351–52; Claeys, 239.

[9] Günther, 258; Bölsche, W. & McCabe, J. (1906). Haeckel, 140; See also Banton (2015), 18; Barzun. 94, 129, 219; Müller-Wille, S. & Rheinberger, H.-J. (2007). Heredity, 17; McLaughlin, (2007). Kant on Heredity, 277–91; Cf. Müller-Wille & Rheinberger, 20 (biology origins date).

hypothesis would become an annexation to German Aryanism.[10] Indeed, no sooner had the ink dried, Origin, than it was promptly translated into German within a year of its publication, where it "*. . . profoundly influenced German thought.*"[11] Weiss (1987b) highlights its considerable impact in Germany, where it was met with widespread acclaim and popular as a philosophical work—

> Darwin's views on the importance of natural selection for . . . social evolution, as well as his casual remarks about the desirability of human breeding, did not fall on deaf ears in Germany.[12]

The criticism by German zoologist Ernst Haeckel (1834-1919) was the intellectual precursor to Darwin's, *The Descent of Man* twelve years later, in 1871. Haeckel's interpretations of evolution profoundly shaped Darwin's views, driving a reformulation of his hypothesis. This persuasion led to a restatement of evolutionary theory that integrated human evolution and social progress, notably shaping the development of eugenics as a proposed solution to human degeneration.[13] Flowers (2022) succinctly acknowledges, "*[i]t was [Haeckel] who introduced Darwin's ideas to German science and further[] popularised them . . . through the Monist League.*"[14]

Worse still, Darwin prophesied that the 'civilised races' would exterminate and replace the 'savage races', referring to all 'low-standing human races', including the Jews, throughout the world.[15] No sooner had *The Descent of Man* appeared than it was appropriated for antisemitic propaganda. Poliakov (1974) unearthed a chilling account in which the German writer Ottomar Beta, in 1876, petitioned the German authorities. Citing the anthropological applications of Darwin's work, Beta framed a "*struggle for existence*" between the German-Aryan race and what he denigrated as the parasitic Jewish race. Consequently, he asserted that legislative measures against Jews were scientifically justified, ultimately inferring their systematic elimination.[16] Darwin based his observations on the animal kingdom and its inherent brutality in nature. His subsequent philosophical reflections on the beastly killings and attendant suffering of lower animals led to the formation of principles with which we are all too familiar: (i) variation, (ii) heredity, (iii) struggle for life, (iv) selection, and (v) adaptation.

It appears that Bölsche & McCabe (1906) emphasises these principles just to demonstrate how profound Haeckel's position was,[17] and much more so than Darwin. Haeckel not only adopted Darwin's Origin, but modified it to apply to human society, for what would later be known as 'Social Darwinism'—

> Special stress is laid on the ever-advancing, ever-uplifting element in evolution. Man is again introduced into the subject. He has "evolved" from the

10 See Weikart (2004), 3–9, for a summarised discourse how Hitler hijacked Darwinism.
11 Weikart (1993), 474; Weikart (2003), 276; Weikart (2004), 10.
12 Weiss (1987b), 30, 33, respectively.
13 Darwin, C. (1876). The Descent, 617–18; Weiss, S. F. (1987b). Race Hygiene, 28; Weiss, S. F. (2010). The Nazi Symbiosis, 23, et seq.
14 Flowers, 66.
15 Darwin (1981). Descent, 1:201, quoted in Weikart, R. (2006). The Impact, 97, 102, 104.
16 Beta, O. (1876). Darwin, Deutschland und die Juden, 11, 32 in Poliakov (1974), 294.
17 Bölsche, W. & McCabe, J. (1906). Haeckel, 135.

> brutality of the animal.[18]

It is not the animal or beast within man; rather, it is the individual who makes the cognitive or volitional choice to lower themselves and become like one. Wertham (1966) draws attention to a noteworthy truism, in comparison to the Darwinian model, that animal behaviour is a higher standard than what Darwinian adherents have acceded to—

> By and large, animals kill for only one reason: survival. Violencc is not identical with hate and hostility. Animals do not kill for hate, spite, revenge, sadism, or greed. They are generally averse to killing members of their own species. Most important of all, they never systematically kill large numbers of the same species. That is a prerogative of man. So when we speak of massacres, extermination camps, mass bombings, and so on, we should not refer to the "bestial" in man. It isn't the beast, it is man himself . . . .
>
> . . . .
>
> Even animals well equipped for attack and predisposed to it may never display destructive aggression unless it is elicited by specific environmental conditions. In contrast to hunger and sex, destructive aggression is not an instinct or even a drive, in the usual sense of these terms.[19]

It is apparent that *New Atlantis* utopia is congruent with Social Darwinist ideals. Weikart (2003) articulated how a trend emerged in writings advocating for racial extermination of those deemed inferior—

> [A] host of leading scientists, physicians, and social thinkers who embraced Darwinian social explanations and eugenics . . . the key to progress was the annihilation of the "lower races," who stood in the way of advanced culture and civilization.[20]

It was common knowledge that *Origin* was not intended to be racially themed with the intention of applying it to human society. Needles to say, Darwin involuntarily bolstered the same racist ideological outcomes as the Aryan/Nordic ideology did. The predominant actor that would revolutionise scientific racism was Haeckel. The manner that he went beyond Darwin was radical. Haeckel's approach differed in that it introduced a continuous grading classification system, ranking various degrees of humanity along a spectrum that included and depicted, "*. . . animal ancestry of humanity the most important aspect of evolutionary theory.*"[21] He was responsible for the birth and popularisation of Darwinism into German intellectual life.[22] Scientific racism has

[18] Bölsche & McCabe, 132, 150–51. See also Graham, L. R. (1977). Science and Values, 1135; Graham, L. R. (1981). Between Science, 24; Haeckel (1868). Natiirliche Schopfungsgeschichte, 5, 93 in Weikart (1993), 475; Weikart (2006), 94.

[19] Wertham, F. (1966). A Sign, 25, 27.

[20] Weikart (2003), *273, 276.

[21] Forti, 13. See also Weikart (2003), 275; Weikart (2004). From Darwin to Hitler, 11, 146–49; Zmarzlik, H.-G. (1972). Social Darwinism in Germany, 442; Zmarzlik, (2011), 10.

[22] Bölsche & McCabe, 167, 172; Friedlander (1995), 9; Gasman (1971); Kelly, 5, 21, 23; Nordenskiöld, 505–27; Schafft, 41; Stein, G. J., 53–54; Weikart (2003), 277.

come a long way since the 1700s, with Kant and Haeckel among the *dramatis personae.* Weikart (1993, 2003) enlightens us how it transformed with the arrival of Haeckel—

> [S]cientific racism based on social Darwinism radicalized racism in the nineteenth and early twentieth centuries by providing a scientific rationale for exterminating non-European races.[23]

The reception of Darwin's hypothesis had a rapid and an "*immediate impact,*" on the German scientific community, to the extent that Haeckel led a debate at the 1863 National Congress of German Scientists in Stettin, advocating Darwinism as a path to progress.[24] Haeckel's address went beyond Darwin's published ideas by refracting biological evolution through a lens of human and social developmental reforms. So much so, Germany "*. . . offered the richest environment for Darwinism to expand beyond the confines of science.*"[25] That same year, Haeckel fused Darwinian evolution with cell biology under the encouragement of microscopic anatomist Max Schultze (1825–1874). According to Weindling (1993), this synthesis was also embraced by the most unlikely of candidates: the staunch physical anthropologist Rudolf Virchow.

However, Zmarzlik (1972, 2011) presents a differentiated perspective. In his account, this extension of Darwinian theory was "*. . . condemned by the leading natural scientists of the day, Rudolf Virchow in the lead.*" Both scholars are, in fact, correct: Virchow largely supported certain scientific applications of Darwin's theory, in particular, biology and medicine. Yet he was sharply critical of the broader philosophical or ideological extrapolations, especially those advanced by Haeckel, which he viewed as speculative and politically dangerous.[26]

For the next four decades, Haeckel became the lead spokesman for scientific racism and, "*. . . the chief apostle for Social Darwinism in Germany,*" advocating that the racial struggle is a significant part.[27] Arguably, many leaders of the Völkisch movement were influenced by, and were connected to, either personally or ideologically to Haeckel, including Hitler, which paved the way for the election of Hitler's party.[28] Gasman (1971) went further to the extent that Nazi biology ideas flowed from Haeckel—

> [N]early all other leading figures in the field of eugenics and racial science in Germany were deeply and consciously indebted to Haeckel for many, if not for most, of their ideas.[29]

As for Hitler himself, according to Gasman's audit trail of Hitler's interactions and associations, Hitler showed his indebtedness to Haeckel in his conversations, speeches, and writings indicative of that they both "*. . . shared a common sense of mission in*

[23] Weikart (1993) 475–77; Weikart (2003), 275.
[24] Gasman, 7–9; Kelly, 5, 22. (See Kelly's opposing point, 93. He asserts Haeckel had many enemies and received little respect within the scientific community). See also Weikart (1993), 470, concerning Kelly's critique. Nordenskiöld, 510; Schleunes, 30–31; Weindling (1993), 29; Weiss (1987b), 33.
[25] Kelly, 5.
[26] Weindling (1993), 38; Zmarzlik (1972), 439; Zmarzlik, (2011), 7.
[27] Schleunes, 30–31, citing Bölsche & McCabe, 150–151, 172; Gasman, xxii, ; Poliakov, 284; Weikart (2004), 186.
[28] Gasman, xxiii, 151, 159–61.
[29] Ibid., 147; Poliakov, 371n.108, quoting Gasman.

regard to man and to his relationship to nature."[30] However, it is worth bearing in mind Weikart's (2004) adverse note on Gasman's thesis: "*. . . Haeckel-to-Hitler thesis ultimately failed in part because he ignored the many areas of sharp disagreement between Haeckel and Hitler.*" He then went on to mention that—

> However, while acknowledging the many differences, we should not ignore the many features of Monist ideology that featured prominently in Hitlers worldview, such as eugenics, euthanasia, and social Darwinist racism.[31]

In fact, Hitler's ideas, solutions and his views on history, politics, religion, Christianity, nature, eugenics, science, art, and evolution, were completely Hacekelian and Darwinistic, even adopting Haeckel's Spartan, beastly approach to purge society. As early as 1920–24, Hitler embraced this Darwinist manifesto, which became central in shaping his thinking during his formative years, prior to his stint in prison.[32] This is consistent with Hitler's Table Talk (2000) discussions, applying Natural Selection ideology. It was recorded that he went as far as to compare the Jewish people with mice (the weak) being killed by a cat (the strong): "*[w]hose fault is it when a cat devouring a mouse? The fault of the mouse, who has never done any harm to a cat?*"

Hitler's advisers, many of whom were anthropologists, influenced his policies to advance his agenda.[33] Haeckel's ideas, grounded in revolutionary Social Darwinism and *Volk* ideology, formed the foundation of this programme. Tenenbaum (1956) draws out this point, emphasising just how deeply the Nazis embedded Natural Selection terminologies into their political systems—

> [T]he [Nazis] specifically incorporated the animal theory of evolution in their political system with nothing left out . . . Their political dictionary was replete with words like space, struggle, selection, and extinction (Ausmerzen) . . . The world is a jungle in which different nations struggle for space. The stronger win, the weaker die or are killed.[34]

Hitler maintained that it is useful to understand the laws of nature in order to uphold them and concluded with: "*Thou shalt preserve the species.*"[35] Hitler's book, *Mein Kampf* [1925] (1971) contained many references to evolution eugenics. Hitler wrote that he had, "*. . . discussed the relationship between politics and natural selection as the basis for a successful German national policy.*"[36] The emphasis on selection is the primary and core idea of Darwinism, not evolution. Haeckel (1900) explained what selection meant when he wrote, "*. . . Darwin created his momentous 'theory of selection' which is what we properly call Darwinism.*"[37] Evolution was part of Darwin's hypothesis; but as Gasman (1971) explained, evolution was describing the outcome of selection. Not the other way round.[38]

---

30 Gasman, 161–62.
31 Weikart (2004), 70.
32 Beyerchen, 95; Kershaw (2000), xli; ; Weikart (2006), 93; Weikart, R. (2013). The Role of Darwinism, 537–38, 541.
33 See Gasman's comparative analysis at *161, 159–169; Jäckel, 90–92; Schafft, 124; Weikart (2004), 215.
34 Tenenbaum, 211–12.
35 Hitler, A. (2000). Hitler's Table Talk, 141.
36 Hitler, A. [1925] (1971). Mein Kampf, 131–37. See also Claeys, 224–25 and Stein, G. J., 51–52.
37 Haeckel (1900). The Riddle of the Universe, 79.
38 Gasman, xxi; Stein, G. J., 53.

Nazi literature, films, speeches, tracts, etc., were "*. . . more faithful to the fullness of Darwin's scientific views.*"[39] Simply put, Social Darwinism is the "*. . . application of the idea of evolution to a higher social type on the basis of social competition between "fit"* and "*unfit*" groups and individuals . . . ," within society where 'selection' and 'struggle,' were closer to the radical insight of Darwin's effort.[40] Darwin was an adherent of Theodor Fritsch (1852–1933) (see analysis at heading Theodor Fritsch, 200) who expressed a similar sentiment in relation to maintaining viability and health of the German race—

> Morality and ethics arise from the law of preservation . . . to raise the species to an ever higher level of physical and mental perfection, that is moral. . . . Whoever seeks to preserve the degenerate and depraved [is false humanity. It] limits space for the healthy and strong, suppresses the life of the whole community, multiplies the sorrows and burdens of existence, and helps rob happiness and sunshine from life.[41]

Accordingly, Hitler believed that Germany was going to face continual increases in population numbers in the 'tangible future.' Therefore, he concluded, drastic measures would be needed to eliminate the "*unfit*" population numbers.[42] Ludmerer & Lappé (1978) loosely framed the practicalities of the unfit in society that they constituted nothing less than a "*major menace.*" The implied solution was to remove medicine provision, the means by which the unfit were increasing—

> [T]hey were surviving in increasing numbers because of modern medicine and charity, and they were procreating far more rapidly than the "fit," thereby diluting the concentration of valuable hereditary characteristics in the population.[43]

Weikart (2004) summarises the Darwinist worldview as the idea that death fosters positive development in evolutionary progress; or, put another way, more deaths lead to greater progress: "*. . . the more death, the better.*"[44] Similarly, the propensity within Darwinist circles to devalue human life and worth, was not a view unique to Haeckel or Hitler, but it was a unique thought process to Social Darwinists who all shared this same optimism for progress.[45] August Weismann's (1834–1914), German cellular biologist and one of the founders of genetics, a die-hard social theorist, made "*. . . significant modifications . . . in the social meaning of Darwinian evolution within the 1880s and 90s.*"[46] David (1909), a Social Democratic member of the German parliament, unravelled the connection between Darwinism and eugenics, and was conversant with Weissman's ideas—

---

39 Stein, G. J., 53.
40 Claeys, 229; Gasman, xxi, xxiii, respectively; Semmel, 114; Stein, G. J., 53.
41 Savage, 413; Weikart, R. (2004). From Darwin, 55.
42 Hitler [1925] (1971), 131–32.
43 Ludmerer & Lappé, 458; Tenenbaum, J. (1956). Race, 212. See also Allen's, 115–20, remarks on Ludmerer's analysis of the history of the eugenics movement, particularly in contrast to interpretations by other scholars.
44 Weikart (2004), *81, 148–49, 170, 187.
45 Ibid., 84, 86, 103.
46 Weiss (1987b), 27.

> This process of natural selection is frustrated through institutions of social assistance, which aim at preserving the life of damaged organisms, allowing them to reproduce and also preserving the lives of their progeny with inferior health."[47]

This rationale is not so abstract as you might think, since its logical basis is extrapolated from 'evolutionary' biology. Weikart (2004) poignantly expresses the practical significance of 'natural selection' in this regard: "*[i]f a particular society has already adopted "unnatural"* institutions that prevent natural selection from operating, it should quickly get rid of them."[48] For example, the application of 'unnatural' is a euphemism for Haeckel's 'artificial selection,' referring to the provision of hospitals and medical care. Modern medical science and humanitarian principles aimed at prolonging life, in turn, preserve the weak and sickly, allowing them to live longer than they might otherwise.[49] The weaker race would have died without intervention; they are two sides of the same coin. Thus, Haeckel positioned this within the context of eradication—

> Now, the longer the diseased parents, with medical assistance, can drag on their sickly existence, the more numerous are the descendants who will inherit incurable evils, and the greater will be the number of individuals, again, in the succeeding generation, thanks to that *artificial 'medical selection,'* who will be infected by their parents with lingering, hereditary disease.[50] (Emphasis added)

Weikart's explanation of 'artificial' selection argues that, rather than abolishing the 'unnatural,' society should preserve the 'higher culture.' He was referring to positive eugenics. In this sense, 'artificial' and 'unnatural' are two parts of the same equation. Haeckel's interpretation of 'artificial,' as referenced in his work *The Wonders of Life* (1905), contextually refers to the undue maintenance and prolonging of the lives of the unfit, who are "*. . . artificially kept alive in our modern communities, and their sufferings are carefully prolonged*" with medical assistance.[51] Stein (1988) elaborated further upon this point, highlighting that Haeckel was not referring to people with illnesses be kept alive, per se, but eliminating those he regarded as genetically imbalanced—individuals who "*. . . bred descendants who polluted the breeding pool and rendered the entire people less fit.*"

His category extended to sickly infants, lunatics, incorrigible, useless people, an umbrella, *inter alia*, for degraded criminals, unemployed, persons who have a weakness of the body and weakness of character.[52] See also Table I-9, 150. This post-1859 *Origin* doctrine gradually consolidated through successive refinements, ultimately reaching its full expression in the 1930s; examined further under Racial Hygiene Ideology Becomes Operational, 141. The terms 'useless,' 'useful,' 'worthless' 'worthy' applied to questions of fitness are framed in explicitly economic terms, denoting individuals as less or more productive assets. Claeys (2000) succinctly interprets this as an ideology of competitive obligation, in which citizens are expected to make a primary contribution—

47 Weikart (2004), 83 et seq., citing David, E. (1909). Darwinismus, 48.
48 Weikart (2004), 44.
49 Bialas & Fritze, 31–32; Hobhouse, L. T. (1911). Social Evolution, 21; Krausnick, 28.
50 Haeckel (1876), 1:173.
51 Haeckel (1905), 118.
52 Haeckel (1876), 1:172; Haeckel (1905), 118–19; Stein, G. J., 56.

> If 'fit nations were those capable of maintaining a competitive edge over their neighbors, "fit" individuals in political economy were those who were productive where others were not.[53]

In *Wonders of Life*, in not so many words, Haeckel painted the new upcoming German *Volk* as the epitome of a "*savage race*" by adhering to the broad measures prescribed under Spartan selection for the improvement of the race.[54] In Mein Kampf, Hitler set out four potential solutions to address overpopulation, framing them through the catchwords of 'selection' and 'struggle.' He deployed these terms allegorically yet with unmistakably direct intent, as noted above, to prepare the ground for their later transformation into national state policy. Indeed, Hitler expressed that the fittest who survive the inclemencies of existence are selected; proved by trials, who demonstrate their fitness by their survival.[55] The inheritance of experience, characteristics and the relative permanence of those adaptations as a result of survival are fixed by selection, guided by reasoned conceptions of social value (capabilities, achievements and quality of interactions); transmit the new genetic information by heredity[56] to form a new, *de novo* better, stronger species and offspring—

> [B]y exposing them with hard trials and deprivation with the result that all those who are less strong and less healthy are forced back into the womb of the eternal unknown . . . [those chosen] are thousandfold tested, hardened, and well adapted to procreate in turn, in order that the process of thoroughgoing selection may begin again from the beginning . . . .[57]

In the Nuremberg Trials, during examination of the defendant, Otto Ohlendorf, the prosecution lawyer, Mr. Heath, quoted to the defendant an excerpt from Hitler's speech from the Nuremberg Party rally in 1933, in which Hitler specified what the rights of the superior race were—

> A higher race, at first higher in the sense of possessing a greater gift for organization, subjects to itself a lower race . . . There thus results the subjection of a number of people under the will often of only a few persons, a subjection based simply on the right of the stronger, a right which, as we see it in nature, can be regarded as the sole conceivable right . . . .[58]

53 Claeys, 234–35.
54 Haeckel (1905), 21, 119; generally, 113–20.
55 Hobhouse (1911), 24.
56 Halliday, R. J. (1971). Social Darwinism, 394; Haeckel (1876), 169–70; Hobhouse (1911), 28–34, 37.
57 Hitler [1925] (1971), 131; Stein, G. J., 51–52, citing Hitler. See also Call, L. (2016), 8–9, 17. Cf. Haeckel (1905), 256.
58 Nurenberg Military Tribunals. (1954). Trials, IV:279; See also Hitler, 132; Tenenbaum, 212, citing Hitler's speech.

ii. The Confluence of German Aryanism and Darwinian Theory: A Synthesis with Anthropology

THE ADVENT OF Darwinian evolutionary thought, the subsequent growing interest in Eugenics, was a complete symbiosis with the Aryan superiority race theory.[59] Weikart (2013) wrote a compelling rebuttal against historians who contend that the Nazis rejected Darwinian evolution. His analysis demonstrates that many historians have missed the mark by (i) failing to examine the scientific underpinnings of the Nazi embrace of Darwinism, and (ii) reflexively resisting the possibility that a Darwinian-inflected Nazi ideology operated in broad harmony with prevailing currents in German scientific thought. Weikart went on to state—

> Indeed the Nazi embrace of Darwinism in their racial ideology demonstrates the influence of science on Nazi ideology. Nazi racial ideology was largely consistent with the scholarship on race taught at German universities. This makes even clearer why so many German anthropologists and biologists supported Nazi racism—they were already committed to it before the Nazis came to power.[60]

Barzun (1938) expresses this harmonious union rather poetically when he states: "*. . . Evolutionism dovetailed beautifully with the notion of [Aryan] superior and inferior races;*" thereby strengthening the Aryan construct, which was as a *sine qua non* central element of "*. . . Darwinism and human evolution with Nazi ideology.*"[61] In fact, Weindling (1993) supports this point with a similar finding, in as many words, with the addition that the Darwinian paradigm provided comprehensive frameworks for the establishment of an orderly and progressive society. Barzun (1938) extends this argument, highlighting

---

[59] There is much debate and ambivalence towards the notion that the Nazis embraced Darwinian human evolution. Kelly (1981:120) criticises Gasman's book and his views on Haeckel. He states that Gasman greatly exaggerated Haeckel's Social Darwinism and his works. Further, Weikart (2013:537) noted Gasman's contradiction, whereby he claimed Social Darwinism was incorporated into the Nazi apparatus, but Gasman then argued that the Nazis rejected human evolution. Weikart (2013:537) posed the question: "*How is it possible to embrace social Darwinism, while rejecting Darwinism and human evolution?*" The answer is submitted that they cherry-picked some Darwinist ideas. The overarching peripheral reason why it is suggested that the Nazis rejected it, is that Darwinist ideas fell foul of their Aryan/Nordic beliefs. German ideologues saw themselves as the elite race, and therefore, it was repugnant and insulting to be compared to animals to determine where they appeared on the hierarchy chain. Thus, they dismissed the idea they evolved from apes. (Mosse's (1964:103); Kurlander (2017a:243)). Mosse states that the Nazis rejected it because they stressed the immutability of the German race.

In rebuttal, let us not lost sight of the simple truth: attempts to impose rational coherence on Nazism are misleading, given that Aryanism was fundamentally irrational from its inception. It was a confused and flawed ideology grounded in myth. The real truth is analogous with the effort to disentangle evolution from Nazism, or eugenics from Nazism through its rebranding as "genetics." In other words, saving the 'hide of evolution' so it is not contaminated with Nazism (Weikart (2004:8). Upon further analysis, their documented Nazi beliefs and behaviour is undoubtedly consistent with Darwin's hypothesis. However, there is evidence to support that they did adopt it. For instance, the Social Darwinian eugenics movement was an extraction from Darwinism which, in turn, propagated a "survival of the fittest" worldview that aligned with their Aryan ideology. Further, Nazism implicitly operated along the same lines of a superiority rationale. It added further weight to Aryan ideals and buttressed them through the inclusion of distorted Darwinist ideas and genetics, producing a synthesis intended to distinguish "superior" and "inferior" elements in order to improve breeding stock (see Banton (1978:94; Hutton 2005:27).

[60] Weikart (2013), 539, 542.

[61] Barzun, 94; Hutton, 9, 28–28, *173–180, 198, 212; Ludmerer & Lappé, 457; Stein, G. J. (1988); Snyder, L. L. (1962). The Idea of Racialism, 26–27; Tenenbaum, 89; Weikart, R. (2013). The Role of Darwinism, 538–39, (547–48, quoting Ernst Lange in *Wille und Macht (1936 Vol. 6, no.5.)*

that Darwin's theory emerged contemporaneously with Adolphe Pictet's (1859, 1863) mammoth two volume treatise on Aryan race-belief, which coincided with the publishing of Darwin's theory in 1859, to which Barzun confirms by noting—

> By an interesting coincidence, in that same year 1859 appeared what is perhaps the most representative monument of nineteenth-century Aryan race-belief, a two-volume work entitled: *Indo-European Origins*, or *The Primitive Aryas, an Essay in Linguistic Paleontology* by the Swiss scholar Adolphe Pictet.[62]

Puzzo (1964) adds that Aryanist ideals existed long before the 1850s that there had been "*hints and suggestions*" about the Nordic characteristics of the Germanic peoples.[63] Accordingly, 1859 was the *mise en scène* synthesis point, marking a perfect wedding between the two ideologies—a seamless integration of evolutionary theory and racial discourse. Gasman (1971) avers that Germany did not just embrace Darwinism partially or on a moderate basis—they adopted it unreservedly, more than any other European country, and transformed it beyond recognition into an comprehensive intellectual endeavour. Darwinism was an all-pervading, logical, closed systematic way of thinking, enhanced with laissez-faire economic competition rhetoric.[64] It was interpreted through Darwinist ideology and then applied to society, Hence, the term 'Social Darwinism.' Weikart (1993) expressed a similar finding—

> For Haeckel economic competition illustrated the universal and inevitable rivalry in human society. He favored free competition among workers because this would foster improvement.[65]

The 1860s and 1870s collective struggle or competition, whilst upholding individual competition was additionally adopted. Although, it was Friedrich von Hellwald (1842-1892) ethnologist, heavily influenced by Haeckel, combined individualism and collectivism. In Hellwald's book *Culturgeschichte* (1875) (trans. History of Culture), he described Haeckel's approach as "*path-breaking,*" crediting him as the first to accurately apply Darwinian theory to politics. In Zmarzlik's (1972, 2011) own words, this socio-political convergence became the "*. . . principal model for social and political thinking.*"[66] Another associate of Haeckel, Wilhelm Preyer (1841–1897), professor of physiology at the University of Jena, had the same Darwinian frame of reference: "*In two published lectures (1869 and 1882) [Preyer,] was also forthright in applying Darwinism to society.*"[67] Other German Darwinist practitioners within the same time period (1860s and 1870s) included, economist, Albert E. F. Schiffle (1831–1903) regarded as the most influential thinker to apply Darwinian thinking to ethical and social theory; Alexander Ecker; Heinrich August Ziegler; Wilhelm Roux; Julius Frobel who applied Darwian processes to political and social discourse, et. al.[68] By the time of 1895, it had grown into a national and racial competition affair.[69]

62 Barzun, 138; See also Poliakov, 259–60.
63 Puzzo, D. A. (1964). Racism, 584.
64 Gasman, xiii; Hofstadter, 5–6; Weikart (1993), 471, 474; Weindling (1993), 27.
65 Weikart, R. (1993), 476.
66 Zmarzlik (1972), 441; Zmarzlik (2011), 9. See also Weikart (1993), 484; Weikart (2004), 34, 168–69.
67 Weikart (1993), 476; Weikart (2004), 91.
68 Weikart (1993), 477–86; Weikart (2004), 28.
69 Weikart (1993), 481–82, 485; Weikart (2003), 276 Weikart (2004), 93.

In 1895, Max Weber (1864–1920), then professor of economics, used his inaugural address at Freiburg to apply evolutionary-biological terminology drawn from natural selection—struggle for existence, selection (Auslese), and adaptation (Anpassung). In doing so, he later made explicit, "*. . . Weber supported economic competition and free trade because he saw these as a part of human struggle.*"[70] Massing (1949) posited that it regulated their political and personal behaviour, and it was uncompromising in its conclusions.[71] It helped to validate German military successes in the Bismarckian wars of 1864–1870, while simultaneously supplying the desired scientific rationale for future aggression, conquest, and destruction, "*. . . as well as threatening (necessitating it) . . . revolving around concepts of hygiene and purity . . . fulfilling psychological necessities [at both] symbolic and populist levels.*"[72] What Savage (2007) meant by these psychological necessities is that they symbolically heightened the morally cleansing aspect of violence, while the populist aspect aimed to appeal to nationalist sentiments to promote violence as justified because the "*. . . nation-state allowed perpetrator groups [to act] . . . .*"[73] No wonder, then, that Germany boasted, drawing on Darwinian terminology grounded in its own Aryan/Nordic racial assumptions, which Pictet's publications appeared to affirm,[74] cemented how Germans understood and perceived themselves—

> In the struggle for survival, the fitness of Germany had been clearly demonstrated. Was not this expressive of a superior spirit or Volksgeist.[75]

Further, it verified for them that their "*. . . prejudices were actually expressions of scientific truth,*"[76] which was what they had been trumpeting all long. This "*superior spirit*" that Scheunles (1970) refers to is exemplified by Haeckel's admiration and praise for the Spartan practice of slaughtering the weak or sickly newborns, as described in Haeckel's book, *The History of Creation* (1876)—

> A remarkable instance of artificial selection in man, on a great scale, is furnished by the ancient Spartans, among whom, in obedience to a special law, all newly born children were subject to a careful examination and selection. All those that were weak, sickly, or affected with any bodily infirmity, were killed. Only the perfectly healthy and strong children were allowed to live, and they alone afterwards propagated the race. By this means, the Spartan race was not only continually preserved in excellent strength and vigour, but the perfection of their bodies increased with every generation.[77]

Haeckel (1876) wrote another similar sounding statement—

---

70 Mommsen, W. J. (1974). Max Weber, (untranslated) 43, cited in Weikart (1993), 478; Weikart (2013), 539 et seq. See Mommsen, W. J. (1990). Max Weber, 39–41, (English version).
71 Massing, 84, 102.
72 Savage, 404–05. See also Chase, 8, 369–70; Claeys, 226; Friedlander (1995), 12; Schleunes, 31.
73 Savage, 405.
74 Already cited in Barzun, 138.
75 Schleunes, 31.
76 Ibid., 32; See also Ludmerer & Lappé, 457.
77 Haeckel (1876), *1:170, 171–74; Burleigh & Wippermann, 30; Stein, G. J., 56; Weikart (2004), 146, 153–54. See also Weiss (1987b), 33–34.

> If any one were to venture the proposal, after the examples of the Spartans and Redskins, to kill, immediately upon their birth, all miserable, crippled children to whom with certainty a sickly life could be prophesied, instead of keeping them in life injurious to them and to the race, our so-called "humane civilization" would utter a cry of indignation.[78]

Far from being one person's expression of individual opinion on how society should function, Haeckel's statement represents a broader vision for societal transformation. His intent was to reconstitute society based on collectivist principles. Haeckel's call for this culling reflects the stance that Burleigh (2002) characterises as a repudiation of liberal-sentimental "*humanitarianism.*"[79] Haeckel advocates for a radical rethinking of societal structures, where such harsh measures were not viewed as extreme but as normative in 'civilised' society. Weikart (2004) elaborates on this by citing Darwin's observation of the savage instinctive hatred and infanticide displayed by queen bees toward rival newborn queens.[80] Friedlander (1995) observed that, after Darwin, evolutionary thinking became so instrumental within the biological sciences that it "*. . . led most scientists to advance theories of human inequality as matters of scientific fact.*"[81]

Physical anthropology, before its merger with Darwinism, centred on examining the anatomical lines of the skull: its shape and the 'facial angle', as putative measures of intelligence, along with attendant features such as 'facial expression' and 'body appearance.' These traits were then used to evaluate individuals and classify their purported "*good*" and "*bad*" mental characteristics.[82] See also biometrics analysis under heading Biometric Technologies, 317. Early anthropology can be concisely described as the study of human beauty, with race serving as the controlling constant. Mosse (2020) argued that these two crucial components, the aesthetic ideals of Enlightenment sciences and human beauty and the intellectual traditions of philosophy and science, influenced one another reciprocally. He then went on to argue—

> [T]he scientific endeavor was directed toward a classification of the human races according to their place in nature and the effect of the environment. The beginning of the new science of anthropology . . . was based upon the attempt to determine man's exact place in nature through observation, measurements, and comparisons between groups of men and animals.[83]

Darwinist ideas began to chip away at the use of physical anthropology to measure racial superiority and inferiority through phrenology (the reading of a person's character from skull measurements) and physiognomy (the reading of a person's character from facial features). These practices were grounded in anatomy and physiology, with the science of craniology emerging as an offshoot of phrenology.[84] Moreover, Schmuhl (2014) quite rightly links "*. . . the genetic foundation for eugenics was laid not in experimental genetics, but in anthropology and psychiatry.*"[85] Consequently, phrenology

78 Haeckel (1876), 1:173. See also Burleigh, M. (1994). Death and Deliverance, 13; Burleigh (2002), 120.
79 Burleigh (2002), 120.
80 Weikart (2004), 74–75.
81 Friedlander (1995), 1, citing Bergmann, A. et al., (1918), 122.
82 Marten, H. G. (1999). Racism, 26.
83 Mosse (2020), 3, 4, respectively. See also Bauman, Z. (2000). Modernity, 218.
84 Bauman, 218; Davies, J. D. (1971). Phrenology, 3; Poliakov, 264.
85 Schmuhl, H. W. (2014). Eugenics and Racial Hygiene, 32.

quickly became an annexation to anthropology. In other words, the development of, and integration of Aryan ideology into anthropology was progressive.

In Poliakov's (1974) analysis of anthropology, from its inception in 1860, the term Aryan was already in circulation. Paul Broca (1824–1880), a French neurologist, anatomist, and leading anthropologist, and founder of the *Société d'anthropologie de Paris* (French Anthropological Society) used anthropometrical data to argue for a "*natural racial hierarchy,*" regarding Aryan races as "*perfectly scientific.*"[86] Ettinger (2011) remarks that he treated biological "*. . . data as an objective basis for the definition of human groups and even for the appreciation of their qualities,*"[87] and established a causal relationship between intelligence and size of the skull—

> [T]he brain is larger in mature adults than in the elderly, in men than in women, in eminent men than in men of mediocre talent, in superior races than in inferior races. Other things equal, there is a remarkable relationship between the development of intelligence and the volume of the brain.[88] (Internal Citations omitted)

Montagu (1974) however, argues that there is no demonstrable relationship, neither "*. . . between cultural and intellectual status and brain size . . . .*" Upon examination of the facts, he concluded—

> The cranial capacity of paleolithic Neanderthal men was, on the average, 1,550 cc. . . . Are we to assume, then, that Neanderthal man was culturally and intellectually superior to the average modern white man? The Negro has an average cranial capacity of 1,350 cc, 50 cc. less than the white . . . .[89]

Darwin relied upon Broca's (1861) study to support his theory of evolution in *The Descent of Man* (1871). Cranium (1909) and Sizer (1888) illustrate the diverse facial distinctions with several facial and skull contours intersecting with the evolutionary development of humans. These included: a snake, a dog, an elephant, an ape, an idiot, a bushman, humans with several phases of human development and, finally, the highest grade, a Caucasian.[90]

---

86 Poliakov, 256, 257 respectively; Savage, 412.
87 Ettinger, S. (2011). The Origins, 228 et seq; Poliakov (1974), 255.
88 Friedlander, 1; Gould, S. J. (1996). The Mismeasure of Man, 115, citing Broca, (1861). Sur le volume, 188, 304; Rushton, J. & Ankney, C. D. (2009). Whole Brain, 693;
89 Montagu, 93, 100–01.
90 Cranium. (1909). The Brain; Sizer, N. (1888). Forty Years in Phrenology, 400.

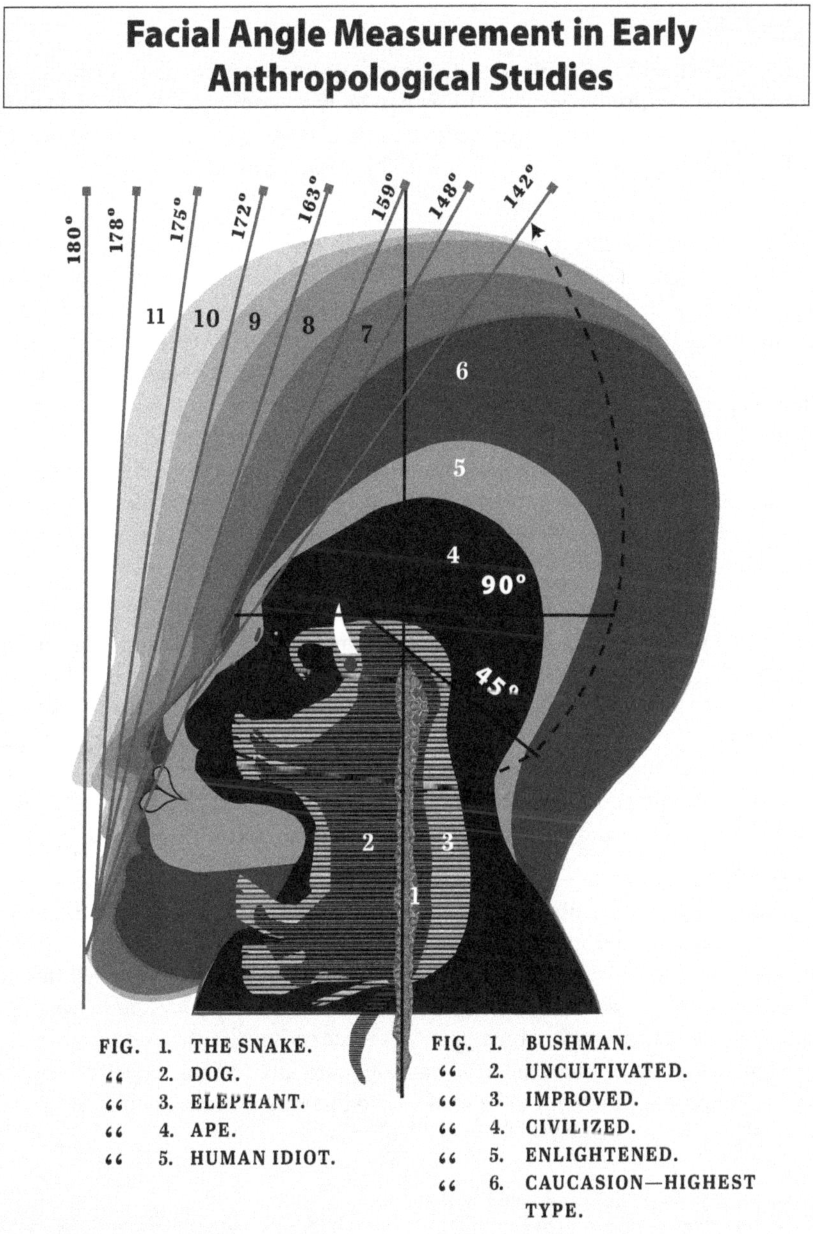

Figure I-2. Facial Angle Measurement and its Role in Racial Classification
Credit: Prof. Sizer (1888)

Davies (1971) draws these points together when he explains that by examining the face and the dimensions of the skull, one can ascertain a person's personality, values, and temperament. Davies submits four different illustrations of faces and skull types, and on the same page he states—

> [A] well-developed region of the head indicates a correspondingly well-developed faculty (propensity) for that region. Consequently, it was thought that a man could make a fairly accurate character analysis by studying the shape of a subject's head in conjunction with his temperament.[91]

In 1888, Francis Galton (1822–1911) was the first to quantify the relationship between brain size and a person's general mental ability, paving the way for his controversial ideas in eugenics and controlled breeding. He reported that more than 1000 male Cambridge University undergraduates that "*. . . obtained high honors degrees had a brain size 2%–5% greater than those who did not.*"[92] The point being that, according to this logic, a person with a large cranium and a favourable facial angle could be identified by a trained Nazi phrenology practitioner as exhibiting the skull patterns associated with good character, high intelligence, and membership in a superior or elite class, whereas those with smaller skulls were deemed inferior.[93] Anon (1909) stated that the skull, the external and internal cranium dimensions, and the size of the brain has an intricate bearing with the measurements of the facial angle—

> It is well understood that in savage tribes, and among the lower classes of men, we find a larger proportion of brain in the middle lobes, where the animal propensities are located, than we find in civilised or highly cultured people. This tends to depress the ear, and by this means the facial angle is enlarged, and the outer end of the lower arm of the angle brought down. The relation of the ear to the face, and the development of the intellectual part of the brain, is the true point of study and the basis of value to the facial angle.[94]

Initially, phrenology furnished the Germans with a narrowed-down, ostensibly elegant process for making distinctions that served to confirm their racial prejudices about who was superior and who was inferior.[95] Mosse (1964) expressed that it was confirmation that "*. . . the skulls of their Aryan ancestors had been long and narrow, . . . soon became a prime criterion for Aryanism.*"[96] Accordingly, if the individual's cranial measurements failed to meet the grade of adequate intellectual capacity, it enabled a rapid, instantaneous judgment to be made, as to the person's fate. Such anthropometric

---

91 Davies (1971), 4.
92 Rushton & Ankney, 693. See also Poliakov, 292.
93 See rebuttal about skull and brain sizes comparative analysis of race in Gould (1996), 115–20. Cf. Galton's paper titled The History of Twins (1875). He explored the relative powers of nature and nurture by meticulously studying twins. By examining the effects of both factors in individuals with similar traits, Galton shed light on the interplay between genetic inheritance and environmental experiences. His groundbreaking work emphasised the impact of life experiences on an individual's development, providing a stark contrast to German anthropology and ideology.
94 Cranium. (1909). The Brain, 208-09. Cf. Chamberlain, H. S. (1899). Die Grundlagen, 1:236, in Mosse (1964), 94–95, citing Chamberlian's aesthetic notation concerning skull measurements.
95 Gould, 106.
96 Mosse (1964), 89.

tests enabled ranking of humans on a linear scale of evolution, and measurements to evaluate mental and moral worth.[97]

Although Gould (1996) documented numerous anomalies and exceptions in Broca's (1861) empirical data, it demonstrated that the science was fundamentally flawed and unstable, leading to its loss of status.[98] However, with the annexation of phrenology into anthropology, it became a far more effective pseudo-scientific research tool—one that was unequivocally used in the examination of thousands of skulls during the Third Reich, during which it too was absorbed into the science of genetics.[99] Thus, the cranium and face biometrics analysis was just one more instrument in the toolbox among other settled practices to form one composite examination protocol, to posit a more definitive and pervasive result.

Whilst Darwin's work introduced the concept of 'Natural Selection' as the mechanism of evolution, the principles of (i) the survival of the fittest and (ii) the struggle for existence, frequently attributed to Darwin, were actually formulated by Herbert Spencer (1820–1903). Spencer coined the phrase 'survival of the fittest' after reading Darwin's *Origin*, and the phrase was later assimilated into subsequent editions of *Origin*.[100] Its bearing on Aryanist interpretations of a racial hierarchy fitted like a glove. Germany rendered the survival of the fitness, not to be understood as survival of the fittest individual, but the "*survival of the fittest race*."[101] In other words, this amounted to the cleansing of the entire German *Volk*. Essentially, it perpetuated the notion of a struggle for existence drawn from the plant and animal kingdoms: the weak and the less fit die, while the stronger, favoured, and fitter are naturally selected to survive. In its human application, this logic held that individuals acquire traits that enable them to survive more effectively, whereas those who lack such traits do not.

Fitness arises when the superior individuals survive—supported by eugenicist measures—adapt, reproduce, and pass their advantageous traits to their offspring. Over time, this leads to the overall improvement of the genetic stock.[102] As a result, a more evolved and superior form of humanity emerges. This philosophy, termed Social Darwinism or Social Evolution, adapts Darwin's hypothesis, although Darwin himself never applied it for the social or political purposes later annexed to it.[103] The supremacy/inferiority aspects of Aryanism ideology was the foundation to a deliberately conceived policy formed by the Nazis, which we know as scientific racism or racial hygiene policy.[104] Social Darwinism ratified Nazi Aryan/Nordic superior and inferior races beliefs, "*. . . which lent an air of scientific legitimisation . . . involving selective breeding, which*

97 Gould, 105.
98 Friedlander (1995), 3; Gould, 120-36, *137-141.
99 Schafft, 232–37; Weiss (2010), 75.
100 Black (2003, 2012), 12; Poliakov, 290.
101 Krüger, A. (1999). Breeding, 42.
102 Call, L. (2016). Anti-Darwin, 8; Stein, G. J., 53. See Claeys, 228, for critique of Social Darwinism and its derivative, survival of the fittest.
103 Banton (2015), 18; Baumslag, 35–36; Barzun, 69; Black (2003, 2012), 12, Burleigh, M. & Wippermann, W. (1991). The Racial State, 28; Kelly, 101; Krausnick (1970), 28; Pearson, K. (1897). Socialism and Natural Selection, 107; Semmel, B. (1958). Karl Pearson : Socialist and Darwinist, 115; Stein, G. J. (1988).
104 Schleunes, 51-52. See critique of this view in review by Schwartz, L. et al. (1982). [The Legacy of Malthus], 502–03. Schwartz et al., derogate from the view that the theories of Aryanism established the scientific foundation for Nazi genocide. They stated that Chase made a assumption that the major forces involved were middle-class, so that they were then able to sway the elite, and in doing so, ". . . exaggerate[d] the importance of hereditarian ideas." A preponderance of evidence in literature (some of this evidence is contained in this paper) written in the early 1800s by anthropology ideologues, do indicate the progressive development towards Aryan hierarchy congruence. Ernst Haeckel's work bridged that gap. See Figure I–1, 28, 'Progression of racist thought.'

had been propounded by, inter alia, Plato, More."[105] Social Darwinism was a splash in the water, in as far as it effected the most significant changes in social life, when it attached itself to racism and antisemitism, Schleunes (1970) outlines—

> [T]he development of Social Darwinism gave to racism and antisemitism and the Völkisch movement a foundation it would not have found for itself.[106]
>
> . . . .
>
> Hitler reshuffled [grossdeutsch, Völkisch, and racist ideas] and cement[ed] them together with a rigid dogma of antisemitism.[107]

Darwin's hypothesis indirectly influenced the formal establishment of eugenics through his cousin, Francis Galton. He coined the term "*eugenics*" between 1883 and 1889 to describe what he conceived as the science of controlled or selective breeding. Allen (1976, 1983, 2002) characterises its more complex formulation as "*. . . a reactionary programmeme for solving social problems through biological technology.*"[108] Ernst Rüdin (1874–1952) later co-founded and advanced eugenics in the early 20th century.[109] Poliakov (1974) cited W. R. Greg's demographic comparison of the Irish and the Scots in 1868, ten years after the publication of *Origin*, which was viewed as the forerunner to eugenics. Greg sounded "*the first childish cry of alarm,*" concluded that the "*. . . dissolute Celts were breeding faster than the virtuous Saxons . . . and the inferior were getting the upper hand because of their [biological] defects.*"[110]

Yet, eugenics was "*. . . a direct logical outgrowth of the scientific doctrine elaborated by his cousin, Charles Darwin,*" in which Forti (2006) asserts that it was totally derived from Plato.[111] Günther (1927) states that Galton was the first to show that it was not the "*. . . environment but heredity [that was] the decisive factor,*" and that he then went on to develop the outlines of the theory of eugenics.[112] This theoretical foundation provided the basis for what later became a global movement. Black (2003, 2012) observed that "*[g]lobal eugenics began in 1912 with the First International Congress of Eugenics in London.*"[113] Race hygiene policy means the practice of eugenics and euthanasia, in which Bock (1983) states—

> [E]ugenics, or, as it was called before and during the Nazi regime and sometimes also in Anglo-Saxon literature, race hygiene which became the core of population policies throughout the Nazi regime. [114] (Footnote call numbers omitted)

Among the genetic considerations that emerged were: (i) the incorporation of eugenics into the goal of preserving the German nation and race, influenced indirectly

---

105 Burleigh & Wippermann, 29; Ehrenreich, 13, 15–16; Müller-Wille & Rheinberger, 12; Schleunes, 32.
106 Schleunes, 32, 48.
107 Ibid., 49.
108 Allen, G. E. (1976). Genetics, 111; Allen, G. E. (1983). The Misuse, 117, 126; Allen, (2002), 13–17 passim; Baumslag, 36; Black (2003, 2012), 13–16; Flowers, 164; Schafft, 43, 39n.6, citing Kuhl, S. (2002). The Nazi Connection, cha 1:1–11.
109 Weindling (1993), 72.
110 Poliakov, 291.
111 Ludmerer & Lappé, 457; Forti, S. (2006). The Biopolitics of Souls, 17–18.
112 Günther (1927), 258, citing Pearson, K. (1922). Francis Galton.
113 Black (2003, 2012), 235, 263.
114 Bock, 401; Schafft, 249.

by Weismann's germ-plasm theory (1904), which provided a broader scientific framework that legitimised Galton's earlier eugenic concepts,[115] and (ii) its role was a contributory vector for the rapid expansion of socialism.[116] In fact, as Weindling (1993) observed, the dividing line between the medical profession, eugenicists, and the State was stripped away, as their symbiotic ideology fused nationalism, eugenics, and social hygiene into a single framework.[117] Furthermore, Ploetz's and the members of his secret society, the Nordic Ring, had formulated unpublished pamphlet entitled "*Unser Weg.*" It documented a much grander plan for *Nordic-Germanic race hygiene*, a broader eugenics programme, aimed at preserving what were conceived as the remaining Nordic elements in the Western hemisphere.[118] Zeidman (2020) more recently argued to the contrary that eugenics is just a euphemism. Racial hygiene is all-encompassing, whereas "*. . . eugenics is euphemistic in that it doesn't capture the amplified fanaticism of Rassenhygiene.*"[119] Weiss (1987b) defined it as a political strategy which consistent with the practicalities political biology—

> "Eugenics" is a political strategy denoting some sort of social control over reproduction. In the interest of "improving" the hereditary substrate of a given population, this supposed science seeks to regulate human procreation by encouraging the fecundity of the allegedly genetically superior groups in society and simultaneously discouraging and even prohibiting so-called inferior types from having children.[120]

Racial policy is *Fortpflanzungshygiene* (procreation hygiene).[121] Similar to Weiss's definition, founding father, Galton (1883) intended eugenics to mean—

> [T]he science of improving stock, which is by no means confined to questions of judicious mating but which, especially in the case of man, takes cognizance of all influences that tend in however remote a degree to give to the more suitable races or strains of blood a better chance of prevailing speedily over the less suitable than they otherwise would have had.[122]
>
> The improvement of humanity by altering its genetic composition by encouraging breeding of those presumed to have desirable genes (positive eugenics), and discouraging breeding of those presumed to have undesirable genes (negative eugenics).[123]
>
> [T]he study of agencies under social control that may improve or impair the racial qualities of future generations, either physically or mentally.[124]

115 Black (2003, 2012), 17–19.
116 Gasman, 147–48; Halliday, 393; Kelly, 6, 101, 103, 105–06, 110; Ludmerer & Lappé, 458; Krausnick (1970), 28; Proctor (1992), 18, 32–33. See also Weismann, A. (1893). The Germ-plasm, and Weismann, A. (1891). Essays Upon Heredity.
117 Weindling (1993), 291.
118 Weiss (1987a), 218 et seq.
119 Zeidman, 319–20.
120 Weiss (1987b), 1.
121 Bock, 401.
122 Galton, F. (1883). Inquiries, 25; Montagu, 235.
123 Freeden, M. (1979). Eugenics and Progressive Thought, 645; Allen (1983), 109.
124 Freidlander, 4; King, R. C. et al. (2014). Eugenics.

Positive eugenics consists of breeding "*good-stock*" humans. Such qualities were understood to include good health, vitality, intelligence, character and psychological robustness,[125] in addition to Lemarck's heavily debated hereditary acquired characteristics, including environmental influences.[126] Thus, the Aryan population must grow, whereas non-Aryans were to be subjected to sterilisation, incarceration, and ultimately extinction. Vecoli (1960) extends this definition, stating that the negative aspect of this programme rested on three policies: segregation, restrictive marriage laws and, as previously noted, sterilisation.[127]

[125] Freeden, 650.
[126] Proctor (1988), 41, 50, *102, 258, 216, 283; Zirkle, C. (1946). The Early History of the Idea of the Inheritance of Acquired Characters.
[127] Schafft, 72–73; Weinreich, 34; Vecoli, R. J. (1960). Sterilization, 194; Weikart (2004), 129–44.

B. Education of Society Matrix to Facilitate Crystallisation of Racial Hygiene Dynamics (1810–1870s)

This section has two parts. The first outlines the intellectual construction of the "*new man*" for future leadership. The second shows how the regime secured the loyalty of lower ranks and mobilised wider societal cooperation. Propaganda provided the infrastructure to reinforce Social Darwinism and entrench it as a self-sustaining worldview. The first subsection covers the general education of German society; the second examines the saturation and influence of the educational establishment.

i. General National Re-Education to Transform and Restructure of Society

THE RISE OF the Third Reich did not materialise overnight. It emerged step by step—each move small enough to seem tolerable, yet together forming the foundations of a totalitarian regime no one fully foresaw. The key was the re-education of the masses. This section shows how the regime used education to prepare society for that incremental transformation, embedding the attitudes that made each step possible.

Generally speaking, it began in earnest from pre-Darwin, c.1810 to the mid-1870s. The emotional and behavioural expectations imposed on German society were underpinned and sustained by systematic re-education efforts, which were prefaced by instilling loyalty and obedience to authority. Haeckel was instrumental. For future application of the role of obedience, see Reengagement of Past Connections, 266 and review Pioneering the Next Phase, 35. It comes as no surprise during the 1860s to 70s, the primary text earmarked for education of the public was the German dictionary. Poliakov (1974) said as much, highlighting that the steps taken were aimed at educating the general public at large, drawing on the ideas of Jacob Grimm (1785-1863). Grimm, by far, was the most influential promoter of the Indo-German Aryan myth, Poliakov wrote. Grimm's famous work, *History of the German Language* (1848) glorified the myth with the view of uniting Germans around their supposed ancestral origins—

> Jacob Grimm, the originator of the famous dictionary, who was considered an oracle by all the authors of literary and historical textbooks. It was from this source, above all, that the myth was to be taught to a large public in the second half of the nineteenth century.[128]

Merton (1938) underlines the locus of successful scientific propaganda: it is most effective when disseminated to uneducated audiences and articulated in technical, medical, or scientific terminology or or in formats that no layperson could understand without specialised training—

> [S]cientific pronouncements of totalitarian spokesmen on race or economy or history are for the uninstructed . . . , the laity is in no position to understand these conceptions or to check their scientific validity and in both instances they may not be consistent with common sense . . . Partly as a result of scientific advance, therefore, the population at large has become ripe for new mysticisms clothed in apparently scientific jargon. This promotes the success

[128] Poliakov, 198 et seq. See also Goodrick-Clarke, N. (2002). Black Sun, 90.

> of propaganda generally.[129]

This is perhaps an over-simplification: re-education was not intended to offset revolution, temper revolutionary impulses, or to preserve the status quo; rather, the intention was to burn the whole state apparatus altogether. It targeted the incoming prospective rank and file leaders first, then the proletariat, second.[130] For the latter, the scope was to educate the general population to inform awareness of the achievements of science.[131] Mosse (1964) expressed a similar sounding statement that—

> Small groups of elitist-oriented initiates were more important for the continuation and propagation of the Volkish ideology [among the proletariat.] Educators, students, and a multitude of such groups attempted to implement the Volkish ideology as a part of the social and national character of the German nation. Through these channels, leadership and aristocratic ideals were exalted above the democracy of the masses; the initiates were called upon to actualize the Volkish 'idealism of deeds' to help diffuse the spirit of the Germanic Volk.[132]

Pause one for moment and fast forward to the Nazi regime. The National Socialist party systematically maintained re-education propaganda. They disseminated a palatable number of composite images through various means, including textbooks, pamphlets, elementary and secondary school curricula, and Nazi-affiliated newspapers. Additionally, specialised medical journals propagated Nazi ideological narratives, contributing to the widespread indoctrination of the Nazi medical community.[133] Noakes (1984a) captures the point succinctly: the purpose of eugenics propaganda was to "*. . . reinforce[ ] deep-rooted forms of prejudice against [designated groups] . . . to devalue individual rights in the name of some collective.*"[134] The Nazis were not content just to see Germans agree with them; they expected society to accept the propaganda without any reservation.[135] However, German society needed no strong persuading. They were already conditioned by their own teachers, rulers and philosophers to accept Nazi propaganda and make the choices they made.[136]

In terms of national antisemitism, the same approach of re-education propaganda was employed. In order to convert "*unreceptive*" audiences of Jewish sympathises into a virulent antisemitic nation, Levy (1975) expressed, "*. . . Germany had to undergo purification and a long period of education before the implementation of sweeping changes.*"[137] Purification, of course, was referring to converting Jewish allies or sympathises into "*sincere*" anti-Semites, accomplished by permanent institutions dedicated to antisemitic education.[138] The following table are principles based upon, and derived from Staub's (1995) psychological-societal analysis—

---

129 Merton, R. K. (1938). Science, 333.
130 Beyerchen, 151; Fest (1970), 40; Kater (1983), 204; Massing, 103; Mosse (1964), 2, 5; Reichmann, 89. See also Le Bon,14.
131 See generally Kelly, 11–16.
132 Mosse (1964), 151.
133 Massing, 75; Proctor (1988), 77; Waller, 225; Weindling (1985), 306–07; Weinreich, 208, 215.
134 Noakes (1984a), 88.
135 Weindling (1985), 19.
136 Browning, 7–8; Kater (1983), 30; Tenenbaum, 211.
137 Levy, 29; Neumann, 127.
138 Levy, 111, 257.

Table I-3. Bystander Cultural Dynamics and Preparation of Society

1. Cultural-Societal Psychological Preparation for Group Violence—
   - Devaluation and Scapegoating
   - Monolithic Cultural Superiority
   - Respect for, and tendency to obey authority
   - Relationship Antecedents to Antagonism
2. Steps Along a Continuum of Destruction—
   - Justification of Mistreatment
   - Moral Equilibration Values Displacement
   - Group Identity Privilege and Differentiation Affirmation
3. Selection of Direct Perpetrators—
   - Acceptance of Propaganda Against Victim Group
   - Acceptance of Group Ideology and Accept Ideological Views
4. The Role of Bystanders : Passivity and Action—
   - Societal Assimilation of Aggression—the New Norm
   - Transformation of Citizens to destructive ideology
   - Create Public Education Propaganda about Psychological tendencies and processes
   - Step Two and Three repeated

Source: Staub, E. (1995). Torture : Psychological and cultural origins, 99–108.

On that point, Weindling asserts that "*. . . mass education would purge societies of ignorance, privilege and superstition.*"[139] Accordingly, Goldhagen (1996, 2009) elaborates that 19th- and 20th-century literature was instrumental in inculcating an eliminationist mindset among ordinary Germans and in identifying volunteers already predisposed to such tasks, for which finding willing Germans was not a problem—

> The unmatched volume and the vitriolic and murderous substance of German antisemitic literature of the nineteenth and twentieth centuries alone indicate that German antisemitism was *sui generis* . . . .[140]

National Socialists portrayed the "*. . . Jews as the penetrators of all the economic and political ills befalling the country. [Even] to the local craftsman and shopkeeper . . . the Jew was a threat[] to their livelihood.*"[141] Jew-baiting in the late 1800s appealed and stirred up existing politically and intellectually acceptable anti-Semitic feeling and negative stereotypes, to legitimise "*National Socialist racist ideology.*"[142] Theodor Fritsch was integral. See heading Theodor Fritsch, 200. Thus, their entire infrastructure of hatred was further radicalised and escalated to the Nazi era, through a constant barrage of re-education propaganda and continued by Nazi leadership and the media.[143] It consisted of the promotion of positive eugenic traits, negative eugenics, and anti-Jewish propaganda. These groups and those invited into their secret conspiracy (ruling classes) were participants in dispensing the Nazi rhetoric. Chrystal (1975) insightfully outlines how the Nazis employed elegant propaganda to ensure that their covert greater goal of

139 Weindling (1993), 3, 26.
140 Browning, 6; Goldhagen (1996), 419; Goldhagen (2009), 94, 96; See also Weikart (2003), 287.
141 Kater (1983), 26.
142 Browning, 7–9; Gerdmar, A. (2009). Roots of Theological Anti-Semitism, 608–10.
143 Browning, 10.

the Final Solution remained concealed from mainstream society for so long. They utilised cutting–edge media control techniques: "*. . . spoken leaflets—were the most volatile of Nazi techniques of persuasion.*"[144] However, leaflets had a fundamental flaw, whereby it lacked the charisma of film which could not rival the impact that film had. Chrystal elaborates on this rationale behind the decision made by the Nazis in adopting film—

> There was another manipulator which outranked the written word . . . [t]he picture in all its forms up to the film has greater possibilities . . . man needs to use his brains even less . . . and thus may more readily accept a pictorial presentation than read an article of any length.[145]

Figure I–3 below, is an example of how the Nazis circulated pro–Aryan pictures in journals and pamphlets encouraging procreation among the "*superior,*" in large numbers to maintain their leadership position in society; whilst extinguishing the threat of racial mixing and a degraded humanity from emerging.[146] It also promoted a subliminal message of hatred for Jews for bringing degenerates into the world.

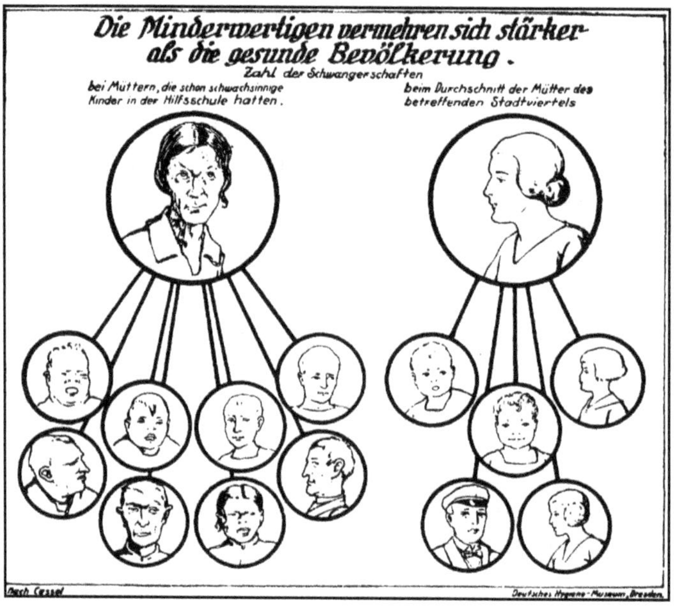

Figure I–3. Eugenic Assumptions of Differential Fertility Across Social Categories in Weimar Germany, 1926
This 1926 illustration exemplifies a central eugenic assumption that so-called "inferior" populations reproduced more rapidly than the "healthy" majority. The chart compares the average number of pregnancies among "mothers with feeble-minded children in schools for backward children" (left) with mothers in the same urban areas (right). Presented as statistical evidence, the figure naturalises socially constructed categories and promotes a biologically deterministic logic foundational to eugenic and later Nazi population policy.
Source: E. Dirksen, "Asociale Familien," Zeitschrift für Volksaufartung und Erbkunde 1 (1926): 11–16, esp. 15; cited in Bock, 407.
Credit: Bock, G. (1983). Racism and Sexism in Nazi Germany.

[144] Chrystal, W. G. (1975). Nazi Party Election Films, 29.
[145] Ibid.; Urwand, B. (2013). The Collaboration, 15.
[146] Bock, 406; Hutton, 10–11.

The precursor to the seizure of educational institutions (to be discussed momentarily) was the re-education of society. From Kelly's (1981) methodical analysis he documented that many German literary developments contributed to widespread reading in the early to mid-1800s, but the three influential factors were, in broad terms: (i) changes in paper-making that paved the way for the mass printing presses; (ii) the price drop of paper per kilogram. "*Newspapers, magazines, and books which were prohibitively expensive to [previous] generations, [had become] affordable . . . ;*"[147] and (iii) "*The widespread use of the petroleum lamp after about 1860 greatly facilitated night-time reading.*"[148]

The most conspicuous success of these reforms came in 1853, when Ernst Keil (1816–1878) founded *Die Gartenlaube*. Kelly noted that his readership was significantly lacking in science education. Therefore, Keil appropriated the role of public educator to correct this deficiency. Keil realised that to educate the masses, the literature had to be "*. . . written in a very easily understandable, elegant, and, where possible, aesthetic form, so that the average manual worker, and especially the women, can understand them,*" rather than in a pedantic, formal or intellectual style.[149] Keil simply dovetailed on the popularity of the magazine, and his flair for being witty and light-hearted was a contributory milestone to raising the general level of education.

By 1890, the demographics of readership had grown astronomically to five million. Although Kelly did not document the year when *Gartenlaube* changed hands, Keil did, however, attract respected names to write science essays.[150] As cited earlier, racial hygiene or Social Darwinism had come to a climax at the *fin-de-siècle*. By continuing to read Part I., this critical intersection also includes various disciplines, the sciences, antisemitism, Aryanism, and occult practices. Kelly (1981) observed that it did not stop with the middle-classes, rather it became the proletariat philosophy in the 1890s.[151] Hence, the link between Keil's accomplishment and the resulting outcome.

The baton was then handed on to Wilhelm Bölsche (1861–1939), the apostle of Darwinism, German author, editor, poet, and publicist. He was widely held to be a "*major cultural phenomenon,*" more than what Haeckel was.[152] Between 1887 and 1901, he transformed "*. . . Darwinism . . . from a tale of bitter struggle to an erotic monism . . . German romanticism is injected as a foreign element into Darwinism.*"[153] Bölsche's voluminous book of *Das Liebeslebenin der Natur: Eine Entwicklungsgeschichte der Liebe* (trans. Love-Life in Nature: The Story of the Evolution of Love, 1898-1901), by romanticising Darwinism with sexual love, elevated it to new heights, such that it is said that Bölsche became the new 'master.'[154] In 1917, it appears that Schallmayer was 'in the know' when he foresaw the future operational Nazi racial hygiene regime. Wegner (2002) quoted a public speech he gave at a doctors forum—

> Speaking before a gathering of doctors in Munich in 1917, Schallmayer emphasized the importance of education in the preparation of public opinion for ever-widening eugenics measures in the population.[155]

[147] Kelly, 15; Weindling (1993), 3.
[148] Kelly, 15, citing Engelsing, R. (1973). Analphabetentum, 127.
[149] Ibid., 16.
[150] Ibid., 16.
[151] Ibid., 23.
[152] Ibid., 37, 135.
[153] Ibid., 42, generally, 36–56.
[154] Ibid., 39–40.
[155] Wegner, 11–12.

If he realised what the forthcoming eugenics objectives were, he would have anticipated that German public opinion, as the recipients of the regime, would also fall in line. This was the process that started in 1925–26. In 1925, the Emergency fund for German Science sponsored a health propaganda week of lectures for the general population, which included a plethora of propaganda methods. In 1926, Administrators, doctors, teachers and politicians from the Reich Ministry of the Interior and the Reich Health Educational Council formed an alliance on the Reich Council for Health Education. This comprised 1000 organisers who conceived the Reich Health Week initiative, intended to bolster nationalist ties, advance Aryan ideals, and promote the German body politic (*Volkskörper).*[156]

ii. Institutional Saturation of Educational Culture and Volkish Indoctrination

FOLLOWING DARWIN'S 1859 hypothesis, Darwinism accelerated and became firmly entrenched through a variety of social, scientific, and cultural factors. This is best explained by Weindling's (1993) observation that the period from the 1860s was characterised as a 'liberal era,' in which medical 'quackery' could flourish with impunity.[157] This liberal era fostered an attitude in which 'anything was acceptable' tout court, enabling an easy break with traditional societal values. In the scientific realm, while these fields were increasingly dominated by privileged professionals, society at large experienced a broad universal education starting in the 1860s.

Initially, this educational outreach, spanning from primary schools to public lectures, aimed to create a larger pool of candidates for entrance into the scientific and medical fields. Moreover, this premise helped inspire the next generation to pursue careers in these disciplines, who became implementers of Nazi medical and racial ideologies.[158] This process thus laid the groundwork for the regime's broader eugenic objectives.

To understand the working strategy behind school reform designed to implement this re-education agenda, a multi-pronged approach was employed. First, a re-education propaganda agenda targeting the youth was first deployed to spread, "*. . . Völkisch thought among the population.*" Second, it was incorporated into school curricula and textbooks, in addition to the "*study of Germanic hero legends and fairy tales to enable students to appreciate more fully the accomplishments of the Volk.*"[159] Macrakis (1993) accentuates the salient point behind the programme for seizing and Nazifying or germanisation of the education system, generally—

> By seizing power at the universities one can educate a new generation of students in the reigning ideology and worldview . . . students played a key role in the transformation of university structures and the spread of National Socialist ideology and worldview.[160]

The subliminal purpose being: young minds are impressionable, so if you win their minds early on, then the conquest for societal transformation has already been won. Accordingly, racial hygiene was imposed on universities for precisely this reason, to transform society along racial and Nazi ideological lines, mobilising educational

156 Mass, S. (2022). The 'Volkskörper'; Weindling (1993), 409–12, 415, 490; Weiss (1987b), 151.
157 Weindling (1993), 14, 21.
158 Chelouche, T. (2022). Teaching Medical Ethics, 16; Weindling (1993), 3, 26.
159 Mosse (1964), 153–55, 157.
160 Macrakis (1993), 74.

institutions as instruments of indoctrination.[161] Macrakis's (1993) point highlights the greater principle that recipients of the new Social Darwinism education in the mid-1850s, they did all the work in spreading it. Doob (1978) poignantly articulated this well-known proverb: "*[i]n order to persuade a good man to do evil, it is not necessary to persuade him to become evil. It is only necessary to teach him that he is doing good.*"[162] This reshaped the rising generation, producing individuals who were "*. . . more Darwinian [than their predecessors, who were] to control German anthropological institutions and research for the next several decades.*"[163] In 1903, a national cadre of educators delivered anthropology lectures that were "*. . . more devoted to Darwinism and other evolutionary tendencies . . . [and] Anthropological studies from an evolutionary perspective [had become] so sophisticated.*"[164] Weikert (2003) documented a case in point how Ploetz, in an effort to recruit German youth into his secret organisation, the Nordic Ring, he wrote a propaganda tract in 1911, titled "*Struggle for existence against the other races.*" This promoted an extreme form of racial eugenics. He advocated fighting in order to win the 'struggle.'[165] Fischer had the same revolutionary idea, who was viewed as an 'active evangelist' to the youth, in the same year, 1911—

> He lectured his young audience on the importance of the 'racial factor' . . . insisting that the race hygienist, like a surgeon for the whole nation, had to be ready to 'cut ruthlessly in where something was rotten' . . . the triumph of National Socialism was still two decades in the future.[166]

A curriculum couched in racial terminology and delivered with euphoria, at a time when racial hygiene was becoming the national norm, made for a highly compelling and persuasive form of teaching. Indeed, Weindling (1993) recorded that the education process was a collaborative effort involving parents, doctors, teachers, and priests.[167] The curricula indoctrination process operated with a dual purpose. First, it was socially relevant because it ensured the inculcation of the mind. If these interactions were maintained, at the appropriate time, the *Volk* ethic could be put into practice. Weindling succinctly states that it was "*. . . based on a socially active and racially-oriented eugenics,*" underpinned by a racially appropriate biological humanism mindset.[168] Thus, this required intentional and deliberate planning. The second aspect of this planning was to transform minds by reorienting them toward an antisemitic worldview, as Steinweis (2006) astutely argues—

> The education system provided a further channel through which "scientific antisemitism" entered the mainstream of German society. Text-books and curricula targeted at students in primary and secondary schools resonated with messages about the racial otherness of the Jews . . . designed to instil in students a visceral revulsion toward the Jews. These educational materials,

161 Kater (1989), 111–12; Macrakis (1993), 57; Massin, 80.
162 Doob, L. W. (1978). Panorama of Evil, quoted in Baumslag, xvii.
163 Massin, 102, 118.
164 Massin, 118.
165 Weikart (2003), 284.
166 Massin, 141–42, quoting Fischer, E., & Pearson, K. (1912). Zur Frage, AfRGB 9: 8-10; Weikart (2004), 115.
167 Weindling (1993), 330, citing Baur, E. et al. (1921). Grundriss der menschlichen.
168 Weindling (1993), 302.

> with their hybrid content of "science" and propaganda . . . embodied what Hitler had in mind when he spoke of the need to combine scientific and emotional antisemitism.[169]

Mosse (1964) fashioned the principle of re-shaping the mind of German youth, in the course of his a commentary on Ludwig Gurlitt (1855–1931), educator and chairman of the advisory council to the Youth Movement—

> [T]he vitality, the spiritual richness and naturalness of the ancients would have to be regained before the German Volk could again assert itself. German youth were ripe for the task.[170]

The inference being that future aspirations for a regime could not rise until they had won the minds of the children.[171] By the late 1870s–90s, the education establishment had profoundly influenced schoolboys and university students, to which Weindling (1993) aptly states, "*. . . who were to provide the germ of the eugenic movement.*"[172] Indeed, the *fin-de-siècle* antisemitic and racial rhetoric had began to penetrate elementary schools, high schools, and universities.[173] The university system by the 1870s and 80s, experienced a boom in student numbers, to which they were being utilised, fundamentally, as 'teaching factories' to saturate the professions with brain-washed students.[174]

This created a crucial interdisciplinary bridge to Darwinism, opening lucrative possibilities within the universities, to augment teaching and research opportunities, and provide huge financial incentives to lecturers and professors, to colour science and medicine lecture teaching with Darwinist propaganda.[175] It was same for the Jews. Scientific racism and antisemitism were concurrent tracks that dovetailed in the 1880s and 90s. Concurrent with the Prussian government's new educational law, which would redirect the future course of German science, Haeckel delivered a 1877 address at a scientific meeting in Munich on the relationship between evolution and scientific inquiry that education should rest on a synthesis of evolutionary biology and the sciences, creating a coherent composite worldview for understanding human existence. Despite ambivalence from his peers, the prohibition of Darwinism in schools, and the issue of caution against inclusion into curricula, Haeckel's ideas prevailed.[176] By the year 1900, state officials were required to distribute Society for National Health journal and booklets containing hygiene propaganda in schools, prisons and other institutions—

> [In an attempt] to arouse popular appreciation for hygiene as a science, and to emphasize the moral responsibilities of citizens of leading a healthy and hygienic

169 Steinweis, A. E. (2006). Studying the Jew, 17.
170 Mosse (1964), 159.
171 For Nazism this would have been the 1890s and 1900s children. Hitler would have been part of this, since he was born in 1889.
172 Weindling (1993), *15, 16, 19.
173 Mosse (1964), 153. Cf. Neumann, 398n.34, quoting Rosenberg, A. (1934). Der deutsche, 11. Neumann highlighted that the education system for 1934 onwards (Rosenberg published in 1934), was run in accordance with the principles laid down by the "ideological oracle, Alfred Rosenberg." Leadership was selected from 70 million men to whom state leadership will be entrusted where world domination was in view, controlled by this envisaged German leadership.
174 Weindling (1993), 66—67.
175 Ibid., 30-31, 33.
176 Nordenskiöld, 521-22.

> life in obedience to the laws of hygiene.[177]

After the year, 1900, it comes as no surprise that education was reformed to the extent that Fritsch's published writings in Der Hammer *inter alia* had become recommended reading for high-school students.[178] See also heading, Theodor Fritsch, 200. The 1920s was purported to be the golden age of health education and propaganda.[179] Weindling (1993) highlighted that in 1925, racial hygiene, in particular, had become mandatory in medical schools—

> The textbook of eugenics of Baur, Fischer and Lenz called for Racial Hygiene to be a compulsory subject in all German medical schools. This aim was restated by the German Racial Hygiene Society in October in October 1922.[180]

This point is crucial, the textbook was authority for establishing the ostensible link between genetics, medicine within eclectic racial hygiene ideologies.[181] Walther Darré (1895–1953) exemplifies this influence. Weindling (1993) further remarks that "*[Darré] blended Nordic racism with human genetics derived from study of the Baur-Fischer-Lenz textbook and Gobineau.*"[182] The intersection between Jewish hatred and indoctrination school curricula is equally illustrative. Weinreich (1999) offers only a brief indication of this, observing that, "*. . . numerous handbooks for teachers on the treatment of the Jewish question contained ample references to the authorities of Baur-Fischer-Lenz . . . ,*" among others.[183] Although Weinreich does not specify precise dates, the 1921 publication of the Baur-Fischer-Lenz volume demonstrates a pre-existing agenda to reform public consciousness. Müller's (1933) review supports this narrative, expressing dismay that the book became less scientifically accurate, and instead was a mouthpiece for the "*. . . crassest kind of popular prejudice,*" twisting records of history and anthropology to favour the authors' egotistic beliefs.[184]

To fill in the gaps left by Weinreich's analysis, it is evident that the 1800s were a critical period in which the accumulation of knowledge through educational indoctrination laid the groundwork for an ideological agenda that predated the Nazi rise to power. This intellectual foundation was so substantial and collaboratively driven that Baur-Fischer-Lenz's manual culminated in a 734-page work. Elsewhere, Weinreich (1999) notes that the consensus among German scholars and other agencies of Goebbels' Propaganda Ministry held that, for such indoctrination literature to generally "*. . . succeed [it] must have in mind [those] who look for scientific 'proof' . . . .*"[185] This approach is consistent with Hitler's early 1920s pretensions in Mein Kampf ([1925] 1971)—

> The crown of the folkish state's entire work of education and training must be to burn the racial sense and racial feeling into the instinct and the intellect,

177 Weindling (1993), 228.
178 Mosse (1964), 163.
179 Weindling (1993), 409.
180 Ibid., 337.
181 Ibid., 473.
182 Ibid., 475, quoting the Ludwig Schemann papers (1925). For commentary on Baur-Fischer-Lenz, see Steinweis, 46–54.
183 Weinreich, 208.
184 Muller, H. J. (1933). Book Reviews, 20.
185 Weinreich, 115.

> the heart and brain of the youth entrusted to it. No boy and no girl must leave school without having been led to an ultimate realization of the necessity and essence of blood purity.[186]

At university level, medical schools was another area that espoused special treatment—the finishing touches to the student's education, especially given that the architects of the Nazi genocide were drawn from the medical profession. Racial hygiene and eugenics had absorbed racial, social, and sexual hygiene, along with human heredity, genetics, psychiatry, anatomy, anthropology, and bacteriology, and, according to Weindling (1993), were being taught in a variety of guises, as set out in the University Lectures Table I-4 below—

Table I-4. University lectures in Racial Hygiene, 1918–1920

Year	University	Professor	Title of Lectures
1918	Göttingen	Rosenthal	National health care (social and Racial Hygiene)
	Berlin	Grotjahn	Racial Hygiene and eugenics
1919	Freiburg	Fischer	Social anthropology
	Freiburg	Nissle	Social hygiene, including Racial Hygiene
	Halle	Drigalski	Social hygiene, politics, biology
	Heidelberg	Dresel	Social hygiene, including Racial Hygiene
	Leipzig	Döllken	Racial Hygiene and criminal psychology
	Munich	Lenz	Social hygiene and Racial Hygiene
	Tübingen	P. Kuhn	Racial Hygiene
1920	Dresden	P. Kuhn	Sexual, racial and social hygiene
	Giessen	Huntemüller	Heredity and Racial Hygiene
	Halle	Anton	Racial welfare and heredity
	Kiel	Kisskalt	Racial Hygiene
	Cologne	Müller	Hygiene, including heredity
	Rostock	Reiter	Population policy, including race
	Tübingen	Basler	Racial physiology

Source: Günther, M. (1982). Die Institutionalisierung.[187]

Accordingly, the German Society for Scientific and Mathematical Education pushed for educational reform linking the sciences and the humanities on one hand, and connecting laws of heredity and eugenics on the other.[188] The new physicians emerging from the 'new humanity' cultivated a novel racial and ethnic consciousness, one that was "*. . . free of the moral restraints . . . .*"[189] See also Romanticism Dynamics, 250.

[186] Hitler, 427; Wegner, 15.
[187] Günther's table reproduced in Weindling (1993), 337, 399.
[188] Weindling (1993), 410–11.
[189] Bialas, W. (2013). Nazi Ethics, 4–5;

C. Manufacturing the Spartan Proto-Superman: Leadership Idealism and the Youth Movement

CAN THERE BE replica supermen? This was precisely the intent for the German Youth movement: to produce compliant, disciplined embodiments of the superman ideal. The strategy sought to create, breed, and train (Züchtung), multiples of youth, shaped into exigent multi-purpose herd-animal supermen or stallions—an extension of Natural Selection. These youths were intended to operate in "*. . . partnership with [Nazi elites regarded as] super-heroes, providing the initial creative idea for super-human powers.*"[190] Like Hitler, the youths recruited into the Hitler Youth movement possessed no superhuman abilities. The background is as follows: Hitler Youth was formally established in November 1901, although its origins extend 100 years earlier; Membership stood at 15,000 by 1911 and expanded to 60,000 by the outbreak of WWI. Jewish participation was not initially excluded, but between 1911 and 1918 Jews became persona non grata, with Aryan members castigating their peers for harbouring "*foreign blood in their midst.*"[191]

Acceptance of this ideology was furthered by utilisation of Romanticism principles. They accentuated loyalty to the concept of the *Volk*, obedience to authority, the use of symbols and badges, an appeal to national pride and defiance, and allegiance to an Aryan identity refracted through the aesthetic ideal of the Nordic man.[192] The Youth Movement was organised by what has been described as the Bund formation structure. According to Mosse (1964) its operational purpose was "*. . . a constant inspiration in the Völkisch search for a means of organising society according [to] their views, a society which was neither capitalist nor socialist.*"[193] This coincided with Ploetz's propaganda campaign and his circle's involvement in Otto Ammon's *Die Gesellschaftsordnung* (1900), which advocated the sterilisation of "*beggars and hoboes,*" while lamenting that the so-called inferior could not be eliminated like animals.[194] Ironically, it was this particular generation of youth, and the next, specifically, that embraced Ploetz's polemic to carry out this programmatic vision of Spartan mass murders in the Nazi regime, that debuted some 33 years later. This was adopted by S.A. (disbanded in 1933), S.S., police, army, and even ordinary Aryan citizens.

The Nazi Party (NSDAP) augmented this theme further in 1919, in order to win the minds of the youth with racial ideas. Consequently, the NSDAP was a youthful party. This was accomplished through deliberately targeted propaganda directed at the youth sector, particularly the Hitler Youth and the National Socialist German Student Association, which was formally established in 1926.[195] German youth was shaped by these movements, including those influenced in the early 1900s and those indoctrinated from 1926 into the post-1940 era. They collectively became participants in the broader machinery of Nazi persecution and mass violence.[196] They were setting Nazi Germany free—isolating the superior from the weaker, inferior races.

[190] Gordon (2016), 229. See also Hannaford, 313; Krüger (1999), 53; Salter, W. M. (1915). Nietzsche's Superman, 426.
[191] Mosse (1964), 184.
[192] Ibid., 172; Reichmann, 196–97.
[193] Mosse (1964) 177–78.
[194] Schafft, 44, citing Ammon, O. (1900). Die Gesellschaftsordnung, 288 Cf. Weikart (2004), 79.
[195] Caplan, J. (1988). Government, 116; Kater, M. H. (1983). The Nazi Party, 139–40.
[196] Mosse (1964), 5–6, 85. See also Mosse (1964), 150-55, 171-89. The Volkish movement transformed top-down the education system from early 1900s and flourished from 1917 onwards. It became national policy to shape racial thought in the next generation of youth. It indoctrinated and infused into the minds of German youth, and "*entered adulthood to spread the ideas they had internalised.*"

The second meaning of "*freedom,*" as Ploetz used the term, referred to winning the Darwinian racial struggle by deploying eugenics as the mechanism for restoring what he believed to be a lost genetic perfection originating before the ice age. Therefore, the Nordic race was considered the 'purest' and the unifying identity for all Germans, while other accounts suggest that the unifying factor was the Nordic-Phalian hybrid, in which Hutton (2005) notes was "*. . . collectively referred to as the 'northern race' (Nordrasse).*"[197] By fixating on Germanic historical ties, this ideology provided a foundation for reclaiming their so-called lost heritage, for articulating a renewed vision of modern Germany, and for inspiring youth to emulate the "*gods*" of their mythic past. The reconciliation of this mythic past and the 1900s formed an essential ideological bridge, made explicit in contemporary currents of occultism and theosophy.[198] Read heading The Superman Doctrine, 72.

The recruiting ground for the S.S., and military was, of course, Hitler Youth, post-1933.[199] What better way to guarantee that the talent you receive will be good stock. By undergoing a meticulously designed training regimen, upon successful completion, for recruitment purposes, you know what you are getting. Consistent with Aryan characteristics, the principal selection criteria, apart from evidence of an Aryan ancestry, was oriented towards athletic prowess—

> [H]ad a long head and blood group A and was tall, strong, blue-eyed, courageous, a superman able to subdue Slavs and Jews.[200]

The youth were proxy supermen for the Third Reich—the hands and feet of Hitler, to make society compliant with Hitler's wishes. Himmler, for example, became the brutal and ruthless commander of the paramilitary forces—the S.A., the federal police, and the S.S.—in large part because he was deeply grounded in racial ideology. He immersed himself in literature on race and anthropology, including the works of Hans F. K. Günther, as well as writings by Friedrich Wilhelm Nietzsche (1844–1900), the German philosopher, essayist, and cultural critic, whose ideas were selectively appropriated and widely circulated within the Third Reich.[201] Himmler and Hitler appraised themselves of Haeckel's books and other Darwinian authors, and its assimilation into Aryan ideology. Webb (1976) elaborates on this point, explaining that the SS training regimen was "*. . . a comprehensive educational plan based on Social Darwinian principles.*"[202]

Consequently, Himmler trained the Hitler Youth and the SS as a terrorizing and brutal military force, and his mandate was to reorganise Germany[203] by funnelling the trained products of his caste structure back into the social order. The ultimate Hitler Youth product was the creation of the Nazi ideal of the "*blond superman,*" obedient, vigorous, brutal, ruthless, merciless, and devoid of remorse—a Spartan. Consequently, only the most exceptional candidates were admitted through a rigorous selection process,

[197] Hutton, 104, citing Günther, H. F. K. ([1922] 1934), Kleine Rassenkunde; 106, 113–114, and 132, citing Hoff, R. v. (1934), 145, and Ruttke, F. (1937), 10; Weikart (2003), 284.
[198] For example, German Druids Wotan Cult, Eastern beliefs, German Kabbalah, Spiritualism, Solar Symbolism (communication with ghosts of a bygone age), Sun worship, practice of magic, German gods worship.
[199] Krüger (1999), 57, 60; Neumann, 82; Peukert, 145.
[200] Krüger, 45, 53. Cf. Snyder, L. L. (1962), 82, purification leading to the super-race.
[201] Mosse (1966), 57; Schafft, 93n.16, citing Breitman, R. (1991). The Architect of Genocide, 34; Shirer, 100.
[202] Webb (1976), 320.
[203] Hale, 27–29; Krüger, 54; Schafft, 92-93.

and "*. . . not even all of the children of the best in one generation were admitted.*"[204] Thus, an obedient Hitlerite was one that acted *sine ira et studio* an emotionless, efficient killing machine transformed by psychological re-education designed to strip away empathy and cultivate absolute obedience.[205] See chapter Institutional Architecture, 250, for a more elaborate discussion.

Grodin (2007, 2010) interprets the psychological dynamics driving the Nazis' commission of evil acts. Typical of closed fraternal orders, Himmler employed oaths, rituals, and initiation rites, separate them from outsiders, cultivate a shared mystique, and establish common values to draw members in. This process fostered camaraderie and devotion to the organisational ideology, which Gordin explains as a form of "*. . . binding prevents members from resisting participation in torture.*"[206] He emphasises that the chief characteristic enabling individuals to commit violence is obedience to authority.[207] See Table IV-4, 277, for indoctrination process to create a killer. Kren and Rappoport (1980) capture the essence of Himmler's unique training schema as a synthesis of iron-hard discipline and a romantic mystique—

> [C]ombining exercises in total Obedience, compulsive attention to detail, dangerous tests of individual courage, and ideological indoctrination-including Himmler's brand of Teutonic elitism as well as standard National Socialist political and racial values. 16 Only after an initial period of testing in this system was the cadet given the privilege of taking the dramatic SS oath of loyalty to Hitler.[208]

Crelinsten (1995), a professor of criminology, argues that the perpetrator ultimately becomes cruel, inhuman, and emotionally detached. Simultaneously, the individual is progressively desensitised and their behaviour normalised. Their actions are further dehumanised through black humour, wordplay, sick jokes, and linguistic twists that objectify victims and strip them of identity, reducing them to demonising labels. Negative attitudes towards victims were maintained and legitimised through propaganda delivered in daily lectures.[209]

Small wonder that these 'supermen' youth were expected to embody the heroic virtues ascribed to their imagined Germanic ancestry: heroism, blind loyalty, honour, racial purity, and unwavering obedience, all in service of reclaiming the greatness of the *Volk*.[210] In fact, SS selections were based upon those who most closely resembled the ideal of the 'Nordic man.' These were the exoteric elements: faith, honour and unconditional obedience. The actual esoteric significance of these attributes was a

---

204 Krüger, 53; Snyder, L. L., 82; See also Koonz, C. (2003). The Nazi, 221–22.

205 The education and training prescribed for SS recruits differs from the previous approach to societal transformation, as illustrated in Table I-3, 57, in that it focuses on the education of society in general.

206 Grodin & Annas (2007), 645; Grodin (2010), 57;

207 Grodin & Annas (2007), 642–46; Grodin (2010), 57; Kressel, N. J. (1996). Mass hate, 162; Staub (1989), 129–30; Staub (1995), 102, 106; Staub (1999), 184.

Ivy League universities in the US and red-brick schools and universities in the UK, such as Eton, are structured on a strict hierarchical obedience system. Communal dining within 'refectories' typically reinforces authoritarian obedience.

208 Kren, G. M. & Rappoport, L. (1980). The Holocaust, 51; Staub (1989), 129, generally, 128–50.

209 Crelinsten, R. D. (1995). In Their Own Words, 39, 48, 52; Crelinsten, R. D. & Schmid, A. P. (1995). Introduction : The Politics, 4; Haritos-Fatouros, M. (1995). The Official Torturer, 136, 142–43; Kressel, N. J. (1996). Mass Hate, 200.

210 Krüger, 53; Mosse (1964), 208.

reference to Hitler, and were verbalised to Hitler under oath of loyalty.[211] Accordingly, Neumann (1944) noted that the reason for the Hitler oath was an act of transference, specifically, the concept of impartation of 'superman' power.[212]

Physical fitness was a central aspect of the "*superman*" ideal, requiring youth to be trained to an optimal level of bodily perfection, an intrinsic element of what the Nazis construed as male beauty.[213] This focus mirrored the superman ideal, who "*. . . exhibit[s] the qualities of bodily strength and beauty, courage, skill, daring, prowess alike in the field of action and in that of thought.*"[214] Mosse (1964) highlighted the thematic significance of the Aryan "*beauty*" connection to the Greek physique as the "*return*" to the source of the Aryan race, namely, the Greek gods' conquest of India (see The Superman Archetyp, 104)—

> Male beauty became the typical Aryan image [in the German Youth Movement]. It combined [racial ideology], an idealization of ancient Germanic strength, and echoes of the Greek ideal of physical beauty . . . [with a] twist of cruelty; while the Greek ideal of bodily proportions supplied an indispensable harmony to the Germanic form.[215]

These qualities, rather than intelligence, were the only attributes required of footmen: a herd mentality suited to carrying out unconscionable Holocaust crimes without questioning the moral character of the orders they received. Rosenberg found support, astonishingly, among the intellectual youth in particular, which would not have been explicable without their grounding in romanticism principles.[216] Essentially, intellectualism is diametrically opposed to Romanticism—the intelligent youth had to undue their intelligence in order to be transformed into a superman—rendering themselves unintelligent! The fusion engendered and elevated the caricature of simple-minded, obedient, power-mad, German supermen. In fact, Wertham (1966) succinctly explains this premise—

> [Long before Nazism], a whole literature had laid the groundwork, hypernationalistic and superman-oriented. The young people especially were conditioned to violence by being told in high-sounding books that they were by birth an elite whose duty it was to expand and conquer. In this way, decent people who loved their homes and children were brought around to picture themselves as heroes. They became willing to kill others who loved theirs.[217]

Poliakov (1974) expressed this notion as the "*ideal of the German man,*" consisting of breeding the youth to adulthood—

> [A] conscience which would stop at nothing to remain unruffled, until it ended in the fantasy of a being totally without conscience—the "blond beast",

211 Neumann, 69, 84–85; Staub (1989), 129.
212 See Neumann, 96–97, for example.
213 Krüger, 46-47.
214 Lindsay, 255.
215 Mosse (1964), 85.
216 Reichmann, 196.
217 Wertham, 77.

> who is a beast, not a man.[218]

Kidd (1924) did not propose a spurious correlation between the superman and a Spartan beast or animal; rather, as noted above, his argument remains entirely consistent with Darwinist precepts—

> The superman—the animal efficient in the struggle for his own interests: 'A new table I set over you, oh my brethren. Become hard'; 'For the best things belong to us, the best food, the purest sky, the fairest women, the strongest thoughts. And if men do not give us these things, we take them.'[219] (Footnote call numbers omitted)

The overlay of Christian and moral qualities onto the superman ideal furnished the perfect motif of blind hero-worship and self-sacrifice cultivated by Hitler.[220] By the time police or military units were deployed, their minds had already been inoculated through training to carry out orders with uncompromising force and without compassion. Such indoctrination instilled the psychological resolve to commit acts of violence, not through inherent strength, but through a cultivated dependence on authority and an erosion of individual moral judgement. Therefore, in the absence of such mental fortitude to kill Jews in particular, one is "*. . . disqualified him from being a Nazi superman . . . .*"[221]

Moreover, without such prior conditioning, the human intellect would be overwhelmed by the psychological strain of carrying out these actions.[222] Accordingly, Lt. Col. Grossman (2017) examined the conundrum of the human inability to overcome the intense resistance to kill, arguing that in doing so, necessarily requires training. Thus, the commission of violence is not the product of a naturally occurring 'passing whim' or 'change of mind.' See The Role of Social Psychology, 256. His analysis demonstrates that such behaviour emerges only under the right environmental conditions—through a "*. . . systemic process of desensitisation, conditioning and training,*" designed to produce a killer.[223] Wertham (1966) notes that an intense drive is required to suppress such resistances initially—

> For murder is never a superficial event. It always requires an enormously strong impulse, an overcoming of resistances and inhibitions, and a building up of rationalizations. There are always counterforces against the negative emotions, death wishes, catalysts, and personality patterns which culminate in the drive to kill. That is why murder is the greatest conflict a man can have with himself or with society.[224]

---

218 Poliakov, 86–87. See also Shirer, 100.
219 Kidd, 61, 67, citing Nietzsche, F. W. (1911). The Twilight of the Idols, and Nietzsche, F. W. (1917). Thus Spoke Zarathustra.
220 Dethloff, 109; Krüger, A. (1999). Breeding, 56, citing Baird, J. W. (1990). To Die for Germany.
221 Goldhagen (1996), 381, 579n.13.
222 Goldhagen (2009), 85–86; Schafft, 163. She documented that nurses and midwives were psychologically damaged by their participation in the euthanasia program.
223 Grossman, D. (2017). On Killing, 143–44, *149.
224 Wertham, 40.

Bialas (2013) highlights how frequent and sustained acts of resistance—"*. . . to suppress all personal feelings of mercy . . . ,*" without becoming emotionally involved—desensitises the conscience, gradually eroding moral agency and "*. . . reliev[ing] even the slightest doubts . . .*" by affirming the belief that reprehensible conduct is justified.[225] Put differently, Nazis exercised their animal passions, the carnality of an animal, in total control of emotion, obeying orders to commit *homo homini lupus* murder: to kill non-Aryans like a dog, without flinching, relentlessly.

Bruns (2014) points out that this represented "*. . . the revival of Spartan customs and practices.*"[226] Such power embodied the superman ideal. A clear example appears in National Socialist population policy, which Neumann (1944) intimates emerged after 1933, rather than during developmental stages of the late 1800s (as discussed earlier). Neumann states that this policy was stripped of compassion, resting on two commandments: (i) for German women, to produce as many children as possible so that the earth might be ruled by the so-called master race; and (ii) for the S.S., to eliminate those deemed "*unhealthy*" or "*unfit to live,*" so that the strong would not be burdened with the care of the weak.[227] Bandura (1999a, 1999b) supports this analysis through his concept of moral disengagement, a matrix of psychological mechanisms with both inhibitive and proactive elements.[228]

Another aspect to disengage the moral propensities was the emergence of the socialist totalitarian political religion itself. Under the aegis ultranationalist press, heightened by peer pressure conformity, is what further gave it teeth to participate in mass executions.[229] Merton (1938) argued that Germany's new totalitarian world, where 'science' was funnelled into a new social context, in which loyalty to the state was dominant, and co-operation with non-Aryans amounted to disloyalty (my paraphrase).[230] Romanticism, through its mystical symbols, folklore, and Aryan beauty myths, fitted seamlessly into this new despotism. It constituted the adhesive binding these ideologies and treatises into a unified restatement of racial hygiene policy. What was lacking was Nationalism. See heading National Awakening, 121. Massing (1949) recorded Adolf Stoecker (1874–1890) and Wilhelm Marr (1819–1904) as saying—

> National Liberalism had [the idea of] national unity, it had the great concept of personal liberty [. . .] Today we must try to re-establish more order on the basis of liberty. [. . .] Democratization of government likewise enhanced the chances of organizing a "mass movement from the Right" with which to offset democratic rule by majority. [231] (Footnote call numbers omitted)

The key takeaway is that whoever first reorganises democracy into a mass movement of national unity, framed as a defence of liberty, will be the first to establish a new racial order for Europe, one grounded in homogeneous political and social control. However,

[225] Bialas (2013), 3, 18. See also Hilberg, R. (1985). The Destruction, 279, 281–82; Waller, 208–09.
[226] Bandura (1999b), 196; Bruns, 219, citing Rohden, A. v. (1934). Verstößt das Gesetz, 8; Crelinsten, R. D. (1995). In Their Own Words, 57–59. Cf. Savage, 413, re lower races were devalued to apes and dogs; O'Flaherty, 85.
[227] Neumann, 112; Shirer, 100 (footnote).
[228] Bandura, "Moral Disengagement," 415–18; Bandura, "Moral Disengagement in the Perpetration of Inhumanities," 194. See also Waller's analysis of Bandura's moral disengagement, 185–90; Bialas (2013, 2014); and Staub (1989, 1990, 1995, 1999).
[229] Poewe, 2; Waller, 218–20.
[230] Merton, R. K. (1938). Science and the Social Order, 326.
[231] Massing, 28, citing Marr, W. (1879). Der Sieg, 43.

whilst Romanticism was a fundamental instrument to subvert the people's emotional expression, it was needful to have some commonalities in beliefs to foster a better homogamy. In this instance, the bonding agent was racism. It was deeply embedded in the masses, and because Romanticism was a nationalist model tailored to satisfy mass desires, it became, *Ipso Facto* the "*. . . pseudo-scientific basis of a mass doctrine.*"[232] See also Romanticism Dynamics, 250.

It is one thing for a belief system to be embraced by groups within society, rallying together to form a religious denomination, political faction, or social movement. It is quite another for any of these groups to consolidate into a single state entity, wherein their doctrines become national policy and transform into autocratic political authority, centralising all political, judicial, and legislative power within the party—or within a single individual. The doctrine of separation of powers, as practised in Western democracies, exists precisely to prevent such a concentration of authority. Hitler assimilated all executive, legislative and judicial powers. Accordingly, Neumann's (1944) explanation about the constitutional function of the National Socialist leader, he expressed of Hitler—

> He combines the functions of supreme legislator, supreme administrator, and supreme judge; he is the leader of the party, the army, and the people. In his person, the power of the state, the people, and the movement are unified.[233]

The antithesis of authoritarian rule is typically a capitalist or free-market system, with a separation-of-powers framework. In these systems, authority was distributed across municipalities, state and provincial councils, and driven by local practices and work ethics that vary regionally and nationally.

[232] Reichmann, 192, 208.
[233] Neumann, 84.n2, citing Koellreutter, O. (1934). Der deutsche Führerstaat; Tolischus, 141.

CHAPTER 3. REFINEMENT AND SYNTHESIS OF ARYANISM, GERMAN NATIONALISM, AND OCCULT INFLUENCES

A. The Superman Doctrine: Theosophy and the Occult Reorientation of German Thought

i. Occult Foundations and the Introduction of the swastika in Germany (1880s–1941)

FOLLOWING THE INTERDISCIPLINARY developments traced thus far—in German science, Aryan doctrine, Social Darwinism, and anthropology ("*Racial Hygiene*")—a unified framework emerged between 1890 and 1900. While the convergence of racial hygiene was nearing completion, the Occult Revival (1880–1910) in Germany was just beginning in 1890. Although antisemitism was already widespread, occultism was a fundamental cohesive force, intensifying and radicalising it. The virulent antisemitism that later defined the Nazi leadership emerged from involvement in occult-political societies such as the *Germanenorden* and the Thule Society, whose core criterion for membership was an explicitly antisemitic worldview. Hitler emerged from this paradigm in 1919. By the 1920s, a further synthesis of occultism and racial hygiene had coalesce, as the following analysis will demonstrate.

The previous section examined the mass education apparatus and the ideological permeation of academia. The next phase in the re-education schema involves the construction of the superman doctrine—a pivotal tenet in the formation of racial and leadership ideals. Originally conceived by Nietzsche, the concept of the *Übermensch* (trans. Overman=Superman) first appeared in his work '*Thus Spoke Zarathustra*' (1883). Interestingly, Nietzsche's Übermensch functions not merely as a philosophical ideal but as a corollary to Darwinian evolution, specifically an extension of Natural Selection toward a higher anthropological type.[1] Lindsay (1916) elaborates, Nietzsche's vision implies a transformation of mankind through selective ascent—a re-imagining of human potential shaped by struggle, will, and transcendence—

> The rule of the strong and the efficient . . . emphasis on force . . . contempt for the weak and for the humbler spiritual virtues may be represented as simply involved in the law of the Survival of the Fittest.[2]

Nietzsche, in crafting his concept of the *Übermensch*, turned to antiquity to derive its defining attributes. He drew heavily on Socrates and the Dionysian era for inspiration for

1 Claeys, 226; Kelly, 109; Lindsay, J. A. (1916). Eugenics, 252; Mueller, R. G. (2019). The Breeding, 186–87.
2 Lindsay, 248–49.

Zarathustra, fusing Dionysus with Zarathustra as an archetype for cultural and spiritual renewal. This vision was not purely philosophical; its Dionysian element gave it distinctly occult[3] and symbolic overtones—tapping into ancient rites, mystical tradition, and mythic regeneration.[4] According to Bein (1964) the term Übermensch itself predates Nietzsche's Zarathustra, appearing early on during the "*. . . 16th century, and is given its new meaning in the sense of a higher ideal stage of [spiritual] development of man . . . .*"[5] What must be understood, however, is how far the concept has drifted in modern popular culture. Kidd (1924) distilled the Nietzschean vision into the pithy expression "*Might is Right*", describing it as a rendering of Darwinism "*. . . in terms of efficiency resting on force.*"[6] In contemporary mass media, the image of the superman has become a simplistic emblem of force, power, and violence—severed from its philosophical and metaphysical roots. This is a reductive return to a modern raw Spartan matrix. This transformation was co-opted and weaponised in the early 20th century, as Nietzsche's vision was selectively distorted to serve totalitarian and racialist ends.

Helena Blavatsky* (1831–1891), a Russian occultist and writer, was the principal architect and founder of modern Theosophy.[7] Together with Henry Steel Olcott (1832–1907), she co-founded the Theosophical Society in New York in 1875. Akin to the recruitment framework of freemasonry, the Society presented itself as a universalist spiritual movement. Staudenmaier (2008) affirms that its stated goal was to promote the brotherhood and unity of mankind.[8] An introductory premise to be explored in this section is that German society embraced what was presented as ancient esoteric wisdom, an amalgam of Satanism with Egyptian and Indian philosophical traditions, harmonising Eastern metaphysics and Western spiritual worldviews. Theosophy synthesised ancient religious concepts with a cardinal belief in Darwinian evolution and the principles of

3 The definition of the terms "occult," "occultism," "esoteric" and "esotericism" in this paper—despite scholarly variation in definitions of esoteric and esotericism (Stuckrad, 2005)—are used synonymously and interchangeably in this paper to refer to 'secret' or 'hidden' knowledge, particularly concerning supernatural, mystical, magical practices, and paranormal energies." These concepts are typically said to involve beliefs or systems of thought that are concealed from the general public or engage with forces that operate beyond ordinary human perception or experience. Occult traditions draw upon or intersect with esoteric currents such as Gnosticism, Neoplatonism, Kabbalah, and the Hermetic treatises on alchemy and magic, and ancient rites, which have historically shaped Western esotericism and differ from broader, cross-cultural forms of alchemy. A central theme running through these occult systems is the belief in the symbolic and metaphysical power of number and geometry. Examples of occult disciplines include Luciferianism, alchemy, astrology, divination, and other esoteric disciplines of symbolic or initiatory knowledge. Importantly, this definition is not applied to hidden financial or political elites systems, or postulations in conspiracy theories.

See Faivre, A. (2010). Western Esotericism; Goodrick-Clarke (2004), 17; Goodrick-Clarke, N. (2008). The Western Esoteric Traditions, 7, 70; Hakl, H. T. (2000). Unknown Sources, 3; Hammer, O. (2003). Claiming Knowledge, 7–22; Hanegraaff, W. J. (2005). Occult/Occultism, 884–89; Owen, A. (2004). The Place of Enchantment, 19–20, 22; Yates, F. A. (1972). The Rosicrucian Enlightenment, 44.

4 Gillespie, M. A. (2005). Slouching, 52–53.

5 Bein, A. (1964). The Jewish Parasite, 27n.71.

6 Kidd, B. (1924). The Science of Power, 40.56, passim.

7 For Blavatsky's background and influence, see generally Hale, 21–30, 51–65; Campbell, B. F. (1980), Ancient Wisdom, for an elaborate analysis and critique of Blavatsky's works; Godwin, J. (1994). The Theosophical Enlightenment, 277–306, 363–65; Goodrick-Clarke (2008), 211–26; Kurlander (2017a), 15–19; Owen, A. (2004), The Place of Enchantment, for her stamp in the UK; Spielvogel & Redles, 85–90, for a synopsis of her impact in Germany; Wolffram, H. (2009b), The Stepchildren, 53–54.

*For additional context on Blavatsky's influence, refer to section, Constructing the Protocols, 195, for similar analysis with particular emphasis on how the occult shaped antisemitism in the formation of the Protocols of the Elders of Zion.

8 Staudenmaier, P. (2008). Race, 6.

modern science.[9] In 1879, the Theosophical Society relocated its headquarters to India.[10] Webb (1976), Staudenmaier (2015), and Goodrick-Clarke (2004) demonstrate that the occult underground, catalysed in part by Blavatsky's time in Germany, secured a major foothold between 1885 and 1886. From a cursory view, her works *Isis Unveiled* (1877) and *The Secret Doctrine* (1888) played a monumental role in reviving and reinterpreting Plato's account of Atlantis, integrating it into a theosophical framework that linked ancient civilisations, esoteric wisdom, and cyclical theories of human evolution. This reinterpretation generated considerable interest and attracted a wide following *sensu lato* accelerated by the publicist Paul Zillmann (1872–1940). Flowers (2022) emphasises, Blavatsky's books were "*. . . often cited in occult literature as the basis of Nazi racial doctrine . . . .*"[11] Blavatsky's subtitle: *A Synthesis of Science, Religion, and Philosophy*, would have piqued the curiosity of onlookers, eager to see how these domains were collectively interpreted. Together with the Society's active promotion of its universalist ideals, a development closely tied to the broader wave of anti-positivism, that came into vogue in both Germany and Austria from the 1880s onward. This revolt against positivism[12] involved a belief in a cosmic life-force, understood through occult means rather than scientific methods.[13]

In 1884, Franz Hartmann (1838–1922) formally joined the Theosophical movement at the invitation of Helena Blavatsky, identifying himself with Blavatsky's Theosophical teachings. Around the same time, Wilhelm Hübbe-Schleiden (1846–1916) founded the German Section of the Theosophical Society, officially established in the "*Occult Room*"—a meeting space in Blavatsky and Olcott's London residence either in 1884 or 1886, according to differing accounts.†[14] He then went on to found and publish the influential occult journal *Die Sphinx* (1886–1896). Among the early contributors were Jörg Lanz von Liebenfels (1874–1954) (born as Adolf Lanz), Guido von List (1848–1919), Friedrich Eckstein (1861–1939), and Lazar von Hellenbach, who conducted scientific experiments with trance mediums.[15]

In 1886, Eckstein had a cordial meeting with Blavatsky, after which he formed an inner circle of Viennese Theosophists. By the late 1880s, both Hartmann and Rudolf Steiner (1861–1925), then serving as Acting General Grand Master of an esoteric

---

9 Goodrick-Clarke (2004), 22; Hammer, 225, 254; Kurlander, E. (2013). The Orientalist, 156; Kurlander, E. (2019). Nazism, 3; Owen, A. (2004), 29. See also remarks by Staudenmaier (2009), 54.

10 Bevir, M. (1994). The West Turns Eastward, 748; Campbell, B. F. (1980). Ancient Wisdom Revived, 1–29; François, 33; Godwin (2010), 69; Goodrick-Clarke (2002), 79, 83; Goodrick-Clarke (2004), 18–19; Goodrick–Clarke, N. (2008). The Western Esoteric Traditions, 211; Meade, M. (1980). Madame Blavatsky; Spence, L. (1920). Encyclopedia of Occultism, entries for *Blavatsky* and *Theosophical Society*, 73, 410–12; Spielvogel & Redles, 85. Webb (1976), 30–31.

11 Flowers, 46, 49; Goodrick-Clarke (2004), 45; Mosse, G. L. (1999). The Fascist Revolution, 122–23, 126; Webb (1976), 30–31, 279–80. Cf. Godwin (2010), 67, 69; and Hammer (2003), 60–61. Blavatsky, as a medium, wrote Isis Unveiled while in a trance, taking dictation from spirits.

12 "Positivism" the methodology that all true knowledge comes from scientific, observable, and empirical evidence. "Anti-positivism," by contrast, embraces subjective, spiritual, or metaphysical explanations of the world.

13 Goodrick-Clarke (2004), 22; Mosse, G. L. (1961). The Mystical Origins, 81; Staudenmaier (2015), 25, 29. See Staudenmaier's (2009, 52–53), for exemplary review of Goodrick-Clarke's (2004) seminal work.

14 Goldstein, J. A. (2011). On Racism, 63; Goodrick-Clarke (2004), 15, 23; Howe, E. (1967). Urania's, 78–79; Staudenmaier (2015), 29, 32, 43; Webb ((1976), 30.

†Myers, P. (2015), An Elite Class, p. 43, states that the German Theosophical Society was established under the leadership of Franz Hartmann and Wilhelm Hübbe-Schleiden during the 1890s, which appears to contradict earlier accounts by Goodrick-Clarke (2004), Howe (1967), Levenda, P. (2002). Unholy Alliance, 42; Staudenmaier (2015) and Webb (1976), who place its founding in either 1884 or 1886.

15 Goodrick-Clarke (2004), 28; Kurlander (2071a), 22, 24–24, 76; Levenda, 42. See generally Kershaw (1998), 49–52, 63.

masonic order noted for its Egyptian-Hermetic influences, had become associated with this growing network.[16] Hartmann later became directly involved with the same order. Levenda (2002) argues this rite was a structural and initiatory bridge between mystical Masonry and the emerging Ariosophical, Armanen, and occult-nationalist movements.[17] The foremost "*magic circle*" of occultists emerging in the pre- and post-fin-de-siècle period were either directly inspired by Blavatsky or were profoundly shaped by her teachings. The following table comprises individuals, arranged in order of importance, influence, and impact—

Table I-5. Occult Architects of Nazism: Theosophical and Ariosophical Influences

Magic Circle	Role / Contribution
Helena Petrovna Blavatsky (1831–1891)	Luciferian and founder of Theosophy. Her root-race doctrine and Aryan mythos established German occultism. Her teachings became an iterative precursor to the Nazi Party (1918) and the Third Reich.
Guido von List (1848–1919)	Originator of Ariosophy. He synthesised Theosophical root-race theory with Germanic mythology and directly influenced Lanz. His ideas indirectly shaped Sebottendorff's occult-nationalist vision.
Jörg Lanz von Liebenfels (1874–1954)	Disciple of List and founder of Ostara. He radicalised Ariosophy into a racial-occult doctrine—the ideological conduit between List and the völkisch societies that followed.
Theodor Fritsch (1852–1933)	Anti-Semitic ideologue who integrated Theosophy and Eastern beliefs into the *Germanenorden*. He heavily influenced Sebottendorff and provided the racial-nationalist scaffolding for the Thule Society.
Rudolf von Sebottendorff (1875–1945)	Founder of the Thule Society. He implemented Fritsch's and Blavatsky's ideas in occult nationalist practice that shaped the embryonic Nazi Party.
Franz Hartmann (1838–1912)	Key Theosophist who connected Blavatsky's ideas to German esotericism. He influenced Zillmann and List, helped revive Rosicrucianism, and transmitted Eastern doctrines into the German occult milieu.
Wilhelm Hübbe-Schleiden (1846–1916)	Early German Theosophical leader who linked occultism with science and mentored Steiner. He laid the intellectual groundwork for integrating science and esotericism in Germany.
Rudolf Steiner (1861–1925)	Theosophical General Secretary who spread occultism and Eastern philosophy across Europe and the U.S. He founded Anthroposophy, revived Rosicrucianism, and published in Zillmann's periodical in 1905.
Paul Zillmann (1872–1940)	Hartmann-inspired publisher and founder of Metaphysische Rundschau. He was the first to publish List and Lanz on esoteric subjects, amplifying the occult revival and racial doctrines in early 20th-century Central Europe.

[16] Goodrick-Clarke (2004), 28; Kurlander (2071a), 22, 24–24, 76; Levenda, 43, 46.
[17] Levenda, 44–45.

Table I-5. Occult Architects of Nazism: Theosophical and Ariosophical Influences

Magic Circle	**Role / Contribution**

Table Note:
This table highlights key esoteric thinkers influenced by Helena Petrovna Blavatsky—mother of modern esotericism and German occultism—who shaped Nazi ideology by fusing Theosophy, Ariosophy, and nationalist mysticism into racial occult doctrine in the 1890s.

Hartmann was concurrently appointed acting president of the Theosophical Society's headquarters while Blavatsky and Olcott travelled to Europe.[18] It is hardly surprising that Hartmann—one of the key signatories of the List Society (see Table I-6, 79)—became closely associated with the organisation, especially because (i) he introduced Guido von List to Blavatsky's works, and (ii) Hartmann translated her writings into German, making him the principal conduit for transmitting Theosophical ideas to List.[19] Hartmann's legacy continued into the First World War era as president of the Austrian Theosophical Society.[20] In Germany, after Blavatsky's death in 1891, Theosophical study shifted from philosophical doctrine to magic or practical occultism, exemplified by List's incorporation of mystical teachings into ritual and nationalist esotericism. The Occult Revival (1880–1910) at the *fin-de-siècle* was gaining momentum giving rise to a range of esoteric outlets in Europe generally, but Germany was "*. . . unique in depth and breath when compared to other European countries.*"[21] In Germany, this revival intersected with the Wilhelmine occultist milieu (c.1871–1918), a subset of the broader Occult Revival. Kurlander (2012) explains, "*[t]he NSDAP [Nazism] retained closer ties to the Wilhelmine occultist milieu than any mass party of the Weimar era.*"[22]

This subculture was coloured by German nationalism, völkisch ideology, racial mysticism, and antisemitism. While still engaging with transnational esoteric currents such as Theosophy and Rosicrucianism, German occultism synthesised mystical ideas with notions of racial purity. It was all-embracing, drawing additionally upon contemporary holistic sciences—vitalism, organicism, and anti-materialist critiques of mechanistic biology—which framed the *Volk*, race, and nation as living organisms whose "*health*" required spiritual and biological purity.[23]

Amid these developments, Hartmann later inspired friend and collaborator Paul Zillmann, who in 1896 founded the monthly periodical *Metaphysische Rundschau* (trans. Metaphysical Review).[24] Contributions covered topics such as yoga, phrenology, astrology, and hypnotism, appearing alongside reprints of German mystics, Rosicrucian-alchemical treatises, and works by the French occultist, Gérard Encausse (1865–1916), better known by his pen name *Papus*. The following year, in early 1897, Zillmann established an occult lodge known as the *Wald-Loge* (Forest Lodge).[25] Goodrick-Clarke (2008) further points out that Theosophy "*cross-fertilized developments in the modern occult revival,*" including ceremonial magical orders and esoteric movements grounded

---

[18] Goodrick-Clarke (2004), 24; Kurlander (2017a), 18, citing Treitel, 95–97; Pokorny, L. K. (2025). The Way, 26. See also Howe (1967), 79, for Blavatsky's critical comments on Franz Hartmann.
[19] Goodrick-Clarke (2004), 24–25; Lovoie (2018), 501.
[20] Kurlander (2017a), 21.
[21] Kurlander (20017a), 14, citing Treitel, 58-9 and Hamann, 7-9, 285-323; Howe (1967), 3–4. For a general discussion of the Occult Revival, see Goodrick-Clarke (2004), 17–31; Howe (1967), 78–103; Owen, A. (2004). The Place of Enchantment, 17–50.
[22] Kurlander (2012), 530, 531.
[23] See Harrington, A. (1996). Reenchanted Science, 3–30, 196.
[24] See also Pokorny, 38n.34.
[25] Goodrick-Clarke (2004), 25–26, citing Zillmann, P. (1898). Die Wald-Loge und Akademie für okkulte Wissenschaften, 1:226–28 and Zillmann, P. (1912). Die Wald-Loge.

in Gnosticism, Hermetism, Neoplatonism, Rosicrucianism, and Goethean science.[26] A key figure in this proliferation was advanced by Zillmann who was the "*. . . first to publish both List and Lanz on esoteric subjects.*" This development coincided with the publication of an article by List in the September 1903 issue of the Viennese occult journal, *Die Gnosis*. In his paper, List placed the triskelion and swastika glyphs at the forefront of his cosmological discourse, presenting a theosophical perspective on the creation of the universe.[27]

Theosophy exerted its influence predominantly through Lanz's Ariosophy (Aryan-gnosis, meaning "*occult wisdom*"), a racist Austro-German offshoot of Theosophy, and through related völkisch-esoteric movements associated with List, the originator of Armanism. Kurlander (2017, 2019) and Lavoie (2018) pithily state that List labelled his movement Wotanism and Armanism, both grounded in racial mysticism and German nationalism, and both of which shaped early Nazi ideology. Indeed, the two Austrian occultists believed they might inspire a reawakening of Aryan civilisation.[28] These currents amalgamated occult philosophy, initially advanced by Blavatsky, with the doctrines of Social Darwinism.[29] Goodrick-Clarke (2004) cements the connection between Theosophy, Ariosophy, and Social Darwinism, observing—

> The central importance of 'Aryan' racism in Ariosophy, albeit compounded by occult notions deriving from theosophy, may be traced to the racial concerns[30] of Social Darwinism in Germany.[31]

List laid the groundwork for Ariosophy; Lanz, building on List's teachings, coined the term 'Ariosophy' in 1915. Yet, Lanz's movement had already coalesced around List's Armanism by the early 1900s, well before its formal naming.[32] Lanz's adaptation, however, was Ariosophy (occultism, Völkisch nationalism, racism, and antisemitism), which intensified these themes, radicalising them with explicit eugenic and conspiratorial elements. Indeed, Goldstein (2011) argues that Ariosophy was "*. . . the outgrowth of the society formed around the mystical thought of Guido von List in Vienna in 1903.*"[33] Lanz popularised these ideas through his periodical Ostara (1905–c.1927), which was the primary vehicle for disseminating his racial-esoteric worldview. The magazine combined elements of Theosophy, Christian mysticism, and fabricated historical narratives to promote Aryan racial supremacy and spiritual regeneration.[34] The content was in

---

26 Goodrick-Clarke (2008), 226.

27 Goodrick-Clarke (2004), 30, 41.

28 Kurlander (2017a), 4–5, 20–22; Kurlander (2019), 3; Lavoie (2018), 499–501, 503–04; Spielvogel & Redles, 88.

29 Blavatsky, H. P. (1888). The Secret Doctrine, vols. I/II; Goodrick-Clarke (2004), 227n.1, defined the term "Ariosophy," to mean occult wisdom; Levenda, 38–40, 66–70, 72, 87 passim; popular fringe accounts such as Pennick, N. (1981). Hitler's, 22–23, later echoed these claims. See also Goodrick-Clarke (2004), chap. 10:123–34, for an analysis on Ariosophy.

30 The "*racial concerns*" referenced by Goodrick-Clarke (2004) should be understood as the pseudo-scientific and spiritual anxieties, hierarchies, and ideologies centred on race—especially ideas of racial evolution, purity, superiority, and decline—as influenced by Social Darwinist and Theosophical thought. These ideas became racialised in the sense that they assigned spiritual worth or evolution to biological-racial categories.

31 Goodrick-Clarke (2004), 14.

32 See Goodrick-Clarke (2004), 164.

33 Goldstein, 68–71; Kurlander, E. (2012). Hitler's Monsters, 529. Kurlander (2017a), 20–22. See also Flowers, 52; Godwin (2010), 121–24; Staudenmaier (2015), 30.

34 Goodrick-Clarke (2004), chap. 8:90–106, for analysis on Jorg Lanz von Liebenfels; Kershaw (1998), 50;

harmony with those of Blavatsky and Steiner; it too presented a synthesis of myth, occult doctrine, pseudo-science, and advanced a number of racial and esoteric beliefs—

i. Theosophy could be used to justify racial hierarchies and the superiority of the Aryan race;
ii. The ancient Atlanteans, conceived as a pure Aryan race, had interbred with inferior beings, creating prehistoric monsters that bred demon hybrids;
iii. A racial split occurred between the noble Atlanteans and their bestial descendants;
iv. Humanity's original gnosis, accessed through the "third eye," was lost due to the corruption of the pineal gland caused by racial and spiritual degeneration.[35]

These beliefs formed the mythic, racial, and esoteric core of what would later be retrospectively labelled Ariosophy—a worldview that fused the Occult Revival with a radicalised vision of racial purity and cultural renewal. In Blavatsky's *The Secret Doctrine* (1888), she believed that humanity progresses through seven root races, each containing seven sub-races, a concept that connects directly to Darwin's evolutionary progression. Lanz's Ariosophy elaborated upon these ideas, asserting a continuity between the advanced sciences of the Atlantean fourth sub-race, destroyed in the legendary fall of Atlantis, and the fifth sub-race, the so-called Aryan race. This latter group was considered the remnant and successor of the Atlanteans, and now is embarking on its ascent toward a higher evolutionary form: the sixth root race.[36]

Furthermore, as discussed in the section Origins of Racist Aryan belief, 17, the impetus for Himmler's expedition to Tibet was driven by a desire to rediscover remnants of alleged Aryan knowledge, imparted by the Atlantean "*god-men.*" Likewise, List, Lanz, and the leadership of the Third Reich embraced the Atlantis (or Thule) narrative: the belief that they were superhumans endowed with occult knowledge and magical powers. These capacities were said to derive from the "*Cyclopean eye*" (also known as "*Dangma's Opened Eye,*" "*Eye of Dangma,*" "*Third Eye,*" "*Eye of Shiva,*" or the pineal gland) and could, once imparted to the Aryans, be recovered by studying antiquity texts and experimenting with their rituals. Blavatsky believed the 'eye' enabled the Atlantean 'god-men' to 'see' "*. . . into the future and read minds.*"[37] According to Blavatsky's root race schema, the sixth and seventh races would return humanity to the original, advanced and enlightened form. These ideas became integral to Hitler's thinking and to the composite ideology of proto-Nazi racial hygiene, later forming the basis of official Nazi doctrine after 1933.[38]

List's work was supported by the Spiritualist medium Fraülein von Rantzau and by Hartmann, who, speaking through the clairvoyant, endorsed List's ideas—specifically, asserting that they were on par with, and bore a striking resemblance to Blavatsky's writings. Regarding the racial dimensions of Aryanism, Goodrick-Clarke (2002, 2004)

---

35 Goodrick-Clarke (2004), 101–02, 106; Kurlander (2017a), 69.
36 Blavatsky, H. P. (1888). The Secret Doctrine, I:108–200 266, 429, 614 passim; Flowers, 50; Godwin (2010), 71, 117; Goodrick-Clarke (2002), 20–21, 80–81; Goodrick-Clarke (2004), 52; Hale, 37, 58; Hammer, 258, 261–67, for critical analysis of Blavatsky's *Secret Doctrine*; Kurlander (2013), 156; Santucci, J. A. (2008). The Notion of Race, 46, 48–49. For a critical view on the concept of the Fifth Race, see Campbell, B. F. (1980). Ancient Wisdom Revived, 65.
For general background on Blavatsky's root races, see Santucci, J. A. (2008). The Notion of Race. Spielvogel & Redles, 85–86; Staudenmaier, P. (2014). Between Occultism, 165.
37 Blavatsky, H. P. (1893). The Secret Doctrine (vol. 1, 45–47, 218; vol. 2, 302–16, 307, 309, 313, 316); Hale, 20–21; Spielvogel & Redles, 86; Staudenmaier (2015), 6–7.
38 Kurlander (2017a), 226, 243, 365n.309, citing Pringle, H. A. (2006). The Master Plan, 79; Kurlander (2017b), 81–83, 91–92; Nagl, M. (1974). SF, Occult, 191; Spielvogel & Redles, 87.

explains that between 1901 and 1908, List's writings adapted and restructured Blavatsky's Theosophical concepts of Atlantis into a racially codified worldview. These thematic parallels with Blavatsky's ideas drew the attention of Hartmann, who, as Mosse (1999) remarks, "*. . . attracted by such parallelisms, became one of List's leading supporters.*"[39] The belief that Atlantis was a highly advanced scientific civilisation with lost knowledge and technologies, is central to this evolving ideological pursuit. The Atlanteans were believed to possess a scientific legacy that imbued them with superhuman abilities and mastery over magical and psychic forces—

> The Atlanteans wielded psychic forces with which our race was not familiar, their gigantism enabled them to build cyclopean structures, and they possessed a superior technology based upon the successful exploitation of Fohat."[40]

The Nazi quest to recover ancient Aryan wisdom was inseparable from the ambition to rediscover and reconstruct Atlantis, and its powerful technologies—the so-called "*destructive miracle*" systems and weapons beyond human comprehension—believed to be key to securing military supremacy and ultimate victory in WWII.[41] Yenne (2010) further notes that List's publication of his runic theory in 1907 or 1908 coincided with the founding of the *Guido von List Gesellschaft* (trans. Guido von List Society), marked by a ceremony embedded within the broader context of Germanic mysticism and occultism.[42]

The signatories of the List Society were, "*. . . drawn from the upper and middle classes of Austria and Germany.*"[43] They held influential positions in politics, publishing, academia, and cultural organisations, enabling them to promote and normalise radical nationalist and esoteric ideologies. As agents of change, they used their authority and networks to shape the sociopolitical landscape of early 1900s Austria and Germany, driving the integration of occultist thought not only into Völkisch and Pan-German nationalist movement,[44] but into the broader fabric of society as a whole by key influential figures, as illustrated below—

Table I-6. Key Signatories of the Guido-von-List-Gesellschaft (List Society)

Name	Title/Position	Affiliation/Occupation	Notable Ideological Alignment or Role
Bernuth, Ludwig von	Chairman	Völkisch health organisation	Völkisch ideology

[39] Staudenmaier (2015), 27; Goodrick-Clarke (2002), 85–86, 148; Goodrick-Clarke (2004), 52–53; Levenda, 46; Mosse (1999), 123; Mosse (1961), 85–87; Spielvogel & Redles, 89. See generally Blavatsky, H. P. (1888). The Secret Doctrine, vols. I/II; François, S. (2023). Nazi Occultism, 32–33; Lovoie (2018), 501–02; Lavoie, J. D. (2021). Theosophical Chronology. For commentary on List and Blavatsky, see Yenne, B. (2010). Hitler's Master, 17–24.

[40] Goodrick-Clarke (2004), 20. See also Kuhn, A. B. (1930). Theosophy, 115–17; Kurlander (2017b), 86, 99.

Fohat is a central concept in Theosophy, described by Blavatsky as the cosmic energy or force that mediates between spirit and matter, functioning as the dynamic agent in the manifestation of the universe. See Blavatsky, H. P. (1888). The Secret Doctrine, I:16, 109; Sellon, E. B. & Weber, R. (1992). Theosophy.

[41] Kurlander (2017a), 263–76.

[42] Yenne, 25–27. See also Goodrick-Clarke (2004), 42 and Levenda, 60–62.

[43] Goodrick-Clarke (2004), 44.

[44] For a general commentary on the Pan-German movement distinctions, see Goodrick-Clarke (2004), 7–16; Levenda, 49.

Table I–6. Key Signatories of the Guido-von-List-Gesellschaft (List Society)

Name	Title/Position	Affiliation/Occupation	Notable Ideological Alignment or Role
Goring, Hugo	Editor	Theosophical literature, Weimar	Occultism, Theosophy
Harpf, Adolf	Editor	Marburger Zeitung	Nationalist press
Hartmann, Franz	Theosophist, Acting President, Author, Occultist	Theosophical Society headquarters	Theosophy, occultism
Hübbe-Schleiden, Wilhelm	Founder, Occultist	First German Theosophical Society	Early Theosophy in Germany
Jostenoode, Harald Arjuna Gravell van	Author	Theosophical writing, Heidelberg	Theosophy, esotericism
Khull, Ferdinand	Committee Member	German Language Club	German cultural nationalism
Lueger, Karl	Mayor of Vienna	Politician	Anti-Semitic, Pan-Germanist
Pfister-Schwaighusen, Hermann	Lecturer in Linguistics	Darmstadt University	Austrian Pan-Germanist, academic
Pickl-Scharfenstein, Wilhelm von	Baron von Witkenberg	Compiler of Anti-Semitic directories	Anti-Semitic propagandist
Polzer, Aurelius	Editor	Nationalist newspapers in Horn and Graz	Völkisch nationalism
Rohmeder, Wilhelm	Pan-German Educationist	Based in Munich	Pan-German educational reform
Schulz, Arthur	Editor	Berlin periodical for völkisch educational reform	Völkisch, nationalist education
Schweiger-Lerchenfeld, Amand Freiherr von	Editor/Army Officer	Stein der Weisen magazine	Popular nationalist press, military
Seiling, Max	Pamphleteer, Philosopher	Munich	Esotericism, popular occult philosophy
Steiner, Rudolf	Grand Master of German Theosophical Society, Philosopher, Occultist	German Section of the Theosophical Society, Berlin	Esotericism, founder of Anthroposophy
Wachler, Ernst	Author/Founder	Open-air Germanic theatre in the Harz Mountains	Völkisch, Germanic cultural revival
Wiegershaus, Friedrich	Chairman	Elberfeld branch, DHV (German Nationalist Commercial Employees' Association)	German nationalist labor movement
Winterstein, Franz	Committee Member	German Social Party (DSP), Kassel	Anti-Semitic party affiliation
Zillmann, Paul	Editor/Lodge Master	Metaphysische Rundschau, occult lodge, Berlin	Occultism, metaphysical publishing

Table I-6. Key Signatories of the Guido-von-List-Gesellschaft (List Society)

Name	Title/Position	Affiliation/Occupation	Notable Ideological Alignment or Role
This is a partial list; membership extended beyond those named here.			

Sources: Goodrick-Clarke (2004), 24-30, 43–44, 58-61; Kurlander (2017a), 19; Treitel, 98-9; Zander, H. (2011). Rudolf Steiner.

Table Note:

This table lists the signatories of the Guido-von-List-Gesellschaft, founded in 1908 to support the occult-nationalist writings of Guido von List.

Hale (2003) adds that the writings of Blavatsky and List inspired a proliferation of cults that emerged in Germany at the *fin-de-siècle*, leaving a discernible imprint upon society.[45] One prominent example is the "*Germanenorden*" (trans. Germanic Order), a secretive, ultranationalist, and antisemitic organisation formally established in May 1912 at the home of Fritsch to indoctrinate German society through a blend of völkisch ideology, racial pseudo-science, and Blavatskian occultism.[46] The *Germanenorden* drew extensively on the writings of both List, and Lanz, whose work exemplified the blend of occultism and racial theory. Interestingly, Goodrick-Clarke (2002, 2005) explains that List's early attempts to revive Wotanism through his writings and his Armanist Order were integrated into the *Germanenorden.*[47]

From its inception, the *Germanenorden* operated as an organisational blueprint for the expansion of additional lodges throughout Northern and Eastern Germany between 1908 and 1918. These lodges were formally constituted through a manifesto comprising nine articles, which outlined the Order's doctrinal objectives, that specified the rights and obligations of its members, and codified its hierarchical and ritualistic framework.[48] Goodrick-Clarke (2004) details in 1911, Hermann Pohl, the chief officer at the Berlin branch in 1916[49]—through correspondence encouraged sympathisers to join the Order and establish their own lodges. Lodges were grounded in racial principles and accompanied by "*. . . racial principles with a ritual based on Germanic pagan tradition,*" and underpinned by antisemitic ideology.[50] In this way, the *Germanenorden* institutionalised the mythic and racial esoteric currents that thinkers like Lanz had articulated, translating occult racial theories into a secretive, hierarchical network aimed at cultural and political influence. In sum, Blavatsky, List and Lanz can be perhaps dubbed as the spiritual fathers and mother of Nazi occultism.

Between 1900 and 1902, Steiner became a committed occultist, a period during which theosophical activity in Germany, already been underway since the 1880s, gained further momentum. However, the German Theosophical Society was not

45 Hale, 132, 161; Flowers, 85–86, 91-103, generally; See also Beaumont, M. (2010). Socialism and Occultism, 228–29; Goodrick-Clarke (2004), Chap. 3:33–48, for a scholarly analysis of Guido von List within the broader context of esotericism and völkisch movements. See generally Flowers, 72–80, on the occult proliferation in Germany.

46 Goodrick-Clarke (2004), *126–27; Goodrick-Clarke, N. (2005). List, 693–94; Kurlander (2017a), 15–22, 35; Kurlander (2017b), 83; Kurlander (2019), 4; Levenda, 73. See also Kurlander (2012), 532, citing Phelps, R. H. (1961). Theodor Fritsch, 442–49, who connects Fritsch's publishing apparatus (Hammerbund) as having "... acted as a political complement to the occult-oriented" *Germanenorden.*

47 Goodrick-Clarke (2002), 258; Goodrick-Clarke, N. (2005). List, 693–94.

48 Goodrick-Clarke (2004), 127, *128–29. He noted that Theodor Fritsch was Grand Master of a grand lodge—the ring leader and chief perpetrator of antisemitism.

49 Goodrick-Clarke (2004), 126 and 142, citing 252n.30, 90-5.

50 Goodrick-Clarke (2004), 127.

Initially, lodge branches included Breslau, Dresden, Konigsberg while lodges at Berlin and Hamburg were already operational prior to 1911.

officially established until 1902, with Steiner assuming the role of General Secretary of the Berlin branch. It was at this time Steiner's leadership marked a turning point in the popularisation of esoteric spiritual traditions within German cultural life. He founded the Anthroposophical Society in 1912, reestablishing it in 1924 following internal divisions. As previously noted, Ariosophy was a racist Austro-German variant of Theosophy, whereas Steiner's Anthroposophy, though structurally similar, adopted a less overtly racialised orientation.

Anthroposophy, as an outgrowth of Theosophy and imbued with Rosicrucian influences, it sought to bridge science, art, and religion—a quasi-spiritual worldview anchored in occult principles that aimed to articulate a spiritual physiology of the human being also referred to as a 'Universal Human.' See futuristic application at heading Class-Based Stratification, 283. Both movements were cut from the same cloth—Blavatsky's books. Indeed, Steiner's approach to Theosophy involved adjusting standard theosophical works, in particular, Blavatsky's *Secret Doctrine*. This adaptation expanded and diversified the scope of Atlantology, presenting a more forthright model customised to accommodate divergent cultural audiences within the broader occult milieu.[51]

Interestingly, Steiner first met Haeckel in 1894, after which he became a vocal defender of Haeckel's Monism. Complementing these esoteric foundations, Haeckel contributed a scientific and philosophical framework that further shaped Steiner's intellectual trajectory. Haeckel was the linchpin in Steiner's evolutionary and racial thought—providing the initial conceptual impetus to link biological theory with social and spiritual frameworks—prior to Steiner's full turn toward esoteric synthesis. Staudenmaier (2008, 2014), Poley (2013), and Kurlander (2013, 2017a) state, thanks to Anthroposophy, it was instrumental in introducing, "*. . . Hinduism, Buddhism, Yoga, and other South Asian spiritual traditions into German [culture].*"[52]

Building upon Steiner's occult framework, both Himmler and Walther Wüst (1901–1993), a Sanskrit scholar and president of the Ahnenerbe from 1937 to 1945, actively pursued the reconstruction of Atlantis, or, in Longerich's (2012) words, the restoration of the Germanic environment through "*the myths of Atlantis,*" indicative of an ancestral Aryan homeland. Kurlander (2017a) points out that certain strands of Nazi esotericism portrayed Lucifer as a positive figure, citing Lewis Spence's *The Occult Causes of the Present War* (1940) to show how Luciferian practices informed Nazi worldviews and the ideological vocabulary later adopted by movement members. Kurlander (2017a) observes specifically that "*[l]ater occultists as well as many Nazis embraced aspects of this 'Luciferian' tradition.*" Once more, a derivative of Blavatsky's Theosophical teachings and the broader Occult Revival, this was the foundation for a reconstructed spiritual worldview.[53]

[51] Faivre, 83, 87, 102. Goodrick-Clarke (2004), 106–21, 192–203; Hammer, 63–65, 103–08, 114–116 passim; Kurlander (2013), 157, citing Glowka, H. J. (1981). Deutsche, 28–32, 50–72, 80–108; Leijenhorst, C. (2005). Anthroposophy, I:82–86; Owen, A. (2004), 50, 228. McDermott, R. A. (1987). Anthroposophy. I:320–21; McDermott, R. A. (1992). Rudolf Steiner, 288–310; Staudenmaier, P. (2008). Race, 4–21; Staudenmaier (2014), 42.

For an expansive commentary on Steiner and anthroposophy, see Staudenmaier, P. (2014). Between Occultism and Nazism, and Staudenmaier's book review of German historian Helmut Zander's study on Steiner's anthroposophy in Staudenmaier, P. (2010), Anthroposophie in Deutschland, 107–16.

[52] Kurlander (2013), 161, citing Wüst, W. (1942). Indogermanisches, 29–39, 86–87, 103; Kurlander (2017a), 186; Poley, J. (2013). Rudolf Steiner, 55–65; Staudenmaier (2008), 6, 9–11; Staudenmaier (2014), 26–29.

[53] Kurlander (2017a), 8, 15, 308nn.34–35; Kurlander (2017b), 94; Spence, L. (1940), The Occult Causes of the Present War, 40–41, 72–73, 81–82; Jacobs, E. (1879), Der Brocken in Geschichte und Sage; Zelle, M. (2012), Externsteine, Detmold: Lippischer Heimatbund.

Indeed, the objective of the later occult architects listed in Table I-5, 75 (Hübbe-Schleiden, List, Lanz, Hartmann, Fritsch, Steiner, Zillmann), was to establish a type of syncretism that harmonised Nordic ideology with Eastern and Western esoteric traditions, drawing from both Theosophy and Steiner's Anthroposophy.[54] Furthermore, as part of this progressive reconstruction, formative Nazi leaders also studied the Quran and drew parallels between Islam, Chinese Taoism, and Japanese Shintoism, while also combining Norse paganism, astrology, witchcraft, and racial ideology—integrating broader mystical and diabolical currents that had emerged from the Occult Revival. All were interpreted as part of a shared ideological harmony, refracted through Germanic Luciferianism.[55]

Both Himmler and Hitler expressed admiration for Eastern mysticism, viewing it as spiritually disciplined and ideologically compatible with their broader worldview. Hitler was explicitly compared to a "*new Mohammed*" because he espoused a worldview akin to Mohammedanism, which teaches discipline, destiny, and, in Jung's words, "*the virtue of the sword.*"[56] Hitler was reportedly vocal in stating that, "*. . . there is no kind of terrorism' in Islam, only 'a promise of happiness.*"[57] Hitler, Himmler, and the Third Reich's interest in Islam, in particular, as part of foreign policy initiatives and Himmler's esoteric beliefs, is well-documented.[58] See Figure IV-2, 310: the conceptual link between the lost Atlantis and its envisioned cultural reawakening.

During Steiner's tenure, Steiner also launched his own journal, *Lucifer*, which was published from 1903 to 1908.[59] Steiner's influence was considerable, helping to stimulate interest in Theosophy among the educated middle classes of Europe and America between 1880 and 1930—an interest that revolved around on the pursuit for answers to the meaning of life—mirroring the predominantly middle-class composition of the List Society's signatories.[60] Indeed, Steiner's lectures had a direct influence on Max Heindel (1865–1919) (also known as Max Grashot), who went on to establish the Rosicrucian Fellowship in California based entirely on Steiner's teachings and

---

54 Poley, 65; Kurlander (2017a), 191, 186–87; citing Williamson, G. S. (2004). The Longing for Myth in Germany, 291–96, Wüst, W. (1942). Indogermanisches Bekenntnis, 29–39, 86–87, 103, and Wüst, W. (1942). Japan und Wir, 3–29; see also Longerich, 285.

55 Herf, J. (2009). Nazi Propaganda, 45; Levenda, 33; Kurlander (2017a), 192; Kurlander (2017b), 95, 97, 99; Kurlander (2019) 2, 15–16; Motadel (2014), Islam, 61.

56 McGuire & Hull (1993), C. G. Jung Speaking, 122–23; Herf, 155, citing a German diplomatic memo from the Germany Embassy in Tehran, which referenced Hitler quoting the Qur'an, Sura 5:82, in Mein Kampf, 69; Kurlander (2017a), 189; Kurlander (2019), 17. The point about Islam's "the virtue of the sword," see also For Hitler's and Himmler's praise and fascination with Islam, see Motadel (2014), 61–63; Picker (2003), Hitlers Tischgespräche, 184; Trevor-Roper (1953), Hitler's Secret Conversations, 319; and Werner (1999), Anthroposophen,74–94.

57 Kurlander (2017a), Hitler's Monsters, *189–92.

58 Motadel. 61–63; Kurlander (2017b), 95, 97, 99; Kurlander (2019), 2, 15–18: Treitel, 213–14, provides analysis on the ideological fusion of Shintoism, Taoism, Islam, and Luciferianism. For the depiction of Hitler as a "new Mohammed," see Spence (1940), Occult Causes of the Present War, 142–44.

59 Godwin, 103–04; Goodrick-Clarke (2004), 26, 28; Goodrick-Clarke (2008), 3, 84, 127, Introvign, M. (2005). Rosicrucianism III:1018; Kurlander (2017a), 18–19; Staudenmaier, 29; McDermott, R. A. (1992). Rudolf Steiner, 292.

Godwin's footnote notation asserts that the journal title *Lucifer* carries common connotations of 'evil,' but it was later renamed to *Lucifer-Gnosis*, retaining the title, to honour its original meaning of 'light-bearer.' However, upon evaluating Steiner's background, his wife, Marie von Sivers, their interest in the occult, and their associations, it is evident that his intent in using the term was far from innocent, but malevolent.

60 Allen, P. M. & Pietzner, C. (1974). A Christian, 455; Poley, J. (2013). Rudolf Steiner, 57; Goodrick-Clarke (2008), 127, 204, 230; McIntosh, C. (1980). The Rosy Cross, 134–44; Owen, A. (2004). The Place of Enchantment, 3–4.

broader theosophical ideas.[61] Hartmann and Steiner were central to the revival of the 1700s and 1800s Rosicrucian movement, whose work was, "*. . . published in Zillmann's periodical during 1905.*"[62]

Goodrick-Clarke (2004) explains that in 1905, Max Altmann entered the expanding occult publishing scene in Germany, eventually launching the influential journal *Zentralblatt für Okkultismus* in July 1907. Its editor, D. Georgiewitz-Weitzer—writing under the pseudonym G. W. Surya—played a significant role in the 1880–1910 modern Occult Revival. Alongside List, Georgiewitz-Weitzer popularised and drew from Rosicrucian themes, alchemy, occult medicine, the military religious orders and freemasonry.[63] Georgiewitz-Weitzer's work bridged traditional Western esotericism with contemporary interests in alternative science, helping to shape the intellectual contours of early 1900s practical German occultism.

Background wise, this revival of Rosicrucianism was a direct link to Germany's legacy as a centre of occult thought, traceable to the origins of the movement in the late 1300s. Its roots are traditionally attributed to Christian Rosenkreuz (1378–1484), whose name means "*rose cross.*" According to legend, Rosenkreuz journeyed through the Middle East, North Africa, and Egypt; he was believed to have returned to Germany in 1403 and founded the Rosicrucian Order. Edighoffer (2005) also found that Rosicrucian societies were subsequently established in the early to mid-1700s across several European states, with Austria emerging as a key centre of Rosicrucian influence between c.1720 and 1780.[64] This Order synthesised elements of Kabbalah, Neoplatonism, Hermeticism, and alchemical doctrine or chemical medicine into what became known as "*secret wisdom.*" Ironically, Yates (1972, 1999) points out, Bacon's *New Atlantis* is "*full of echoes of the Rosicrucian manifestos,*" postulating a deeper alignment between utopian science and esoteric thought than is typically acknowledged.[65]

While this origin narrative is compelling, Spence (1920) disputes its historicity, asserting that it has been "*. . . proved to be wholly fabulous,*" though he concedes that the movement did indeed originate in Germany, albeit not necessarily authored by Rosenkreuz.[66] The Rosicrucian tradition gained momentum in the early 1700s, with the publication of its foundational manifestos: Fama Fraternitatis (1614), Confessio Fraternitatis (1615), and The Chymical Wedding of Christian Rosenkreuz (1616). These texts, rich in allegory and spiritual symbolism, blended mysticism with calls for intellectual and religious reform, firmly embedding Rosicrucianism within the esoteric currents of early modern Europe. In contrast to Spence, McIntosh (2011) argues that is "*. . . a powerful legend had taken root . . . and has spawned a great variety of movements.*"

From 1918 onwards, in Staudenmaier's (2013) carefully explicated analysis of the question: "*Why did a wide range of occultists succumb to the allure of Nazism, at least initially?*" This heading has explored thus far the vector through which occult ideas

---

61 Goodrick-Clarke (2004), 243n.48; Howe (1967), 84–*85; Kurlander (2012) 534n.39; Wachsmuth, G. (1955). The Life, 190–91.
62 Goodrick-Clarke (2004), 60, 235n.15, citing Anon. (1905). Geheime Figuren der Rosenkreuzer aus dem 16ten und 17 ten Jahrhundert', 41-88, 92-88; Moran, B. T. (1994). Alchemy, 112; Newman, W. R. & Grafton, A. (2001). Secrets, 238; Stavish, M. (2007). freemasonry, 143; White, R. (1999). The Rosicrucian, xiv, 8, 40, 41, 263.
63 Goodrick-Clarke (2004), 27, 58; Owen, A. (2004), 225–26, on how British occultism viewed German Brotherhood developments.
64 Edighoffer, R. (2005). Rosicrucianism, II:1016.
65 Yates, F. A. (1999). The Occult Philosophy, 203–04; Yates, F. A. (1972). The Rosicrucian Enlightenment, 118–29. See generally Churton, T. (2002). Golden Builders; Goodrick-Clarke (2008), 107–27, for a concise analysis on the subject.
66 McIntosh, C. (2011). The Rose Cross, 27; Spence, entry for 'Rosicrucianism,' 340–342.

gained cultural traction. Yet the broader question remains: What was it about the occult that initially attracted people? In his preliminary analysis, Staudenmaier (2013) partly answers his own question by postulating that, "*. . . several of the more zealous Nazi opponents of occultism had come from occult backgrounds themselves.*"[67]

As outlined earlier, a pragmatic answer to the broader question, as Staudenmaier (2015) points out, is that (i) Blavatsky ignited a surge of interest esotericism, and (ii) more importantly, the Theosophical Society propelled that momentum by deliberately marketing and publicizing its activities to piqued interest. This initiative could not be deliverable by one influential person alone; rather it was a concerted effort by those embracing Blavatsky's teachings, such as the occult architects of Nazism listed in Table I-5, 75, and the key signatories of the List Society in Table I-6, 79. Goodrick-Clarke (2002, 2004) avers that during the period from 1914 to 1918, encompassing WWI and the pre-Weimar era, "*. . . theosophical thought lent itself to völkisch adoption.*"

Goodrick-Clarke further noted that, "*. . . early (pre-Nazi) völkisch and Pan-German groups before and after the First World War, were definitely influenced by esoteric ideas.*"[68] Kurlander (2017a), drawing on Gerhard Szczesny's 1940 dissertation, emphasises that post-WWI Germany had fostered an unusually large "*occult general public*" compared to other Western nations. Although Szczesny's dissertation focused on occult journals and periodicals, these publications revealed broad cultural receptivity to esoteric interests among the general German public. This was so conspicuous that Wolffram (2009a) observes that, by May 1921, it was even reflected in police investigations, where telepaths, mediums or spiritualists were enlisted to shed light on unsolved crimes.[69]

Small wonder, then, that Goodrick-Clarke (2002) asserts that the postwar appeal of occultism's *raison d'être* stemmed from a psychological need to restore national pride, mitigate the humiliation of military defeat and retribution stimulated numerous utopian Weimar fantasy writers to herald a coming saviour.[70] This complements Fisher's (1984) findings that the cultural climate of Weimar Germany was marked by a desire for retribution, invigorating a wave of utopian fiction envisaging a redemptive, saviour-like messianic figure.[71] Black & Kurlander (2015) links this cultural stereotype to the societal impact of "*. . . deaths unredeemed by battlefield victory—had a devastating effect on German society.*"[72] And, as Szczesny reported, the general public were quite "*. . . susceptible to occult ideas well before 1914,*" a susceptibility aligned with the Occult Revival.

In 1918, Rudolf von Sebottendorff (1875–1945) (born Adam Alfred Rudolf Glauer), originally initiated into Masonry in Egypt, founded the proto-Nazi Thule Society (Atlantis) in Munich. Conceived as a direct offshoot of the *Germanenorden* lodge, the Thule Society inherited its ideological framework, grounded in List's esoteric racial

---

67 Staudenmaier, P. (2013). Nazi Perceptions, 26, 29.

68 Goodrick-Clarke (2002), 126. Goodrick-Clarke (2004), *31, 33; Yenne, 16-17, 39; See also Tourlamain, G. (2014). "Völkisch" Writers, 9.

This "influence" in the quote is most likely attributable to the prominence of List, whom Goodrick-Clarke (2004), 40, describes as a ". . . celebrity amongst Pan-German groups of Austria," as early as 1897, citing newspaper article, Ostdeutsche Rundschau. (1895, April 9). 3, and the Wiedener Sängerbund programmes from the Wolfgang Heinrich Collection (Linz).

69 Wolffram, H. (2009a). Crime, 581–601.

70 Goodrick-Clarke (2002), 129; Kurlander (2017a), 65. Wolffram (2009a, 591) poses a further cause that the 'occult wave' was ". . . a complex combination of mass grief, economic deprivation and perceived moral and philosophical decline . . . ."

71 Fisher, P. S. (1991). Fantasy and politics; Kurlander (2017a), 89, 94.

72 Black & Kurlander, 6.

doctrines and reinforced by quasi-masonic ritual practices.[73] According to Goodrick-Clark (2004) "*[w]ithout this man . . . Germanenorden and Ariosophy would have been condemned to oblivion.*"[74] Another splinter group, the German Order Walvater of the Holy Grail founded in 1913, was less politically focused than Thule and the *Germanenorden*, it instead emphasised mystical and esoteric themes. Closely associated with the *Germanenorden*, the Thule was shaped by the antisemitic activism of Fritsch and Theosophical preoccupation with lost Atlantis. In 1933, when Sebottendorff returned to Germany, the Third Reich permitted him to revive Thule.[75]

It was a covert cover name for the *Germanenorden* until their official merger in March 1919. Fritsch helped popularise concepts originally derived from Blavatsky's Theosophy, where they were adapted into a racial-nationalist framework, which were then further adapted into a racial-nationalist ideology, solidifying the link between occultism and the forthcoming Nazi Party (keep reading).[76] See also heading, Theodor Fritsch, 200. Kurlander (2013) remarks that Fritsch and Sebottendorff were instrumental and Steiner (as previously discussed), to incorporate Eastern beliefs as part of *Germanenorden's* ongoing development. This included Hinduism, Islam, and Kabbalistic practices within its initiations and liturgy, in addition to elements of Tibetan Buddhism, all of which carried over into Sebottendorff's successor organisation, the Thule Society.[77] Phelps (1963) and Kurlander (2012, 2017a, 2017b) detail the complex transactional evolution of the Nazi Party. Sebottendorff's occult-inspired Society and its parent organisation, the *Germanenorden*, were fundamental in forging the ideological framework, organisational blueprint, and propaganda apparatus for its embryonic formation.[78]

First, Sebottendorff secured de facto control of the Münchener Beobachter (Munich Observer)—later renamed the Völkischer Beobachter (Racial Observer, "VB"). He effectively purchased the paper through proxy ownership by leveraging investments made by his companion Käthe Bierbaumer and her sister Dora Kunze in Franz Eher Verlag, the official Nazi publishing house, headquartered in Munich. This undertaking was carried out through Franz Eher Verlag, and was eventually acquired by the Nazi Party in 1920. Once the VB relocated its operations there, Sebottendorff assumed editorial control, transforming the paper from a disguised sports journal, originally a covert front for Thule's activities, into a primary propaganda organ of the emerging Nazi Party. Further, Thule itself operated as a secret front for the *Germanenorden*. Under Sebottendorff's leadership, the paper evolved into the primary propaganda organ of the emerging Nazi Party.[79]

Second, in October 1918, under Sebottendorff's direction, VB editor and Thule member, Karl Harrer, together with fellow Thule member Anton Drexler, co-founded

---

[73] François, S. (2023). Nazi Occultism, 9–11; Goodrick-Clarke (2005), 694; Redles, D. (2005). Hitler's, 53; Kurlander (2019), 4; Levenda, 72–79; Pennick, 24.

[74] Goodrick-Clarke (2002), 258; Goodrick-Clarke (2004), 29, 39, 49f, 129–30, *133f, 200. See also Howe (1967), 86; Howe, E. (1983). *Germanenorden*, 4:1081–82; Kurlander (2012), 529; Kurlander (2015a), 500; Levenda, 33, 73, citing Gilbhard, H. (1994). Die Thule-Gesellschaft, 10–15; Redles (2005), 53. See generally Goodrick-Clarke (2004), chap. 11:135–52, for a meticulous analysis on Rudolf von Sebottendorff.

[75] Kurlander (2017a), 39, 163, 179; Kulrander (2017b), 88; Phelps (1963), 247.

[76] Goodrick-Clarke (2002), 114; Koehne, S. (2015). Were the National Socialists, 787; Kurlander (2017a), 33–42, 48 et seq; Phelps, R. H. (1963). "Before Hitler Came," 247, 249, 250, 251, 252.

[77] Kurlander (2013), 157.

[78] Kurlander (2012), 533; Kurlander (2017a), 15; Kurlander (2017b), 84. See also Howe (1967), 78–90; Howe (1983), 4:1082; Phelps, R. H. (1963). "Before Hitler Came", 247, 255–61.

[79] Kershaw (1998), 139; Kurlander (2017a), 41, citing Sebottendorff, R. v. (1933). Bevor Hitler kam, 47–49; Redles (2004), 54.

the Politische Arbeiter-Zirkel (trans. Political Workers' Circle) as a parallel body to the Thule ring, which soon reorganised into the Deutsche Arbeiterpartei ("DAP").

Third, in 1919, the DAP evolved into the Nationalsozialistische Deutsche Arbeiterpartei ("NSDAP"), now known as the Nazi party. Within weeks, Hitler joined. By September of the same year, his vehement antisemitism cultivated by the coterie membership of both the *Germanenorden* and Thule, essential criterion and unifying factor, became glaringly evident. Hitler's antisemitic ideology, which would later become a core tenet of Nazi policy, was starkly outlined in his foundational speech on 13 August, 1920 at a Nazi party meeting in the Munich Hofbräuhaus. This was a several-hour-long oration, titled "*Why Are We Antisemites?*", marking a pivotal moment in which Hitler explicitly declared his goal to eliminate the Jews. By 1920, Hitler had emerged as de facto leader of the NSDAP and had gained control of the VB—the party's official newspaper. NSDAP party membership skyrocketed, quadrupling from 1,100 in June 1920, to 3,300 by August 1921.[80]

Mommsen (1996) explains that this surge in growth was "*. . . less to do with [the NSDAP's] political programme, but more to do with the unique political style that Hitler and his close supporters*" introduced.[81] In July 1921, Hitler was officially appointed party leader and chairman, in which he asserted a personal claim to unlimited power, Mommsen posits.[82] This sequence illustrates the 'magic circle' transmission of Blavatsky's teachings, ultimately found their home in the Nazi Party. Kurlander (2017a) summarises this matrix succinctly—

> Without Sebottendorff's Thule Society, without Harrer's Political Workers' Circle, and without the infamous newspaper they purchased to promote their worldviews, the Nazi Party would almost certainly not have been born.[83] (Footnote call numbers deleted)

By 1933, as the Third Reich ascended, Hitler's entourage from the 1920s were already steeped in the occult and esoteric ideologies distilled from the Occult Revival period, refracted through Luciferianism, as previously outlined. Kurlander (2017a) illustrates how these influences, ranging from Luciferian themes to werewolf lore, continued to suffuse the worldview of the regime's inner circle.[84] This elite preoccupation both reflected and extended a broader cultural trend: the unusually large "*occult general public*" identified in the 1920s, as mentioned earlier. Kurlander (2017a) contends that this widespread fascination with the occult persisted into the 1930s, was absorbed by the Third Reich, and was "*. . . consumed by the thousands.*" He attributes this persistence largely to the NSDAP's occult-themed propaganda, which he argues, "*. . . illustrates affinities between supernatural thinking and the rise of the Third Reich.*"[85]

---

[80] Spielvogel & Redles (2011), 84, 95n.24, citing Deuerlein, E. (1959). Hitlers Eintritt, 204; Mommsen, H. (1996). The Rise, 77, 156, 163; Stachura, P. D. (1980). The Political Strategy, 263–64.

[81] Kershaw (1998), 139; Mommsen (1996), 160; Steinweis, 8.

[82] Kurlander (2017a), 40–47; Kurlander (2017b), 84; Phelps, 256–58, citing Sebottendorff, 74. See also Childers, T. (1983). The Nazi Voter, 43–49; Mommsen (1996), *160–61.

[83] Kurlander (2017a), 46.

[84] Kurlander (2019), 12; Kurlander (2017a), 173, 163, 276–81, 283, and also 50–60, citing Picker, H. (1968). Hitlers Tischgesprache, 94–95, Piper, E. (2005). Alfred Rosenberg, 15–17. See also Fest (1970), 188–92; Heiden, K. (1935). A History, 45, 67; Steigmann-Gall, R. (2003). The Holy Reich, 142–43.

[85] Kurlander (2017a), 84, 87 et seq.

ii. From Mystical Origins to Political Emblem: The Swastika's Occult Trajectory

WEBB (1976) **ARGUES** that List's symbolic interpretation of the Armanen Futhark runes, including swastika-related symbols and in particular the swastika, was pivotal in List's *Das Geheimnis der Runen* (1908) (trans. The Secret of the Runes) linking the runes to symbols of ancient Aryan knowledge. List, inspired by Eastern wisdom from Tibet and the solar symbolism of the sun-wheel, popularised and associated the swastika, found among ancient Hindu symbols, with 'the discovery of the Aryan secret priestly caste, known as the 'Armanen.'

In addition, the exoteric religion of Wotan or Odin worship was a belief he claimed to have held since childhood.[86] It is noteworthy, that Goodrick-Clarke (2004) and Koehne (2015) point out that these runes, in particular, the vowels, "*. . . provided List with a model for an hierarchical priesthood of [the sun or the sun-Wotan cult] whose power derived from initiation.*" Among these runes, which stemmed from Blavatsky, the swastika held particular prominence.[87] Interestingly, despite what my present analysis demonstrates, the origins of these cultic beliefs do not lie with List or Lanz, as might initially appear. Earlier sources reveal that their antecedents lie in the work of Gobineau, widely regarded as the father of racism, as previously discussed.

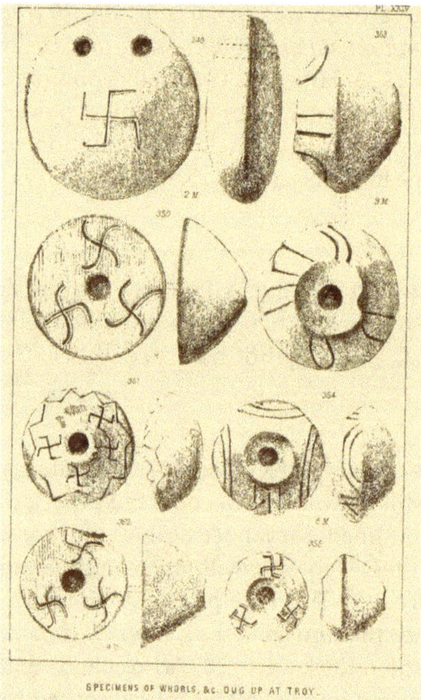

Figure I-4. Specimens of Whols finds at the Troy Excavation site, 1875.
Source: Schliemann, H. (1875). Troy and Its Remains, Plate XXIV

Accordingly, Butler's (1941) treatise, praised as an "*engrossing and scholarly*" contribution, unearthed a letter from Gobineau to his sister, dated 1876. In it, Gobineau expressed personal adherence to the cult of Odin and felt dignified to assume the role of "*high-priest of Odin, Thor, and Frey.*"[88] In the same letter, he referred to the cult as "*the religion of our fathers,*" highlighting his romanticised view of Norse mythology and its connection to ancestral and racial traditions. This is a significant observation, given Gobineau's posthumous status among German ideological circles as the father of racism. It further underlines his belief in cultural and racial superiority, which he embedded in an imagined lineage of ancient Germanic religious identity—linking myth, race, and

[86] Flowers, 63–64; Goodrick-Clarke (2002), 62; Goodrick-Clarke (2004), 34, citing List, G. v. (1913). Deutsch-Mythologische Landschaftsbilder, II:592; Hale, 323; Jung, C. G. (1970). "Wotan"; Kershaw (1998), 50; Spielvogel & Redles, 89; Webb (1976), 280; Yenne, 30.

[87] Flowers, 96–97, 315–25; Goodrick-Clarke (2004), 39, 41–42, 56–57 passim; Koehne (2015), 766, 768; Levenda, 52, 55, 58. See generally Goodrick-Clarke (2004), chap. 4: 49–55; Levenda, 49–53.

Stephen Flowers is widely regarded as a leading authority on runic esotericism. See Flowers, S. E. (2022), *The Occult in National Socialism*, as well as other works more specifically focused on runes.

Also note that Levenda, 56, provides the 1875 ritual detail without citation. If supporting sources exist, they are not identified.

[88] Butler, R. d. O. (1941). The Roots, 139–40; Flowers, 33–37, 55; Pulzer, P. (1988). The Rise, 52; Wedgwood, C. V. (1941). The Roots.

spiritual hierarchy. Goodrick-Clarke's (1998, 2002) meticulous study of Savitri Devi, is referred to as 'Hitler's priestess.'[89] He also conducted research on the famous German archaeologist Heinrich Schliemann (1822–1890), to which we now turn. Schliemann (1875) conducted excavations at Hissarlik, the ancient city of Troy (modern-day Turkey), between 1871 and 1875.[90] Among the hundreds of discoveries made, he unearthed a treasure trove; Figure I-4 shows this included a considerable collection of pottery fragments and terra-cotta whorls, many of which bore repeated depictions of the swastika, then spelled "*suastika*" in Schliemann's writings.[91]

Goodrick-Clarke (1998) observed that developments in archaeology and modern occultism contributed significantly to the evolving Aryan myth, particularly "*. . . through the identification of the swastika as a racial symbol.*"[92] This early phase, rooted in intellectual and esoteric circles, laid the ideological groundwork for the symbol's racialised reinterpretation in Europe. However, the swastika did not enter broader public consciousness until 1892, when Blavatsky's Theosophical Society began actively disseminating the symbol in the West. After encountering it in India, Blavatsky promoted the swastika through translated sacred texts that fused Eastern spirituality with Western occult philosophy. François (2023) and Goodrick-Clarke (2004) affirm its wider dissemination through the Theosophical Society. Goodrick-Clarke (2004) explains—

> From 1892 translations of Indian sacred texts and were printed in his periodical, Lotusbliithen [Lotus Blossoms] (1892–1900), which was the first German publication to sport the theosophical swastika upon its cover.[93]

This publication marked a turning point in the symbol's visibility in the West, particularly within esoteric, spiritualist, and eventually nationalist currents. Its growing appeal coincided with the widening Occult Revival in Europe, which not only drew increased public interest but also expanded the thematic range of esoteric exploration.[94] This context supports the finding of François (2023) who maintains that the swastika was more commonly adopted by occultists than political movements at the time of the Occult Revival—highlighting its popularisation through modern occultism and, in particular, Blavatsky's role as the key agent of its popularisation in the West.[95]

Coinciding with the same timeframe as the Occult Revival (1880–1910), Wilson (1896), then Curator of the Department of Prehistoric Anthropology at the United States National Museum, presented a comprehensive analysis of the swastika's antiquity.[96] Quinn (1994) documented that Polish librarian Michael Zmigrodski addressed the Tenth International Congress of Anthropology and Prehistoric Archaeology on the subject of the swastika. Walker (1939) and Quinn both assert that proponents of the swastika—

---

89 See Goodrick-Clarke, N. (1998). Hitler's Priestess for full biography of Savitri Devi; Goodrick-Clarke (2002), 88–106.

90 See Quinn, M. (1994). The swastika, for a comprehensive analytical analysis on Schliemann's findings.

91 Goodrick-Clarke (1998), 23; Quinn, 24–25; Schliemann, 98–121. See letter from Max Müller to Schliemann issuing him a warning about annexing Indian cultural swastika symbol. Cited in Quinn, 1, 45. See also Walker, N. (1939). Real History of the swastika, challenging Schliemann's Aryan interpretation quoted in Quinn, 44.

92 Goodrick-Clarke (1998), 33–34.

93 Goodrick-Clarke (2004), 25; François, 45.

94 See Goodrick-Clarke (2004), chap. 2:17–31.

95 François, 44; Morrisson, M. S. (2007). The Periodical Culture of the Occult Revival.

96 Wilson's publication appeared in the report of the Museum for 1894,

Alexandre Bertrand, Heinrich Schliemann, Émile Burnouf, and Ludwig Müller—claimed it was the "*. . . emblem of the supreme god of the Aryan race [in ancient India].*"[97]

Schliemann, at the time of writing in 1875, wrote that the swastika was, "*. . . the most sacred symbols of our Aryan forefathers.*"[98] Recognised as an oriental motif deeply embedded in Indian culture, religious and pictorial traditions, according to Quinn (1994), it was "*. . . fetishised within an Aryanist discourse.*"[99] Reich (1970) illustrated an Indian Swatika-Cross that was discovered with a decorative three dots at the end of each prong.[100] Consequently, through Schliemann's contacts, the "*. . . swastika was launched as the Aryan symbol in the European mind*"—a process which, according to Quinn's documentation of Schliemann's correspondence with contemporaries such as Ludvig Müller—can be traced to the period between 1887 and 1889, culminating in the 1889 congress meeting.[101] Indeed, Schliemann asserted that the swastika appeared frequently in Hindu temples and notably in ancient warfare, which scholars date between c.1500–1000 BCE—

> [T]he ships of King Rama in which he carried his troops across the Ganges on his expedition of conquest to India and Ceylon — bore the swastika on their prows. Sanscrit scholars believe that this heroic epic (the Ramayana) was composed at the latest 800 years before Christ . . . .[102]

Not surprisingly, the rationale for placing the swastika prominently on the "*prows*" was its supposed intrinsic power to bring victory and conquest. This reflects the same ideological mentality later adopted by Nazism, where the symbol was displayed at the forefront of marches, banners, and regalia as a means to project dominance and to actively pursue the defeat of the Allies. Goodrick-Clarke (1998) went on to mention that Schliemann's obsession with the swastika led him to adorn his Athens residence with decorative swastikas. This continuity is later evident in the activities of Lanz. According to Fest (1974), Lanz flew the swastika from his castle in Austria, which was the headquarters of the Aryan League as early as 1907.[103]

However, evidence suggests that the swastika predates its classical associations. The symbol, also known as the Gammadion, is derived from the Greek uppercase block letter gamma (Γ), mirrored along both axes to form a hooked cross. It is constructed by arranging four gammas at 90-degree rotations—upward, downward, leftward, and right—forming a swastika-like figure. Variations of the Gammadion have been found in archaeological contexts dating back as early as 3,000 BCE (thirteenth century).[104] Drawing on iconographic and numismatic evidence from global antiquities, Wilson (1896) argues and d'Alviella (1894) shows, represented in Figure I-5 below, that the symbol long predates Indian civilisation and Buddhism. They both show that the swastika is one of the oldest and most widely disseminated symbols in the ancient world, migrating across cultures and appearing in various forms and under different names in civilisations including America, in "*. . . as much separated by time as by space . . . ,*"[105] and

97 Goodrick-Clarke (1998), 24, 34; Quinn, 23, 53.
98 Goodrick-Clarke, 23–24; Schliemann, H. (1875). Troy and Its Remains, 102, 105.
99 Quinn, M. (1994). The swastika, 8.
100 Reich, W. (1970). The Mass, 101–03.
101 Goodrick-Clarke (1998), 24, 34; Quinn, 23.
102 Schliemann, 102–03.
103 Fest, J. C. (1974). Hitler, 36, 38; Goldstein, 69n.29; Levenda, 85.
104 d'Alviella, E. G. (1894). The Migration, xii. See also Cirlot, J. E. (1971). Swastika, 322–323.
105 Wilson, T. (1896). The swastika, 981.

as diverse as the United States, Assyria, Egypt, India, Greece, Celtic Europe, Japan, and Korea, among others.[106] Goodrick-Clarke (1998) adeptly synthesises these historical strands, demonstrating how the swastika's symbolic Indian origins specifically pinpoint the true origin in Greece "... *ancient Teutons, the Homeric Greeks, and Vedic India.*"[107]

[106] d'Alviella, chap. II:32–83; Wilson, *767–68, 770–983; Kurlander (2017a), 1283; Quinn, M. (1994). The swastika, 7, 44,141n.12, citing Walker, N. (1939). Real History of the swastika, 44. See also criticism from Quinn, 17–18, on the terms 'symbol' and 'migration' as contradictory.
[107] Goodrick-Clarke (1998), 24; Kurlander (2013), 157.

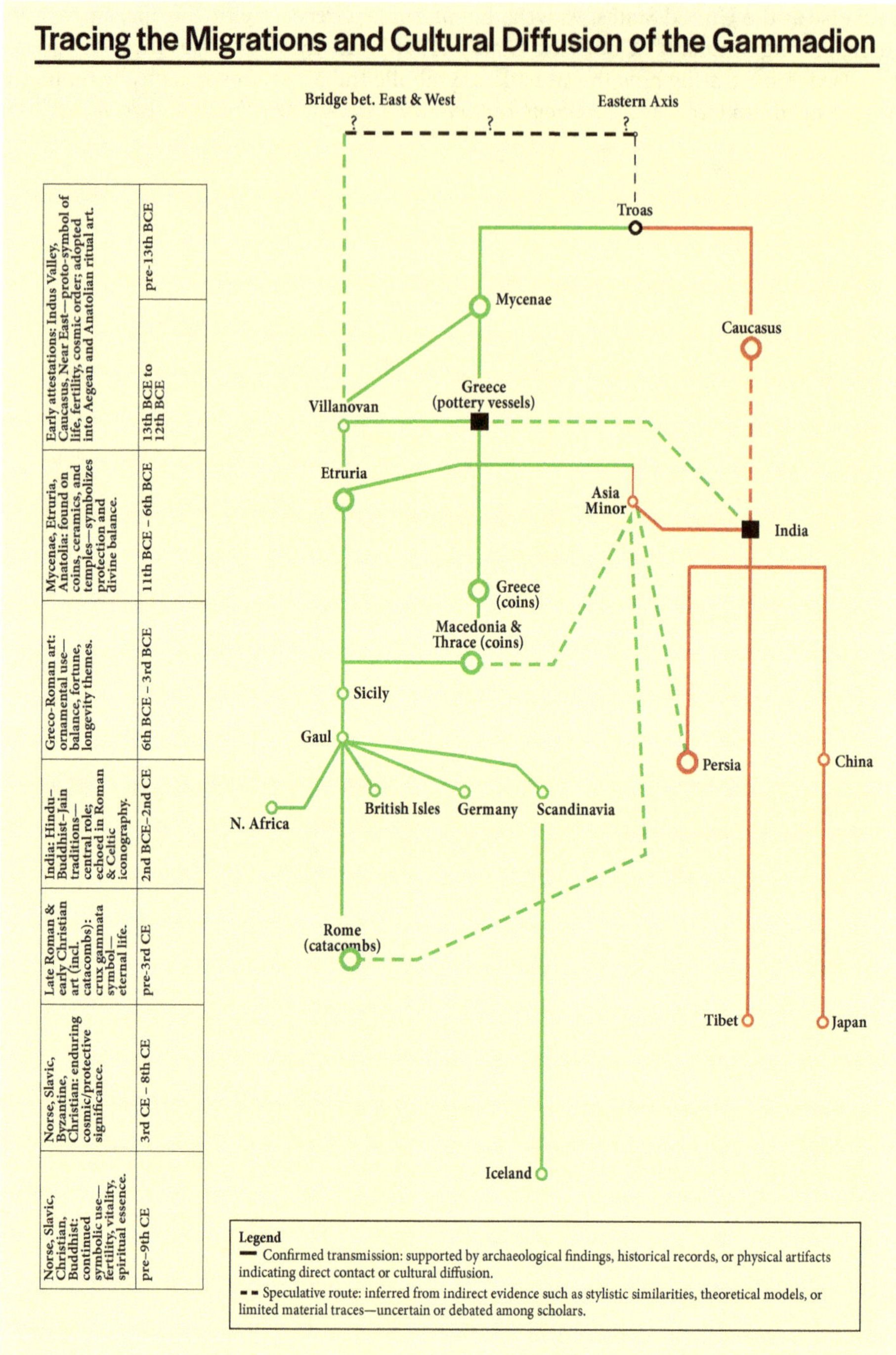

Figure I-5. Historical Diffusion of the Gammadion (Swastika) Motif.
This figure summarizes the historical diffusion of the gammadion (swastika) motif across regions and centuries, tracing its early appearance in Troas and its subsequent circulation throughout Europe, Asia, and North Africa. Adapted from É. G. d'Alviella, The Migration of Symbols (1894).

In Tibet, Hale (2003) found it was the "... *ancient symbol of good luck which represented the Wheel of Life....*"[108] Similarly, in China, it was regarded as a symbol of good luck and health, consistent with its traditional auspicious connotations within Indian culture, while in England and France, it appeared on small bronze ornaments.[109] Based on its global presence and deep antiquity, Wilson (1896) classified the swastika as a prehistoric icon of early Aryan symbolism. By the time of the *fin-de-siècle*, the swastika became an established Aryan racial symbol and was adopted as the official seal of both antisemitic and Theosophical societies.π[110] It is noteworthy that the swastika entered the public domain more explicitly as a symbol of antisemitism in the early 1920s, partly due to Hitler's claims in Mein Kampf, but it only fully assumed this meaning later, during the Third Reich.[111]

Mosse (1961) maintains that the swastika was a prevalent artistic motif to depict the Aryan emerging from the sun, while Cirlot (1971) asserts that, "... *the most general interpretation was that it symbolized movement and the power of the sun,*" and an ancient mystical symbol of the sun. The painter Fidus (born Hugo Höppener) intricately linked the swastika with the "... *German youth movement, us[ing] this motif constantly.*"[112] Mosse (1961, 1999) makes good on this argument by demonstrating Eugen Diederichs' mesmerisation with this symbolism, and subsequently created "... *a red and golden flag with the sun as its centrepiece,*" further intertwining the sun's symbolism with the cultural identity of the German people.[113]

Mosse (1961, 1999) and Spielvogel & Redles (2011) illustrate the connection through Prana (trans. power of the sun), a German monthly devoted to applied spiritualism and occult sciences, published by the Theosophical press in Leipzig. Both highlight that Prana championed vegetarianism and abstinence from alcohol as key to understanding nature and the cosmic life-force. Its editor, Johannes Balzli (1864–1939), also served as secretary of the List Society, with Hartmann, a honorary member and frequent contributor. Steiner contributed to Prana, and, in addition, content based on his work appeared regularly, though it was edited and adapted by Balzli. Ultimately, this all circled back to List.[114]

In both of List's books, *Das Geheimnis der Runen* (1908) (trans. The Secret of the Runes) and *Die Bilderschrift der Ario-Germanen* (1910) (trans. The Hieroglyphs of the Ario-Teutons), he maintained that the swastika was a secret occult "*sign of salvation.*" Echoing this idea, a 1923 editorial column in Sebottendorff's newspaper, the VB (formerly Münchener Beobachter), titled "*A United Decision*" (No. 198, 26 September 1923) and cited by Mühlberger (2004), declared—

108 Hale, 9.
109 Goodrick-Clarke, N. (1998), 4, 23–24, 44–45, 62, 66, 69, 85, 107, 110; Goodrick-Clarke (2002), 94; Pennick, 32; Wilson, 855.
110 François, 44; Goldstein, 69n.29; Goodrick-Clarke (1998), 34, 44; Samuel, R. H. (1963). The Origin, 74, citing List, G. (1910). Die Bilderschrift, Webb (1976), 1–29, 262.
π Note: Spielvogel & Redles, 89, state that Blavatsky state that Blavatsky "... had used the [swastika] in the seal of the Theosophical Society." However, Blavatsky died in 1891. While she wrote extensively about the swastika as a powerful occult symbol in The Secret Doctrine (1888) and vol. II, published posthumously in 1893, it seems an oversight to claim that she directed the seal's use. She could not have; she was dead. As argued here, it was likely her disciples who adopted the symbol at the *fin-de-siècle*.
111 Hitler, A. ([1925] (1971)), 497; Reich, 100.
112 Cirlot, 322–23; Mosse (1961), 86; Mosse (1999), 124; Pagel, A. (1992). Jugendstil and Racism, 100–01.
113 Mosse (1961), 87; Mosse (1999), 124; Quinn, M. (1994). The swastika, 25; Schliemann, 118–19.
114 Mosse (1961), 87–88n.31, citing Prana, Organ für angewandte Geisteswissenschaften, VI, 1-2 (1915), 348–349; Mosse (1999), 124–125; Lavoie (2018), 500–06; Spielvogel & Redles, 89; Staudenmaier (2014), 85–86; Staudenmaier (2015), 27, 29.

> The eternal struggle between Ormuzd and Ahriman . . . has once again ended in the victory of the sun, whose symbol is the ancient Aryan sign of salvation: the swastika![115]

Building on the earlier discussion of the NSDAP's esoteric roots in Thule and *Germanenorden*, the party adopted the pagan sun-wheel swastika as its symbol in December 1922. Key ideologues, Gottfried Feder (1883–1941) and Alfred Rosenberg, refracted the swastika through Christian iconography, blending it with imagery from Egyptian solar deities and Eastern fertility cults. Gottfried Feder (1883–1941) described it as the "*. . . symbol of reawakening life,*" while Rosenberg saw it as a timeless emblem of cultural renewal. Their amalgamation drew on antiquity—most notably the Chi-Rho (or khi-rho), which appeared on Ptolemaic coinage as early as 230 BCE and in Roman-era iconography.[116]

Readers will recall the discussion of the developmental stages of nature worship in Crystallization of Racial Hygiene, 110, spanning the 1850s to the 1880s. Mosse (1964, 1999) astutely elucidates how List's romanticised nature mysticism—rooted in Völkisch ideology and infused with Theosophy—added vigour to their pseudoscientific pretensions—

> Taken by themselves, theosophy, sun worship, and the Nordic ideal of beauty seem outrageously irrational. But taken as a whole, they provided an explanation of the world and man's place in it. Combined, they crystallized the yearnings of those classes who were sinking in the social scale.[117]

The "*ideal of beauty*" that Mosse speaks of was embraced by movements like Lebensreform (life-reform), which championed a holistic return to nature—a trail that leads back to Hartmann.[118] This entailed a fascination with Greek antiquity and emphasised health, vegetarianism, veneration of the human form, and athleticism as essential to both personal and societal renewal. De Jonge (1979) underlines this aspect, noting that nudism was not connected to sex at all, but rather was "*. . . central to the cult of the body, sun worship [emphasizing] the freedom of physical movement . . .*"—a philosophy rooted in pantheistic ideals and an embrace of nature's transformative power—that also intertwined with avant-garde art, progressive politics, scientific innovation, geopolitics, philology, anthropology, history, archaeology, astrology, mythology and the occult.[119] As demonstrated by Goodrick-Clarke (2004), Kurlander (2017a), Koehne (2015), and Redles (2005), occultist practices were not repudiated, nor was paganism excluded from

[115] Mühlberger, D. (2004). Hitler's Voice, 89–*90; See also Koehne (2015), 771, 787n.144, citing Döllinger (1920), 182, in a comparable context; Kurlander (2017a), 49, 180; Kurlander (2017b), 49, 180; Kurlander (2019), 6; Spielvogel & Redles, 89. Cf. Staudenmaier (2008), 12, citing Steiner linking Lucifer and Ahriman together.

[116] Hastings, D. (2009). Catholicism, 127–28; Koehne, S. (2014). The Racial Yardstick, 587, 595n.70; Koehne (2015), 585–86; Kurlander (2017a), 168, 179–80; Maitland, C. (1846). The Church, 168–70; McGovern, J. B. (1904). Concerning, 5–9; Petrie, W. M. F. (1898). Religion, 45–47; Reden, S. v. (2007). Money in Ptolemaic Egypt, 67; Society of Antiquaries of Newcastle upon Tyne. (1884). Archaeologia, 48:241–43.

[117] Mosse (1964), 106; Mosse (1999), 121.

[118] Levenda, 45.

[119] Cornwell, J. (2003). Hitler's scientists, 192; De Jonge, 169, *174–76; Flowers, 53; Kurlander (2013), 158, 159; Levenda, 44; Mosse (1961), 86–87. For further Lebensreform discourse, see Bernard, L. (2024). Selling Natural Laws; Flowers, 46, 180–81; Kennedy, G. (1998). Children of the sun; Warren, H. C. (1933). Social nudism; Williams, J. A. (2007). Turning to nature.

the expressionist visual arts, literature, film, and theatre during the NSDAP's ascent throughout the Weimar period.[120] In particular, Eisner (1969) and Kurlander (2017a) emphasise that the Nazis exhibited a notable predilection for killing and for horror-inflected expressionism, especially within visual art and film, where occult themes and macabre imagery permeated cultural production.[121]

iii. Counterarguments to the Established Narrative

IMPORTANTLY, ACCORDING TO Kurlander (2017a, 2019), Schertel (2009), and Timothy Ryback (2010), Hitler's occult interests spanned paganism, pseudoscience and gnosticism. His interests went well beyond Ernst Schertel's 'Magic: History, Theory, Practice' (2009; originally published 1923), a heavily annotated volume, replete with marginal notes and heavy underlining, from Hitler's personal library, as well as other occult-themed works.[122] Particularly striking are the passages concerning 'magic' that Hitler underlined, including those that articulate its disciplinary demands and its promise of self-transformation, including—

> Satan is the beginning . . . Satan is in everything that lives and appears . . . Out of this, which is unreasonable (irrational), reason in the true sense is born. Without this preceding gloom, creation would have no reality; darkness is its necessary heritage.[123] (Omission in original)
>
> [C]entral tenets of parapsychology: the power of the will in appealing to the collective unconscious, including the marked line: "He who does not have the demonic seed within himself will never give birth to a magical world."[124]

In this light, Hitler's behaviour itself was peppered with the tropes of a magician: ritualized speech, performative charisma—a deliberate emulation of magical power. So much so, in magical theory, like a magician or shaman pronouncing an incantation with rising intensity, Hitler believed in delivering the same message in constant repetition, a message that could be transformed into political reality.[125] Cavendish's (1983) encyclopaedia entry on 'incantation' is consistent with Hitler's documented practices and lends substantial support to the view that he practiced a form of real magic—

> Words are weapons of power in magic . . . The magical theory is that just as someone of powerful personality can use forceful words to dominate other people, so the magician armed with the lightning of concentrated will-power and the thunder of overwhelming language can dominate anything in the universe, natural or supernatural.[126]

---

120 Goodrick-Clarke (2004), 168–171; Koehne, 763–64 passim; Kurlander (2017a), 47–49, 76, 186, 50n.125, citing Bullock, A. (1991). Hitler, 79–80; Redles (2005), 56–59.
121 Eisner, L. H. (1969). The Haunted Screen, 95–97; Kurlander (2017a), 76–77.
122 Kurlander (2019), 4; Kurlander (2017a), 182, 352nn.206, 207, 208, citing Bärsch, C.-E. (1998). Die politische Religion, 293-96, 354-57: Ach, M. (1977). Hitlers "Religion," 52-54, 76-77, 108, 140-43: Koehne (2015), 789-90; Gugenberger, E., & Schweidlenka, R. (1993). Die Fäden, 146-50; Spence (1940), 66-69.
123 Schertel, E. (2009). Magic, 118–19, quoted in Kurlander (2017a), 182, 352n.204.
124 Kurlander (2012), 534; Kurlander (2017a), 56–57, 69–75; Ryback, T. W. (2010). Hitler's Private Library, 160, citing Schertel, 17, 69, 82.
125 Kurlander (2017a), 72, generally, 69–74; Sickinger, R. L. (2000). Hitler, 116.
126 Cavendish, R. (1983). Incantation, 5:1418.

Waite (1978) notes that Hitler did not learn from the intellectual giants of German history; rather, his beliefs were first shaped by the pamphlets of List and Lanz, and again between 1919 and 1923, when his political ideas came under their renewed influence, to which we now turn. This timeline parallels with the trajectory of his involvement in the occult. Kershaw (1998) and Kurlander (2017a) cite Lanz's 1951 memoir, which recounts that in August 1909, while in Vienna, a 20-year-old Hitler visited Lanz's Ostara office in Rodaun and requested back issues. Impressed by his earnestness, Lanz reportedly gave him back issues free of charge.[127]

While scholars have heavily disputed the reliability of this meeting, Goodrick-Clarke's (2004) meticulous analysis, together with Kershaw's (1998) record about Hitler's conversion to antisemitism, indicates that Hitler is likely to have collected Ostara, which would substantiate the connection between Lanz's racial-occult writings and Hitler's early worldview.‡ [128] Sebottendorff's own account in Bevor Hitler kam (1933), as documented by Phelps (1963), recounts Thule's formative role in the early Nazi movement. According to Sebottendorff, in 1919, Hitler initially turned to Thule members for support—and they were the first to align themselves with him.[129] It is all the more revealing that, from the occult-infused foundations of the NSDAP and Hitler's early contact with esoterically inclined ideologues of List, Lanz, Fritsch, Sebottendorff, following Thule's merger with the *Germanenorden*, this occult milieu coalesced around Hitler as its emergent political leader. He was in many respects one of their own.

Esoteric belief was not exclusive to Hitler's magic circle leadership, but included societal elites and middle classes. Indeed, Mosse (1975) maintains that its appeal extended beyond the upper echelons of power, attracting members of the higher social classes, the educated bourgeoisie, and participants in the broader nationalist movement.[130] Manjapra (2014) corroborates this perspective, explaining that "*... occult rituals appealed to a segment of the respectable middle class and served as a conduit for the popularisation of Schopenhauerianism in Germany.*"[131] Mosse (1970), Stern (1987), and Staub (1989) advance this argument further, contending that Hitler's rise was not solely attributable to the elites and the middle classes, who played a substantial role, but was also facilitated by parliamentary majority rule amid a paralysed party system. In addition, this dynamic was reinforced by decisive grassroots support from millions of voters who underestimated and misperceived him.[132]

How is this paradox even conceivable? In response to a similar question, Stern (1987) argues that it emerged from two intertwined processes: the collapse of the

[127] Kershaw (1998), 49–50; Kurlander (2017a), 3 56n.199, citing Daim, W. (1958). Der Mann, 17–48, documenting Lanz's reported 1909 encounter; Pennick, 25; Waite, R. G. L. (1978). The Psychopathic god, 90–*91n., 92–95.

[128] François, 20; Goodrick-Clarke (2004), 99–118, *196 passim; Kershaw (1998), 50–51; Kurlander (2012), 529, 534–*35, citing Hamann, B. (1996). Hitlers Wien, 285–323; Kurlander (2013), 157, 167n.17, citing McGetchin, D. T. (2009). Indology, 171–6; Levenda, 80–82; Mosse (1961), 83–96; Mosse (1975), *197–98.

‡ Weikart (2016, chap. 7) and Hamann (1999:221–22) urge strong caution, dismissing the alleged visit as exaggerated and unsupported by verifiable evidence. Hamann maintains that eyewitness Elsa Schmidt-Falk reported no knowledge of any connection between Hitler and Lanz. For a rebuttal of Hamann and a detailed examination of the evidence linking both Lanz and List to Hitler, see Goodrick-Clarke (2004:194–200). Goodrick-Clarke's position is further supported by Kurlander (2017a), 4, 31–32.

[129] Phelps (1963), 245; Levenda, 46–47.

[130] Flowers, 301; Goldstein, 62; Mosse (1975), *15–16.

[131] Manjapra, K. (2014). Age of Entanglement, 64, citing Gerhard Klamp, "Schopenhauertradition und Forschung im zwanzigsten Jahrhundert," Zeitschrift für philosophische Forschung 14 (1960), 438– 452; Ulrich Linse, Ökopax und Anarchie: eine Geschichte der ökologische Bewegungen in Deutschland (1986), 125–152.

[132] Mosse (1970), 214–15; Staub (1989), 46; Stern, F. (1987). Dreams and Delusions, 125.

Weimar Republic and the rise of Hitler.[133] To this, it may be added that propaganda directed at the general public and the educational establishment constituted central modes of transference. Stern, however, offers a more nuanced answer, specifying that the crucial dividing line lay in the nature of this support: middle-class supporters, Völkisch intellectuals, and political elites knowingly endorsed Hitler as one of their own, while the broader public were largely ignorant. Kurlander (2012, 2013) cements this understanding further, demonstrating that the diffusion of occult mode of thought permeated multiple layers of society, driven top-down and absorbed across Germany's more eclectic population. Kurlander (2012, 2013) commented—

> These complex and contradictory attitudes towards the supernatural make perfect sense when one considers the diversity of the Nazi movement. Incorporating virtually all elements of the German middle classes (and a significant minority of the working classes), Catholics as well as Protestants, former liberals and conservatives, Nazism can hardly be viewed as monolithic in its intellectual antecedents and political praxis. Many Germans embraced aspects of the occult . . . .[134]

One case in point is highlighted by Mosse (1964), who examines the activities of novelist and Thule member Ellegaard Ellerbek. He worked tirelessly to cultivate public interest in the occult, particularly among young adults, by travelling across Germany to deliver lectures in schools. His presentations, frequently themed around questions such as, "*'Is there a world of spirits?'—with an answer that always rang out in the affirmative,*"[135] reflecting an agenda of indoctrination aimed at normalizing occult ideas among impressionable audiences. In one especially striking lecture, he even exhorted his student audience that they were *gods.*[136] See also Reengagement of Past Connections, 266 for quotation on Ellerbek. The antecedents of occultism were clearly established prior to 1933, and its practice especially after WWI, persisted after Hitler's assumption of power.[137]

These Nazified occult concepts bring us full circle to the Hitler-List connection (the Lanz–Hitler linkage already established). On August 13, 1920, only a few months after joining the NSDAP, Hitler delivered a lengthy speech that portrayed German prehistory through an Atlantean framework. Mees (2008) rightly points out the irony that, despite his affirmative position on Atlanteanism, Hitler held a clear disdain for Wotanism, the *Germanenorden*, and the Thule Society. He suggests that Hitler sought to distance himself from these movements, though he stops short of explaining why.[138] Hamann (1999) and Waite (1978) affirm his ideological connection to, and influence by List, noting how Hitler reiterated the teachings of List and Fritsch. Hitler emphasised the symbolism of the swastika and opined—

> We know that all of these people held one sign in common: the symbol of the sun. All of their cults were built on *light*, and you can find this symbol, the

[133] Stern (1987), 121.
[134] Kurlander, (2012), 547; Kurlander (2013), 165, citing Treitel, 213–16. See also Goodrick-Clarke (2004), 202; Hamann, 7–9, 285–323; Howe (1967), 4.
[135] Mosse (1964), 83. See also Goodrick-Clarke (2004), 46; and 154 for comments about Ellerbek.
[136] Mosse (1964), *82–83.
[137] Godwin (2010), 124; Goldstein, 66.
[138] Mees, B. (2008). The Science, 113.

> means of the generation of fire, the Quirl, the cross. You can find this cross as a swastika not only here [in Germany], but also exactly the same [symbol] carved into temple posts in India and Japan. It is the swastika of the community (Gemeinwesen) once founded by Aryan culture (Kultur).[139] (Emphasis in original) (Alteration in original)

Even the VB, as previously discussed, reported in 1920 that the NSDAP openly proclaimed its affiliation with List. The party asserted that the revolutions of 1918–19—marking the collapse of the German Empire and the transition to the " . . . *Weimar Republic had been 'prophesied' in "the Edda and the teachings of [List's] Armanen.*"[140] Phelps (1954) draws attention to a key detail regarding Hitler's personal affinity for List's Armanen teachings. In 1921, Hitler received a copy of an essay on Nationalism by Rabindranath Tagore as a birthday gift, bearing a personal dedication from B. Steininger: "*To Mr. Adolf Hitler, my dear Armanen brother,*" followed by the runic invocation "*logapore wodan: wigiponar.*" According to an interpretation by Prof. Dr. Moser of Tübingen, the inscription explicitly links both Steininger and Hitler to List's Armanen circle.[141] There is little doubt that Hitler shared a harmony for List's ideas and vocabulary during his early years in Vienna, and later infused them into his political speeches.[142]

In sum, by 1921, the NSDAP had integrated itself with List's occult circle and Hitler, then serving as party leader and chairman, was personally connected to List's Armanen ideology. Furthermore, Thule (Atlantis) was a front or covert designation for the *Germanenorden*, within which List's Armanen Order was subsumed—a lineage ultimately tracing back to Blavatsky. Moreover, as previously noted, the occult journal *Prana* was under the editorship of Johannes Balzli, a member of List's occult society. Hartmann, one of the early occult architects, was a frequent contributor. Vegetarianism and teetotalism were central themes of Prana, and Hitler had adopted both practices by the 1920s. Taken in isolation this is trivial, but given that Prana was an esoteric-occult journal published by the Theosophical Society, and considering the wider set of antecedents, it is unlikely to be an isolated coincidence. Accordingly, Kurlander (2012) asserts that the cultural environment that Hitler was exposed to in the 1920s was prevalent with occult practices—

> Hitler's prewar Vienna and Munich were immersed in occult and esoteric doctrines, from Theosophy, Anthroposophy and Ariosophy, and World Ice Theory to astrology, clairvoyance, paganism, [and] witchcraft.
>
> . . . .

139 Hamann, B. (1999). Hitler's Vienna, *210–213, quoting Jäckel, E. & Kuhn, A. (Eds.). (1980). Hitler, 186–87; Kurlander (2017b), *86–87; Waite, R. G. L. (1978). The Psychopathic, 96–98. See also Cocks, G. C. (2012). The State, 46; Koehne (2015), *774–75, 776; Mees, B. (2004). Hitler, 268 (also citing Jäckel & Kuhn, 1980, 186–87).

140 Kurlander (2019), 6. See also Koehne (2015), 777.

141 Goodrick-Clarke (2004), 199; Kurlander (2012), 534–35; Phelps, R. H. (1954). Die Hitler, 925; See also Koehne (2015), 778 and Lavoie (2018), 507, et seq.

142 Cocks, 46; Waite (1978), 95–97. See opposing view by Spielvogel & Redles, 90, who cast doubt on a direct, personal relationship between Hitler and List in Vienna from 1907–13. However, scholarly authorities suggest that List's teachings shaped the development of Hitler's beliefs. It is known that Hitler frequented circles connected to List. At some point, he formed an association with List, though not necessarily during the specified period—whether prior to 1907 (when Hitler was 18 years old), in the intervening years, or post-1913.

> All these cultural discourses and social phenomena were remarkably widespread across interwar Germany, compatible with many aspects of modernity [143]

Given these collective insights, it is somewhat difficult to maintain that practical occultism played no role in Hitler's inner circle. What cements this point further is the composition of the grassroots Nazi Party in the early 1920s.[144] The following outlines Hitler's principal cabinet members, ordered by their influence and significance within Nazi ideology and occult associations (a separate discussion of Himmler follows)—

Table I-7. Key Occult-Influenced Figures in Hitler's Circle (1920s–Post-1933)

	Position/Role	Notable Esoteric / Ideological Associations
Dietrich Eckart (1868–1923)	Hitler's mentor; co-founder of the DAP; early Nazi propagandist	Early occult-nationalist mentor; key in launching Völkischer Beobachter
Alfred Rosenberg (1893–1946)	Chief Nazi ideologue; Reich Minister	Central architect of Nazi racial mythos; integrated pagan mysticism
Walther Darré (1895–1953)	SS Race and Settlement Office Chief	Influenced by Anthroposophy and Armanism; shaped Nazi race theory
Hermann Wirth (1885–1981)	Co-founder of the Ahnenerbe	Advanced foundational mytho-historical racial theories
Rudolf Hess (1894–1987)	Hitler's Deputy Führer	Deeply involved in astrology and Eastern esotericism
Anton Drexler (1884–1942)	Founder of the German Workers' Party (DAP)	Founding member of Thule Society; pivotal in channelling esoteric-nationalist ideas into early party doctrine
Ernst Röhm (1887–1934)	Co-founder and leader of the SA	Early Hitler ally; interested in astrology and mystical ideas
Erich Ludendorff (1865–1937)	WWI Field Marshal; Beer Hall Putsch co-defendant	Promoted nationalist-occult beliefs post-war
Wolf-Heinrich von Helldorff (1896–1944)	SA leader; Berlin Police Commissioner	Linked to occultist Erik Jan Hanussen; active in occult circles

Table Note:
This table outlines selected members of Hitler's inner circle and ideological affiliates whose beliefs, writings, or associations reflected occult, mystical, or esoteric influences. It emphasises the overlap between Nazi leadership and spiritual worldviews that shaped the racial, mythological, and nationalist ideologies of the early Third Reich.

This continuity of occult influence, traced from Blavatsky through List, reached the highest levels of Nazi leadership (see also Table I-6, 79), particularly under Hitler and Himmler. A disciple of Lanz, Eckart—along with Rosenberg, Drexler, and Hess, each affiliated with the Thule Society—either directly mentored or associated with Hitler; all can be linked back to List, Lanz, or both. Others listed in Table I-7 also held significant occult affiliations. Far from a fringe curiosity—esoteric beliefs were deeply embedded

[143] Kurlander (2012), 534n.39, 549, respectively. See also Howe (1967), 4–7; Spielvogel & Redles, 91.
[144] Kurlander (2019), 4–6.

in the early Nazi movement. These individuals not only occupied key positions within the regime, but decisively moulded and reinforced Hitler's evolving ideological and political convictions.[145]

Wiligut emerged as a key proponent in channelling these Luciferian influences within Nazi ideology.[146] After Wiligut joined freemasonry in 1899, he became a disciple of List and Lanz. He authored several occult-pagan texts, and "*. . . was drawn to Liebenfels' Order of the New Templars.*"[147] In November 1933, Wiligut met Himmler and impressed him with his profound esoteric knowledge, leading to his appointment as head of the SS Department for Pre- and Early History.

His mandate was to guide the SS using ancestral clairvoyant insights into ancient Germanic religion and tradition. He later worked for the *Deutsches Ahnenerbe Studiengesellschaft für Geistesurgeschichte* (trans. German Ancestral Heritage Studies Society for Spiritual Prehistory; hereafter "Ahnenerbe"), a SS research institute founded by Himmler in July 1935.[148] The Ahnenerbe sponsored fringe, border scientific research, conducted occult investigations, and carried out anthropological studies—all aimed at advancing Himmler's vision of a purely Germanic society. Wiligut soon earned the nickname "*Himmler's Rasputin*" for his spiritual authority and influence, in which he a counsellor on magical and religious matters.[149] Goodrick-Clarke (2004) explains that Wiligut's esoteric influence was central to the SS's Aryan-theosophical architecture shaping Himmler's imaginative world.[150]

Within the Nazi regime itself, interest in the occult was neither pervasive nor uniform, but rather concentrated among specific leaders and factions. Indeed, François (2023) and Kurlander (2013) noted, "*. . . it was the SS, through its leader, and not Nazism in general, which took an interest in occult doctrines and neo-paganism . . . ,*" a category that included Steiner's Anthroposophy.[151] Occultism was not a central driving force of Nazism, but, as François explains, it was more of "*a symptom or precursor*" of the broader ideological landscape.[152] Nonetheless, figures outside the SS, such as Julius Streicher, occasionally engaged in occult-linked practices; this fascination was far from widespread among the broader Nazi membership.[153]

However, occult practice apparently flourished indirectly, dovetailing on Kurlander's (2017b) argument that the Third Reich embraced occult groups because they propagated the "*. . . idea of a lost, Ur-Aryan (Atlantean) civilisation,*" which, of course, carries occult significance.[154] This enduring presence of occultism is exemplified in a handful of

---

145 Koehne (2015), 779, 780–82, for general commentary on Eckart; Kurlander (2017a), 173, 163, 50–60, citing Picker, H. (1968). Hitlers Tischgesprache, 94–95, Piper, E. (2005). Alfred Rosenberg, 15–17. See also Fest (1970), 188–92; Heiden, K. (1935). A History, 45, 67; Levenda, 92–94, 107; Mommsen (1996), 158–59; Steigmann-Gall, R. (2003). The Holy Reich, 142–43.
146 See Longerich, 282–85, for analysis on Wiligut.
147 François, 37; Kershaw (1998), 50.
148 For general overview about the founding of 'Ahnenerbe' see Longerich, 275–79.
149 François, 37–38; Flowers, 104–120, 286–88; Godwin (2010), 143–46, 149; Goodrick-Clarke (2004), 160, 162, 183, 240n.1, citing Daim, W. (1958). Der Mann, der Hitler die Ideen gab; Kurlander (2017a), 175–79, 182; Kurlander (2019), 13; Laqueur, T. (2006). Why the Margins Matter, 115–16; Yenne, 98–101.
See also Godwin (2010), 118–21; Flowers (2022) 52, point that theosophy laid the groundwork for List and Lanz. See generally Goodrick-Clarke (2004), chap. 14:177–91, for meticulous analysis on Karl Maria Wiligut. See Goodrick-Clarke (2004), chap. 9:106–22, for analysis on Lanz and his Order of the New Templars.
150 Goodrick-Clarke (2002), 135–36, 327n.17, citing Mund, R. J. (1976). Jörg Lanz v. Liebenfels und der Neue Templer Orden;
151 François, 37; Goodrick-Clarke (2002), 126; Kurlander (2013), 162.
152 François, 22.
153 Ibid,, 37.
154 Kurlander (2017b), 88.

characters, to which De Jonge (1979) posits, when "*. . . Germany was literally full of persons describing themselves as spiritual leaders,*" which included the occult.[155] The Nazis pragmatically utilised occultism as a tool in their broader pursuit of power, blending ideological commitment with ruthless strategic opportunism. Thus, the tension between eradication of occult practices and its utilisation was not a contradiction, but rather a calculated strategy.

Treitel (2004) foremost expert on German occultism, partially agrees with this perspective, acknowledging the cultural affinities between the occult and Nazi ideology. However, she advances a more divergent interpretation, contending that these affinities, "*. . . never translated into a sociopolitical alliance of occultists with the state.*"[156] Kurlander (2015a, 2015b) does not entirely dismiss Treitel's stance that the occult had no 'special role' in shaping the history of the Third Reich, but instead states that it is a "*more complex picture.*"[157] Although Treitel downplays the significance of occult beliefs, other historians have demonstrated that such ideas played both symbolic and practical roles within the Nazi regime. Kurlander (2015a) expresses a similar point that, "*. . . historians not only play down the Nazis' occult proclivities, but also argue that the larger story of the Nazi regime and the occult movement is one of escalating hostility.*"[158]

Again, Treitel maintains that Himmler, as the SS head, "*. . . presided over a twelve-year attempt to eradicate the occult movement from Germany.*"[159] Both Staudenmaier (2014) and Kurlander (2017a) acknowledge that the Third Reich's war on the occult commenced only on 9 June 1941.[160] The timing of this initiative, under the direction of both Himmler and Hess, the second-in command, complicates the notion of a consistent twelve-year suppression of occultism and instead points to a delayed and selectively targeted campaign. Kurlander (2012, 2015a, 2015b, 2017a) cites the striking case of Hess, himself guided by astrological counsel, who in May 1941 undertook an unauthorised solo flight to Scotland with a romantic diplomatic notion to negotiate peace with Britain. He writes—

> Nazi regime employed occultism "pragmatically" throughout the war, recruiting astrologers and clairvoyants to provide military and diplomatic intelligence both before and after the May 1941 Hess affair.[161]

Kurlander (2012, 2015b, 2017a) highlights a key paradox in Nazism's ties to the occult. He criticises 'reactionary modernism' or revisionist accounts like Treitel's downplay of Nazi occult involvement, arguing instead that Nazism selectively incorporated elements of the occult to complement and reinforce their racial sciences and Aryan supremacy,

155 De Jonge, 177–81.

156 Treitel, C. (2004). A Science, *211, 248. See also Kurlander's (2017a), 14, 16, 23 passim, for an opposing view; Laqueur's (2006), 112–135, objective analysis of Treitel; Flowers (2022), 185, brief remarks; Staudenmaier's (2009), 54–56, review and critique.

157 Kurlander (2015a), 501; Kurlander, E. (2015b). Hitler's Supernatural Sciences, 132.

158 Kurlander (2015a), 501; Treitel, 209-10.

159 Treitel, 211.

160 *Kurlander (2017a), chap. 4:99–130; Staudenmaier (2013), 47–50; Staudenmaier (2014), 214–47, for extended analytical discussion.

161 Kurlander (2012), 542; Kurlander (2015a), 499, 512–13; Kurlander (2015b), 132, 136; Kurlander (2017a), 53, 118–30.

"*. . . than [what] Treitel and other revisionist accounts suggest.*"[162] Kurlander (2012, 2013, 2015a, 2015b) identifies this dynamic as a clear case of Himmler's hypocrisy: while the regime publicly denounced and persecuted occultists, the regime selectively and simultaneously exploited them for opportunistic purposes when politically or ideologically expedient.[163] This duplicity is further supported by Varshizky (2019) who argues that Himmler and, to a lesser extent Hess, were "*. . . known for their affection for esoteric and occult ideas . . . ,*" supporting the view that Himmler maintained a deliberate double standard.[164]

On one hand, Himmler justified the suppression of occult movements in the name of "*. . . political control and 'scientific' professionalization*"—persecuting them as threats to the Nazi vision of a rational, scientifically ordered society.[165] On the other, the regime actively employed occult practitioners for their perceived utility in military and diplomatic contexts. For example, Schellinger et al. (2015) document the recruitment of various occultists in Berlin to investigate Mussolini's whereabouts in 1943.[166] Kurlander (2017a) similarly avers that the occult interests of Himmler and Hess shaped the SS's engagement with occultists and border scientists after 1933, and increasingly after 1941, as part of the war effort.[167]

The background outlined by the occult architects in Table I-5, 75, together with the key esoteric figures in Hitler's inner circle presented in Table I-7, 99, helps contextualize the ideological lineage that made such recruitment plausible. In addition, individuals such as Edgar Dacqué, Karl Maria Wiligut, Ludwig Ferdinand Clauß, Friedrich Bernhard Marby, and Siegfried Adolf Kummer received institutional support elsewhere in the regime—revealing a broader receptivity to occult and pseudoscientific ideas than is commonly acknowledged.[168]

Kurlander (2015a, 2017a) situates this dual approach within the uneven timing of state action, noting that, "*. . . during its first four years of rule, the Nazi regime showed little consistency in the way . . .*" it policed occultists.[169] This timeframe corresponds with what Staudenmaier (2013, 2014) documented, namely, that several early actions taken against esoteric groups between 1935 and 1937, including the closure of the Association for Occult Science, the dissolution of the Society for Esoteric Studies in Leipzig, and other similar measures. Affiliated circles in Berlin were also placed under surveillance.[170] The term 'policing' is strategic usage, insofar as the emerging dichotomy

[162] See Kurlander (2015b), 135–43, 149; Kurlander (2017a), xiii–xiv. See also Kurlandrer (2012), 545–46, for a broad rebuttal of revisionist accounts; Longerich, 77–78. For actual examples of revisionist work, see Treitel, Science; Harrington, Reenchanted Science; Owen, A. (2004). The Place of Enchantment; Rabinbach, Catastrophe; Gregory, Nature Lost; Paul J. Croce, Science and Religion in the Era of William James, vol. 1: Eclipse of Certainty (Chapel Hill, 1995).

[163] For example, see Kurlander (2015a). The Nazi Magicians', 498–522; See also Kurlander (2012), 542–*44; Kurlander (2013), 155–56; Kurlander (2015b), 2, 149. See also comments in, Schellinger, U. et al. (2015). Pragmatic Occultism, 157.

[164] Kurlander (2012), 542, 544–45; Kurlander (2015a), 511;Varshizky, A. (2019). The Metaphysics, 277, citing Franz Wegener, Heinrich Himmler: Deutscher Spiritismus, französischer Okkultismus und der Reichsführer SS (Gladbeck: Kulturförderverein Ruhrgebiet, 2014). Varshizky also references *The Kersten Memoirs* (1957), a source of limited reliability, alongside more substantive works such as Wegener's.

[165] Kurlander (2012), 534; Kurlander (2015b), 138.

[166] Kurlander (2017a), 220; Schellinger, U. et al. (2015). Hitler's Supernatural Sciences, 169–71, quoting Walter Hagen [i.e., Wilhelm Höttl], Unternehmen Bernhard: Ein historischer Tatsachenbericht über die größte Geldfälschungsaktion aller Zeiten (Wels-Starnberg: Welsermühl, 1955), 135–37.
See also Schellinger's, U. et al., 170, annotation that Höttl's accounts were biased.

[167] Kurlander (2017a), 220, 222; Schellinger et al., 164. For a general discussion of border science, see Kurlander (2017a), 197–230.

[168] Kurlander (2017a), 87, 200.

[169] Kurlander (2015a), 502; Kurlander (2007a), 100, 101, 111.

[170] Staudenmaier (2013), 33; Staudenmaier (2014), 223.

lay in eliminating borderline or fringe sciences, while preserving forms of occultism considered scientifically and methodologically sound. Kurlander (2015a) sharply distinguishes between the "*vampire of superstition*" on one hand, referring to fringe paranormal practices, and the respected "*scientific*" occultism on the other, that sought to diffuse scientific method. He further shows that Himmler instructed Arthur Nebe, chief of the criminal police (Kripo), to develop criteria for differentiating between the two—

> [T]he investigation of all the occurrences that are inaccessible through natural human perception (scientific occultism)" versus "every activity that rests on and exploits superstition.[171]

Thus, emphasising how the Nazi regime's approach to occultism was not uniformly repressive but strategically bifurcated. Archival material, particularly from the *Bundesarchiv*, documents Nazi efforts to distinguish between "*border science,*" "*charlatan science*" or "*occult swindlers,*" and "*scientific occultism.*"[172] These first two categories were framed as commercial occultism under the trope of 'border science.' This exploited "*. . . unscientific and pseudoscientific occultism . . .*" and therefore "*. . . requir[ing] the sharpest police oversight and countermeasures . . . .*" By contrast, the boundary lay in the investigation of "*. . . scientific phenomena that conventional science were unable to detect,*"[173] yet still "*. . . follow the established rules of scientific inquiry.*"[174] Nebe carved out a space for "*. . . selectively appropriating 'scientific occultism,'*" whether for military technology development, intelligence gathering, or political prognostication, while actively weeding out the 'vampire of superstition.'[175] Returning to Himmler's duplicity outlined above, Kurlander (2015a, 2017a) highlights that Himmler was "*. . . deeply invested in occult and border-scientific thinking and saw no incompatibility . . .*" between selective appropriation and its policing.[176]

Still, not all scholars agree on the extent or significance of these links. Strube's (2017) review of Staudenmaier's (2014) work, Strube contends that popular narratives of Nazi occultism are exaggerated and overly sensational. Accordingly, Strube questions the evidential basis for asserting any significant connection between occultism and Nazism. Strube writes—

> [T]he criticism of sensationalist narratives sometimes seemed to have resulted in corresponding simplifications. Consequently, it would be necessary to pay more attention to the historical evidence showing actual connections between occultism and National Socialism.[177]

While Strube calls for more careful historical inquiry, his argument underestimates key developments. He overlooks the Occult Revival of 1880–1910, and fails to fully

171 Kurlander (2015a), 504–*505.
172 Kurlander, E. (2015a), 514n.107; Kurlander, E. (2017a). Hitler's Monsters, x, 22–31, 101–02, *chap. 5:131–61, border science analysis; Schellinger et al., 157–66.
173 Kurlander (2015a), 511, citing Kröner, W. (1938). Die Wiedergeburt des Magischen, 12, 14–16.
174 Kurlander (2017a), 22–23. See for example, Sommer, A. (2009). From Astronomy to Transcendental Darwinism : Carl du Prel (1839–1899).
175 Kurlander (2015a), 505–06.
176 Kurlander (2015a) 506. See also Kurlander (2015b), 135–43, for background to Himmler's hypocrisy; *Kurlander (2017a), chap. 4:99–130.
177 Strube, J. (2017). Between Occultism and Nazism, 254.

engage with Himmler's well-documented double standard. Furthermore, Hitler's extensive personal library contained numerous books on occultism—among them, a heavily annotated volume by Ernst Schertel. Hitler also invoked incantations in his speeches and maintained associations with Lanz; as previously discussed, Lanz's occult journal *Ostara* exerted a formative and provocative influence on Hitler.

However, Treitel's more ambivalent view is valuable, particularly when considered alongside Heilbronner's (2015) concept of the "*Wewelsburg Effect*."[178] Heilbronner argues that the postwar fascination with Nazi horror—which persists into contemporary times, especially in literature, film, and intellectual discourse—blurred the line between myth and fact. This cultural tendency, he claims, has led many to exaggerate Nazi occultism into something fantastical, creating a macabre mystique that overshadows more grounded historical realities. While acknowledging the ongoing scholarly debate, the discussion highlights Hayton's (2015) observation that Nazi wickedness, including their involvement in occultism, is typically rationalised and explained away through less malevolent acts, thereby overshadowing the regime's engagement with the occult.[179]

Taking this present analysis together, these perspectives reveal that occultism in Nazi ideology was neither central nor marginal. It operated on multiple levels—elite and popular, symbolic and strategic, sincere and cynical. A nuanced understanding demands moving beyond both sensational myths and outright denial. Stripping the occult from Nazism risks degrading the historical record of the Holocaust. While revisionist scholars aim to correct sensationalism, overcorrecting by tidying up or erasing the occult narrative, it blurs the ideological machinery that made genocide imaginable, justifiable, and ultimately executable. This approach undermines our understanding of the Holocaust not merely as a crime, but as the product of a belief system in which occultism provided a narrative scaffold for Nazi ideology, shaping power, identity, and violence in supernatural and esoteric beliefs.

iv. The Superman Archetype in Western and Eastern Antiquity

LIST, AS DID Steiner, developed a syncretic system integrating karma, reincarnation, and the teachings of freemason Johann Baptist Kerning. Staudenmaier (2015) documented fifteen distinct esoteric practices in a 1910 poster issued by the '*Zentrale für praktischen Okkultismus*' in Germany (Headquarters for Practical Occultism).[180] Little wonder that, according to Mosse (1964) German philologists—

> [F]used a large number of mystical Indian symbols with the gods of ancient Germans. . . . Willibald Hentschel, in 1923, gave his order for the development of true Aryans the name Artamanen, after the Indian sun god Artam.[181]

Just as the swastika was repurposed by occult organisations, Mosse rightly affirms that Germanic societies were also designated with the names of the same Indian pantheon.[182] This, of course, conflates the sun-god archetypes: Horus, Dionysus, Apollo,

[178] Heilbronner, 273–74.
[179] Hayton, J. (2015). Beyond Good and Evil, 249.
[180] Staudenmaier (2015), 28.
[181] Mosse (1964), 89–90.
[182] Ibid., 90.

and the old Norse deity Baldr (or Baldur), a derivative of the Greek Apollo. Kershaw (1998) pertinently adds that List took the swastika as a sign of the "'*unconquerable,' the Germanic Hero, the 'Strong One from Above*'"—a title originally ascribed to Apollo.[183] All of which were used to objectify and sacralise völkisch ideology, particularly within groups like the *Germanenorden* and Germanic Druids obsessed with racial purity. Their ideas were subsequently disseminated through theosophical publications.[184] During the Occult Revival (1880–1910) the swastika itself transitioned: from an esoteric symbol within the occult, to a marker of Aryan identity, then a leading symbol of antisemitism, and by 1933, it became synonymous with German political power.[185]

The aforementioned sun worship aspect reinforces the earlier point that Germans perceived themselves as the progeny of the god-men, specifically identifying with the sun-god Apollo. Simply put, they emerged from the sun. Given this connection, the ideological thread extends to Atlantis, well-known for its connection to solar veneration, where Blavatsky was the foremost authority in occult Atlantology. Ignatius Donnelly (1882) contemporary of Blavatsky, affirms that "*[s]un-worship [was] the ancient religion of Atlantis* . . . ."[186] India has a long and complex history of invasions and conquests. Among the earliest were the incursion of Alexander the Great and the earlier expansion of the Persian Empire into the region. In this context, the notion of an invasion by the armies of Dionysus, while speculative, is not entirely implausible. Furthermore, some groups in Punjab, India, have traditionally claimed descent from Dionysus.[187]

However, Taylor (1906) dissents with the Horus/Dionysus/Apollo sun-god view. He argued that the Indian origin of the Dionysiac myth was debunked by a comparison of Dionysus with Assyrian sun-god who was named Dianisu. Upon review, Taylor (1906) offers no authority for his finding, and no corroborating sources substantiate his viewpoint. Indeed, The Saturday Review (1890) critiques Taylor's assertions as questionable, with many considered disputable.[188] As for Apollo, Donnelly believed that none of the myths of Apollo resemble any of the Indian sun-gods, because no trace of him could be found in the Vedic hymns.[189] It is widely acknowledged that Apollo was known as the sun-god. A further irony from Jung's (1970) analysis is that Wotan, the Germanic god of storm and frenzy, has ancient connections with Dionysus and in turn with Nietzsche's superman.[190]

Dionysus/Apollo were also known by various names and epithets, varied across regional dialects, but the same person. Indeed, according to Smith & Anthon (1851), Donnelly (1882), Strabo (1854), Herodotus (1912), Fabre d'Olivet (1915), Evola (1995), and Godwin (2010), the Dionysus Egyptian pharaoh was believed to be the manifestation, usually identified under various names of Zeus or son of Zeus, Hades, Bacchus, Horus, Apollo, Jupiter, lacchus, Daksa, or Vaivasvata Manu or Ram by the Celts, and Rama by the Hindus. Furthermore, various deities from antiquity—including those from Greek, Roman, and Hindu traditions—were identified as manifestations of the Egyptian Pharaoh, Osiris, viewed as one and the same.

Strassler (2007) highlights the oscillating distinctions between Horus, son of Osiris, in Egypt, whom the Hellenes equated with Apollo, and Osiris, whom the Hellenes

183 Kershaw (1998), 50.
184 Mees (2008), 147; Mosse (1964), 67–87.
185 Webb (1976), 262; Goldstein, 62n.12.
186 Donnelly, 284; Godwin (2010), 64–87.
187 Smith & Anthon, 562, 588; Mallory, 63.
188 Cook, J. D. et al. (Eds.). (1890). The Origin, 22–23.
189 Taylor, I. (1906). The Origin, 304.
190 Jung, C. G. (1970). "Wotan," 180–82, 185. See also Goodrick-Clarke (2002), 179.

referred to as Dionysus. Another inversion posits that Apollo is the son of Osiris and Isis. While Strassler appears not to definitively endorse this inversion, the preceding discussion illustrates how different cultural traditions render diverse names, with these 'gods' embodying similar archetypal roles. Cultural exchanges and syncretism allow for gods to merge and transform, enabling them to manifest in multiple forms. In Latin tradition, Hercules also known as *ἰδαῖος* (Idaios), was associated with the sun and shared many attributes with Apollo. In other cultural contexts, he was identified as Krishna in Hinduism; as Herakles and Pan in Greek mythology; and as Harikrishna, Harikesa, and Hrsikesa within the Hindu tradition.[191] Thus, it is not another, different sun-god; they are the manifestations of the same solar entity, known by different names across various languages, cultures and periods.

Prior to the destruction of Atlantis, Blavatsky and Donnelly (1831–1901) pioneered the Atlantis-related theories, suggesting that the kings of Atlantis became the gods of the Greeks. Aryanism began with these Greek demi-gods invading India, specifically Dionysus and Hercules.[192] Following their conquest of India, this belief extended to the idea that they founded a dynasty of god-men who ruled in their wake. Interestingly, the superhero god-men ideology was a symbiotic part of Aryan/Nordic belief referring to, "*. . . Übermensch, 'an Apollo-like creature who jumps over skyscrapers.'*" This is a variant of the common saying for Apollo, depicted as 'riding his Quadriga across the sky.'[193] One of the most famous landmarks of Germany, is the statute of Viktoria (Victoria) riding the Quadriga with four horsemen on top of Brandenburg Gate, Berlin. He/she was known for being androgynous. See heading, Crystallization of Racial Hygiene, 110, for discussion on the national spirit of Germany and inclusion of mysticism. Viktoria is Roman-Germanic for the Apollo-Dionysus archetype; a personification of victory and divine state power.

Drawing from the earlier swastika analysis, a composite picture emerges. You will recall that the swastika on the prows of King Rama's ships represented power, victory and conquest. In tandem with these ideas, Kershaw (1998) found that List adopted the Swastika as a sign of the "*. . . 'unconquerable,' the Germanic Hero, the Strong One from Above.'*"[194] Although speculative, List arguably saw Hitler as the embodiment of the unconquerable Germanic hero, given Hitler's familiarity with List's writings and the positive influence they exerted upon him. Hitler, in turn, interpreted List's swastika belief as a reference to, or prophecy about himself.

To complete the composite, drawing from the Origins of Racist Aryan belief, 17 heading, this prophecy notion is not an exclusive to Hitler alone. As teenagers, Hitler and Kubizek reportedly had a mystical experience on Freinberg mountain after watching Richard Wagner's opera Rienzi. During this experience, Hitler is held to have prophesied his future as a political leader. Hitler later regarded Wagner as a mentor, idol, and his

[191] Donnelly, 284; Evola, J. (1995). Revolt, 8, 213, 224n.16, 257, (See generally Goodrick-Clarke's (2002) commentary on Julius Evola, 52–71); Fabre d'Olivet, A. (1915). Hermeneutic Interpretation, 30, 118–19, 168–69, 184–85, 192; Godwin (2010), 41–42; Goodrick-Clarke (2002), 58, 60; Hislop, A. (1903). Two Babylon's, 48; How, W. W. (1912). Herodotus, vol. I, bk. 2:186, para.42.2; Jaiswal, S. (1967). The Identity of Herakles; O'Flaherty, 86; Schliemann, 121; Smith, W. & Anthon, C. (1851). entry for 'Dionysus', 264, 706; Strabo. (1854). The Geography of Strabo, 238; Strassler (Ed.), 2:42–2.44, pp.136–38; 2:2.144, p.185; 2:2.156, p.192; Stafford, E. (2012). Herakles; Thapar, R. (1965). A History of India, 1:32–33, 44.
[192] Brown, T. S. (1955). The Reliability; Donnelly, 283–307; Flowers, 46; Godwin (2010), 1–35, 64; Karmarkar, R. D. (1950). The First Greek Conqueror; O'Flaherty, W. D. (1980). Dionysus; Smith & Anthon (1851), 265. See Mosse's (1964, 210), passing comment on Dionysus re the Munich circle.
[193] Dethloff, C. (2011). Coming Up to Code, 109. See also Goldstein, 77; Keith, A. B. (1933). The Origins, 817; Smith, W. (1842). Currus, 310, with the notation, ". . . who is crowned with rays as he rises in a quadriga."
[194] Kershaw (1998), 50.

hero, who played a critical role in shaping his worldview. Webb (1976) further explores this oscillating connection between Hitler and Wagner, showing that Wagner was deeply involved in occultism, holding a position as a Grand Master in certain esoteric circles. Wagner's blending of myth and esoteric interpretations strongly linked the key figures to Hitler, and in turn, to Hitler's inner circle, to occult networks.†[195] Indeed, Kurlander (2017a) found that Wagner's Ring cycle—steeped in the same heroism embodied by figures like Siegfried and Wotan—"*. . . constituted a central moment in the popularisation of German national mythology . . . ,*" and was pivotal in shaping Hitler's own vision of an Ario-Germanic ideology.[196]

As previously mentioned, the retrospective examination of the ancient world of Atlantis (Thule) was rooted in the Aryan belief that the German people, as children of the gods, were of demi-god extraction. The Aryan race was said was to be derived from and identified with one individual—Dionysus, revered as the Indian sun-god hero.[197] Thus, they were considered the original race of supermen, believed to have inherited the immense knowledge and magical powers of the lost world of Atlantis.[198]

Arguably, the superman doctrine, Aryan ancestry and the German connection were intended to evoke connotations with the Greek and Mesopotamian (Sumerian) gods, also known as the Anunnaki, by drawing parallels with classical antiquity (Hercules, Gilgamesh, Perseus, Achilles, Thor, Hermes et. al., dubbed as mortal gods), with a view that modern-man would "*. . . take on the trappings of these divine figures.*" The physiognomy of ancient Greeks, in turn, according to Günther (1927), is described as resembling that of Germans.[199] Hale (2003) presents a variant perspective that challenges the conventional view that an "*elite priesthood*" of survivors, literally held to be giants—sought refuge in the Himalayas and established themselves in Shambhala, Tibet.[200] There, they regarded Shambhala as a repository of ancient wisdom and Aryan heritage. This narrative can be reconciled with the proposition that these gods fled from Atlantis to both India and Greece. Chamberlain believed that a "*race of supermen*" would return through a "*long historic process*" of "*pure*" races.[201] This "*long process*" is Francis Bacon's *New Atlantis*, which is still yet future.

Germany's enduring connection to occultism is illustrated by three key elements. First, its nexus with Blavatsky's Theosophy; second, the swastika, popularised through her Theosophical Society; and third, occultism fused with the völkisch worldview and persisted into the Third Reich.[202] In fact, Blavatsky's philosophical framework profoundly affected List, who co-opted her concepts and reinterpreted them, combining völkisch ideology with occultism and theosophy to underpin Nazi racial ideology. This was a subtle process. Staudenmaier (2015) found that occultists absorbed scientific

[195] Webb (1976), 43–44 and 76n.37, citing Otto Julius Hartmann, Die Esoterik im Werk Richard Wagners (Freiburg, 1960).

† Webb (1976) suggests that Wagner was a "Grand Master" in occult circles. I urge caution and scepticism. While Wagner's involvement in esoteric practices and views is acknowledged, the claim that he held such a position remains speculative, with no reliable historical evidence to support it beyond Webb's reference to Otto Julius Hartmann's work.

[196] Kurlander (2017a) 6–7, citing Gugenberger, E. & Schweidlenka, R. (1993). Die Fäden, 97-99.

[197] Jaiswal, 111; Karmarkar, 242; Puskás, I (1990). Magasthenes; Rawlinson, G. et al. (1880). History of Herodotus, 223–24; Stafford, E. (2012). Herakles.

[198] Davies, 25; Hale, 40, 51; Kurlander (2017a), 176; Kurlander, E. (2017b). One Foot in Atlantis.

[199] Gordon, J. (2016). When Superman, 212-13, 228; Günther (1927), 2, 142, 152–72; See also Haslem, W. et al. (2007). Super/heroes : From Hercules to Superman, for example.

[200] Hale, 59. See also Flowers, 65.

[201] Neumann, 107–08.

[202] Black, M. & Kurlander, E. (2015). Revisiting the 'Nazi Occult', 8, citing, Michael Hagner, "Bye Bye Science, Welcome Pseudoscience?," in Rupnow et al., Pseudowissenschaft, 50; Flowers, 84–85; Kurlander (2017a), 13; Staudenmaier (2015), 31.

themes and liberal ideas, anchoring various occult doctrines to a plethora of concepts about race and assigning them special spiritual and esoteric significance.[203]

v. Elevation of Hitler as Superman

HITLER WAS, IN part, viewed as superman. He idolised the ground that Nietzsche walked on. Shirer (1960) noted that "*Hitler often visited the Nietzsche museum in Weimar . . . posing for photographs of himself staring in rapture at the bust of the great man.*"[204] Shirer posited that these chosen supermen were described as the "*lords of the earth,*" which struck a responsive chord with Hitler that he was the fulfilment of Nietzsche's prophecy, namely, he was superman, and these "*lords*" would be appointed by him to rule society—

> Society is not entitled to exist for its own sake but only as a substructure and scaffolding, by means of which a select race of beings may elevate themselves to their higher duties . . . .[205]

Hitler was taken *sui generis* by the German people, regarded as both a superman and a god-man, with "*[S]uperhuman qualities [that] emanate from him . . . pervade the state, party, and people.*"[206] This perception was not merely an outcome of political strategy but, as François (2023) explains, had ties to his engagement with esoteric ideas. François quotes historian François Delpla, who asserts that—

> That Hitler borrowed from magic, both to seduce the masses and to articulate his ideology is undeniable; that he resorted to esoteric belief to help himself believe in it, plausible.[207]

Supporting this perspective, Mosse (1999) observed that Hitler maintained a fascination with the secret sciences until he died—

> [W]ho to the end of his life believed in "secret sciences" and occult forces. It is important to unravel this strand of Nazi ideology because this mysticism was at the core of much of the irrationalism of the movement . . . .[208]

Mosse (1999) elaborates that the secret sciences were not conventional sciences as understood in modern terms. Instead, they were rooted in occult wisdom, drawing upon ancient and medieval sources, and were prominently revealed in Blavatsky's

[203] Staudenmaier (2015), 31, citing Helmut Zander, "Sozialdarwinistische Rassentheorien aus dem okkulten Untergrund des Kaiserreichs," in Handbuch zur 'V.lkischen Bewegung' 1871–1918, ed. Uwe Puschner, Walter Schmitz, and Justus Ulbricht (Munich: Saur, 1996), 224–51; Julia Iwersen, "Rassismus und Antisemitismus in der Esoterik," in Wege der Esoterik: Ideen und Ziele (Freiburg: Herder, 2003), 205–13; Ansgar Martins, Rassismus und Geschichtsmetaphysik: Esoterischer Darwinismus und Freiheitsphilosophie bei Rudolf Steiner (Frankfurt: Info3, 2012). See also Treitel, C. (2004). A Science, 190.
[204] Shirer, 100.
[205] Ibid., *100–01.
[206] Neumann, 85; Tolischus, 136.
[207] François, 44.
[208] François, 20–21; Mosse, (1999), 40, 117.

Isis Unveiled.[209] Yet, despite the almost mythological aura attributed to him, Hitler did not embody the actual "*superhero*" abilities and powers traditionally ascribed to figures of classical antiquity. Instead, his so-called superhuman qualities were carefully constructed and projected through a combination of political theatrics, propaganda, and ideological mysticism. Despite this, Hitler was viewed as a 'strong man' an anticipation of the superman, and a redeemer and saviour of 13 million Germans. Kershaw (2000) framed how he was received by the nation upon his arrival—

> The pseudo-religious strains of the cult built up around him . . . portray[ed] him as a secular 'redeemer.' A lost war, national humiliation, profound economic and social misery, lack of faith in democratic institutions and politicians, and readiness to look to a 'strong man' able to overcome through force the apparently insurmountable acute political chasms prevailing in a comprehensive state crisis, had all contributed to drawing large sections of the masses towards seductive slogans of national salvation.[210]

209 Mosse (1999), 123.
210 Kershaw (2000), xlii. See also Kershaw (1987), 263.

CHAPTER 4. CRYSTALLIZATION OF RACIAL HYGIENE: CONVERGENCE WITH SOCIAL DARWINISM AND GERMAN NATIONAL SOCIALISM

A. Social Darwinism gives birth to National Socialism (1881–1918)

HOW THE NAZIS consolidated scientific racism on a national scale is a matter to which we now turn. Mosse (1961) argues that historians have largely ignored National Socialism ideology in modern German history from the 1890s to the first decade. In particular, because it was "*. . . too outré to be taken seriously.*" Mosse posits the question, who can take seriously, "*. . . an ideology which drew upon the occultism of Madame Blavatski, rejected science in favor of 'seeing with one's soul'. . . . ?*"[1]

Gasman (1971), Mosse (1964, 1975) and Stein (1988) summarised the "*revolution*" ethics of the new socio-political synergy as "*. . . a pseudo-scientific [national] religion of nature worship and nature-mysticism combined with' notions of racism,*" "*with a literal, and not analogical, transfer of the laws of biology to the social and political arena.*"[2] Hitler held the same view believing that natural selection revealed how deeply rooted humanity was in nature—how far humankind had travelled from its animal past, and how close it was to its animal forbears.[3] The popularity of both German Social Darwinism and racial hygiene were such that, they integrated nicely among the educated classes—the military milieu, nobility, aristocracy, the middle class, professionals, and most Germans including scientists, et al.—who considered it to be scientifically true.[4]

It is worth making the point, sooner rather than later, that racial hygiene was shaped by class bias. Conversely, Weiss (1987a) angles her point towards the more chilling truth that Germ an eugenics, both before and after 1933, was not at all concerned with improving Nordic stock, but was "*. . . more preoccupied with class than with race.*"[5] While both perspectives co-exist, the former maybe argued that race was just a subset, in which it targeted ethnicity for culling, and preserved the aristocracy and the middle-

---

1 Mosse (1961), 81. See also Flowers, 88–120, 178, for a meticulous and extensive analysis of the relationship between the occult and National Socialism, as well as Goodrick-Clarke (2004), 217–23.

Flowers, 89, critiques Goodrick-Clarke's failure to examine the occult form of Christianity, noting: "most people pass right over the most obvious fact about National Socialism: its occult Christian foundation. Although Nicholas Goodrick-Clarke proposed to explore 'the occult roots of Nazism,' he neglected to address the occult form of Christianity that underpins Nazism, a form explicitly articulated in the NSDAP's program under the term 'Positive Christianity.'"

2 Gasman, xxiii, 154; Mosse (1961), 82–83, 93; Mosse (1975), 71–72; Stein, G. J., 53. See also Kelly, 84. The key text for mysticism and nature worship is Mosse (1994, 1998) The Crisis of German Ideology passim. See also Gordon, S. A. (1984). Hitler, Germans, 25.

3 Gasman, 11, 13, *154–55.

4 Graham (1977), 1143; Graham (1981), 230; Friedlander (1995), 1; Kelly, 8; Schleunes, 31; Stein, G. J., 51; Weindling (1993), 470, 471.

5 Weiss (1987a), 195.

classes. In accordance with the latter, German eugenicists "*. . . equated the "fit"* with the educated and socially useful elements in society . . . ."[6]

Haeckel's most successful book, for example, *The Riddle of the Universe* (1899), appeared to adjust Darwinism again. He called for a new monistic or non-dualistic religion fusing science and the broad spectrum of religions, which, in turn, would be based on evolutionary Natural Selection or evolutionary naturalism, to explain and guide all society.[7] Haeckel as a nationalist was instrumental in synthesising these constructs together under the banner of National Socialism—a branch or sub-division of Social Darwinism that garnered much support post–1870s period. His influence extended beyond this fusion, encompassing the intersections of science, paganism, religion, and other cultural paradigms. See also counterpart heading, German Racial Hygiene subordination of Jewish hatred, 187. This formed the basis for the framework for the 'national spirit' of Germany. It incorporated music, popular songs, mythologies, legends, holy flames, flags, banners, emblems, and ceremonies to inculcate loyalty, "*. . . but, above all, national monuments in stone and mortar.*"[8] This new German "*religion*" would be a potpourri replete with Aryan myths, heroes and other Aryan ideals to foster national regeneration.[9] Appropriately, Otto Tolischus, distinguished New York Times foreign correspondent and analyst of German affairs from 1926 to 1942, employs a satirical tone in his book, *They Wanted War* (1940) to highlight the absurdity of the National Socialist appropriation of mythical and occult themes. The German people, he remarks—

> [I]s dominated by Richard Wagner—not the Richard Wagner of the incomparable though still debated melodies, but the Richard Wagner who brought back to life the dismal, pitiless and forgotten world of German antiquity, the world of fighting gods and fighting heroes, of dragons and demons, of destiny and pagan epics, which presents itself to other peoples as mere Wagnerian opera, but which has become subconscious reality to the German masses and has been elevated to the inspirational mythos of the National Socialist movement that rules the Third Reich.[10]

Although Haeckel's modifications were not strictly grounded in science, he was instrumental in the unification of Germany by promoting an extreme and distinctive form of Darwinism in which evolution, mysticism, and science were fused into a single monist (pantheistic) worldview. This worldview incorporated scientific ideals while simultaneously evoking the patriotic spirit of the German people and the pagan German past. Haeckel was viewed as a national prophet, and a god.[11] Haeckel's influence was extensive. Accordingly, as early as 1866–68, with the publication of Natürliche Schöpfungsgeschichte (trans. Natural History of Creation), Gasman (1971) traces the continuity of this intellectual lineage, arguing first: "*[w]hen Germans refer to Darwin and Darwinism [they saw] through the distorted lenses of Ernst Haeckel . . . and his monist philosophy.*" Second, Hitler's numerous references of indebtedness to Haeckel,

---

6 Weiss (1987a), 217.

7 Haeckel (1900), 350–64; Flowers, 67, 156–64; Gasman, 6–7; Graham (1981), 24; Stein, G. J., 54; Weikart (2004), 66.

8 De Jonge, A. (1979). The Weimar Chronicle, 177; Mosse (1975), 2, 6–*8; Marten, 27; Schuman, 82, 83, 92–93; Slight, 846–47.

9 Wegner, 25.

10 Tolischus, O. D. (1940). They Wanted War, 10–11. See also Trachtenberg, J. (1983). The Devil, 5.

11 Arvidsson, 332; Gasman, 15–17; Kelly, 56; Schafft, 123; Weindling (1993), 40–48.

and parallels between their respective ideas, are indicate affinities in both rhetorical style and worldviews.[12]

Essentially, just as Aryan belief found symbiosis with Darwinist belief, so too, Darwinist doctrines and scientific racism found a permanent home in National Socialism. In other words, the origins of National Socialism was sourced from Haeckel's Social Darwinism.[13] This is no strange thing that the above were indissolubly combined. In fact, in socialist leader August Bebel's (1910) discussions on socialism and Darwinism, a key point is his remark that Darwinist philosophies, "*. . . may be consciously and expediently applied to all human beings. But that will only be possible under socialism.*"[14] The National Socialist revolution, Mosse (1966) wrote, was achieved "*. . . through an appeal to tradition and sentiment.*"[15] Weindling (1993) succinctly expresses the purpose for the amalgamation to become a weapon, ultimately—

> Science was to weld socialism and nationalism into a coherent ideology of biological social reform, and be the weapon in the struggle to regenerate the race.[16]

In 1879, Karl Pearson (1857-1936), British mathematician, Social Darwinist, and socialist, was one of the first, apart from Haeckel, to synthesise Darwinian ideas into nationalistic stereotypes.[17] In 1881, Eugen Karl Duehring's book, *Die Judenfrage als Frage der Racenschädlichkeit für Existenz Sitte Und Cultur der Völker* (trans. The Jewish Question as a Problem of Race Morals and Culture), Tenenbaum (1956) commented that for that time, Duehring presented "*. . . an almost complete Nazi programme.*"[18] This entailed the annulment of mixed marriages in order to protect Germans from blood contamination; a concept deemed incompatible with the Nordic spirit.[19] Essentially, from the 1860s, with the rise of the Darwinian paradigm, physical anthropology was already on its way out. As efforts intensified toward the *fin-de-siècle*, scholars worked assiduously to construct a new caste system. The graph in Figure I-6, 136, illustrates how the architecture of racial hygiene was progressively infused into society, a process that would later crystallise into scientific racism.

In Otto Ammon's paper, for example, '*The Social Order and Its Natural Foundation*' (1895) he socialized and humanized the concept of Natural Selection among humans, adding altruistic value.[20] Ammon broadened the remit of Social Darwinism by adding in "*Anthropo-Sociology,*" obfuscating the demarcation between physical anthropology and Social Darwinism by enriching it with social applications, edging towards amalgamation. Ammon and his collaborators approached it from the physical anthropology perspective, not only by examining the cranium for measurements, but also by incorporating social

[12] Banton (1978), 94; Gasman, 160–61, 178–79n.66, citing Maser, W. (1970). Hitler's 'Mein Kampf', 77, and Rauschning, H. (1940). The Voice of Destruction, 50. Gasman's citation to Rauschning, as evidence for Hitler referencing Haeckel offers little probative value. See also Staub (1989), 99, Hitler's intention to unite socialism and nationalism.
[13] Massin, 80; Kater, M. H. (1989). Doctors Under Hitler, 112; Proctor (1992), 17–19.
[14] Kelly, 131; Bebel, A. (1910). Woman and Socialism, 254.
[15] Mosse (1966), 17.
[16] Weindlng (1993), 64.
[17] Alexander (2019), 71; Provine, W. B. (1971). The Origins, 25–34; Gasman, xxvi; Phillips, D. (2015), 380; Semmel, B. (1958). Karl Pearson : Socialist and Darwinist.
[18] Tenenbaum, 12.
[19] Ibid.
[20] Graham (1977), 1135, citing Ammon, O. (1895). Die Gesellschaftsordnung und ihre natürlichen Grundlagen (The Social Order and Its Natural Foundation), 25–26; Sorokin, P. A. (1928). Contemporary, 247. See also Kelly, 106 and Weikart (2003), 280.

factors, contrasting, "*. . . according to their, and their parents', social position and according to their country or city origin.*"[21]

British scientist and philosopher, in Julian Huxley's (1887-1975) book, *Religion Without Revelation* (1927, 1957) was a derivative work and extension of Haeckel's book *The Riddle*. He was an ardent follower of Darwinism, to the extent, just as Haeckel adapted Darwin's hypothesis, Huxley too recommended a radical modification of Social Darwinism. It paralleled Haeckel's monistic fusion of science and religion, a context in which Huxley urged scientists to develop a secular religion for a "*new humanity*." Huxley described religion as "*scientific humanism*," an advanced form of Social Darwinism that he later developed further in the 1957 edition, renamed "*evolutionary humanism*."[22] He advocated the amalgamation of science and religion, as well as the creation of a new human species, although the latter falls outside the scope of this section. This represents an extraneous application to Nazism, since the necessary scientific framework did not exist at the time, but it remains highly relevant to emerging or future technological regimes. See Scientific Ideologies and Governance Technologies, 282.

i. Broadening the Scope of Racial Hygiene and its Synthesis with Nationalism

SOCIAL DARWINISM HAD taken on many eclectic forms, amassing many emphases, appropriating various ideologies, integrating antisemitism with extreme nationalism, assimilating racial ideas, embracing art depicting the ideal of Aryan beauty, and introducing new monographs. Thus, Nazi nationalism final form was not an impulsive edifice—it was developmental, and an ongoing process of formation.[23] Reichmann (1951) poetically encapsulates this concept about altering principles to accommodate new developments—

> [D]octrines are extremely flexible in the hands of those who wish to transform them, and no great difficulty would have been involved in giving even "scientific Socialism" some antisemitic accessories.[24]

Such tinkering paved the way for the rise of a scientific racism caste, reaching its zenith in reshaping society. This process aimed to create the *Volk* ethical nation, uniting humanity under a single ideology, which, as aforementioned, was fundamentally rooted in the principles of Social Darwinism.[25] Nazi crimes against humanity were the inevitable corollary of such adjustments.[26] By 1880, racial theory begun to become part of the national mood.[27] However, a coherent nationalistic political modality was still lacking. In order to carry Social Darwinist philosophy forward for inculcating the masses with racial hygiene ideology, required a complete, national and social integration.[28]

Place an asterisk at either side of the *fin-de-siècle*. It marked a pivotal milestone, heralding the advent of modern genetics. Between 1890 and 1903, this period marked a crucial turning point in the sciences, firmly entrenching the development of racial hygiene and Social Darwinism policies. See also the counterpart heading at the *fin-*

21 Banton (1978), 92–93; Barzun, J. (1938, 1965); Sorokin, 244–*45, 246; Stark, W. (1961).
22 Huxley, J. (1927). Religion Without Revelation, 9; Huxley, J. (1957). Religion, 203–39; Phillips, T. (2007). One World, One Faith; Tenenbaum, 211.
23 Le Bon, 14, 16; Mosse (1964), 52, 78–84; Pagel, 97; Volkov, S. (1978). Antisemitism, 31, et seq, 33–34.
24 Reichmann, 100.
25 Mosse (1964), 48; Krausnick, 32; Reichmann, 83.
26 Baumslag, 35.
27 Massing, 76, *80, 83.
28 Ibid., 76.

de-siècle, The Protocols of the Elders of Zion, 187. It was during this transformative period according to Cornwell's (2003) comprehensive survey of German scientific and technological achievements that "*. . . Germany had become an international Mecca of science,*" where it, "*. . . stood pre-eminent in many areas of the natural sciences, mathematics, and technology . . . .*"[29] Furthermore, this was contemporaneous with the transformation of nationalism in which Wiese (2005) aptly observed—

> [This transformation] that started in the Bismarck era (1862–1890), culminated in the mid-1890s in the prevalence of an 'integral nationalism' often tinged by antisemitic convictions.[30]

This form of nationalism transcended the goal of mere unification. Instead, it defined the nation in exclusionary and repeatedly hostile terms, targeting Jewish communities and other minorities. This period, emblematic of the *fin-de-siècle*, conflated scientific progress and troubling social ideologies, setting the stage for the racial and political upheavals of the 1930s. These ideologies coalesced into a pivotal fusion point within the apparatus of National Socialism.[31] Indeed, Reichmann (1951) astutely observed that this precipitated the "*. . . fatal [amalgamation] of imperialism and racialism.*"[32] *Volk*ov (1978) complements this amalgam, elaborating further that—

> By the mid 1890s . . . the Reich was absorbed by a society predisposed to it . . . . What was forged in the heat of passion in the 1870s became a matter of course by the 1890s. Antisemitism . . . became a part of a whole culture towards the end of the century.[33]

Simply put, prior to this critical time, these ideologies were moderate; after the *fin-de-siècle*, they became radical and more pervasive. See also heading, German Racial Hygiene subordination of Jewish hatred, 187. New branches of 'hygiene' disciplines started emerging, and new approaches to hygiene within existing modules of medical practice (e.g., bacteriology, epidemiology) became prevalent—particularly in the social hygiene domain, such as public health legislation, social welfare measures, and social medicine insurance. All of these disciplines acquired greater depth after 1900, functioning with greater symbiosis.[34] It was a Darwinian revolution, wherein social problems had to be expressed in conjunction with evolution and genetics. This framework allowed the notion of future exterminations to be intellectually rationalized and accepted *in toto*.[35]

While Blut und Boden (trans. Blood and Soil, 1933) was a vital component of the ideology of National Socialism, in accordance with Mehrtens's (1987) *dicta*, it actually comprised a large collection of sporadic pieces that had to be reined-in, and organised under one head.[36] Indeed, Nazi Germany and widespread support for racial hygiene, did not arise spontaneously or by chance. It was not 1933, nor even the 1920's.

[29] Cornwell, 7, 38–46 passim; Flowers, 85.
[30] Wiese, C. (2005). Challenging Colonial Discourse, 42, 47–48.
[31] Kater (1989), 111; Mosse (1964), 92; See also Weikart (1993), 470; Weikart (2003), 279; Weindling (1993), 94.
[32] Reichmann, 83.
[33] Volkov, *43–44, 45–46, respectively. See also Tal, U. (1975). Christians, 266.
[34] Weindling (1993), 220–21, 297–98.
[35] Voegelin, 314; Zmarzlik (1972), 442; Zmarzlik (2011), 10; Weikart (2003), 286; Weikart (2004), 203.
[36] Mehrtens, H. (1987). The Social System, 162.

Reichmann (1951) aptly explains that, "*. . . it was the outcome of certain trends in the social development.*"[37] One irregular social trend was engagement in the occult, for example. See The Superman Doctrine, 72. Langbehn believed that the linkage between science and the occult was symbiotic—

> [M]ysticism . . . transformed science into art was precisely this life-force defined in terms of the occult. The ideology of this movement had direct ties with those occult and spiritualist movements which were in vogue toward the end of the century [namely, the 1890s]. Such ties were especially fostered by Theosophy.[38]
>
> Such a linkage between theosophy and the volkish world view will remain throughout the movement's history. This can be conclusively demonstrated through Prana, which called itself a German monthly for applied spiritualism and which was published by the theosophical publishing house at Leipzig.[39]

The final form encompassed an eclectic worldview, assimilating mystical and spiritual elements with an occult vision of national unity, grounded in blood and soil. It also involved the wedding of anthropological theories with the incessantly revised adaptations of Social Darwinism,[40] continually being adjusted to accommodate practical political aims, forming the basis for subsequent political reforms.[41] The single most important component of National Socialism however, was virulent antisemitism, the sealant that held this heterogeneous socialist movement together. These are the facets which passed into the National Socialist movement to which it joined hands quite fittingly with Social Darwinism.[42] As previously mentioned, Social Darwinism generally found its strongest support among the nobility and the respectable, educated classes, and the same was true of National Socialism and antiemitism.[43]

Furthermore, this synthesis highlights how societal predilections and scientific advancements interwove to create a fertile ground for radical ideologies. The founding of genetics enabled theoretical eugenic interests to be mobilised into an organised movement "*. . . combined characteristics of a national association, a scientific society and a professional pressure group.*"[44] The integration of public health, social welfare, and medical practices provided a legitimising veneer to otherwise extreme views, making them not only intellectually defensible but also seemingly indispensable for societal progress. This period marked a convergence where ideology and science mutually reinforced one another, cementing a trajectory towards radicalisation and systemic violence opened the door to Pandora's box and foreshadowed the Europe-wide genocidal violence between 1939 and 1945. Hitler [1925] (1971) was merely adopting Haeckel's Spartan specimen approach. In fact, Weindling (1993) cited Hitler showing a clear ideological affinity with Haeckel's views, a point noted at the beginning of this section—

37 Reichmann, 172.
38 Mosse (1961), *85, 88. See also Webb (1976), 282.
39 Mosse (1961), 87.
40 Cecil, 69–70; Krausnick, 27–28; Schafft, 121; Tal, 76 ff.
41 Krausnick, 28, 32; Mosse (1964), 91.
42 Bracher, K. D. (1970). The German Dictatorship, 13; Gordon (1984), 26; Krausnick (1970), 27–35; Mosse (1961), 93; Reichmann, E. G. (1951). Hostages of Civilisation, 83.
43 Mosse (1964), 1; Browning, 8.
44 Ludmerer & Lappé, 458; Weindling (1993), 125.

> In 1928 Hitler wrote of his admiration for how the spartans killed newborn weaklings. He associated selective killing with the type of Spartan military prowess that was needed to wage a war for Lebensraum?[45]

According to Haeckel (1876, 1900, 1905) and Hitler, those who were too weak to care for themselves should be corporately denied care and left to die. Haeckel, who believed in negative eugenics, referred to this as 'artificial selection,' following what he saw as the Spartan custom.[46] In other words, it was deemed ethically justifiable to murder the weak, sick, those affected by impairments, and the elderly for the sake of the healthy. Haeckel (1876) wrote, "*. . . perfectly healthy and strong children were allowed to live, and they alone afterwards propagated the race.*"[47] Remember, the Spartan doctrine was still being developed, approximately 30 years before Hitler's rise to power. Essentially, medical assistance by virtue of hospital care and/or the prescription of life saving drugs and treatments are only for the hereditary fit.

Haeckel's propositions advocating medicine and hospital treatment denials and just being left to die, became institutionalised in policy from 1890 onward. Weindling (1993) states that "*[a] movement for criminal biology arose during the 1890s seeking to prevent crime by scientific rather than moral and educational means.*"[48] Moreover, commonplace physical anthropology applications surrendered to the "*. . . second stage of Social Darwinist thinking . . . ,*" as Zmarzlik (1972, 2011) termed it. This assisted the introduction of eugenics, which was reoriented with a "*. . . new set of political attitudes and values.*"[49] By the turn of the century, the older, Virchowian form of physical anthropology was dead. Weikart (1993, 2004) inferred as much in the context of the 'old guard' German anthropologists. For example, Virchow who rejected Social Darwinism and biological racism, saw the passing of the baton "*. . . to a younger generation of anthropologists shortly after 1900, both Darwinism and scientific racism came to the fore.*"[50]

Weindling (1985) argued that, "*[b]efore 1914, Racial Hygiene had been decisively rejected by the imperial administration as a violation of ethical codes and personal liberty.*"[51] See also Crystallization of Racial Hygiene, 110. Ploetz's organisation was in a state of flux—it was still evolving and was gaining ground, insofar as it was highly "*. . . influential amongst physicians, but he never appealed to the masses, since he only favoured an elitist approach to organising the eugenics movement.*"[52] The reason, according to Weindling (1993), was that Ploetz's racial hygiene programme at this stage was confined to the middle-classes and the elite—

> [I]n line with the idea of a breeding elite. The Society was now prepared to admit as members official bodies, associations and institutions [in order to expand.][53]

---

45 Weindling (1993), 545.
46 Gallagher, H. G. (1990). By Trust Betrayed, 52; Haeckel, E. (1876). The History of Creation, 1:170; Haeckel, E. (1900). The Riddle of the Universe; Haeckel, E. (1905). The Wonders of Life.
47 Haeckel (1876), 1:170; Bialas, W. (2014). Nazi Ethics, 32.
48 Weindling (1993), 82.
49 Massin, 118–122, 130. Zmarzlik, H.-G. (1972). Social Darwinism, 442; Zmarzlik (2011), 10; Weindling (1993), 134–35.
50 Weikart (1993), 488; Weikart (2004), *115, 116–17, 126.
51 Weindling (1985), 304.
52 Weikart (2004), 68; Weindling (1993), 146.
53 Weindling (1993), 300.

Furthermore, it required the additional passive participation of the masses, for them to make such '*altruistic value*' assessments. Weiss (1987b) draws out this point further, indicating that Germans themselves had a duty to make these distinctions—

> Strong lines of demarcation between the strong and the weak, the talented and the stupid, the moral and the immoral, the "healthy" and the "pathological" had to be made by all Germans.[54]

Although Haeckel first introduced the concept of "*race hygiene*" (*Rassenhygiene*) in 1895, the intellectual structure of the movement was developed by Ploetz, its principal architect, who formally coined the term.[55] Black (2003, 2012) and Proctor (1992) make the distinction that it should not be translated as "*race hygiene*," but rather as eugenics or human germ plasm, as the two concepts are essentially the same.[56] Weiss (1990) asserts that Ploetz intended for it to signify—

> [W]hat approximately Anglo-Saxons had in mind when they employed the word eugenics. In other publications Ploetz explained that *Rassenhygiene* had a much larger scope than the English term eugenics . . . .[57]

Wilhelm Schallmayer (1857-1919) and Ploetz are widely held as co-founders of race hygiene in Germany;[58] their contributions, however, diverged significantly in focus. While Schallmayer emphasised eugenics, Ploetz was instrumental in establishing the organisational foundations of German eugenics. Weiss's (1987b, 1990) analysis highlights this distinction, noting that Schallmayer, unlike Ploetz, never actively participated in the institutional development of *Rassenhygiene* and refrained from mentioning the term because he adamantly opposed its racial connotations. Thus, Weiss (1987a, 1990, 2010) asserts, *Rassen* is an inherently ambiguous term to "*. . . denote [anything from a] small ethnic community, a nation, an anthropological race, or the entire human race.*" Weiss extends this critique to Ploetz's newly coined term, explaining that *Rassenhygiene* was equally broad and vague, referring to the hygiene of any and all of these groups.[59]

By the early 1890s, Schallmayer's "*. . . numerous books and articles gave eugenics its theoretical foundation,*" and he is therefore, credited with authoring the earliest eugenic tract.[60] In 1891, Schallmayer, published the first eugenics tract in Germany titled *Uber die drohende körperliche Entartung der Kulturmenschheit und die Verstaatlichung des ärztlichen Standes* (trans. Concerning the Imminent Physical Degeneration of Civilised Humanity and the Nationalization of the Medical Profession). This seminal work, later reissued in 1895 and 1898, laid the foundation for his subsequent achievements.[61] Between 1900 and 1902, purportedly the same period as the birth of genetics, Friedrich

54 Weiss (1987b), 56, (no source given for the point that "all Germans" had these distinctions).
55 Black (2003, 2012), 262; See Weiss (1990) for full Ploetz and Schallmayer analysis. See Weiss, S. F. (1987b). Race Hygiene and National Efficiency : The Eugenics of Wilhelm Schallmayer for extensive discussion.
56 Black (2003, 2012), 262; Proctor (1992), 18.
57 Weiss (1990), 17.
58 Graham (1977), 1139n.11, citing Ploetz, Die Tüchtigkeit, 5; Graham (1981), 224; Weikart (1993), 488; Weikart (2004), 118; Weiss (1987b), 52; Weiss (2010), 25.
59 Weiss (1987a), 201–202; Weiss (1990), 16–17; Weiss (2010), 27.
60 Weiss (1987b), 3, 52; Weiss, S. F. (1990). The Race, 14, 18, 33.
61 Weiss (1987b), 3.

Albert Krupp (1854–1902), arms manufacturer and dilettante biologist, sponsored an essay competition titled "*What can we learn from the principles of Darwinism for application to inner political development and the laws of the state?*"[62]

The judging panel was chaired by the leading Darwinist, Haeckel—whose influence by that point had broadly extended across scientific and ideological spheres.[63] Schallmayer's cornerstone magnum opus to German eugenics was *Vererbung und Auslese im Lebenslauf der Völker* (trans. Inheritance and Selection in the Life History of Peoples), published in 1903, 1910, and 1918 as a revised and expanded version of his earlier tract, which became the definitive German eugenics textbook.[64] Schallymayer's treatise, in particular, was utilised as a framework to inform German national laws, political reforms, and policy development. It was elevated to the status of a world philosophy, and influenced laws and regulations later enacted under the Third Reich hegemony.[65]

Indeed, the competition provided the impetus for the formation of German eugenics movement. What followed, was exactly that. Germany, established the first eugenics society and journal.[66] See heading Weimar Era Refinements, 131, for a discussion on the realisation of Schallmayer's plan. Weiss (1987b) identified the crucial link in the transmission of Darwinist ideology to Schallmayer, insofar as Haeckel was the foremost influence on his intellectual development.[67] The point being that retention of Haeckel's radical Darwinist views, informed his writing for Schallmayer's winning essay! By the 1920s, eugenics had been annexed to racial hygiene, although over time racial hygiene became a subdivision of the broader eugenics movement.

Between 1902 and 1904, two national journals were founded. Ludwig Woltmann (1871–1907), one of Haeckel's principal pupils, regarded as among the first to employ the term "*Social Darwinist*" and to apply it systematically to social thought and legislation. Mosse (1964, 1998) wrote that "*. . . Woltman produced a merger of Darwinism and Völkisch organicism, and argued that the leaders of the Italian Renaissance were descended from the Goths and Longobard.*"[68] This development can be understood as an outgrowth of what Weiss (1987b) characterises as Woltmann's broader attempt to synthesise Darwinism with neo-Kantian philosophy, grounding racialised historical narratives within a seemingly coherent scientific-philosophical framework.[69] Woltmann established the journal, '*Politisch- Anthropologische Revue,*' in which he noted that "*. . . every page was influenced by the spirit of Darwin.*"[70] Pulzer (1988) maintains that the journal was also devoted to questions of racial vitality and degeneration, and to proving the virtues of racial purity and Germanic superiority.[71]

In 1904, Ploetz founded the journal, '*Archiv für Rassen-und Gesellschafts-biologie*' (trans. Archive for Racial and Social Biology), excluding genealogical periodicals established between 1870 and 1918, for which he received approval and acclaim

---

62 See Weiss (1987b), chap. III:64–89, 'The Krupp Competition of 1900 and Schallmayer's Award-Winning Treatise.'
63 Field, 212; Gasman, 150-51; Graham (1977), 1135; Graham (1981), 220; Kelly, 107–08; Mosse (1964), 99; Poliakov, 294–95; Weikart (2004), 35, 50, 65; Weindling (1993), 116.
64 Weiss (1987b), 3–4, 74.
65 Ehrenreich, 15, 18, *34–57; Friedlander (1995), 19. Krausnick, 28; Wegner, 11.
66 Weiss (2010), 26–27.
67 Weiss (1987b), 33. See Weiss's analysis on Schallmayer re eugenics views, in chap. II:38–63.
68 Mosse (1964), 100–02; Mosse (1998), 102–03, cited in Hutton, 175.
69 Weiss (1987b), 96, 186n.22, citing Steinberg, H.-J. (1972). Sozialismus, 54-55.
70 Weikart (2013), 545; Weiss (1987b), 96–97.
71 Pulzer, 230.

from his mentor Haeckel.[72] Weindling (1993), quoting Ploetz's foreword from January 1904, stipulated that—

> [Salvation of the German race] could be attained through science. Such a science should combine logic, experiment, and racial and social categories. . . . and should establish optimal conditions for the maintenance and improvement of the race.[73]

Racial hygienists regarded science as salvation for the German race.[74] Ploetz's intention for *Archiv* was to—

i. Expand racial hygiene by optimal preservation and development of the race to promote diverse views of Darwinism;
ii. Inform its readers of awareness about developments in eugenics through its numerous book reviews and announcements.
iii. Serve as a platform to make racial hygiene reforms and as a means for reforming lifestyle.[75]

On 22 June, 1905, Ploetz founded and organised in Berlin what he called the world's first professional eugenics organisation—*Gesellschaft für Rassenhygiene* (trans. Society for Racial Hygiene) later renamed *Deutsche Gesellschaft fur Rassenhygicne* in 1911. Haeckel and Weismann were appointed honorary chairmen, although Haeckel never engaged in the Society's activities. Ploetz openly advocated for the elimination of so-called "*weaklings,*" while Aryan ideologist, Lenz, developed a theory of racial inequality on Aryan lines,[76] solidifying the link between eugenics and Darwinian principles. In 1907 and again between 1911–1912, social hygiene gained recognition as an academic discipline in Vienna, Munich, and Berlin.[77] See heading Weimar Era Refinements, 131. Indeed, Schmuhl (2014) documents that "*[i]n 1913, eugenics was formally accepted into the medical section of the Society of German Natural Scientists and Physicians.*"[78] Until 1914, Ploetz played a dominant role in shaping the Society's development and leading the eugenics movement.[79] The Society was elevated to international status, with satellite branches established across countries in both the Northern and Southern Hemispheres. However, in 1916, it was stripped of its status, and, as one historian observed "*. . . became a thoroughly German organisation.*"[80] Ploetz claimed that racial hygiene was essentially—

> [T]he ground between socialism and social hygiene on the one side, and Aryan ideologies on the other. Intellectually the aim

72 Field, 212; Gasman, 148; Graham (1981), 221; Kelly, 109; Poliakov, 295–96. See Mosse (1964), 99–104, for background on Woltmann's views. See also Schmuhl (2014), 31–32 ; Schmuhl, H. W. (2008). The Kaiser Wilhelm; Schmuhl, H. W. (2005). Grenzüberschreitungen; Weikart (1993), 472, 488; Weikart (2003), 279; Weikart (2004), 119–20; Weindling (1993), 132; Weiss (1990), 22; Zmarzlik (1972), 458, 460; Zmarzlik (2011), 26, 28.

73 Weindling (1993), 129–*30.

74 Ibid., passim.

75 Ibid., 131, 137 et seq; Weiss (1987a), 206–07.

76 Allen (2002), 32–34; Baumslag, N. (2005). Murderous Medicine, 37; Massin, B. (1996). From Virchow to Fischer, 93, 135; Proctor (1992), 18; Weindling (1985), 306; Weiss (1990), 22–24; Zmarzlik (1972), 459, 473n.38; Zmarzlik (2011), 27, 41n.38.

77 Weindling (1993), 225.

78 Schmuhl, H. W. (2014). Eugenics and Racial Hygiene

79 Proctor (1988), 33; Weindling (1993), 142; Weiss (1987a), 207, 209.

80 Weiss (2010), 27–32, 40, respectively.

> was a synthesis of anthropology and medicine. . . . By this combination of science and a personal sense of responsibility . . . an elite with supreme physical and scientific qualities would arise.[81]

Pearson's Socialism and Natural Selection (1897) further consolidated Ploetz's position by integrating racial-nationalist and collectivist themes that did not emerge *ex nihilo* in 1920, but were already developing in embryonic form and would later be taken up within National Socialist ideology, creating a synthesis with Social Darwinism—

> [W]e are endeavouring to throw the net [(Social Darwinism—National Selection)] which shall draw talent into useful social activity over the widest possible area [(National Socialism)]; but, at the same time, we are restricting the power of special ability to crush out the less competent [(struggle for existence and elimination)].[82]

This is despite the fact that Haeckel was negative towards socialism.[83] Haeckel's monograph devolved into a full-bodied biopolitical ideology and governmental form underpinned by Romanticism principles that founded National Socialism, and was the prelude to it.[84] See heading Romanticism Dynamics, 250. They set in place the necessary groundwork for the 1933 Nazi race hygiene policy, sixty years later (from the time of Haeckel's modifications in 1863). Indeed, Social Darwinist theory of biological laws of development, Natural Selection and inheritance became the standard upon which to found legislation and socio-economic planning.[85] See also heading, Broadening the Scope of Racial Hygiene, 113.

Schallmayer advanced a new philosophical ethics of "*rational selection*" within the doctrine of *Rassenhygiene.* The first measure he adopted was negative eugenics justified on medical grounds for the eradication and elimination of disease.[86] He made this explicit, envisaging, "*. . . future medical [institutions empowered to] exercise permanent eugenic control over the German people.*"[87] Weikart (2004) found that Schallmayer ardently believed that Darwinism would be basis for the new ethics, and then went on to quote Schallmayer (1903)—

> Theoretically the theory of evolution leads undeniably to the demand for the continued development of ethics in the sense of evolutionary ethics.[88]

Weikart's annotation on the text is that, the function of 'evolution ethics' would provide an idiosyncratic advantage to adherents, inasmuch they will triumph in the 'struggle for existence' between societies, over-and-above those retaining traditional

81 Weindling (1993), 135, 141.
82 Pearson (1897), 120–21, 123.
83 Gasman, 147–48.
84 Ibid., xiv–xxiii; Krausnick, 27; Reichmann, 186.
85 Weindling, P. J. (1985). Weimar Eugenics, 306; Weiss (1987b), 26–28.
86 Weiss (1987b), *52–53, 54.
87 Poliakov (1974), 295. See also Weindling (1993), 118; Weiss (1987a), 203–06, for background how Schallmayer formed this view of the future from his internship in the university hospital's psychiatric clinic.
88 Schallmayer, W. (1903). Vererbung und Auslese, ix–x, quote at 243, in Weikart (2004), 50.

morality values. Hygiene was the basis for morality, *tout court.*[89] Weiss (1987b) posited Schallmayer's second argument for the "*. . . adoption of a policy of 'rational selection' on the inherent utility and cost-effectiveness of eugenics.*'"[90] Effectively, his 'rational selection' policy was the negative eugenics pursuit of biological efficiency, which embodied his aspiration to identify and cull all impediments from the population, which Schallmayer maintained could, "*. . . only be adequately removed by embarking on a eugenics programme.*"[91] He hoped this would become the new "*moral code,*" which he termed "*generative ethics*" and "*biomedical technocracy.*"

It was for this reason that, in the aforementioned prize-winning treatise, he advocated the nationalisation of the medical profession: (i) to accomplish his overarching goal of increasing the hereditary efficiency of the nation; (i)(a) through the imposition of generative ethics as its new manifesto, intended to realise his expansive vision for Germany and secure the biological vitality of the nation on a national scale; (i)(b) through the systematic deployment of programmes designed to achieve this objective; and (ii) through education and propaganda, by which individuals would be encouraged to internalise and adopt the new moral code.[92] In 1911, another competition emerged titled, "*What Do the Inferior Elements [Minderwertigen] Cost the State and Society?*" Although it did not engender half as much criticism as the Krupp competition did, although its themes were realised during the Weimar period.[93]

Fast-forward to the Nazi regime momentarily. Dr. Bernhard Rust (1883–1945), Reich Minister of Science, Education, and Culture, used his opening speech in May 1937 to signal the trajectory that scientific racism would take—outlining the organisation of genetics through a retrospective lens tracing back to the nineteenth century.[94] Rust expressed that, "*[t]he National Socialist awakening has called science to where the decisive battle will be fought.*"[95]

ii. National Awakening: Annexation of Racial Hygiene to Authoritarian Morality

RACIAL POLICY EXPANDED through increasingly refined anthropological criteria, accentuated with the inclusion of Darwinism/Haeckelian principles, an approach Woltmann had attempted, but failed to realise. Contemporaneously, Semmel (1958), quoting from Pearson's (1897) rebuttal against early Darwinists, observed that Pearson believed their application of Social Darwinist principles did not go far enough, since it was limited to the struggle between individuals. Pearson contended that it was on a much grander scale than that of the individual, specifically, through the police state's control of individual conduct—

> The truly elevating struggle was not that between individuals but 'the struggle of tribe against tribe, of race against race.' . . . [T]he herd exists owing to its social instincts, and that human sympathy and racial and national feelings are strong natural forces controlling individual conduct. . . . A nation . . . was 'an organized whole', which was 'kept up to a high pitch of external efficiency by contest, chiefly

89 Weikart (2004) 50–511 Weindling (1993), 346.
90 Schallmayer, W. (1895). Die drohende, 10, quoted in Weiss (1987b), 53, 177n.44.
91 Weiss (1987b), 114.
92 Ibid., *58, 62, *79–80, 87, 91.
93 Kaup, I. 1913. Was kosten die minderwertigen, 723, quoted in Weiss (1990), 25–26.
94 Harwood, J. (1987). National Styles 390–91.
95 Rust quoted in Macrakis (1993), 91. Cf. Massin, 79–80, re scientific racism continuity.

> by way of war with inferior races . . .[96] (Emphasis added)

In order to transform it from an individual or localised belief into a national racial state, the Völkisch ideal was conceived as a racially and ethnically homogeneous whole, rather than a mixture of superior and inferior races.[97] In the first phase, the Nazis sought to construct a political framework grounded in national unity. They consisted of racial and cultural unification, framed as a national awakening. They claimed a new order was necessary—"*. . . a strategy to ensure the stability of the state by preempting a social revolution . . . .*"[98] This rationale became the vector through which the prospective regime could penetrate every German household and organisation, enabling it to solve what it defined as Germany's social problems.[99] See also chapter, Scientific Ideologies, 282.

The concept of this unification is analogous to the famous debate about the unification of church and state. The Nazi state subordinated church beliefs, ethics, and values to state control.[100] Accordingly, in Hitler's speech to the Reichstag on 23 March 1933, speaking about morals in public life, he said it was necessary to create and secure "*. . . the conditions necessary for a really profound revival of religious life.*"[101] In other words, not even orthodox religious persons were excluded from the Nazi totalitarian regime; church authorities moved in lockstep with it and provided an inclusive channel through which its reach extended to the non-religious as well.[102] Accordingly, Fest (1970) in the sense of national community, elaborates on how this infiltration manifested—

> Socialism in the Third Reich . . . was everything that served the interests of the German people; and since their interests were consistently identified by the National Socialist leaders with their own personal power aims, the rigid totalitarian permeation of every sphere of life was indeed 'socialism.'[103]

What is still missing is the escalating Jewish hatred component and its integration into the broader socialist-totalitarian structure. See Historical Anti-Semitism, 180. It was the agency of nationalism (the majority) that made this possible. Arendt (1973) expressed a pithy statement highlighting the parameters for installing such a tyrannical movement, best summed up as 'organising the masses.' Arendt writes—

> In order to establish a totalitarian regime, terror must be presented as an instrument for carrying out a specific ideology; and that ideology must have won the adherence of many, and even a majority, before terror can be stabilized. [104]

---

[96] Semmel, 115, citing Pearson, K. (1905). National Life, 49, 55, respectively. Haeckel accepted Pearson's position: glorifying war and the state as the means to regulate and reduce unfit stock. See Sammel, 123. Cf. Pearson (1897), 108–09, rebuttal against Haeckel re the competition aspects; Krausnick (1970), 30; Cf. Baynes, N. H. (Ed.). (1942). The Speeches of Adolf Hitler, 1:369–72, re Hitler's speech about the foundation of morals.

[97] Hitler, A. [1925] (1971), 403-04; Schafft, 72n.39, citing Hitler. Original source emphasis added.

[98] Weiss (1987b), 13.

[99] Helmreich, E. C. (1980). The German Churches, 85, 483; Jäckel, E. (1972). Hitler's Weltanschauung, 15n.19, citing Rauschning, H. (1939). Die Revolution Des Nihilismus; Weiss (1987b), 8–14.

[100] Helmreich, 11, 72; Schafft, 67–68.

[101] Baynes, 370–71. See also Helmreich, 131–32, citing Baynes, 370–71. See also Baynes, 488n.43.

[102] Schafft, 68.

[103] Fest (1970), 43.

[104] Arendt, H. (1973). The Origins of Totalitarianism, *6, 308, 310–26.

The sinister and covert intent behind nationalism was to transform Jewish hatred into a state-sanctioned ideology, embedding it at the heart of Nazi scientific racism. Goldhagen (1996) mapped this link, demonstrating how Germany's antisemitism became pervasive and institutionalised within civic life—

> No other country's antisemitism was at once so widespread as to have been a cultural axiom, was so firmly wedded to racism, had as its foundation such a pernicious image of Jews that deemed them to be a mortal threat to the Volk . . . . [and it] more or less governed the ideational life of German civil society in . . . pre-Nazi history cemented into place the hegemonic racist antisemitic ideology, from which few dissented.[105]

The second phase required that the proposed national socialist state had to become the new standard for political governance, national policy, and social reforms, going forward. One can identify the skeleton elements of this intent for an integrated, cohesive state with the following statement: "*The National Socialist state . . . allowed a new cooperation to develop among science, industry, and the state through ideological concord.*"[106] The socialist state became the arbiter of morality: dictating policy, its application, and remit. The principal objection has always been that the state-wide imposition of authoritarian rule upon society, constitutes an appropriation of rights, freedoms, and morality of the individual. Yet, having inherited total power, the state sought to "*control individual conduct*" quoted above, in order to enforce adherence to its prescribed moral code.[107]

Proctor (1992) observed, "*. . . without a strong state to back [the new caste], Racial Hygiene was relatively impotent.*"[108] Essentially, individual German states were independent and fragmented. The significance being, from the perspective of racial hygiene, it was dead and ineffectual. Nationalism was the mechanism that enabled racial hygiene to be swiftly deployed, and its rules and protocols universally applied.[109] For precisely this reason the "*. . . early Racial Hygiene movement was primarily nationalistic and meritocratic rather than antisemitic or Nordic supremacist,*" with the view that its system of ideas would percolate down through the provinces.[110] A collective (Nationalism) that is in agreement is better than one in which there is a variety of opinion.[111] A reformulation of this principle is that a professional body of opinion is more effective in delivering a service, product or instrument, when there is only one approach and method, with all others excluded, even if they are valid.

Germany's quest of national unity was repeatedly attempted, and as Mosse (1964) revealed, such unity might have emerged at several historical junctures: the Congress of Vienna in 1814, and again in 1817, 1821, 1848, 1871, 1873, and even on the eve of the First World War. Each attempt, however, ultimately failed.[112] Pearson (1897) highlighted

---

105 Browing, 6; Goldhagen, D. J. (1996). Hitler's Willing Executioners, 106, 419.
106 Summary of Peter Thiessen's lecture, quoted in Macrakis (1993),104n.36, 233.
107 Kershaw, I. (1987). The 'Hitler myth', 49.
108 Proctor (1992), 20; See also Neumann, 47, 49, 51; generally, 41–61.
109 Hobhouse (1911), 186.
110 Proctor (1992), 18. See also Mosse (1964), 5.
111 Hobhouse (1911), 187.
112 See Browning, C. (2004). The Origins, 6; Craig, G. A. (1978). Germany, 13–14, 19–20; Gasman, 1–5; Mosse (1964), 1–7, 91, 219; Mosse (1975). Nationalization, 17–18; Pagel, A. (1992). Jugendstil, 104–05; Semmel, 115.

on point that the militant, regimental character of a nationalised morality model relied on the exercise of state power to eradicate the "*unfit*"—

> Socialism which proposes to use the power of the State in order . . . to 'organise' society or some part of it. That is to say, this "regimental" Socialism proposes to interfere with the freedom of the individual to whatever extent the sovereign may dictate, for the purpose of more or less completely neutralising the effects of the innate inequalities of men.[113]

Mosse ([1966] 1968, 1975) focused on the years 1933–1939 of the Nazi regime, yet he also captured the broader objective behind Pearson's 1897 formulation, tracing how nationalisation evolved from the late nineteenth century through the Weimar period. He further explained how Hitler's concept "*nationalization of the masses*" meant this evolution—

> All individuals and all organisations in Germany had to be 'nationalized' in the sense of making them subject to party control; for the party was the guardian of the Germanic world view and through the will and power of its chief, the Führer . . . .[114]

The crux of Mosse's ([1966] 1968) and Pearson's (1897) arguments lies in their claim that this ideological matrix enabled a swift and seamless mobilisation from the existing state apparatus to the Third Reich, ensuring a controlled transition rather than a violent revolution. Put another way, committed social-Darwinists preferred a peaceful form of racial competition in which the "*inferior*" would yield to annihilation, rather than a ruthless bloodshed, in order to secure the "*. . . racial struggle for world supremacy.*"[115] Remember from our earlier analysis of Bacon's *New Atlantis*? This was a partial realisation of that utopia. Aldous Huxley labelled George's Orwell's '1984' a brutal version of *New Atlantis*; in contrast, Huxley's 'Brave New World' (1932) presented a softer vision—not a "*boot smashing down on the face*" as Orwell depicted.[116] Paradoxically, the Nazi regime embodied a *homo homini lupus* ethos—barbaric and predatory toward non-Aryans, but markedly softer and protective toward its own Aryan stock.

The objective was to forge a state-wide community in line with the new constructed Germanic identity, through intensive Nazi propaganda. Consequently, the new racial order, presented as egalitarian, elite-bound, romantic, and familial, found its fullest expression in the SS. According to Segev (1987) this 'all in one' nature encapsulates the regime's dual characteristics.[117] Welch (1993, 2004) similarly remarks that the population was obliged to "*. . . put 'the community before the individual' (Gemeinnutz vor Eigennutz) and to place their faith in slogans like 'One People! One Reich! One Führer!'*"[118] It was endowed with seductive altruistic notions of harmony, emotion and feeling, respect, comradely and brotherhood—"*. . . a great German family, bound to each*

[113] Pearson (1897), 112.
[114] Mosse (1966), xx; Mosse, G. L. (1975). The Nationalization of the Masses. See also Massing, 76–83. He emphasises nationalism in context of the 'fatherland', and how antisemitism became a vital component of nationalism.
[115] Weikart (2004), 185–87, 192.
[116] See Staub (1989), 83, for commentary and analysis.
[117] Massing, 82, 102; Waller, J. (2007). Becoming Evil, 31; Segev, T. (1987). Soldiers of Evil, 62.
[118] Welch, D. (1993). Manufacturing a Consensus, 4; Welch, D. (2004). Nazi Propaganda, 217–*218.

other through [genetics and] blood ties."[119] Thus, for those who shared the prescribed blood ties, the regime promised love, happiness, peace, and equality. In return, it demanded each individual's acceptance, devotion to *Volk* and fatherland, obedience to the new authority, unwavering loyalty to the regime, and commitment to fulfilling its nationalistic ambitions.[120]

Mosse (1975) now expresses the purpose for nationalism—

> [T]he careful efforts of nationalist movements were directed toward disciplining and directing the masses in order to avoid that chaos which defeats the creation of a meaningful mass movement.[121] (Emphasis added)

Nationalism espoused its organic concomitant, namely, the "*. . . demand that all should become "'good human beings', herd animals, blue-eyed, benevolent, beautiful souls . . . .*"[122] The Nazis dictated what the ubiquitous standards of morality were to be for the Menschen (trans. new man), or the so-called "*good human being.*"[123] See future application under Class-Based Stratification, 283. According to Bialas (2013, 2014) this moral code presented as inherently 'good,' was just a veneer, ensuring that Germans would acquiesce, more often than not, conform to its prescribed dictates—

> [I]mmorality as the morality of a new order to make sure that people are willing to fall in line. It was important for the Nazi perpetrators to appear to themselves as decent people acting in a moral fashion.[124]

Another aspect of the "*new man*" was that the individual would be born from a new myth of life.[125] Thus, even in the absence of direct participation in barbaric killings, a person could still be expected to contribute to or aid eliminationist ends, as Goldhagen (2009) puts it. Simple conformity with the regime from bystanders amounts to indirect influcnce and participation. He thematically delineated these distinctions—

> [i.] If he orders or organizes or has a ministerial or command role in institutions that take part in the eliminationist programs, then he is a perpetrator of the overall mass murder or eliminationist program.
>
> [ii.] If he kills or facilitates the killing of many people, then he is a perpetrator of mass murder.
>
> [iii.] If he helps to drive people from their home and country, then he is a perpetrator of eliminationist expulsion.

119 Ehrenreich, 39, citing Karl Förster, "Deutsche Ahnengemeinschaft (DA). Ein Aufruf," Thüringer Heimatspiegel 8 (1931): 226–28. See also Mosse (1964), 114; Mosse (1966), 3; Schafft, 123, quote in International Military Tribunal, 29:123.

120 Ehrenreich, 38-39, 48; Fest (1970), 40, 42; Kershaw (2000), xxxvii; Massing, 76; Mosse (1975), 1–21; Mosse (1964), 93–94.

121 Mosse (1975), 15, *16.

122 Call, L. 16. Cf. Claeys, 235.

123 Bialas (2013), 3–25; Bialas (2014), 16–30; Gasman, 172; See also Kater (1983), 204.

124 Bialas (2013), 9, generally, 3–30; Bialas (2014), 17–18.

125 Arvidsson, 344.

> [iv.] If he beats and tortures people but somehow manages to do nothing to contribute to people's deaths, then he is a perpetrator of assault and torture.[126] (Emphasis added)

The formation of the final product of the creation of the *Menschen* was necessary for the Nazis to be uniformly brutal, cruel and unsympathetic, within which, the Nazis "*. . . had already practiced future morality of strength and ruthlessness,*" to ensure the genetic degenerate would be destroyed before it endangered the Aryan race.[127] Thus, the "*humanist*" values they mandated masked the brutality of Haeckel's Spartan-savage doctrine.

German intellectuals rediscovered Tacitus's (c.56 AD—c.120 AD) classical text, *Germania.* On the basis of two hypothetical remarks within it, they constructed a new social milieu that fostered the myth of proximity to soil and earth—leading Germans to conclude they were autochthonous, *tout court.*[128] These were the component parts that established a fixed sphere of identity. As such, the propaganda film Blood and Soil became a nationalist symbol, conveying notions of community, rootedness, and antiquity—motifs that permeated Nazi vocabulary.

The symbol portrayed the idea that Nazi Germany's national body was a pure "*Aryan/Nordic*" superior race, and "*. . . one of the chief reasons was its closeness to the soil.*"[129] This was the secular religion: the people worshipping themselves, transformed into worship of the nation. Mosse (1975) elaborates further, noting that it was specifically the "*general will.*"[130] Mosse went on to draw out its significance by distinguishing the 18th and 19th centuries politics—

> The nation in the eighteenth century was now said to be based upon the people themselves, on their general will, and was no longer symbolized solely by allegiance to established royal dynasties. The worship of the people thus became the worship of the nation, and the new politics sought to express this unity through the creation of a political style which became, in reality, a secularized religion.[131]

Essentially, Social Darwinists inaugurated a new nationalised morality for the emerging new man within the *Volksgemeinschaft.*[132] Massing (1949) characterised Aryanism as an extraction from organised myth, one that became integral to German nationalism.[133] Hitler was regarded as the chief figure of the National Socialism, leading the emerging socialist elite who believed in the new "*. . . dominate racial community of all German people.*"[134] Not so unsurprising that National Socialism was known as Hitlerism, a construct that enabled the deployment of a unified ideological system tied to the ethics of the new Nazi-Darwinist order. Moreover, beyond formally nationalising

---

126 Goldhagen (2009), 91–92, 130, *131–32.
127 Bialas (2014), *23, 27, 30. See also Bialas (2013) 3–6, how the new man behaviours were nurtured; Cocks, 56–58; Kershaw, I. (2000). Hitler, 39n.203, citing Domarus, M. (1968). Der Reichstag 717; Schafft, 225.
128 Isaac, B. (2006). The Invention, 56; Poliakov, 80, 86.
129 Gasman, xxii; Lovin, C. R. (1967). Blut Und Boden, 284; Mosse (1964), 6, 16, 54–55, 67, 113, 218.
130 Mosse (1975), 1–20.
131 Ibid., 2.
132 Bialas (2014), 16–30.
133 Field, 200; Massing, 15, 80n.11, and 233; Snyder, L. L.,50.
134 Cecil, 65; Mosse (1964), 61.

antisemitism, it offered followers a "*. . . sense of togetherness, community, and belonging to those who followed Hitler*"—qualities central to Hitlerite identity.[135] This new order unified Germany's faction groups into one composite mould, as Gasman (1971) wrote—

> [T]he trends of racism, imperialism, romanticism, antisemitism, and nationalism which were floating around among various dissatisfied and frustrated groups in German society, especially among the lower middle-classes.[136]

The conventional socialism model was founded upon egalitarian principles, whereas Social Darwinism was quite simply *ipso facto* the politics of discrimination and violence.[137] Therefore, the new socialist caste rested upon a "*national edifice*" that rejected both the survival of, and equality with, the unfit.[138] Cecil (1972) observed accordingly "*[t]here was nothing egalitarian about Nazism at any level.*"[139] His point being that Germany's mould was not strictly socialist, but it was egalitarian among its own—the *Volk* racialists. Rather, it was Haeckel's mould.

This enabled them to discriminate on two obvious grounds. The first, in due course, post-1933 finished product, the Nazi National Socialist movement fostered a corporate hostility. This, in turn, inaugurated the catastrophe for German Jews, and a new mindset governed by a 'belonging' test. It proceeded not on a citizenship basis, but on colour (racial), purity of language and shape. These were the grounds, to "*. . . categoriz[e] humans according to their 'overall' genetic value . . . ,*" against those outside of the *Volk*, especially the Jews, in addition to non-Aryans, who were considered "*at the bottom of the heap*" of society "*. . . was balanced by the improvement of those at the 'top end.'*"[140]

The second ground made assimilation impossible and facilitated easy identification, segregation, or quarantine of non-Aryans—individuals who were initially compelled to assimilate.[141] The "*mass*" of "*good human beings,*" the general public was compelled, not by coercion but by wilful consent, to watch, listen, and report any suspicious or undesirable behaviour exhibited by those who did not conform or fit in, even when the targets were their own family members[142] Arguably, the Hitler salute was deliberately introduced to easily identify deviant behaviour among those who failed to conform to the totalitarian regime.[143] Hall (2009) classified these informants, both paid agents and the general public as—

> Spies, agents, snitches, snouts, snoopers, busybodies, gossipers, denouncers, nosey parkers, informers, Greifer (literally translated as 'grabbers' and 'catchers' - Jewish informers working for the Gestapo) . . . .[144]

[135] Beyerchen, 11, 84, respectively; Jäckel, *19, 22, quoting Görlitz, W. & Quint, H. A. (1952). Adolf Hitler, 627.
[136] Gasman, xxiii. See also Massing, 37.
[137] Halliday, 404.
[138] Ibid, 399; Mosse (1964), 34.
[139] Cecil, R. (1972). The Myth, 64.
[140] Graham (1977), 1158; Graham (1981), 248; Hannaford, 14; Krüger (1999), 53; Reichmann, 172.
[141] Cecil, 71; Ehrenreich, 44; Gasman, 158–59.
[142] Gellately, R. (1996). Denunciations, 931; Johnson, E. A. (2000). Nazi Terror, 254, generally, 253–61, 283–301; Johnson, E. A. (2011), Criminal Justice, 602.
[143] Bettelheim, B. (1952). Remarks, 90–91; Gellately, R. (1988). The Gestapo, 684; Hall, C. M. (2009). An Army of Spies? Johnson (2011), 603–05; Schafft, 69; Cf. Kershaw (2000), 133, re the Gestapo was helped by eager citizens to turn in those suspected of being friends with Jewish people.
[144] Hall, 253–54.

As for public institutions, Macrakis (1993) writes that "*. . . paid informants were spies planted in many public institutions . . . to be the 'eyes' of the National Socialists to find people who they perceived as undesirable.*"[145] Another metric of recognition was the degree of involvement, or lack thereof, in the nationalist construct. Essentially, this made it straightforward to identify anyone who failed to participate in the collective "*awakening.*" Citizen or not, non-participation stamped the individual as the "*elephant in the room,*" paving the way for systematic targeting and elimination.[146] Just as Peukert (1987) emphasised a "*. . . system of terror which made it dangerous for people not to look fired with enthusiasm,*" within Hitler's aforementioned *esprit de corps* harmonious community.[147] In today's parlance for a person who's personality and/or work ethic is not consistent with the company's culture, is generally typecast or branded as a 'poor cultural fit.' In the opening statement of Gellately's (1996) article, he writes "*[one] of the duties of the 'good citizen,' . . . was to inform the authorities in order to hinder the commission of crimes, track down criminals, or uphold the existing order.*"[148]

This "*existing order*" involved offences that targeted anyone deemed undesirable or a threat to Nazi ideology, who were treated as enemies of the state, and was enforced through the '*For the Defense against Malicious Attacks*' (1933) law. This was further entrenched by the Treachery Act of 1934, which made denunciations, inflammatory remarks, or critical utterances against the police, the State, or party, a criminal offense. Making negative comments about Hitler or expressing criticism of the Nazi revolution was deemed high treason. In less severe cases, offenders might be released by the Gestapo with a warning; however, more serious offenses were punishable by imprisonment, just disappeared, or worst-case, death.[149]

What best captures this regime's ideology is Evans' (2005) concept of 'self-surveillance,' which Gellately initially, and Hall (2009) later framed as a "*'self-policing' state operating within a 'consensus dictatorship.'*"[150] What Hall meant was that ordinary citizens themselves became part of the surveillance network, an "*. . . intrusive instrument of surveillance and control . . .*" through which the regime's surveillance expectations were internalised and enacted.[151] Gellately (1996) supports this notion that it was structurally impossible for the Gestapo to sustain all-encompassing surveillance without the voluntary assistance of German society, which played a crucial role.[152] Within this cohesive spy apparatus, Arendt (1973) defines these facets as totalitarianism, characterised by—

> Totalitarian movements are mass organisations of atomized, isolated individuals, where collaboration [and collusion] of the population is integral. In totalitarian countries, these movements are so well organized that in a system of ubiquitous spying, everyone may be a police agent, and each individual feels himself under constant surveillance.[153]

[145] Macrakis (1993), 60. See also Gellately, R. (1988). The Gestapo; Hall, C. M. (2009). An Army of Spies?, 247–49.
[146] Goldhagen, D. J. (2009). Worse Than War, 59.
[147] Peukert, D. (1987). Inside Nazi Germany, 49, 50, respectively.
[148] Gellately, 931.
[149] Arendt (1973), 434–35; Dams, C. & Stolle, M. (2014). The Gestapo, 77–78; Johnson (2000), 11; Moore, P. (2024). It isn't a book, 209–10.
[150] Evans, R. J. (2005). The Third Reich, 114; Gallately, 931; Hall, 247, 250, 252–53.
[151] Hall, 253.
[152] Gallately, 935 and *933–34, citing Koehl, R. L. (1983). The Black Corps, 159; Johnson (2011), 601.
[153] Arendt (1973), 323, 431, where two related points concerning totalitarianism are stitched together. See also Johnson (2000), 11, 15; Johnson (2011), 600.

Thus, the blunt reality was that society itself was doing all the donkey work for the authorities—an annexation of the general public to the state's policing apparatus. However, let us not forget the underlying *raison d'etre* for the entire new apparatus in the first place: to identify and remove Jews to "*. . . preserve as much as possible the purity of the German tribes from alien-type elements.*"[154] The upsurge of nationalism and its fusion with scientific racism ignited antisemitism, as aforementioned, as a matter of course, Jews were marginalised and earmarked as an easy target for isolation and persecution.[155]

Parallels can be drawn with the Japanese National Party, the Kokutai, in the manner in which the state addressed domestic civil unrest and social revolution. As with Germany, so too the Communist fifth column was menacing Japanese society. With the spectre of social revolution, Janice Matsumura (1996) described it as a "*. . . 'self-perpetuating cycle' of fear of threats to the nation.*" Suspected subversives were arrested by the "*thought police*'," just as Nazi Germany did for having 'dangerous thoughts.'[156]

This led to a national state of alarm and the possibility of domestic unrest. To counter this, dissidents were charged with trumped-up offences to quell the social uprising and were killed.[157] Similarly, non-conformists were eliminated by Hitler's 'mobile death squads' which carried out—

> [S]pecial operations . . . of kill[ing] large numbers these enemies of the Reich as fast as possible. Lists were prepared of government leaders, nobility, clergy, professionals, and intellectuals. "[158]

It was apparent that the Nazis intentionally delivered a boot-smashing, overtly intimidating, coordinated regime designed to root out such dissenters, as Pearson (1897) compellingly argues—

> The only true contradictory of Individualism is that more common kind of Socialism which proposes to use the power of the State in order . . . to "organise" society or some part of it. That is to say, this "regimental" Socialism proposes to interfere with the freedom of the individual to whatever extent the sovereign may dictate . . . It is militarism in a new shape, requiring the implicit obedience of the individual to a governmental commander-in-chief, whose business it is to wage war against natural inequality, and to set artificial equality in its place.[159]

There is also a third dimension to consider. German unity transcended the mere notion of citizenship; it reflected Haeckel's monism, a secular religion for humanity that exerted considerable influence beyond Germany.[160] The influence of Haeckelian monism significantly impacted the publisher and author Eugen Diederichs (1867–1930), further embedding these ideas within the cultural fabric of the time. He helped bring Völkisch ideas into mainstream respectability and, in 1912, founded *Die Tat*,

154 Ehrenreich, 28; Kershaw (1987), 60, 170.
155 Fest (1970), 40; Krausnick (1970), 20, 22.
156 Gellately (1988), 655, citing Hans Buchheim, "The SS-Instrument of Domination," in Anatomy of the SS State, by Helmut Krausnick et al., trans. R. Barry et al. (London, 1968), 147.
157 Yellen, J. A. (2013). The Spectre of Revolution, 206–07, quoting Matsumura, J. (1996). Internal Security, 407.
158 Schafft, 92n.14, citing Nicholas, L. H. (1994). The Rape of Europa, 61.
159 Pearson (1897), 112.
160 Webb (1976), 65.

contributing to the aesthetic framing of National Socialism under the rubric of "*New Romanticism.*"[161] Such propagandist ideals created *en bloc* a new community of thought underpinned by German mythology, characterised by an irrational, emotional, and mystical worldview, sustained by feeling—a force subsumed within the Geist.[162] See also Romanticism Dynamics, 250.

It comes as no surprise then that Hitler was held to be the "*National Socialist par excellence . . . and also the protagonist of the movement.*"[163] With the nationalisation and unification of all institutions and agencies, it is little wonder that this process consolidated Hitler's political power. Furthermore, new employment positions were created to ensure that the national system operated like a well-oiled machine, controlling all sectors of society. For instance, Schaftt (2004) outlined that some of these duties included identifying groups deemed suitable for Germanisation or Aryanization, liaising with museums, advising on the compilation of school materials, and overseeing the Germanisation of banks and other institutions.[164] See also heading Adoption of Darwinism, 35, regarding the purging of professions. Arguably, while the de facto purpose of these initiatives was to stimulate economic growth through research and development, their actual effect tightened the regime's authoritarian grip over society and facilitated control.

Similarly, candidates selected for "*refolking-Germanisation*" were judged "*. . . racially inadequate, or were too refractory to Germanisation, and were killed.*"[165] Forti (2006) argues that, in terms of political governance, race in its Natural Selection form, "*. . . functioned as an effective vector for a 'naturalization' and 'biologization' of the object of power.*"[166] This enabled Hitler and his inner circle to make life-and-death decisions between those whose lives must be preserved and those who must die.[167] Friedländer (1995) captures the implications of this logic—

> "Aufartung durch Ausmerzung," which can be translated as "improvement through exclusion." But this translation does not fully transmit the perversion and brutality of the phrase. A better translation is "physical regeneration through eradication . . . ."[168]

Hitler embraced the eugenics thinking, which attracted support from eugenics proponents such as Fritz Lenz.[169] A single universal standard enabled the enforcement of rules across the German population and ensured broad compliance. As already mentioned, this facilitated the withdrawal of "*medicine and charity*" with ease, restrictions

[161] Gasman, 154–55; Mosse (1961), 82–96; Mosse (1964), 17, 39, *52–66.
[162] Gasman, xxiii; Mosse (1964), 55–57, 72, 77; generally, 67–87; Mosse (1975), 7, 41–42.
[163] Fest (1970), 4. Cf. Weiss (1987a), 202 and Weiss (1990), 17, citing Ploetz, A. J. (1895). Die Tüchtigkeit, 5. Ploetz posits his differentiation between the Jews with whom he says are Aryan—the Jewish race is ". . . is the cultural race *par excellence*." Yet, as against the Germanic race, they represent the best portion of the Aryan race. (Emphasis in the original). Ploetz placed the Jewish race on the same level as the Aryans.
[164] Schafft, 100, 102-03, 142; Tenenbaum, 136–49.
[165] Tenenbaum, 33–34.
[166] Forti, 11.
[167] Ibid., 11–12.
[168] Friedlander (1995), 20.
[169] Lenz, F. (1931). Die Stellung des Nationalsozialismus zur *Rassenhygiene*, 302, cited in Weiss (1990), 32.

on rationed food, denial of services, and limits on participation in social, cultural, and recreational activities.[170]

B. Weimar Era Refinements: From Concept to Practice (1918–1932)

THE CRYSTALLIZATION POINT for eugenics and Nazi racism/Aryanism was the 1920s, the prelude to the Third Reich and the Holocaust.[171] During this period, Hitler was already being referred to as 'Führer' within the NSDAP (Nazi Party) by 1920–21. For background on how the Nazi Party developed cumulatively. See The Superman Doctrine. By July 1921, when he became the leader of the party, he was addressed as 'our Führer,' synonymous with 'Führer of the NSDAP,' which replaced the more conventional title of 'Chairman.'[172]

This era marked the culmination of 200 years of racial hygiene policy development, whose synthesis was further refined through minor adjustments in the 1920s. Its final formulation unfolded piecemeal. In 1920, empirical justifications for compulsory sterilisation and euthanasia began to emerge from prominent professionals such as jurist Karl Binding (1846-1920) and psychiatrist Alfred Hoche (1865-1943), both were key players in the early 1920s. They authored *The Release of the Destruction of Life Devoid of Value* (1920), a work grounded in medical and scientific argument, followed by a second edition in 1922. This text was highly influential, and Wertham (1996) asserted that it, "*. . . Crystallised the thinking of a whole generation.*"[173] As the title suggests, it advocated the killing of mental patients classified under Division II, corresponding to the taxonomy of disorders listed in Table I-9, 150.

By the close of the 1920s, "*Rassenhygiene*" had attained widespread popularity in Germany since its introduction by Ploetz. The seminal book titled *Human Heredity*, published in 1931 by Baur-Fischer-Lenz became the cornerstone of Nazi eugenics. In other words, the skeletal infrastructure was already in place by 1933; it merely required form. Weindling (1985) provides a detailed examination of the transformation in physical anthropology during the late 1920s, which was reoriented toward biological heredity and racial mapping applications. This included the subdivision of disciplines such as serology and the study of hereditary diseases, while traditional anthropometric methods were phased out *in toto*.[174] Weindling, in his focus on the relationship between eugenics and genetics, argues that German eugenics "*. . . passed through a series of distinct phases, each having a profound impact on German genetics and hereditarian biology.*"[175] Furthermore, as for Hitler's reception himself in 1933, Rosenbaum (2014) and Waller (2007) reiterate this author's refrain that—

> [E]liminationist antisemitism in Germany was already 'pregnant with murder.' Germans were of "one mind" about the Jews . . . It was not Hitler's willingness to murder the Jews . . . It was the willingness of the German people. . . . Hitler was more a facilitator of an irresistible compulsion rather than a charismatic instigator. The German people were the ventriloquists; Hitler was their dummy.[176]

170 Cf. Claeys, 229.
171 Weiss (1990), 9.
172 Kershaw (1987), 21–22.
173 Weindling (1993), 394–96; Wertham, 161–62.
174 Weindling, (1985), 313.
175 Ibid., 303–04.
176 Rosenbaum, R. (2014). Explaining Hitler, 350; Waller, 38.

By the time the WWI began, German anthropology was gaining pace in many universities, supported by six regularly published journals. Among these were the prestigious *Sozialistische Monatshefte* and *Die Gesellschaft*. The intent was to raise public awareness of the practical implications of emerging genetic science and to counter the more biologically determinist interpretations that began to circulate. These moderating efforts, however, declined toward the end of the Weimar period as the field became increasingly dominated by extremist racial applications.[177] It was logistically vital for the finalisation and implementation of national race policy, since Germany's rebirth was interpreted through biological laws.[178] Kershaw (2000) elaborates on this linkage that Jewish eradication was a basis for "*national renewal*" resting on racial 'purification.'[179] The eugenics of the 1920s had a profound effect on social welfare policies. The epithet of the welfare lobby was 'positive health' lifestyle activity, a subdivision of positive eugenics propaganda. Inculcating a new eugenic mentality among the populace was intended to cultivate disciplined behaviour and cement a heightened sense of national duty. [180]

Following the post-WWI years, racial hygiene expansion had come to a halt, but it was revived in 1922 as a scientific forum. With funding from the Rockefeller Foundation around 1924, German scientific research was revitalised.[181] Furthermore, from the early 1900s, the ideology of racial hygiene (eugenics) gained increasing influence across Europe, UK, and Russia. Within this context, countries with high illiteracy increasingly attracted Rockefeller funding for public-health science—investments framed as humanitarian, yet closely tied with emerging population-policy concerns.[182] See also the future application at Biometric Technologies, 317.

In 1923, a population policy, organised along eugenic lines, was added to the Rockefeller portfolio.[183] Weindling (1985, 1993) noted that the defining characteristics of the Weimar period were national social hygiene problems. For example, the 1920s marked the beginning of a national science policy, to which biology and eugenics were viewed as crucial to social reform and moderation.[184] This period was a special time of refinement and "*. . . redefinition of the scope of scientific activity.*"[185] Indeed, eugenicists had already posited that, due to increases in low-quality birth rates, one-third of the population was degenerate—a claim interpreted as indicative of societal emasculation and a warning that population culling was imminent.[186]

---

[177] Graham (1977), 1141–42, 1144; Graham (1981), 226–27, 230; Schafft, 45.
[178] Ehrenreich, 15, 18, *34–57; Gasman, 172.
[179] Kershaw (2000), xliii–xliv.
[180] Weindling (1993), 366–67, 405 (Weindling included a financial basis reason additionally).
[181] Weindling, P. (1988). The Rockefeller Foundation; Weindling (1993), 324; Weiss (1990), 36.
[182] See Balinska, M. A. (2000). The Rockefeller Foundation; Black (2003, 2012), 243–44 passim; Kevles, D. J. (1985). In the Name of Eugenics, 63, 76, 200, 202; Löwy, I. & Zylberman, P. (2000). Medicine as a social instrument; Murard, L. & Zylberman, P. (2000). Seeds for French health care; Palló, G. (2000). Rescue and cordon sanitaire; Rodrı et al. (2000). Foreign expertise; Weindling, P. (1988). The Rockefeller Foundation and German Biomedical Sciences; Weindling, P. (1997). Philanthropy and World Health; Weindling, P. (2000). An overloaded ark; Wilkinson, L. (2000). Burgeoning visions.
[183] Weindling (1993), 315–19.
[184] Weindling (1985), 305–05; Weindling (1993), 321, 342.

See Weindling (1993), 341–42, 356–57, for a partial background discussion on population policy, the result of social hygiene problems. Views were varied between political parties, all passed through their respective ideology funnels, that converged into one funnel, which come forth as an united population policy, the "*unifying bond between political parties.*" Weindling posits the push for the policy was collectively because Germany was in a state of a 'racial biological' crisis of the nation. This comprised war causalities, psychiatric disorders, starvation, high disease rates, were deemed as issues of national concern.
[185] Weindling (1985), 306.
[186] Weindling (1993), 344–45, 385, establishment of registry and record offices for data collation.

The mid-1920s marked another crucial period of preparatory, pre-1933 racial hygiene formalisation—an essential phase in the development of subsequent state policy.[187] The primary characters were Alfred Ploetz and his pupil Fritz Lenz.[188] Racial hygiene developments mushroomed in 1925. For example, the 'new moral code,' (discussed earlier) fulfils Schallmayer's 1909 proposal on 'generative ethics,' a subdivision of his broader concept of 'rational selection.' Weiss (1987b) documented that Schallmayer proposed the first step in implementing the strategy of rational selection involved creating an index-card system to record and codify the nation's stock—

> According to Schallmayer's plan, all individuals . . . would be issued an official two-sided Krankenpaßkarte (health passport) by the state. Side one was to contain an exact description of the holder, considerably more detailed than an ordinary travel passport. . . . On the reverse side ample room would be provided for an individual's personal physician to enter any and all cases of a disease suffered by the passholder . . . The physician also had the duty to record on side two, cases of any illness which he believed might lead to the confirmation of a significant hereditary trait. . . .
>
> [I]ndividuals with entries on their card would be compelled to turn in their passports once a year to their personal physician. He, in turn, had the obligation of delivering the passports to a local health office to be stored and processed. . . . These statistics, once compiled, would then be analyzed and evaluated. The entire process of rationalizing German health records would be directed and supervised by Germany's medical practitioners.[189]

These measures were extracted from Schallmayer's (1909) work, summed up by Weiss (1987b) as the "*. . . direct' hygienic means by which Schallmayer sought to monitor human selection.*"[190] Bruns (2014) clarifies the intent and application of these ideas, by drawing on Schallmayer's work—

> Since these [medical and racial criteria of inclusion and exclusion] ethics were aimed at a supposed natural selection within society and the supposed greater good of the people and future generations, the term used was "generational ethics."[191]

Because parish registers held only the original birth certificates (in the era before photocopying), Germans had to obtain a stamped form from the church parish, the *Standesamt* (trans. State Vital Statistics Office), or synagogue—

> [This] form confirmed the accuracy of the information found in the original register. The accumulated stamped forms would accompany the Aryan Affidavit (the Aryan Pass), which was a cover sheet summarizing the racial

[187] Graham (1977), 1134, 1138, 1157; Graham (1981), 224, 245. Proctor (1992), 18–19.
[188] Graham (1977), 1136; Graham (1981), 221; Krausnick (1970), 32–33; Weiss (1990), 30–33.
[189] Weiss (1987b), 55, 58. See generally Hertz, D. (1997). The Genealogy.
[190] Weiss (1987b), 58n.56, citing Schallmayer, W. (1909). Generative Ethik, 6:177; See also Ehrenreich, 42–43, 45; Weindling (1993), 293; Weindling, J. (1985). Weimar Eugenics.
[191] Bruns (2014), 214, citing Schallmayer (1909), 199-231.

> heritage of the individual.[192]

Another interesting development that occurred in 1925 was the merger of racial hygiene with bacteriology,[193] discussed further at Racial Hygiene at its Zenith, 160. Between 1925 and 1926, following intervention by the Rockefeller Foundation, they issued a landmark donation of 2.57m to support German eugenics. It fostered a more proactive role that continued until Hitler's rise to power by "... *facilitating the expansion of German hereditarian science into areas of social relevance,*" to put in place the '*Human Biology Programme*.' It provided greater integration with the sciences based upon a modernised form of eugenics.[194] In the same year, 1925, *The German League for National Regeneration and Heredity* was founded in Berlin, but later merged with the Racial Hygiene Society in 1932.[195] Its remit prior to amalgamation was the health of future generations, in which Weindling (1993) expressed that it had a closer connection with the welfare state apparatus,[196] than what Ploetz's grass roots 'Racial Hygiene Society' did.

Hutton (2005) asserts that this fusion enabled learning and scholarship to "... *be in step with the wider goals of the movement and the Volk.*"[197] Another vector that cemented this fusion was the structuring of university medical curricula around racial hygiene, with the aim of churning out graduates who would carry its principles into the medical profession. Weiss (2012) points to a fine example of how the curriculum played out in practice—

> [D]uring their preclinical years medical students would attend an introductory lecture in human genetics, followed by more specialized courses as part of the clinical curriculum. Finally, at the end of their studies, all medical students would take part in a capstone course in the institute's genetic clinic. Here they would go beyond theory and experience, firsthand, myriad types of genetic deformities that plagued the Volk. Such experiential learning would prove especially important for those who would eventually go on to serve on the hereditary health courts or as genetic counselors.[198]

Further entrenchment came through the mandatory subject of medical historiography, which was deliberately skewed to legitimise Nazi medical ethics as historically grounded. According to Bruns (2014), the intention was to "... *constantly reassure doctors that they were on the morally correct path.*"[199] Bruns & Chelouche (2017) affirms that this "... *revised curriculum ... was intended to explicitly create a 'new type of physician.*'" In tandem with the 1925 developments, the role of the 'new physician' shifted the focus from individual care to the "*general welfare of society,*" one that intricately linked the goals of the Nazi State with the broader goals of German health, as Weiss (2012) outlines in the context of genetics and racial hygiene.[200] See also Class-Based

192 Hertz, D. (1997). The Genealogy, *58, 61.
193 Weindling (1993), 336.
194 Black (2003, 2012), 364-365; Weindling (1993), 432–33.
195 Weindling (1993), 500.
196 Ibid., 406.
197 Hutton, 193.
198 Weiss (2012), 641.
199 Bruns (2014), 225, 227, 231.
200 Ibid., 225; Bruns & Chelouche, 592; Weiss, S. F. (2012). The Loyal Genetic Doctor, 631, 635.

Stratification, 283 heading for future application. This policy crystallised in August 1933, when the university curriculum was formally restructured—

> Lenz called for race-hygiene to become the main subject of the curriculum for medical students, and in January of 1934 was appointed to the first Chair of race-hygiene at the University of Berlin.[201]

However, the wider objectives of these revisions were the mobilisation of racial hygiene across all sectors of German society, despite the fact that it rested on psedoscience.[202] A compelling suggestion was put forward to make the medical profession more accountable, to which Berlin medical officer, Richard Roeder, returned a startling denial announcement in 1925—

> Today everyone from a minister to a worker is subject to controls, and rightly so. Only the doctor is uncontrolled, and not even by colleagues, and the sick are expected to offer up their greatest possession, their health.[203]

The transfer of power to doctors was arguably preparatory, allowing them to establish the system that would elevate them to 'gods' of the nation in the Third Reich. Kater (1987) adds that a respectable number of doctors had "*. . . joined precursors of the Nazi Party,*" which later merged with the Hitler movement by 1925–.[204] Worse still, some graduate Nazi physicians who had received medical diplomas struggled to grasp basic principles of medicine or anthropology. Yet they were assigned to Nazi camps to conduct the so-called scientific research and experiments on prisoners, having been conditioned to feel no remorse or guilt and to believe their actions were morally justified.[205] This process is commonly conceptualized by Bandura (1999a), (1999b), Staub (1989, 1990, 1995, 1999) and Bialas (2013) as moral disengagement, whereby individuals cognitively restructure harmful conduct to appear justified as illustrated in Figure I–6 below. This research was not confined to typhus studies alone but encompassed a wide range of infectious disease experiments.[206] See Table I–10, Table IV–1, Table IV–4.

---

[201] Allen (1983), 124.
[202] Friedlander (1995), 21–22; Hutton, 193, 205.
[203] Hansen, E. (Ed.). (1981). Verschüttete Alternativen, 180, in Weindling (1993), 355.
[204] Kater, M. H. (1987). Hitler's Early Doctors, 29–30.
[205] Bialas (2013), 17–25; Bialas (2014), 16–30; Bruns & Chelouche, 592.
[206] Baumslag, 129.

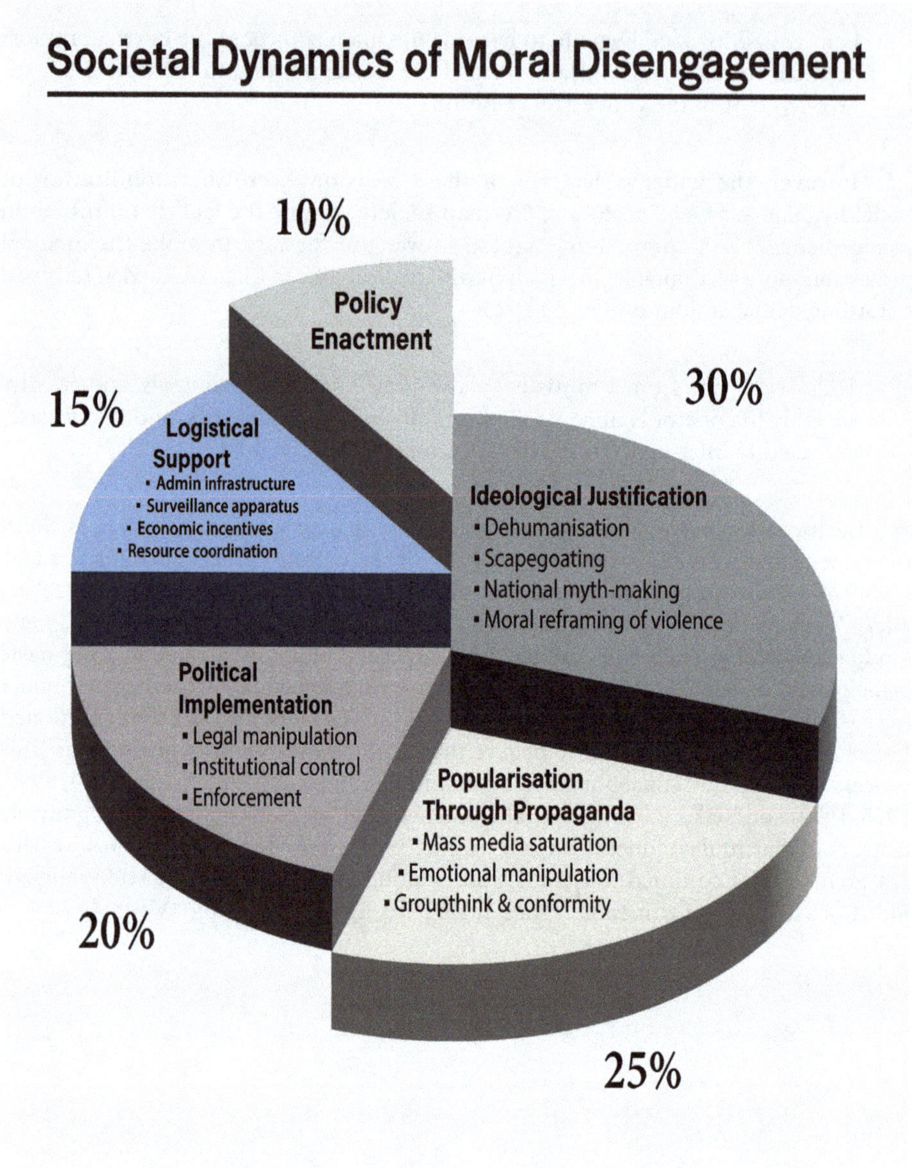

Figure I–6. Societal Dynamics of Moral Disengagement
Conceptual model of societal moral disengagement illustrating how belief systems, messaging, political structures, and logistical mechanisms interact to normalize systemic harm and diffuse ethical responsibility. Unlike individual or institution-centred models, this framework emphasises the interdependence of social, ideological, and operational components in sustaining collective disengagement. Terminology is adapted from Bialas (2013:3). Percentage allocations represent conceptual estimates of each component's relative influence based on initiating function, structural necessity, and cumulative impact.

Moral disengagement is relevant because these specialised, esoteric developments in genetics, medicine, and racial hygiene were predominantly invisible to the wider community, masking the incremental measures that were constructed a tyrannical society just round the corner. Rather, from 1925-29, German industry was booming. The Weimar epoch was aptly described as the "*national rebirth*" of the German people, and the 'golden twenties,' referring to Germany's economic boom.[207] Grunberger's (1971) integral discussion on Germany's economy, he noted, "*. . . Germany experienced internal stability and an economic revival for the first time in over a decade.*"[208] Gasman (1971) elaborated upon this "*national rebirth*" point, that it intersected with the "*. . . pseudo-scientific myth of the racial unity of the Aryans.*"[209] Hans Schemm, founder of the National Socialist Teacher league, delivered a telling speech in a meeting before Munich professors, in which he declared—

> From now on, the question for you is not to determine whether something is true, but to determine whether it is in the spirit of the National Socialist revolution.[210]

In other words, the Third Reich embodied the 'spirit of lies'—the bonding agent that held the regime's machinery together. Germany's apparent prosperity, unbeknown to its citizens, was the calm before the storm—the Great Depression coinciding with the unleashing of the German New Order in 1933. Weindling (1985) highlights that "*[d]uring the 1920s eugenics had a profound influence on social welfare policies. . . . during the phase, from 1927 to 1932, when geneticists assumed leadership and control of the eugenics movement.*"[211] This period witnessed simultaneous transformations in social attitudes, medical practice and legislation, which collectively steered Germany toward a society founded upon eugenics and racial ideology. The restrictions, elimination policies and regulatory measures introduced after 1933 merely piggybacked on this already-entrenched caste framework. Ploetz's 1895 book, *Die Tüchtigkeit* (trans. The fitness of our race and the protection of the weak) for example, set out the formative principles of a skeletal negative eugenics programme—one that endorsed infanticide, the withdrawal of welfare support, and the denial of medical care to the weak—

> [T]he white race was confronted with a profound moral dilemma: within the foreseeable future it would be possible for society to shield all its members, no matter how unfit, from the Darwinian struggle for life. . . . Crudely put, humanitarianism would save individual cripples at the expense of crippling the entire race. . . . All children would be examined at birth by a team of doctors, and those found unfit would receive a fatal morphine dose.[212]

---

207 Peukert, 26.
208 Grunberger, R. (1971). The 12-year Reich, 4.
209 Gasman, xviv.; Krausnick (1970), 32; Mosse (1964), 52, 99. See generally Bialas & Fritze, 16–100. See also Schafft, 15–34, how 'pseudoscience' research played out during the ghettos. See heading, Racial Hygiene at its Zenith, 160.
210 Beyerchen, 52, citing the speech of Hans Schemm in Niekisch, E. (1953). Das Reich, 197. Cf. Krausnick, H. (1970), 32 National Socialism was predicated on the "so-called scientific evidence" of Racial Hygiene.
211 Weindling (1985), 303–04.
212 Kelly, 106–07n.18, citing Ploetz, A. J. (1895). Die Tüchtigkeit, 144-47; Savage, 413; Weiss, S. F. (1990). The Race, 18, citing Die Tüchtigkeit, 226.

The killing of the newborns is reminiscent of Bacon's Atlantis: the "*technological and scientific rationality,*"[213] a genetic-engineering programme that sanctioned lawful murder. Racial hygiene initiatives in Ploetz's era were grassroots institutions, and Nazi science had not yet been nationalised, as the lethal components of eugenic practice were still evolving. Even so, concomitant racial hygiene research was gathering momentum within social medicine, intensified by the inexorable growth of perceived racial impurity.[214] Indeed, the psychologist Gerhard Pfahler (1938, 1943) recast *Volk* ideology as a practical socio-political premise. He named 'race' and 'heredity' as the twin pillars of the new social order for German society—the New Reich, fortified the *Volk*, which, in turn, rests upon the "*science of biology,*" without which life itself could not exist.[215]

Essentially, all of the state components and eugenics measures' were the *mise en scéne* for the appearance of the Third Reich, "*. . . for which the change in political control in 1933 permitted full implementation.*"[216] The Weimar Republic was a crucial interval for the laying of the infrastructure,[217] the normalisation of eugenic policies in the public eye, and empowering state officials, geneticists, and doctors to codify arbitrary eugenic sterilisation laws. These measures were justified as solutions to social problems, upholding pernicious ideals of racial purity[218] Weiss's (1987b) concluding remarks on the Schallmayer eugenics discourse offered an iron-clad corroboration for the view that "*[r]ace hygiene under the swastika witnessed the complete synthesis of the two disciplines of Weimar eugenics and Aryan ideology.*"[219] In other words, overhauling the old political administration was no great leap, given that the Nazi racial hygiene apparatus was far better suited to its racist philosophy—was already in place during the Weimar era and ready to be implemented when the opportune moment arrived.[220]

Between the mid-1920s until 1932, a barrage of proposals ensued from various actors for mass sterilisation as a solution. No sooner had the ink dried on the racial hygiene laws of the early 1930s, prior to the formal debut of the Third Reich than further revisions, transformations, and continual adjustments were already underway.[221] It was an ongoing work to make it compatible with Nazism's new objectives and to subordinate it to the needs of Nazism's system of medicine.[222] For instance, during the Weimar period, there was a cascade of innumerable scholarly racial journals, registers, and criminal biological indices of the German population for future practical eugenic measures. Race hygiene centres and genealogical societies proliferated at this time.[223] The Kaiser Wilhelm Institute for Anthropology, Human Heredity, and Eugenics, directed by Eugene Fischer, was established in 1927 to support scientific ventures and research.[224]

---

213 Mehrtens, 163.
214 Ehrenreich, 38; Kershaw (2000), xlii; Weindling (1985), 307.
215 Barzun, 27–28; Hutton, 137–38, citing Pfahler, G. (1938), 127, 162–3 and Pfahler, G. (1943), xiii. Krausnick, H. (1970), 30; Mosse (1964), 100. See also Hutton, 205, citing Spelter, J. (1937), 4; . Cf. Bialas & Fritze, 25.
216 Graham (1981), 224; Weindling (1985), 305.
217 Ehrenreich, 45; Krausnick (1970), 30-31.
218 Goldschmidt, R. (1960). In and Out of the Ivory Tower, 231; Krüger (1999), 52, citing Friedlander, H. (1995). Origins of Nazi Genocide; Noakes (1984a), 84–85; Weindling (1985), *304–05, 316; Weindling (1993), 489.
219 Weiss (1987b), 104.
220 Massing, 98.
221 See generally Weindling (1993), 462–71.
222 Weindling (1993), 493, 514.
223 Proctor (1992), 20; Weindling (1993), 469.
224 Proctor (1992), 20; Schafft, 45–56; Weindling (1985), 305 citing Graham (1977), 1136–37 and Graham (1981), 221–23; Weiss (1990), 36.

By 1929, there were eight racial hygiene branches. By 1931, six more branches were founded, with six more in the pipeline, doubling membership.[225]

Between 1929 and 1931 an economic crisis had trickled in. This period was marked by authoritarianism, reversal of separation of powers doctrine with a dictatorial system of emergency decrees, spiralling fiscal revenue, high welfare benefits expenditure, massive unemployment, and mass poverty. These factors, individually and collectively, contributed to the nation's degeneracy and pushed Germany to the brink of bankruptcy. These were some of the casual factors, though not all; the mood led to "*. . . demands for corporate and authoritarian management of social problems.*"[226] Initially, the authorities resisted these proposals, but Germany's failing economy and the resulting social ills made sterilisation measures appear increasingly feasible, in particular, for those receiving welfare.

What began as a narrowly framed initiative quickly became a slippery slope, as the scope of sterilisation expanded to address perceived problems of poverty, with poor families increasingly regarded as a burden on the state. See the future application under heading Class-Based Stratification, 283, onwards. In 1931, two years before Hitler's rise, he renewed calls for sterilisation, adding hereditary diseases and the chronically ill to the expanding list of conditions deemed to warrant it.[227] Table I-9, 150 illustrates the exponential growth of these classifications. Politically, the fiscal crisis heightened demands for a 'superman' to resolve the nation's escalating problems. According to Kershaw (1987) when Hitler came to power, one of the constructs of the "*Hitler myth*" created by the Reich government was to foster the perception that—

> [T]he government was energetically combating the great problems of unemployment, rural indebtedness, and poverty, and the first noticeable signs of improvement in these areas, gave rise to new hopes and won Hitler and his government growing stature and prestige.[228]

The people had been socialised and conditioned to accept any viable solution to manage the burgeoning ills of society. Thus, the national mood was one of collective acceptance, to the extent that Alexander (1949) asserts, the guise of "*sickness,*" the first round of exterminations became permissible—

> [In the interests of] utility, efficiency and productivity. . . . The killing center is the *reductio ad absurdum* of all health planning based only on rational principles and economy and not on humane compassion and divine law.[229]
>
> . . . .
>
> It is rather significant that the German people were considered by their Nazi leaders more ready to accept the exterminations of the sick than those for political reasons. It was for that reason that the first exterminations of the latter group [was] carried out under the guise of sickness.[230]

---

225 Noakes, J. (1984a). Nazism and Eugenics, 83; Proctor, 19–20.
226 Weindling (1993), 441, *442–46, 448–49.
227 Alexander, L. (1949). Medical Science, 38; Weindling (1993), 444–56.
228 Kershaw (1987), *61–62.
229 Alexander (1949), 41, 46.
230 Ibid.

There is a modicum of truth to the generalisation that a number of Darwinist adherents, having collaborated covertly to establish a statewide racial-hygiene infrastructure, must have recognised, over the seven years from 1925 onwards, the approaching emergence of a Nazi Party totalitarian order. While maintaining a degree of caution, one could be forgiven for thinking that a state government, explicitly premised on a constitutional racial hygiene institutional structure had simply materialised overnight, in March 1933. In reality, the Weimar period, characterised by institutional consolidation, mergers, and takeovers, played a key role in accelerating the institutionalisation of racial hygiene.

As the 'old guard' of grassroots founders stepped aside, a new approach took shape.[231] This required extensive education networks, supported by modern media, museum exhibits, health exhibitions, and literature distribution.[232] See also Education of Society Matrix, 55. These efforts, coupled with government-backed racial hygiene initiatives, furthered the movement's reach and influence. Consistent with this author's reframe, much like the boiling-frog syndrome, the regime advanced through a gradual build-up—a constant barrage of piecemeal adjustments to racial-hygiene policy.

[231] Weindling (1993), 501.
[232] Ibid., 409–10.

CHAPTER 5. RACIAL HYGIENE IDEOLOGY BECOMES OPERATIONAL, MARCH 1933–1945

A. Outline of Eugenics Progression: the framework for Segregation Policy

i. Nazi Racial Theory: Aryanisation and Societal Displacement

HITLER'S YOUTH YEARS were profoundly influenced by the matrix of racial-scientific ideologies and technological advancements epitomized by Germany's status as the "*science Mecca*" that defined the *fin-de-siècle* in Germany and Austria. Accordingly, Cornwall (2003) posits that "*[w]hen Hitler came to power in 1933 science, medicine and technology were pressed together into the service of a new regime.*"[1] Haeckel's intellectual influence upon Hitler was instrumental, further reinforced by his meticulous study of "*Human Heredity,*" a work that profoundly convinced him to support these ideas and translate them into practical measures.[2] See heading Adoption of Darwinism, 35.

Moreover, 1933 marked the critical juncture when the institutionalised final form of racial hygiene became operational. It was the point at which a, "*. . . symbiotic relationship [was] forged between German biomedical scientists and functionaries of the Nazi state.*"[3] Proctor (1992) takes the argument a step further, asserting that "*[b]iomedical scientists played an active and leading role in the initiation, administration, and execution of each of the major Nazi racial programmes.*"[4] In an earlier study, Proctor (1988) elaborates on the scope of biological science in this context, citing Dr. Hermann Boehm (1884–1962), a honorary professor of racial care. Boehm broadly summarised the remit of racial hygiene, a view that closely accorded with the perspectives of many of his contemporaries—

> [M]edicine [examines the health of] 'German genetic streams.' [It is] preventative rather than curative, medicine . . . whereas vitamins had proved valuable for curing illness, genetics was valuable primarily for finding ways to prevent illness . . . [T]he short-sighted 'individualism' of the Weimar Republic had blinded people to the value of this longer-term preventative science (genetics) and its practical application in Racial Hygiene.[5]

Boehm's expression of 'practical application' signalled a hands-on approach to racial science. Rudolf Hess referred to this notion as "*applied biology,*" casting Nazi Germany

1 Cornwall, 8, 28–29.
2 Muller, H. J. (1962). Studies in genetics, 541-544, in Glass, B. (1981). A Hidden Chapter, 357–58, 359, respectively.
3 Weiss, S. F. (2010). The Nazi Symbiosis, 20, 67.
4 Proctor, R. N. (1992). Nazi Doctors, 18.
5 Proctor (1988), 84.

as the "*biological will of the German people.*" In this view, National Socialism became "*political biology*" or the "*medicalization of politics.*"[6] It is not unsurprising that Hitler was often referred to as the "*great doctor*" or "*great physician of the German people*" and "*father of physicians*" within this medicalised political movement.[7] And, as our earlier analysis reminds us, he was also portrayed as the chief of the socialist movement.

The distinction between applied biology and political biology is now clear: the former dealt with eugenic practice, while the latter converted those ideas into state policy and enforcement. Accordingly, Stein (1988) framed the politics of applied biology as, "*[t]he business of the corporate state was eugenics or artificial selection.*"[8] Indeed, Weiss (1987a) drew out a further symbiotic relationship point: after 1933, it was extremely difficult to "*. . . separate the goals and activities of 'professional eugenicists' from the rhetoric and racial policies of Hitler and high-ranking Nazi Party members.*" The underlying logic was that it was intended as a weapon of objective evidence to support Race hygiene initiatives. Doctors were repurposed under a new head of activity that elevated the doctor as a Gesundheitsführer (health leaders), tasked with aligning medical practice with Nazi healthcare policy.[9]

The so-called "*inferior classes*" encompassed diverse groups deemed unfit for the regime's imagined *New Atlantis*.[10] The programme unfolded in two phases: first, eugenic identification; second, mass elimination. Nazi racial hygiene operated on a clear logic. Party leaders planned the framework, "*. . . under the rule of an exceedingly well organised dictatorship.*" This structure supported their pursuit of a long-term goal: the destruction of Jews and other groups labelled genetically inferior—"*. . . to purify the Aryan peoples and to create a racially pure and physically perfect nation.*"[11] An incidental aim was a disease-free society achieved by removing those classified as dysgenic. The ultimate, covert ambition was the creation of a new world—an Aryan *New Atlantis* populated by a supposedly superior human type.

We digress briefly to return to 1931 and observe the movement's name change and its subsequent regression by 1933. After the merger of the *Deutscher Bund für Volksaufartung* with the *Deutsche Gesellschaft für Rassenhygiene* (trans. German Society for Racial Hygiene), the Berlin Racial Hygiene Society made a minor adjustment to its name, by tacking on *Eugeniks* (Eugenics) to draw in more members.[12] As a result, it procured more power and became more centralized than it had been previously.[13] Weindling (1985, 1993) observed that racial hygiene as a concept was flawed—a taboo held in disrepute among officials eager to distance themselves from the national ideology

---

6 Bruns, F. & Chelouche, T. (2017). Lectures on Inhumanity; Cocks, 46; Gallin, S. & Bedzow, I. (Eds.). (2022). Bioethics and the Holocaust, 15; Gasman, 172, citing Krausnick, H. (1968). The Persecution of the Jews, 15n.143; Grodin (2010), 51; Kreniske, J. (1989). [Racial Hygiene] Review, 412; Proctor (1988), 64, 79–82; Stein, G. J. 56; Weinreich, 27, 30, 208, 215; Weindling (1993), 309.

See also Krausnick, H. (1970) edition, 31. Poliakov (1974), 197; Proctor, R. (1982). Pawns or Pioneers?, 37, in Weiss (1987a), 226; Proctor (1992), 19; Savage, 404; Schafft, 225; Stein (1988), 56; Weindling (1993), 505.

7 Burleigh, M. (2002). The Legacy, 117; Grodin (2010), 51, 53; Proctor (1988), 64; Proctor (1990), 23; Proctor (1992), 19; Weyers, W. (1998). Death of medicine, 152.

8 Stein (1988), 56; Weindling (1993), 506. See also Grodin (2010), 51.

9 Bruns (2014), 231, 233; Arendt (1944), 52; Weiss (1987a), 226.

10 Freidlander, 17.

11 Gittelman, M. (2006). The Holocaust, 11; Mosse (2020), xxiv; Spielvogel & Redles, 84; Wegner, 8.

12 Weiss (1987b), 156, and Weiss (1987a), 221, Weiss (1990), 36, Weiss (2010), 68, proffers a slight variation to the record that the Society was renamed. Not a complete name change. The change consisted of adding the word Eugeniks onto the Society's name—'Deutsche Gesellschaft fur *Rassenhygiene*.'

13 Weiss (1990), 36.

of racial purity. Ploetz and Fischer supported the shift.[14] Weiss (1987a, 1990) expands on this rationale: the amendment aimed to fracture society by driving a wedge between Aryan right-wing racists and the declining Weimar left-wing non-racist eugenicists.[15] The new terminology conferred a scientific rigour that racial hygiene lacked; an accelerant for the rapid enactment of legislative reforms in 1933.[16]

However, in 1933, the word *Eugenik* was removed from the Society's official name, and Ploetz's original name, *Deutsche Gesellschaft fur Rassenhygiene*, was reinstated—reversing the synthesis made in 1931. Weiss (1987a) identified the key distinction: the 1931 change had separated the Society from its racial applications in an effort to present itself as "*pure science.*" In 1933, racial hygiene doctrine was recalibrated to satisfy the "*. . . racial policies of Hitler and high-ranking Nazi party members.*"[17] After the National Socialists assumed political power in March 1933, the medical profession became the first institutional sphere to be reorganised, a process systematically chronicled by Proctor (1988).[18]

ii. Stages of Development: Local Physician to Becomes Gods of the Nation

The following table outlines the reforms implemented through a blend of legal, economic, and ideological instruments. The regime aimed to forge a unified Reich by subordinating all institutions and individuals to National Socialist doctrine, advancing through an iterative sequence of stages toward the consolidation of a totalitarian state[19]—

Table I-8. Stages in the Nazi Regime's Progressive Persecution and Extermination of Jews[20]

Stages of Nazi Persecution of Jewish People

Stage/Tier	Description	Principles
1. **Führer Principle**	Vertical chain of command reporting—restructuring government agencies and society.	Indoctrination; Centralisation of control.
2. ***Gleichschaltung* (Coordination)**	Reorganization and alignment of racial hygiene to subordinate to National Socialism: ▪ Replacing old ethics with new ones. ▪ Dissolving satellite organisations. ▪ Stripping autonomy of educational, scientific, and welfare institutions. ▪ Unification of various agencies. ▪ Abolishing independence of states and municipalities by statute. ▪ Converting of private businesses to serve Nazi causes	Legal transformation and alignment of institutions.

[14] Graham (1977), 1139–40n.11, citing Ploetz, A. J. (1895). Die Tüchtigkeit unsrer Rasse, 5; Graham (1981), 224–25; Poliakov, 294; Proctor (1992), 18; Weindling (1985), 304; Weindling (1993), 436, 462, 482;
[15] Weiss (1987b), 156; Weiss (1990), 10–11.
[16] Friedlander (1995), 17; Noakes (1984a), 75; Weindling (1993), 432.
[17] Weiss (1987a) 226–27; Weiss (1990), 40–41.
[18] Proctor (1988), 64–94.
[19] See Bracher, K. D. (1972). Stages of Totalitarian, 109–28.
[20] See also Staub (1989) 117–18. Cf. Kressel's 155, application of the 'foot-in-the-door' principle reflects the systemic process of marginalization outlined in the above table.

Table I-8. Stages in the Nazi Regime's Progressive Persecution and Extermination of Jews[20]

Stages of Nazi Persecution of Jewish People		
Stage/Tier	**Description**	**Principles**
3. **Racial Hygiene Process**	Exclusion of Jews and non-Aryan classes from Employment: ▪ Dismissing from public and private sector jobs ▪ Displacing from medical and scientific professions ▪ Restricting entrance to the medical profession ▪ Excluding from educational institutions	Racial Cleansing; Exclusion from employment.
4. **Economic Displacement**	Takeover of Jewish Businesses	Financial Marginalisation
5. **Law for Restoration of Civil Service**	▪ Exclusion of Jews from German life and society. ▪ Legal enactment making it illegal to employ and reemploy Jews and non-Aryans.	Legal and Societal Exclusion
6. **Constant Amalgamations of Institutions**	Consolidation of Power: ▪ Continuous amalgamations narrowed the number of agencies. ▪ Merging the German Medical Association and *Hartmannbund* into a centralized regulatory body. ▪ Establishing the Association of German Health Insurance Physicians and the Office for Public Health to further centralize authority.	Centralization; Continuous Mergers
7. **Earmarking non-Aryans for isolation**	Identifying non-Aryans for isolation or sterilization: ▪ Establishing health offices to collect, analsze, and store medical data. ▪ Creating inventories of individuals deemed bad stock. ▪ Enforcing Nuremberg Laws to prevent marriage or sexual relations with Jews. ▪ Conducting blood tests to establish genetic ethnicity.	Initial Identification and Isolation

In the medical sector specifically, what began as a polycratic system of governance[21] gradually progressed through ongoing restructuring and mergers. By the time of tier six, this process had consolidated authority, narrowing the system into a monocratic structure by unified, centralised leadership. The first tier involved a comprehensive reorganisation of government and society, establishing a new chain of command in accordance with to the "*Führer or leadership*" principle. The medical industry was likewise reoriented into a hierarchical, unified state apparatus consistent with this

[21] Kaminsky, U. (2014). "Mercy Killing", 246–47.

leadership structure.[22] Grodin (2010) specifies what may be broadly characterised as the background to the Holocaust, arguing that the explanation lies in the new medicine reforms—a condensed restatement of the themes discussed thus far—

> [T]he historical accounts of the role of medicine and physicians in relation to Racial Hygiene theories, the medicalization of social ills, and the meshing of medicine with nationalist socialist ideology may provide answers.[23]

Dr. Rudolf Ramm, head of the medical establishment in 1943, explains that, "*. . . regulations concerning the profession were no longer to be made by majority decision but rather by the leadership principle.*"[24] Effectively, the Führer principle deployed a top-down system of authoritarian socialist medicine. The majority of physicians who joined the Nazi Party between 1933 and 1936 (around 50 percent), were predominantly young, mirroring an earlier trend in which nearly 3,000 doctors, or six percent of the profession, had already joined by January 1933.[25] See also connections with SS at heading, Reengagement of Past Connections, 266. Grodin (2010) observed the Nazi Party's deliberate portrayal of its ideology in a medicalised way was a critical factor in attracting physicians.[26] The Führer principle emphasised that every aspect of healthcare and society rested on one single health leadership authority. These physicians were to become the new leaders, accountable to a single superior under the Führer principle,[27] which Weyers (1998) elucidates that, in matters of discipline, it mandated that "*. . . physicians practice their profession in the interest of the community, as defined by National Socialist ideology.*"[28]

The second tier comprised reforms aimed at transforming both the public and private sectors of economic activity and trade. The Führer principle was an indoctrination and disciplinary mechanism to maintain control over the rank and file. This vertical chain of command culminated in the formation of the National Socialist Physicians' League, which "*. . . was in turn subordinated to the National Socialist Party.*"[29] The organisation continued to publish a dedicated journal, *Racial Medicine*, to keep scientific and medical practitioners up to date with ideological and policy developments.[30] This was Germany's policy of "*alignment,*" "*synchronization,*" or "*coordination*" as part of the general process of *Gleichschaltung*.

The significance, as Cornwell (2003) emphasises, and Macrakis (1993) specifically, that *Gleichschaltung* "*. . . meant the use of 'legal' means to transform and align institutions into one unified Reich.*"[31] For example, in June 1933, racial hygiene was reorganised, and its satellite organisations were consequently dissolved. This process stripped educational, scientific, and welfare organisations of their autonomy, facilitated the centralised control

22 Berenbaum, M. (2022). The Question, 17; Beyerchen, 55; Kater (1989), 21–22; Proctor (1988), 72; Schafft, 59; See also Kershaw (1987), 45, re leadership principle and Macrakis (1993), 102; Weiss (1987a), 227; Weiss (1990), 41.
23 Grodin (2010), 51.
24 Ramm, R. (2019). Medical Jurisprudence, 41, quoted in Chelouche, 17; Proctor (1988), 174.
25 Chelouche, 17; Kater (1989), 56–60; Proctor (1992), 19.
26 Grodin (2010), 59. See also Cocks, 52.
27 Berenbaum, 17; Neumann, 83–84; Proctor (1988), 72; Cf. Jäckel, 82.
28 Weyers, W. (1998). Death of medicine, 46.
29 Proctor (1988), 70. See also Kater (1989), 56–57; Macrakis (1993), 51; Proctor (1992), 19. Cf. Kater (1983), 204–05.
30 Proctor (1988), 77.
31 Cornwell, 8–9, 129, 164; Macrakis (1993), 51–52, 222-23n.2. See also Barnett, V. (1999). Bystanders, 18; Neumann, 51, 55.

and unification of additional agencies, and granted the regime authority to abolish the independence of states and municipalities by statute.[32] Consequently, every organisation and institution became subordinate to National Socialism, placing absolute power in the hands of a single, committed Nazi party-appointed official.[33] Proctor (1988) argued that the scope of duties in the new medical industry was not exclusive to family practice alone, but encompassed a wide range of matters that would continue to evolve over the following months—

> The task of medicine was thus transformed: the traditional doctor of the individual would be replaced with genetic doctors. The genetic doctor was not a new speciality but an entirely new kind of doctor: one who cared for the future of the race . . . .[34]

Unsurprisingly, Hitler expressed that, "*. . . Racial Hygiene was to be the task of the German Physician,*" a declaration upon which anti-Jewish philosophy and eliminatory policies were to be built.[35] Physicians were selected by Hitler for their commitment to racial hygiene, and in return they enjoyed economic incentives and professional advantages which were readily welcomed.[36] Zeidman (2020) captures the two main reasons why Jewish doctors and scientists were "*cleansed,*" because first—

> Nazi officials were charged with the task of purging non–Aryans, and Jewish doctors/scientists were viewed as 'racially foreign' themselves. . . . Those who were inferior themselves . . . could never reliably cleanse the asylums and clinics of hereditary diseases.[37]

In addition to acquiring new policing powers, many doctors and medical officers were, for the most part, military experienced and retained ranks as reserve officers. Weindling (1993) explains that this enabled the medical establishment to be mobilised for military purposes, including the enforcement of marshal law and lockdowns.[38] Their authority inoculated them against conscience, so that their medical crimes were justified as morally conscionable, defensible, and scientifically sound along eugenic lines. This, they believed, ensured the continued biological survival of the nation.[39] Many ideologues, philosophers, social theorists, and physicians believed, in reality, that their moment had arrived—victory for the German *Volk.* They cultivated gratis, a sense of loyalty and social responsibility, gaining an "*advantageous position*" of power. The far-reaching implications of this practical application proved tantamount to genocide—a reality that, unbeknownst to them, would ultimately materialise.

The third tier emerged from the implementation of the first and second tiers, instigating the persecution of Jews. Essentially, the racial hygiene programme operated on a twofold basis under the architecture of the Führer principle and *Gleichschaltung.*

32 Beyerchen, 55. See also Peukert's analysis on '*Gleichschaltung*', 97–98, 100; Weindling (1993), 499.
33 Beyerchen, 55. See also Neumann, 400–03 (National Socialist Principles of Organization), re the takeover of all organisations and brought under control.
34 Proctor (1988), 105. See also Baumslag, 47.
35 Proctor (1988), 64 Weinreich, 17, 44.
36 Baumslag, 47–48; Kershaw (2000), 133; Proctor (1988), 105.
37 Zeidman, L. A. (2020). Brain Science, 320.
38 Weindling (1993), 162.
39 Savage, 419.

First, it targeted the displacement of Jewish physicians from the medical and scientific professions, including medical schools and universities. Second, it restricted entry into the medical profession, ensuring that admission was limited to individuals sympathetic to the National Socialist cause.[40]

The fourth tier involved the takeover of Jewish businesses, followed by the expulsion of individuals of non-German heritage (primarily Jews) from every industry sector and, in particular, the medical field. As to why large numbers cooperated in displacing the Jewish presence from the workplace, Staub (1989) dismisses fear of the Nazis as the primary motivation. The general population opposed the euthanasia program, yet they paradoxically cooperated passively with the persecution of Jews.[41] As established in section National Awakening, 121, this author argues that such cooperation stemmed from deeply rooted nationalist antisemitism, which created a shared cultural and ideological framework of tacit approval. This union of nationalistic sentiment and antisemitic ideology normalised participation in the cleansing and persecution of Jews, for many, rendered it a patriotic duty. Thus, it was a shared mutual approval.

The fifth tier was societal exclusion. In April 1933, the German government enacted legislation preventing Jews and political undesirables from obtaining or retaining employment, effectively barring them from participation in German life. Gellately (2001) intimates that the removal of Jewish and Polish civil servants from government posts was conducted covertly, as the Nazis "*[c]leverly called [this measure the] 'Law for the Restoration of a Professional Civil Service' [established in 1933] to avoid the impression that the Nazis were tampering. . . .*"[42] The enactment presented such measures positively, lending them legitimacy, facilitating a seamless transition, and thereby masking the coercive logic of racial policy. It barred anyone deemed Jewish or politically unreliable from employment in the civil service and universities.[43] A supplemental law extended its reach, making it increasingly difficult for Jews to practice medicine. Other sectors affected included the judiciary, law enforcement, the education system, the arts, the press, and other independent professions. This process began in earnest during the Weimar period, as early as 1920, and eventually encompassed the "*. . . entire corps of officials all the way down to the village level.*"[44] See also heading Education of Society Matrix, 55, for discussion on Germanisation.

The sixth tier concerns the ongoing amalgamation of Germany's medical establishments. By April 2, 1933, the two leading societies, the German Medical Association and the Hartmannbund, merged into a single, centralised regulatory body for the profession. Weindling (1993) traces the rallying of the medical profession around collectivist ideologies to c.1900,[45] when doctors transitioned from "*physicians of the individual*" to "*physicians of the nation.*"[46] However, these developments may more plausibly be traced to the Revolution of 1848–1849, when Weiss (1987a, 1987b) argues

40 Beyerchen, 45; Friedlander (1995), 20; Kater (1989), 177–221; Merton, 322–23; Proctor (1988), 64–70, 75, 91; Staub (1989), 117–18; Zeidman, 7, 33; Weinreich, 18–19. See also Kelly 71–74, for background and concluding discussion how the school/university curriculums and textbooks were revised, 30-40 years previous to the Nazi regime, in the 1890s, incorporating Darwinism.

41 Staub (1989), 118.

42 Friedlander (1995), 1; Gellately, R. (2001). Backing Hitler, 25; Macrakis (1993), 77–78; Noakes (1984a), 87; Schafft, 70.

43 Proctor (1988), 91; Weiss (2010), 190.

44 Beyerchen, 11–13; Gellately (2001), 25; Macrakis (1993), 51–53, 74. Kershaw (2000), 42. Kershaw also noted that measures came into effect in 1938, specifically, banning all Jewish doctors from medical practice.

45 Weindling (1993), 217.

46 Bialas (2014), 33; Berenbaum, 17; Proctor (1988), 73.

that medical professionals first saw themselves as "*custodians of national health,*" a role that continued to evolve over the following decades.[47] In 1899, Schallmayer published the same vision in an article titled "*A Ministry of Medicine,*" advocating for the careful appointment of organisational leaders to ensure that the ministry functioned efficiently and was staffed by those who were—

> [B]iologically-trained social scientists and government officials willing and able to examine all social, political, and economic institutions from the standpoint of their effectiveness . . . ."[48]

Weindling (1993) underlines the radical nature of the new medical profession by highlighting its transformative changes—

> Medicine was transformed from a free profession, as it was proclaimed by the German Confederation in 1868, to the doctor carrying out duties of state officials in the interests not of the individual patient but of society and of future generations.[49]

Weiss (2010) expanded on Weindling's point by clarifying what the scope of these new duties were. She cited Otmar Freiherr von Verschuer (1896–1969), a leading advocate of the Sterilisation Law in Germany, who explained in discussions with Fischer how the newly established professional ethics norms would shape expectations for incoming doctors from 1933 onward—

> The medical role of the physician in today's völkisch national socialist state is completely different, and much broader [than before] [. . . .] The new task of medicine today is: care of the national body through the preservation and advancement of healthy genes, the elimination of unhealthy genes and through the conservation of the specific racial character of our people [. . . .][50] (Alteration in original) (Omission in original)

By April 27, Germany's local and regional medical establishments had been seized.[51] Celebrating the victory, the National Socialist Physicians' League proclaimed on banner headlines: "*We take command.*"[52] Many agencies were dissolved or merged,[53] consolidating absolute authority in the hands of Nazi doctors—equivalent in power to medical universities rectors and endowed with Führer prerogatives. These mergers also created additional regulatory bodies, further centralising control over the profession.[54] Doctors had become gods. Proctor (1988) said as much, within eight months of the Nazis coming to power—

47 Weiss (1987a), 229; Weiss (1987b), 16, 17–18; Weiss (1990), 13.
48 Weiss (1987b), 114.
49 Weindling (1993), 6, 545.
50 Weiss (2010), 100.
51 Proctor (1988), 70–72; Weindling (1993), 519–*520.
52 Proctor (1988), 71–72.
53 Ibid., 72–73; Tenenbaum, 87.
54 Kater, M. H. (1989). Doctors Under Hitler, 111.

> By the end of 1933 the Führer of the National Socialist Physicians league was clearly Germany's most powerful medical men.[55]

Between 2 August 1933 and 3 July 1934, Germany's medical associations were unified into a single, autonomous, quasi-political body: the Association of German Health Insurance Physicians. They were subsequently restructured to establish the Office for Public Health, organised in accordance with Nazi ideological principles.[56] Following the Führer principle, lower-echelon leaders were tasked with overseeing pharmaceutical, dental, and health affairs, including medical insurance, thereby guaranteeing that regional health administration complied with national policies.[57] The intent behind these successive mergers and reforms was to repurpose the medical profession as a national policing authority of public health—a concept first articulated in 1855 by Max von Pettenkofer (1818–1901), a Munich chemist.[58]

In sum, Fein (1979) identifies this tiered process, as "*functional sequence*" processing. This is further substantiated by Hannaford (1996), who outlines the sequential stages as: "*... assimilation, naturalization, evacuation, exclusion, expulsion, and finally liquidation.*"[59] This pattern is mirrored in the 1920s through takeovers and mergers, in which broad institutional unification became progressively narrower and more intrusive. Similar to the initial measures in the Weimar period: sterilisation, institutionalization, and euthanasia were mild and non-mandatory. Under Nazism, however, these policies were incrementally tightened or repealed, and escalated into extinction practices (Ausmerzung) against the Jews, sanctioned with "*... an air of scientific authority, about it ...,*" marking the final stage.[60]

This sequence echoes Orwell's dystopian vision in '1984.' Milieu control was maintained through a persistent barrage of propaganda and lies, aimed at rewriting history in a language called 'newspeak,' thus shaping public perception and compliance. Classical texts, fiction, scientific works were repeatedly revised, simplified, and withdrawn from circulation.[61] The process served a dual purpose: to popularise racial hygiene and to keep Nazi physicians abreast of new developments in racial biology.[62]

The seventh tier involved identifying the general population to earmark non-Aryans for isolation or referral for sterilisation. This aligned with the role of physicians as "*physicians of the nation,*" endowed with "*... powers and responsibilities equal to the other Reich ministries ... directed and staffed by physicians.*"[63] This phase marked the initial implementation of Schallmayer's rational selection screening strategy, as discussed under, Broadening the Scope of Racial Hygiene, 113.

55 Proctor (1988), 73. See also Baumslag, 75.
56 Baumslag, 48; Proctor (1988), 71–73.
57 Proctor (1988), 73.
58 Weindling (1993), 155,157–58, 519.
59 Fein (1979), 21; Hannford, xvi, 15.
60 Bock, 409; Friedlander (1995), 17; Müller-Hill, B. (1998). Murderous science, 23–24; Proctor (1988), 115, 117, 177–222; Weindling (1985), 304–05; Weinreich, 30. Noakes (1984a), 75. Noakes highlighted that the implementation of the sterilization was concurrent with other measures against the Jews and culminated with the 'euthanasia' programme.
61 Moustaki, N. & Borman, G. (2011). CliffsNotes on Orwell's 1984, 59; Reznikov, A. (2001). George Orwell's Theory, 53.
62 Proctor (1988), 77.
63 Weiss (1987b), 61–*62, additionally quoting Schallmayer, "Ein Medizinalministerium," Das neue Jahrhundert 1 (1898): 393.

B. Eugenics Developments: Surveillance Society and Social Control

i. Academic Mandates and the Role of Universities in Nazi Eugenics Policies

UNIVERSITY INSTITUTIONS FROM 1933 onwards were mandated to support the government's objectives of enhancing the German race and preventing the reproduction of those deemed genetically inferior. This approach differed from the university lectures delivered between 1918–1920 (outlined in Table I-4, 64). The 1920s, frequently referred to as the 'silent years,' were the preparatory phase for students, laying the groundwork for the radical medical measures introduced in the Nazi era. This period emphasised mental training, and fostered an environment conducive to antisemitism and to the formation of leaders in racial hygiene. Students of the 1920s became the young leaders of the 1930s, such as Walter Gross, who rose to prominence at age of 29.[64] Initially, lectures in racial hygiene were exclusive to university students; however, by 1933, professors were also required to deliver external lectures on these subjects—

> [E]ugenics, genetics, and anthropology courses to state-employed physicians and SS doctors; to help carry out the sterilization law by providing *Gutachten* (expert testimony) in cases coming before the genetic health courts . . . .[65] (Emphasis in original)

Three distinct categories of 'undesirables' fell within the purview of racial hygiene application—

Table I-9. Racial Hygiene Divisions

Merged Categories		
Division I Ethnicity Classes	**Division II** Scientific–Genetic–Medical Classes[a]	**Division III** Social classes[b]
• Jews	• Genetic/Congenital Conditions	• Alcoholics/Drug Addicts
• Poles	• Neurological and Psychiatric Disorders (One Million mentally ill)[66]	• Asocial Individuals[b] (est. 1 Million)
• Orientals	• Cognitive/Intellectual Impairment[d] [67]	• Beggars/Vagabonds
• Negroes	• Chronic/Systemic Diseases	• Political/ideological nonconformists
• Gypsies	• Sensory and Physical Disabilities	• Criminal Dispositions
• Asians	• Infectious/Reproductive Diseases	• Religious or cultural nonconformists
• Half-Casts	• Susceptibility to infections/diseases	• Homosexuals

64 Bruns (2014), 232; Koonz, 106-08.
65 Weiss (1987a), 233.
66 Ibid., 230; Weiss (1987b), 153–54; Weindling (1993), 533.
67 Weiss (1987a), 230.

• Mixed German offspring		• Laziness • Prostitutes • Idiots (170,000 in the Social Interest)[68] • Welfare Benefit Recipients • Elderly and Infirm • Dysfunctional or "socially unfit" family units • Single Mothers and Orphans • Illegitimate Children • Unproductive Employees
	▼	▼
Division I. confined to isolation facilities (Jewish Ghettos, for example), gassed and bodies burned. **Division 11**. underwent sterilization procedure then killed at special euthanasia centres disguised as normal working hospitals		Identified via administrative registers; subjected to **Division II.** procedures, resulting in starvation and/or execution.

[a] Anything framed as innate, clinical, or diagnosable—however dubious the "science." Serious disorders and less serious conditions such as nervousness and exhaustion, were lumped together as manifest evidence for a dysgenic or weak hereditary constitution.[69]

[b] Anything defined by conduct, morality, or social status

[c] Deviant Antisocial behaviour.[70]

[d] One million individuals were designated as "feeble-minded" in medical classifications, a category which included traits such as lying, querulousness, laziness, and susceptibility to influence.

Within Division I., Jewish people were the principal targets of ethnic cleansing,[71] and were constructed as the central focus of these policies in the first place. This category also embraced other groups branded as "*inferior*" by the Nazi regime, including Gypsies and Black people.[72] Nazi negative eugenics extended well beyond "*cleansing*" of professions or the targeting of Jews deemed "*guilty of "racial degeneration.*" To legitimise this agenda in genetic terms, Schallmayer supplied a definition of degeneration intended to clarify its meaning—

> Degeneration . . . is a generational hereditary development accompanied by a deterioration of the functional efficiency of one or more physical or mental organs, which results in future generations becoming less adapted to their living conditions. . . . National degeneration [Volksentartung] means a decline in the average hereditary qualities of a people such that its overall fitness [. . .] is diminished.[73] (Omission in original)

68 Burleigh, M. (1994). Death and Deliverance, 17–19; Weiss (1987a), 230; Weiss (1990), 44.

69 Weiss (1987b), 19, 25; Weiss (1990), 13, 43–44; Wertham, 158–59.

70 Weindling (1993), 533.

71 Kershaw (2000), 234; Proctor (1988), 78, quoting Stähle, E. (1934). Blut und Rasse, 4–6; Peukert, 210–12; Schafft, 96, 103; Weinreich, 22, *31–32, 43.

72 Milton, S. (1991). Gypsies 378, generally, 375–87.

73 Kultur und Entartung." Monatsschrift für soziale Medizin und Hygiene 1 (1906):483–84 quoted in Weiss (1987b) 120.

Division II. comprised those classified as genetically degenerate, in which heredity was considered a predisposing factor in disease.[74] The ethnic and genetic divisions were treated as equivalent. In practice, Nazi racial "*science*" sought to collapse Divisions I. and II. into a single category—a process realised in 1938.[75] This intent is made explicit in Professor Lenz's *Human Heredity* (1931, third edition), where he asserted—

> We must remember that no clear biological distinction can be drawn between the heredity factors of race, on one hand, and other heredity factors including the heredity factors of disease, on the other.[76]

Individuals within the scope of Division III., the social category, were cast as social reprobates—candidates for compulsory sterilisation. Steinweis (2006) stresses the Nazi regime's systematic programme of social engineering, designed to redefine and police the boundaries of German society through "*. . . stigmatizing social nonconformism as a hereditary abnormality, and forcibly sterilising disabled Germans . . . .*"[77] Seidelman (1992) remarked that this process culminated in institutional integration whereby, "*. . . hospitals had to provide the facilities (operating rooms, anaesthetics, beds, nursing staff) to support the [sterilisation] programme.*"[78]

Ploetz envisaged state registers to identify those in Division III. who were "*roaming free*" and had evaded detection, given that they displayed no visible physical or mental impairments. His model operationalised Schallmayer's concept of "*generative ethics.*" He proposed a direct classification measure to identify them through biological censuses and surveys, designed to capture comprehensive data. This administrative architecture was realised with the establishment of health offices tasked with collecting systematic medical information on the entire population.

This new code encompassed genealogy tracing through genetic testing regimens, genetic registers for newborns, health passports, marriage certificates, medical genealogies, hereditary health papers for students, and civil registrars recording genealogical information. In an ironic turn, Proctor (1992) observed that these same "*genetic registries*" were "*. . . later used to round up Jews and Gypsies.*"[79] According to Weiss (2012), when blood tests failed to determine biological paternity, anthropological racial tests were employed. These involved a 32-page family-history questionnaire, followed by a somewhat arbitrary physical examination assessing traits such as height, head shape, nose size and shape, and ear length, although Weiss emphasised that phenotype alone was not definitive—

---

74 Bock, 405; Chase, A. (1980). The Legacy of Malthus, xvii; Friedman, L. J. (1977). [The Legacy of Malthus]. This has also received mixed reviews. See Schwartz, L. & Schwartz, M. (1982). The Politics of Scientific Racism; Friedman, L. J. (1977).

75 Bock, 408; Gittelman, 11; Proctor (1988), 73; Zeidman, 1.

76 Baur et al., 623–24, 681.

77 Steinweis, 24.

78 Seidelman, W. E. (1992). Racial Hygiene, 150. See also, Allen (1976), 111, 120, Allen (1983) 122–23 and Allen (2002), 29, although Allen's analysis focuses on the American eugenics movement, he states that the same regimen uncomfortably mirrored Nazi Germany's racial hygiene protocol. See also Burleigh (2002), 113, 117, where he notes that the stench of medicalized criminality extends from the concentration camps and the asylums implicated in the "euthanasia" program to institutions such as the Kaiser Wilhelm/Max Planck Society. See also Burleigh, M. (1994). Death and Deliverance : 'Euthanasia' in Germany.

79 Proctor (1992), 20.

> [P]henotype does not deviate from that of the average southwest German population. . . . the subject lacks obvious Jewish traits. Striking is only the large convex nose that can be seen in profile owing to the flattening of its tip. As the mother's nose is large, however, and convex noses can be found among German-blooded individuals, this trait does not lead one to conclude the existence of an alien racial taint.[80] (Alteration in original)

Weiss (2012) later remarked that the "*exact racial ancestry*" based on phenotype alone was not definitive.[81] Physicians meticulously analysed and evaluated this data to create inventories of individuals considered as "*bad stock.*" This information was stored in colour-coded card indices at centralised data banks within each regional health office, detailing patients' families, ancestry, and health across four generations. If unfitness was discovered, they were candidates for sterilisation.[82]

Once these practices had become standardised, no clear distinction was made between the genetically fit and unfit, or between individuals of good or poor health. Germany's rising crime rates and the growing number individuals classed under Division III. as "*degenerates*" exacerbated administrative pressures, created a backlog of referrals and detention costs running into millions annually.[83] First, this environment facilitated the identification, localisation, and isolation of the "*elephant in the room.*" It also fortified the case for an active race hygiene policy engineered to reduce waste in the welfare budget.[84] Simultaneously, it increased the capacity of such individuals to blend in and become indistinguishable among the wider population. Second, population registers were introduced to "*combat asocial mentality*" and prevent procreation, thus safeguarding German genetic stock.[85] The 1933 stock market crash triggered the Great Depression and pushed six million people into unemployment, adding a further layer of economic and administrative complexity.

ii. Legislation and Genetic Health Courts: Instruments of Social Extermination

THE INTERSECTION WITH the mobilisation of racial hygiene highlighted a social and economic vulnerability in employee productivity: those fortunate, yet unfortunate enough to secure new employment, were arbitrarily and potentially placed on the radar as non-productive. This positioned them within Division III. of Table I-9, the social category of individuals considered useless and unworthy of life. It heightened awareness of racial hygiene as *a fortiori* urgent solution to address this growing problem and reduce the financial burden of the unfit.[86] Friedlander (1995) records that the Italian physician Cesare Lombroso (1835–1909), the father of criminal anthropology—

> [N]ot only attributed atavistic criminality to individuals from the lower classes who committed crimes but also depicted entire groups as criminal. The handicapped were one such group. He thus defined epilepsy

80 Weiss (2012), 653–*54.
81 Ibid., 659.
82 Weindling (1993), 526–27, 530–31; Weiss (1987b), 155–56.
83 Weiss (1990), 37; Weiss (2010), 23.
84 Weiss (1990), 37.
85 Müller-Hill, 61; Weindling (1993), 234, 238.
86 Weiss (2010), 60, 133.

> as a mark of criminality . . . .[87]

It was later extended to include anyone whose behaviour was *gemeinschaftsfremd* (trans. alien to the community), in particular, those "*. . . seen as sullying the image of the new society by begging, refusing work, or any sort of 'antisocial' behaviour.*" This engaged Division III., the social category, whose members were described as exhibiting "*hereditary social inadequacy.*"[88] Individuals were labelled *Minderwertigen*, considered to have no hereditary value or social usefulness.[89] Thus, Division III. encompassed the vulnerable, elderly, illiterate, and those affected by socio-economic deprivation: non-workers, unproductive people, welfare claimants, the "*lazy,*" asocials, and the visibly ill. The Nazi aim was to replace these perceived 'misfits' with artificial selection in place of natural selection.[90] They were either killed, worked or starved to death, including children.[91] In essence, Weiss (1990) captures the notion succinctly: middle-class prejudices and eugenic motivations equated Division III. with "*. . . fitness with productivity and achievement, and degeneracy with asocial behaviour and the inability to contribute to society . . . .*"[92]

What other German racial enactments had failed to achieve was accomplished by the 1935 Nuremberg Laws, comprising the Reich Citizenship Law, Blood Protection Law, and Martial Health Law. Marriage between the "*genetically ill*" (Jews and non-Aryans) and the "*genetically fit*" (German subjects) was prohibited. Such miscegenation was regarded as a "*public health measure*" and a "*public health hazard.*"[93] Thus, negative eugenic measures— "*. . . to discourage the dysgenic from procreation by means of marriage controls, confinement in asylums, or sterilisation*"—were perceived as a timely solution to contain the growing problem.[94]

As for displaying humanity and sympathy, Noakes (1984a) quoted Professor Kauup of the Seminar for Social Hygiene in Munich, who "*. . . insisted that the 'question of isolation' of the inferiors there is no place for sentimentality. Our healthy offspring have a right to protection from corruption from the genetically damaged . . . .*"[95] This construct accords with Chase (1980), who observed that it legitimised, and implicitly justified, the creation of centres to contain a purported epidemic among the unfit—

> The purposes of scientific racism, on the other hand, were the hereditary endowments 'in the blood' of the ill, the uneducated, and the poor for their chronic (and eminently environmental) infectious and hunger diseases, for their illiteracy, and above all else for their poverty.[96]

Moreover, the genetic aspects of Division II., outlined in the early 1933 sterilisation law draft of the *Gesetz zur Verhütung erbkranken Nachwuchses* (trans. Law for the

---

87 Friedlander (1995), 3.
88 Bock, 417; Chase, xx; Noakes (1984a), 76.
89 Bock, 417; Chase, xx; Friedlander (1995), 17–18; Ludmerer & Lappé, 458; Weiss (1990),
90 Bock, 404; Chase, 3; Caplan, A. (1994). The Relevance of the Holocaust, 11–12; Gallagher, 52; Hooton, 31; Hutton, 27; Pellegrino, xvi; Weindling (1993), 55; Weiss (1987a), 229.
91 Wertham, 144, 180–81.
92 Weiss (1990), 10.
93 Proctor, 131–132, 141; Savage, 418, 427.
94 Noakes (1984a), 80; Proctor (1988), 58, 103, 130–76 passim; Proctor (1990), 33.
95 Pellegrino, E. D. (2005). Foreward: Nazi Medicine, xiv; Gallagher, 52; Noakes (1984a), 79.
96 Chase, 3; Proctor (1988), *69–70; See also Weinreich, 28. He states that Jews were to be blamed. They were the embodiment of everything to be resented.

Prevention of Genetically Diseased Offspring), applied to German citizens. Responsibility for implementing this policy fell to physician, Walter Gross, who, three months earlier (April, 1933), had been tasked to create the National Socialist Office for Enlightenment on Population Policy and Racial Welfare. This office inaugurated the Third Reich's temporary best-kept secret: sterilisation law.[97] Section one set out a non-definitive list of hereditary diseases which read as follows—

1. (i) Anyone who has a hereditary illness can be rendered sterile by a surgical operation if, according to the experience of medical science, there is a strong probability that their offspring will suffer from serious hereditary defects of a physical or mental nature.

   (ii) Anyone is hereditarily ill within the meaning of this law who suffers from one of the following illnesses:

   a) Congenital feeble-mindedness.
   b) Schizophrenia.
   c) Manic depression.
   d) Hereditary epilepsy.
   e) Huntington's chorea.
   f) Hereditary blindness.
   g) Hereditary deafness.
   h) Serious physical deformities.

   (iii) In addition, anyone who suffers from chronic alcoholism can be sterilized.[98]

While draft law never attained legal standing, just shy of six months after Hitler assumed office, it became mandatory in July 1933. See Division II.[99] Weiss (1987b) stressed that their moral duty was to prevent the diseased and infectious from entering society—

> [T]he prevention of degeneration was not merely desirable; it was an ethical imperative. The state . . . had a moral duty to protect its biological vitality.[100]

For this precise reason, it also was mandatory for Jewish women, conceived as "*racially inferior,*" to undergo abortions and sterilisations, and were forced to accordingly. Becoming pregnant was a criminal offence, and being visibly pregnant was an "*immediate death sentence.*" Jewish men, however, were regarded as the greatest threat.[101] Conversely, Aryan women were encouraged to bear more children for the future of the German race and were prohibited from undergoing abortions or sterilisations.[102] This initiated a process that the Nazis believed would reverse the degeneration of German progeny

97 Koonz, 105; generally, 103–30.
98 Baur et al., 213–426; Bock, 418; Chase, xvii, 5; Gittleman, 8; Proctor (1988), 96–97; Weindling (1985), 309–10. Weiss (2010), 140. See also Weikart's (2004, 89 -102) discussion on the historical development of generative ethics for elimination of the disabled from 1895 onwards.
99 Bruns, 218; Proctor (1988), 96–97, 175, passim; Proctor (1990), 29; Schmuhl, H. W. (2005), Grenzüberschreitungen, 145–48, cited in Weiss (2010), 66; Weiss (1987a), 224–25; Weiss (1987b), 153; Weiss (2010), 66, 99, 139.
100 Weiss (1987b), 54.
101 Bock, 401; Lovin, 286; Waxman, Z. V. (2010). Towards an Integrated History, 313–14.
102 Ludmerer & Lappé, 59; Waxman, 313.

and restore the Aryan race.[103] What ratified the Nazi rationale was the assertion that "*. . . if there is little [that can be done] for a sick individual, we must help the coming generations by preventing the sick genes from being spread in the future.*"[104] Consequently, the improvement of the human race involved careful selection of mates from the intellectually elite genetically good stock to avert biological decline.[105]

German citizens judged genetically unfit were prohibited from marrying under the 1933 *Lebensunwertes Leben* policy, later renamed the Law for the Prevention of Hereditary Diseased Offspring, if physicians in the Genetic Health Courts judged their lives as "*not worth living.*"[106] This point was emphasised by Erwin Baur, director of the Kaiser Wilhelm Institute for Breeding Research, when he wrote—

> Every farmer knows that should he slaughter the best specimens of his domestic animals without letting them procreate and should instead continue breeding inferior individuals his breeds would degenerate hopelessly . . . As a recompense for our humanness of today, we must see to it that these inferior people do not procreate . . . no one approves of the new sterilization laws more than I do, but I must repeat over and over that they constitute only a beginning.[107]

However, Bock (1983) highlights a dichotomy that she characterises as a "*double-edged essence*": on one hand, among Germans there was a decline in the birthrate, "*racial suicide,*" because they preferred to have fewer children; on the other hand, the unfit, the poor, and the socially inferior were multiplying faster than the talented and fit, "*racial degeneration.*"[108] Racial policy was conceived as the remedy to reverse this trend. Germans were offered financial and social incentives to procreate, whereas the perceived inferior were subjected to sterilisation.[109]

Schafft (2004) explained the mandatory reporting of Divisions I. and II. disorders and diseases by healthcare personnel to the Genetic Health Courts, with incentives to comply.[110] As a result, social misfits representing a subproletarian class[111] estimated at one million were targeted. To catch those who slipped through the net, registry offices were established throughout the Reich to record sicknesses and other irregularities. Consequently, between 2,000 and 3,000 courts were erected (one in each major city and county) to process the "*valueless*" individuals within the nation.[112] Indeed, the Third Reich followed Ploetz's 1895 schema, dismantling welfare institutions and replacing them with sterilisation legislation.[113] Weiss (1987a) cites a 1933 article quoting the physician Friedrich Maier (1833–1906), in which he advocated scrapping the welfare system, "*. . . which generally only served the weakest and asocial individuals . . . .*"[114]

Negative eugenics was deployed through social policy, employing non-standardised metrics such as IQ testing to classify individuals along a spectrum of exceptional, average,

---

[103] Proctor (1988), 69; Weinreich, 34–35.
[104] Zeidman, 245, citing inaugural speech of Hugo Spatz on April 1, 1937.
[105] Proctor (1988), 17, 19; Kreniske, J. (1989). [Racial Hygiene], 410; Weiss (1987a), 199, 203, 207.
[106] Baumslag, 41; Binding, K. & Hoche, A. (1975). The Release; Bock, 408, 412–5; Gittelman, 11; Müller-Hill, *31, 32–35; Noakes (1984a), 84. Proctor (1988), 73, 96.
[107] Weinreich, 31.
[108] Bock, 405–06; Proctor (1988), 15, 19.
[109] Bock, 405–06.
[110] Schafft, 76.
[111] Weiss (1987b) 5.
[112] Weiss (1987a), 229–30, 236.
[113] Weindling, (1985), 315, 318, 509.
[114] Weiss (1987a), 228–29.

or mentally unfit as a preliminary selection filter. While not determinative alone, Proctor (1988, 1990, 1992) records that IQ test results informed evaluations submitted to the Genetic Health Courts.[115] Montagu (1974) demonstrates that the supposed link between race and IQ, and the ideological cohesion of these claims, was a "*muddled myth.*" He then offers a definition of race in this context—

> [T]he myth of "race," the belief that physical and mental traits are linked, that the physical differences are associated with rather pronounced differences in mental capacities, and that these differences are measurable by IQ tests and the cultural achievements of these populations.[116]

Schafft (2004) documented the invasive Aryan genetic examination process of 1935, which supplemented IQ evaluations. To obtain a racial certificate, assessments of race and ancestry "*. . . consisted of a blood test, a look at the shape and physiology of the eye and the shape of the head, and taking a photograph from the front and in profile.*" However, Schafft concluded that these tests were effectively useless.[117] Anthropology departments across Germany routinely made decisions based on their own judgement, and the tests they employed were not regarded as conclusive criteria for establishing ancestry.[118] For example, if a child's continuous assessment indicated a low IQ or failed to meet developmental standards, the "*. . . application of human reason to human selection*" was invoked.[119] This involved a cost–benefit analysis weighing both eugenic considerations and the child's education.

Noakes (1984a) asserts that the rise of the "*new activist*" doctors created a fertile climate in which eugenic ideals could be operationalised, providing a basis for applying a "*. . . 'scientific' solution to social problems; social hygiene . . . [was] supplemented with Racial Hygiene*'" as the answer.[120] In other words, the criteria of racial science were fluid and could be invoked at will and justified accordingly. This social group was thus resigned to the same fate of sterilisation, institutionalisation, and, ultimately covert plans for murder (euthanasia) through 'legalised killings' rationalised as 'preventive psychiatry.' Medical ethics were distorted to argue that such killings were desirable for the collective good.[121] See also the 'moral balancing act' discussed under Enlightenment Legacies, 282. Grodin (2010) analyses the psychological motivations underlying the vulnerability of Nazi doctors to becoming perpetrators and summarises the following factors—

115 Proctor (1988), 111–12, 115; Proctor (1990), 29; Proctor (1992), 21.
116 Montagu, A. (1974). Man's Most Dangerous Myth, for general discussion, but specifically, 5, 10.
117 Schafft, 42, *73–74, 77. See also Proctor, 139, for different grades of tests to obtain a certificate. Proctor, 112-13, 115, further notes that Jews were never specifically targeted with IQ tests.
118 Ehrenreich. Nazi Ancestral Proof. Friedlander (1995), 2–3; Gould, 118.
119 Weiss (1987b), 53n.44, 177, quoting Schallmayer, W. (1895). Die drohende physische Entartung, 9–10.
120 Noakes (1984a), 78.
121 Binding, K. & Hoche, A. (1975). The Release of the Destruction of Life; Friedlander, 18; Gripp, R. F. (1981). [The Release of the Destruction of Life], 454–55; Mueller, T. & Beddies, T. (2006). The Destruction, 97–98.

Table I-10. Why Physicians are Vulnerable to Becoming Perpetrators

1. Compartmentalization
2. Tendencies toward sadism, voyeurism
3. Healing through hurting, repressing awareness of violence
4. Use of science to objectify violence
5. Use of metaphors and euphemisms
6. Justifying and rationalizing
7. Medicalizing

Source: Grodin, M. A. (2010). Mad, Bad, or Evil, 58.

The realities of reporting responsibilities for physicians, teachers, police officers, civil servants, social workers, and even church designated liaison officers, are best illustrated by the following scenario.[122] If one had a common cold or had a natural propensity towards biological imbalances that made one more susceptible to disease,[123] the poor wretch who made a routine visit to a GP would be in for a rude awakening. Unaware of the GP's newly expanded responsibilities, the patient was examined, and the doctor discovered that he/she was suffering from a disorder or illness. The criteria from the social division within Table I-9, are thus effectively met. Now, suddenly, the individual would potentially be flagged for sterilisation. The GP had to report this immediately to the district public health officer.

Any visit to the GP for a minor ailment, further down the line, after the sterilisation process, could result in the individual being selected as a candidate for the euthanasia killing centres, more commonly referred to as the "*life unworthy of life*" programme. In the early stages, Genetic Health Courts with an appellate division, ordered the sterilisations before they were eventually dismantled, to make way for the killing centres.[124] See next heading Racial Hygiene at its Zenith, 160, how it played out. In a lecture delivered by Lenz on February 20, 1935, titled: "*Problems of Practical Race Hygiene.*" Weiss (2010) quoting Lenz's talk, reports that if the law were "*. . . vigorously applied, the number of feebleminded could be cut in half in a matter of years,*" in Germany, as well as in occupied and friendly countries.[125] Accordingly, Table I-11 sets out sterilisation figures from the initial stages following the 1933 enactment—

Table I-11. Grounds given for sterilization in the adjudication of the 1933 Sterilization Law

Reason for Sterilization	Study		
	K.H. Bauer[a]	Mikulicz-Radeckib[b]	Verschuer[c]
Congenital feeble mindedness	42.9	66.8	29.6
Schizophrenia	25.4	22.6	34.4
Hereditary epilepsy	13.4	8.1	24.6
Severe alcoholism	5.0	0.5	6.8
Manic-depressive insanity	1.6	0.5	2.8
Hereditary deafness	1.4	1.5	0.6
Hereditary blindness	1.2	1.5	—
Huntington's chorea	0.2	1.5	0.3

[122] Koonz, *105-06; Proctor (1988), 73, 104-06; Weindling (1993), 526.
[123] Weiss (2010), 76.
[124] Lewis, A. (1934). German Eugenic Legislation, 183-85; Proctor, 41, 101-04, 106-08, 296; Schafft, 75.
[125] Weiss (2010), 192, 217.

Table I-11. Grounds given for sterilization in the adjudication of the 1933 Sterilization Law

Serious bodily malformation (hereditary)	0.2	1.5	0.9
Diagnosis not given	8.7	—	—
	100.0%	100.0%	100.0%

Source: Otmar Freiherr von Verschuer, Erbpathologie: Ein Lehrbuch für Ärzte und Medizinstudierende (1934), 2nd ed. (Dresden, 1937), 178, in Proctor, R. N. (1988). Racial Hygiene, 107–08. See also Mosse, (1978), 91–92; Mosse (2020), 83–84 and Proctor (1992), 20–23.

[a] Based on a sample of 6,052 men.
[b] Based on a sample of 6,032 women.
[c] Based on a sample of 325 men and women.

A case in point is Lenz, who wrote a telling statement in the fourth edition of *Human Heredity* (1931) regarding future intentions: "*. . . in paving the way for sound selection in the sense of purification, improvement and elevation of the race.*"[126] This is especially pertinent given that ghettoisation, segregation, and deportations were still nine years away, beginning in 1939. Sterilisation law was therefore viewed as an instrument of isolation and population-control, facilitating the identification and rounding up of those earmarked for the next phase of "*medical*" murder by gas and injection. Peukert (1987) estimated that by 1945 between 200,000 and 350,000 had been sterilised.[127] Lenz's statement affirms the dual aims of racial hygiene: he intended "*upgrading*" and cleansing of German society of Jews and genetically unfit, using both positive and negative eugenic measures.

[126] Weinreich, *31-32, quoting Baur, E., Fischer, E., & Lenz, F. (1936). Menschliche Erblebre und Rassenbygiene, Band I Menschliche Erblebre (München, Lehmann, p. 733). See also Browning, 172–73.
[127] Peukert 218–19 (sterilization numbers source not provided).

C. Racial Hygiene at its Zenith: Diseases as Justifications for Isolation

i. Discoveries in Bacteriology Transference of terminology onto Jews

We have traced the development of racial hygiene policy, in which eugenics ideology was entrenched from its inception, culminating in a blueprint for genocide. The 1880s and 1890s marked a period of "*rapid scientific advances*" in bacteriology. Many German discoveries arose from the isolation of causal organisms of bacterial diseases, including anthrax, chorea, Malta fever, plague, and dysentery between 1873 and 1900.[128] This was accomplished through microbiological breakthroughs, "*. . . applying standard methods of cultivation and replication . . . ,*" bacteriological techniques derived from industrialisation, and serum therapies and immunisation.[129]

Weindling (1993) argues that post-1880s bacteriology increasingly adopted military analogies to describe microbial behaviour. Robert Koch (1843–1910), a principal founder of modern bacteriology, identified the causative agents of tuberculosis (1882) and cholera bacilli (1883), established new standards for diagnosis, treatment, and prevention.[130] It is noteworthy that Weiss (1987b) stresses the impact of Koch's discoveries, noting that they "*. . . heralded a new dawn in the classification and treatment of infectious diseases.*"[131] So much so, racial hygiene was regarded as incomplete without its extension into bacteriology—

> [O]nly in the 1880s, largely as a result of Koch's discoveries and the subsequent rise of bacteriology, did hygienic practitioners really possess the necessary theoretical and scientific knowledge to adequately fulfil their duty as custodians of the nation's health.
>
> Only after the germ theory of disease was firmly established could the link between epidemics and poverty be demonstrated.[132]

In Otis's (1999) analysis of Koch, she writes, "*. . . [under the microscope] his techniques for identifying microorganisms . . . , he developed systematic processes to eliminate diseases from [given geographical areas permanently], first by identifying bacteria; then, elimination (Ausrottung) and destruction (Vernichtung).*"[133] Thus, it is not a stretch to maintain that this was an early conceptual blueprint, already visible in the 1880s–1890s, for defining biological "*enemies*" and the subsequent transposition of this structure onto the Jews—beginning with blood testing, followed by ghettoisation, and culminating in the death camps.

Moreover, the turning point came when bacteriology was reoriented through a new vocabulary of health and medicine.[134] In around July 1885, Koch began expressing bacteriological concepts in a monocausal, war-metaphorical manner.[135] He described how hoards of "*alien parasites*" had invaded the body, and chemotherapy drugs were

---

128 Weindling (1993), 160.
129 Ibid., *159–64.
130 Otis, L. (1999). Membranes, 28–29, 33; Weiss (1987b), 17.
131 Weiss (1987b), 18.
132 Ibid., 115, 190n.107, respectively, citing Gottstein, Adolf. "Die Entwicklung der Hygiene im letzten Viertel- jahrhundert Zeitschrift fur Sozialwissenschaft 12 (1909):74.
133 Otis, 34–35. For commentary on Koch, see Otis, 8–36. See also Savage, 417–18.
134 Cocks, 50–51.
135 Weindling (1993), 167n.39, 167.

referred to as "*magic bullets.*"[136] Bacteriology itself had become 'infected' as antisemitic radicals started to draw upon its terminology of disease to formulate jokes and pathogenic metaphors, such as "*Jewish bacillus,*" while disease-causing bacteria was inflected with invisible enemy parasites or dirty Jews.[137]

Interestingly, the term "*parasite*" was already in prolific use in writings as early as 1883,[138] prior to the emergence of bacteriology as a distinct field. Moreover, its actual origins can be traced back to the medieval era. See analysis in Constructing Robust Preparedness, 232. Its appropriation by antisemities transformed it into a contagion of language. These new metaphoric applications continued to proliferate, especially by the 1930s, when viruses gained scientific prominence. This coincided with the rise of Nazism, which added further disease rhetoric and contributed to the consolidation of "*medical microbiology,*" with bacteriology representing a narrower, microscopic focus.[139]

Bacon's *New Atlantis* was constructed as a biological utopia of perfection in which Weindling (1993) explains that state socialism, sanitary reform, and Darwinian principles were combined. This triad is highly relevant here, given that Ploetz relied on bacteriology and hereditary science to rationalise systems of social regulation, public-health management, viewing "*. . . Racial Hygiene [as] preventive therapy against alien bacteria . . . .*"[140] Of course, during viral outbreaks, public outcry directed at a designated outgroup and driven by fear could escalate rapidly. Within this climate, Ploetz's approach—encouraged the elimination of pathogenic threats in the name of collective well-being—intensified demands for state intervention. Calls for strict quarantines, accompanied by exigent appeals for "*. . . incarceration of the infectious [to be imposed] as priorities,*" were only a short step away from legitimising the eradication of the presumed source.[141] The state then becomes the arbiter and "*. . . guardian of public morality*" as a consequence.[142]

Across 1851, 1886, and 1892, Lagarde and Langbehn advanced what might be called a bacteriological register, openly advocating the extermination of Jews and comparing them to "*the cancer of our entire existence,*" "*vermin,*" "*a passing pest,*" baneful germs, and "*cholera.*"[143] In 1902, the physician Alexander Pilcz (1871–1954) asserted in an Austrian medical journal that Jews suffered from acute psychosis and insanity.[144] These antisemitic metaphors simply never took hold at that time, as Social Darwinism, nationalism, and the emerging Nazi genetic caste system had not yet reached their zenith. These developments constituted rudimentary forms of knowledge transmission that defined the embryonic stages of racial hygiene development in the mid-1800s.

Fast forward to the Nazi era. The bacteriological analogies of the 1800s were no longer figurative. Under the regime, such metaphors were treated as literal truths

136 Ibid., 167.
137 Dundes, A. & Hauschild, T. (1983). Auschwitz Jokes; Hughes, E. C. (1962). Good People and Dirty Work; Linke, U. & Dundes, A. (1988). More on Auschwitz Savage, 414–418; Weindling (1993), 170. Cf. Gilman (1991), 227, 231. Gilman observed that in 1980s Germany, public discourse around AIDS began to echo ". . . the format of the vocabulary of disease in the Nazi period." It framed marginalized groups—such as foreigners, drug users, and sex workers—as sources of contamination, thus invoking historical associations that had previously linked Jews as agents of infection. See also Hagemeister, M. (2008). The Protocols, 94.
138 Bein, 19.
139 Savage, 415; Tomes, N. (1998). The Gospel of Germs, 245.
140 Weindling (1993), 131.
141 Ibid., 346.
142 Ibid., 77–78, 80, 155.
143 Chamberlain, H. S. (1912). Foundations, 2:228; Gutteridge, 21; Pulzer, 236;Tatz, C. M. et al. (2006). Genocide Perspectives III, 87, quoted in Savage, 417. See also Bein, 12, 32; Wegner, 9–10. For a general discussion about Langbehn's vitriolic antisemitism, see Pulzer, 233–38.
144 Procter (1988) 196.

and embraced *in toto* as reality. Hitler adopted these tropes, insisting that Germans must avoid all contact with carriers of this "*poisonous bacillus.*"[145] For the cultural and historical scaffolding from which these disease pathotropes evolved, see heading Hatred transmission, 181.

ii. From Disease Legitimisation to Isolation and Elimination

By the Nazi era, Gerhard Wagner (1888–1939), leader of the German medical profession, taught in his lectures that racially "*bastardized*" Jews constituted a "*diseased*" race and that Judaism itself was "*disease*" incarnate.[146] Metcalfe (2002) and Proctor (1988, 2000) observe that depicting Jews as cancerous growths or tumours, a disease, plagues, or carriers of pestilence was neither new nor unique to Nazi Germany or the Holocaust, even within academic discourse.[147]

Savage (2007) captured this dynamic succinctly: "*[i]t is not the metaphor itself, but the 'reality' which this metaphor represents, the perception of intrinsic essentialised threat. . . .*"[148] The implication is clear: Jews were cast as a disease and as a cancerous growth, a portrayal that helped construct a narrative demanding action. In a political debate on March 6, 1895, Hermann Ahlwardt labelled Jews as "*cholera bacilli*" or "*cholera germs,*" claiming they were contagious.[149] Gerhard (1998) found in the stenographic record that Ahlwardt's smear was met with ". . . *a call for separation of the Jews from the Germans . . . the need for segregation and exclusion . . . using the model of hygienic defense against epidemics.*"[150]

First, it amounted to incitement to violence, urging drastic measures to excise the "*cancer growth*" from the German body politic[151]—an alleged pandemic attributed to the extraneous presence of millions of Jews in Europe. Second, its widespread and repeatedly publicised use by ideological actors served both (i) as an indicator of the scale of violence anticipated, genocide *ab origine*, and (ii) as an implicit summons to 'hygienically' cut away the disease from the national body.[152] Furthermore, Goebbels equated Germans with "*surgeons,*" declaring: ". . . *our task here is surgical, not humanitarian . . . drastic incisions, or some day Europe will [die] . . . .* "[153] On this basis, the same rationale was extended to justify eliminating other groups deemed inferior.[154] Goebbels asserted that they represented a societal pathogenic threat: ". . . *as bacteria, as pathogens for the nation. Such pathogens had to be eliminated.*"[155]

In accordance with Bandura's (1999b) account of 'moral disengagement,' one key mechanism is dehumanisation, which provides a warrant for genocidal action.[156] He argues that humane convictions, social obligations, and ". . . *self-censure, for cruel conduct . . .*" can be neutralised, suspended or disengaged by stripping targets of their human qualities. In the Nazi case, once the Jewish identity and humanness were eroded,

---

145 Bein, 32–*33; Gutteridge 21.
146 Hannaford, 320; Proctor (1988) 196; Savage, 422, 437n.139.
147 Metcalfe, N. (2002). A Short History, 272, quoted in Baumslag, 204; Proctor (1988) 196; Proctor (2000). The Nazi War on Cancer, 8, 11, 46.
148 Savage, 424.
149 Baumslag, 44; Bein, 106; Gerhard, U. (1998). The Discoursive, 89; Mosse (1964), 44; See generally Mosse (1975), 31–51; Langbehn, J. (1892). Der Rembrandtdeutsche, 184; Pagel, A. (1992). Jugendstil, 109n.12; Savage, 417; Stern, F. (1974). The Politics, xiii, 142.
150 Gerhard, 89; Savage, 435n.89.
151 Grodin (2010), 60 et seq; Savage, 409, 415; Wegner, 17.
152 Proctor (2000), 46; Savage, 419–20, 425–26; Sontag, S. (2001). Illness, 83–84.
153 Hochhuth, R. (1980). A German Love Story, 18, quoted in Lifton, 477, quoted in Savage, 423.
154 Proctor (2000), 46. Savage, 426, 437n.144.
155 Gittelman, 8; Savage, 418.
156 Bandura (1999b), 200; Savage, 425–26.

they were reconceived as objects to be managed, manipulated, or reduced to subhuman status in pursuit of some abstract 'good.' Once established, they were expelled from the moral community, "*. . . no longer deserving treatment as moral human beings.*"[157] Accordingly, Bandura depicts how they were subsequently viewed—

> [M]indless 'savages,' 'gooks,' . . . despicable wretches . . . 'satanic fiends,' 'degenerates. . . . It is easier to brutalize people [when they have been reduced to such base forms].
>
> . . . .
>
> [It gives perpetrators] satisfaction and build[s] their sense of self-worth, and are less burdened by stress.[158]

Consequently, the 1800s analogies evolved into doctrine, generating what Lifton (1986) clichéd as, the "*healing-killing paradox.*"[159] It signified in a simple sense, "*. . . healing the [body of the] Aryan race by killing those seen as threatening it.*"[160] This was compatible with the primary objectives of Nazi racial hygiene: the elimination of the biologically weak. This was not construed as inhumane conduct, but as a therapeutic mission—a belief that administering death constituted "*. . . purely a healing treatment and a 'healing work.'*"[161] In essence, this is Haeckel's Spartan-animal behavioural standard. Thus, any physician who helped the weak in any capacity was cast as 'the persistent enemy of progress.'[162] After two centuries of development, Lenz (1931) finalised the codified doctrine of racial hygiene, incorporating terms such as infectious, diseased, and, inter alia, filthy, into official policy—

> If people were dirty in their ways and backward in their civilisation, those who make so much of the influence of environment would not be slow to find the cause in the parsimony of nature . . . the Nordic is . . . animated by a love of order and by cleanliness of mind. Bodily cleanliness too, is rightly regarded by Guther as a characteristic of the Nordic races . . . Lice, which are still common in eastern and southern Europe, are very rare in the north and north west.[163]

This paper's thematic direction, traced from the publication of Darwin's Origins, frames animalistic conduct as central to understanding why Hitler and the Nazi regime conceived humans as animals, and the Jew "*. . . remains the eternal parasite, a sponger (Schmarotzer) [is] like a pernicious bacillus . . . .*" to whom genetic laws derived from

---

157 Biaslas (2013), 19, 22–*23; Opotow, S. (1990a). Moral Exclusion; Opotow, S. (1990b). Deterring Moral Exclusion; Staub, E. (1990). Moral Exclusion; Staub, E. (1999). The Roots of Evil, 183; Waller, 244–49; Weiss (1987a), 236; Weiss (1990), 49.
158 Bandura (1999b), 193, 200; See also Savage, 428–29, *425, citing Bandura, A. (1999a). Moral Disengagement. In Charny, I. W. (Ed.), Encyclopedia of Genocide; Tenenbaum, 105; Weiss (1987a), 236.
159 Lifton, 150, 413, 482–483. See also Grodin (2010), 52; Staub (1990), 51, 58; Waller, 113, general analysis of Lifton, 113-23; Weyers, 153.
160 Savage, 428–30.
161 Bialas (2014), 23; Grodin, M. & Annas, G. (2007). Physicians and torture, 642; Grodin, M. A. (2010). Mad, Bad, or Evil, 51; Lifton, 46, citing Binding, K. & Hoche, A. (1920). Die Freigabe; Proctor (1988), 73, *177–223; Savage, 405, 422; Zeidman, 371–72.
162 Claeys, 226; Hobhouse (1911), 22.
163 Baur, E. et al, 660.

livestock breeding could be applied.[164] The rationale follows that, because the camps were ostensibly free of epidemics, the SS doctor was "*. . . neither killing nor healing, . . . but [was] achieving the necessary healing-killing balance,*" as Eduard Wirths (1909–1945), chief SS doctor at Auschwitz, explained.[165] Savage (2007) refined Lifton's account, maintaining that the SS doctors fulfilled their mandated role by, "*. . . destroying the source of disease, in order to heal the organism.*"[166]

The first step was initially the requirement to identify the 'inferior races' for spatial separation and quarantines, the source of disease and contamination, "*. . . that could endanger German personnel or reach beyond the ghetto walls and afflict the Aryan population . . . .*"[167] Of course, this was justified as a hygienic necessity, exemplified by the ban on Jews from railway travel as indicated in the headline: "*Germ carriers banned from Railways*" to keep the Jews isolated and to keep Aryans out.[168] Such dehumanising tropes stamped the Jews, leading directly to preliminary measures, beginning with the initial isolation measure of forced confinement in the ghettos. Concerns about disease transmission and separation were part of a covert agenda with a far greater objective. Steinweis (2006) encapsulates the point with precision, observing that "*. . . Nazism's preoccupation with race distinguished it from previous antisemitic movements . . . and endowed it with an iron logic of exclusion and separation . . . ultimately leading to the Holocaust.*"[169]

Eugenics operated largely as a non-lethal instrument, aside from a few unpublicised secret killings rather than mass exterminations, designed to identify the "*diseased.*" This process smoothed the path toward the eventual eradication of what the regime regarded as the unmanageable and excessive numbers of genetic misfits and social undesirables that had accumulated by 1938. In October 1938, 17,000 Polish Jews were expelled across the border into Poland. In November 1938, these groups were rounded up and sent to concentration camps Buchenwald, Dachau, and Sachsenhausen. Between February and October 1939, Jews in Germany were given seven months' notice by the authorities to depart. The remaining 163,889 who did not leave were gathered into ghettos in Theresienstadt, Czech Republic, and Lodz, Tarnów and Warsaw, and into the smaller ones of Cracow, Lublin, Radom, and other parts of Poland.[170]

Schafft's book, *From Racism to Genocide* (2004), examined the records of the Kaiser Wilhelm Institute Anthropologists. She reported that they were conducting research on living Jews from the Jews of the Tarnów ghetto, with the intention of analysing their anthropometrical data. She demonstrates that such investigations were just another pretence to legitimise ethnic cleansing and the evacuations. She concluded that the "*. . . anthropologists' work in the Tarnów ghetto [represented] a damning picture of the pseudoscience*" that characterised the overall work in the Third Reich."[171] For instance, despite the positive framing, investigators reported that they observed none of the claimed cases of infectious disease. Furthermore, the data yielded no meaningful

[164] Bein, A. (1964). The Jewish Parasite, 7, *19. See also Hitler ([1925] (1971)), 305; Hitler, A. (1939). Mein Kampf, 252–73, quoted in Poliakov, 2; See also; Schafft, 123.
[165] Lifton, 202.
[166] Savage, 421.
[167] Berenbaum, M. (2022). The Question, 7. See also Bauman, 223; Goldhagen (1996), 136–37; Schafft, 124–25, 153.
[168] Baumslag, N. & Shmookler, B. M. (2014). Typhus Epidemic, 41; Proctor (1988), 199.
[169] Steinweis, 9.
[170] Proctor (1988), 199; Schafft, 80, 91–92.
[171] Schafft, 34; generally, 15–34. See also Fuchs, 345; For an opposing view, Weindling (2017) considers pseudoscience as a real science, but characterises it as unrestrained science that released ethical restrictions and inhibitions. See Weindling (2017). Intrroduction, 4, 8; Weindling, P. J. (1993). Health 574. See also Chelouche, 14.

conclusions, and the anthropologists "*... did not know how to do the simplest statistical procedures.*"[172] In light of this, was there any time that Aryan science was ever anything other than pseudo-science? Remember from our previous discussions, anthropology predated the Third Reich by 200–300 years before later being amalgamated with German Aryanism, giving form and coherence to an already emerging Darwinist racial ideology. Consequently, it developed as a form of pseudoscience and it continued to develop along those lines.

Coincidently, the Nazis stepped up their annihilation programme against all of the inferior, both Jewish and non-Jewish; converging on the year 1938, one year before the outbreak of WWII in 1939. The fusion of racial hygiene Divisions I. and II. in Table I–9, also occurred in 1938. This additionally coincided with the late-1938 decision to process the non-Jewish genetically diseased for elimination, while the relocation of Jews likewise began that same year. These developments converged on 1939. The establishment of the Łódź ghetto, and its continued confinement, were justified as forms of medical quarantine, publicised as necessary to prevent the further spread of contagion. Hitler further justified the liquidation of the ghettos and their "*evacuation*" by arguing that, since Germany had to evacuate its own civilian population, why should we not do the same with the Jews (my paraphrase).[173]

The point is this: ghettoisation was a provisional, transitional, quick-fix quarantine zone for control and containment, not a long-term solution. Rather, it was necessary to prepare subsequent stages of Nazi Jewish policy, both by expediting the establishment of the ghettos and facilitating their eventual elimination.[174] Nazi medical authorities employed the threat of disease as a pretext for a series of repressive measures of confinement within the ghettos, cloaking these policies in "*objective science,*"[175] piggy-backing on claims that they were "*scientifically 'proven'*" by Social Darwinist ideology.[176] However, the underlying philosophy behind the deliberate intention construction of a disease environment remains to be fully examined. It was multi-layered and stems back to the Revolution of 1848–49. In Joseph Goebbels diary, entry for 7 August, 1941, in the context of reports that typhus was increasing in the Warsaw Ghetto, he wrote—

> The Jews have always been carriers of infectious diseases. They must either be cooped up in a ghetto and left alone, or liquidated; otherwise they will always infect the healthy population of the civilized nations.[177]

In Dalton's (2019) commentary on Goebbels, he argued that the phrase 'liquidation' was a "*... nonlethal reference to dissolving the Warsaw Ghetto.*"[178] Dalton supports this interpretation by citing Goebbels's diary entry of 11 August 1941 (Hitler's prophecy),[179]

---

[172] Schafft, 27, 32.
[173] Dalton, T., Ed. & Trans. (2019). Goebbels on the Jews, 11 August 1941 (II.1.213); Hitler (2000), 24.
[174] Baumslag, 51, 49, quoting Trunk, I. (1972). Judenrat, 143; Bein, 106; Browning, 116, *111-12, quoting National Archives Microfilm T175/239/2728524–28. Conference between Heydrich's division heads and Einsatzgruppen leaders, September 21, 1939. See Browning, 111–68, general ghetto discussion.
[175] Weiss (1987a), 235; Weiss (1990), 49.
[176] See Ehrenreich, E. (2007). Nazi Ancestral Proof, 16, 28; Hutton, 125, citing Gruber, M. v. (1927), 247.
[177] Dalton, T. (Ed. & Trans.). (2019). Goebbels, entry for 7 August 1941, vol. II.1.189; See also Glass, J. M. (1997). "Life unworthy of life", 63–64; Kershaw (2000), 960n.71; Proctor, 199.
[178] Dalton, II.1.189.
[179] Hitler's "prophecy": he cast himself in the role of a prophet, proclaiming that he would become Führer of Germany and foretelling the annihilation of European Jewry. See Weinreich, 81–82, 168–69, 210.

as well as discussions recorded in Hitler's Table Talk (2000). Trunk (1989) emphasises the calculated nature of the policy, noting that it banked on the expectation of an epidemic outbreak in the ghettos—

> In their campaign of annihilation against the Jewish population of Poland the Nazis employed not only the well-known technique of "deportations" to the death camps, but also used bacteria. . . . It is . . . clear . . . that this technique occupied a prominent place in Hitler's project to exterminate the Jews. By deliberate design the German authorities created conditions in the ghettos which made the outbreak of epidemics inevitable.[180] (Alteration in Lindenthal's quote)

Of course, this formed the embryonic foundation for manufacturing a medical crisis as a pretext for sterilisation and the initial confinement of individuals within asylums, psychiatric institutions, and other segregation centres.[181] Tenenbaum (1956) redirects our attention to this paper's central argument, namely, that the Nazis harboured an obsessive fear that their Nordic superiority might be contaminated by even a trace of non-Aryan or mixed blood. Therefore, they utilised quarantine centres, camps, and ghettos as instruments that "*. . . acted as huge filters separating the pure from the impure.*"[182] The ostensible purpose was to contain the contrived communicability of disease. Yet, as Roland (1992) notes, there was no evidence to support such a claim—

> There is little evidence that such epidemics existed to any serious degree before the ghetto was created and no evidence that Jews were any more likely to suffer from, or to transmit, typhus than any other group.[183]

Lindenthal (2014) expands on this perspective, citing various historical epidemiological studies from 1919 to 1936, and concludes that, in many cases, Jews, "*. . . were in a healthier state compared to the non-Jews, demonstrating a lower incidence of disease.*"[184] Ironically, once confined to the ghettos, Jews were deliberately forced to live in unsanitary, overcrowded conditions designed to accelerate mortality. Following Trunk's (1953) point, they consequently contracted typhus in these environments. The Third Reich was well aware that rancid housing conditions were implicated in the aetiology of typhus, given that bacteriological science had already highlighted such links by the 1890s. Weindling (1993) makes the point sharply that "*Housing surveys, WohnungsEnquête, linked TB to poor ventilation, damp and over-crowding . . . ,*"[185] underlining the inevitability of increased pathogenicity.[186] See Figure I-7 below.

180 Lindenthal, J. J. (2014). The Epidemiological Status, 18, 36n.28, quoting Trunk, I. (1953). Epidemics and Mortality in the Warsaw Ghetto 1939–1942. YIVO Annual 8, 82–122.
181 Berenbaum, 7; Pellegrino, xiii; Schafft, 152–62.
182 Tenenbaum, 90. See also Hertz, 55–56.
183 Baumslag, 22, 52–53; Roland, C. G. (1992). Courage Under Siege, 130–31. Cf. Fishberg, M. (1911). The Jews, 325. Dr. Lazar Solomonovich Minor's conclusion was that Jews are not predisposed to nervous diseases anymore than what Christians would be. See also Gilman (1991), 63.
184 Lindenthal, 16–*17.
185 Weindling (1993), 165, 167. See also Lindenthal, 25.
186 Beinfeld, S. (2014). Health Care, 130, 202; Baumslag (2005), 28, 54, citing Trunk, 144; Baumslag & Shmookler (2014), 40; Goldhagen (1996), 289, 291; Roland, 4–5. See Baumslag, 61, re a similar approach was adopted for the death camps.

This supports Lindenthal's methodical analysis, which exposed the deplorable living conditions within the ghettos.[187] As a result, when inmates contracted dysentery, cholera, typhoid, tuberculosis, or typhus, the Nazi authorities retrospectively referenced their original diagnoses, citing their "*prehistoric claims*" and prophecies as evidence[188]—that the Jews *ipso facto* had always been disease carriers. This, in turn, furnished the rationale for disinfection, quarantine, and ultimately mass murder, extending even to those who had merely come into contact with them.[189] Beyond the use of typhus as justification itself, Roland (1992) duly noted that—

> By setting up the ghetto and forcing so many people to live there, the Germans made their contention a self-fulfilling prophecy: it seldom took much time before epidemics arose in a ghetto and turned it into a death trap."[190]

If they did not die from these conditions, they were murdered daily by occupying soldiers, or, upon deportation, subjected to mass shootings.[191] For example, focusing solely on the Polish territories annexed to the Reich, including, the ghettos, Bauer (1981, 2017) recorded that the initial population of 692,000 Jews had fell to 460,000 between September 1939 and February 1940, even before the establishment of the death camps.[192]

---

[187] Lindenthal, 17–28.
[188] Baumslag, 55, quoting Hagen, W. (1973). Krieg, 117; Fuchs, B. (2008). German Anthropology; Proctor (1988), 200; Schafft, 19.
[189] Baumslag, xxiv, 4.
[190] Beinfeld, 106, 110. See also Baumslag, 35, 49; Proctor (1988), 198; Roland, 131.
[191] Baumslag, 52; Schafft, 18, 20, 31.
[192] Browning, 116; Bauer, Y. (1981). American Jewry, 69.

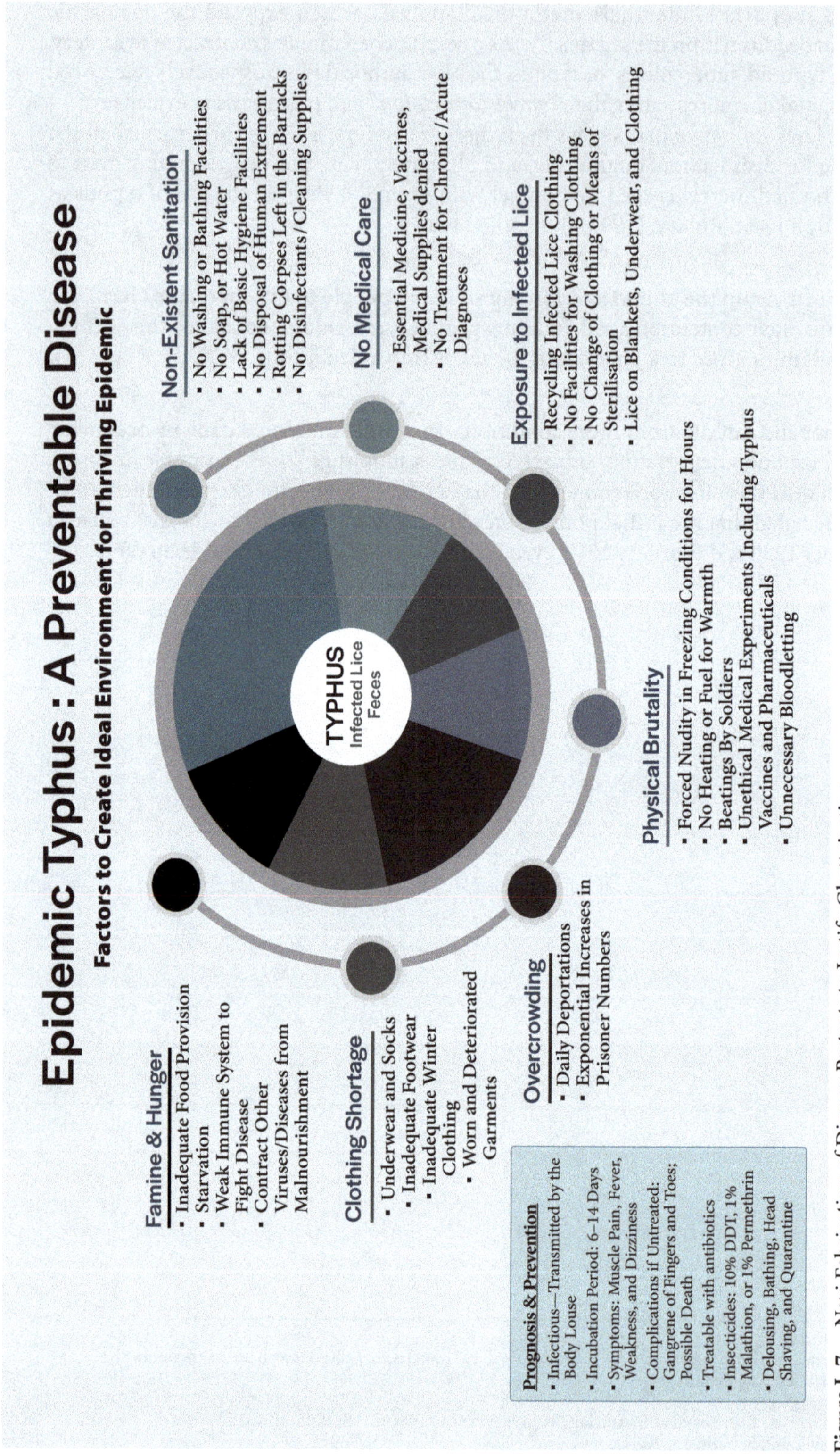

Figure I-7. Nazi Fabrication of Disease Pretexts to Justify Ghettoization and Mass Extermination.
This figure reveals how the Nazis fabricated disease threats to justify isolating Jews in ghettos and later deporting them to death camps. By framing Jews as disease carriers, they created a false rationale for quarantine, disinfection, and ultimately mass murder. (Source: Baumslag, 91).

The continual influx of new arrivals into the ghettos only intensified the crisis. Overcrowding became a major contributory factor, reaching epidemic proportions: "*. . . soon after concentration . . . Jews in these ghettos did begin suffer from higher rates of infectious disease.*"[193] Moreover, the authorities were fully capable of breaking the vicious circle of hunger-induced epidemics simply by providing sufficient food to restore health. The British Medical Journal (1940) reported that the Nazis created an additional pretext by installing numerous disinfection facilities, presenting the appearance of disease control, yet in practice they did nothing—

> [T]o eradicate the source of the disease by clearing out the worst part of the ghetto, where tens of thousands of Jews are confined under pestilential conditions.[194]

Beside the fact that the Nazis invoked disease metaphors and concocted a non-existent disease for the creation of the ghettos "*. . . under the guise of 'quarantine': ghettoes were a 'hygienic necessity' . . . ,*"[195] this narrative not only stigmatised Jews as carriers of contagion but also relied dehumanising descriptors such as viruses, bacilli, microbes, cancerous growths, and other metaphorical labels.[196] A further irony lies in the reality that, following deportation to the death camps, Jews were subjected to typhus experiments designed to test the communicability of the disease through a series of procedures. A wider array of extreme medical experiments was also undertaken as part of eugenic research programmes, spanning clinical and institutional settings, large-scale camp-based experimentation, and smaller trials carried out under the guise of 'health care.'[197] The ghettos were employed as a *cordon sanitaire*, purportedly to protect the Aryan gene pool from communicable diseases and to prevent those viewed as genetically diseased from infecting or procreating with Aryans.

Taking inspiration from the English and French public hygiene movements, German medical men Rudolf Virchow, Solomon Neumann, and Rudolf Leubuscher "*. . . during the revolutionary year of 1848, they joined forces to secure overdue health reforms.*"[198] They undertook studies to determine the causal relationship between poverty, housing, squalor, and disease. Virchow conceived the scope of public health as broadly as possible, focusing on the living conditions of social groups and their effects on health.[199] Sontag (1989, 2001) highlighted the significance of metaphors of physical weakness, character weakness, and moral decay, and corruption. Under the Nazis, these were reinterpreted as literal markers of disease, so that "*[t]he disease itself becomes a metaphor.*"[200] Thus, to be located within the ghettos were automatically constructed as "*diseased.*" The ghetto itself was conceptualised as a pathological entity to be eradicated from the German political body. By extension, impoverished towns and cities were likewise designated as a cesspits of 'disease,' even though Aryans were physically healthy, and without any illness.

The whole matrix of the racial science programme was a top-down deception enabled by propaganda. Thus, creation of contrived narratives from inception, paved the way for genocide a process that was both multi-faceted and multi-layered. They

193 Proctor (1988), 200. See also Baumslag, xiii–xiv; generally, 2–29; Browning. 160;
194 British Medical Journal. (1940). Medical News, 2:36; Proctor (1988), 200n.88, citing the BMJ, 36.
195 Savage, 418.
196 Ibid., 404, 415–17.
197 Langbein, H. (1972). Menschen, 378, cited in Lifton, 255, *262, quoted in Baumslag, 150; Weindling (1993), 495–97; Weindling, P. J. (2017). Intrroduction, *4–25.
198 Rosen, G. (1958). A History, 254–255.
199 Ibid., 255; Weiss (1987b), 16.
200 Sontag, S. (1989). Illness, 58; Sontag (2001), 60, quoted in Savage 414.

were framed within both a positive and negative contexts. One particular layer was the guise of public health. Baumslag & Shmookler (2014) and Weindling (2014) write, "*. . . through obsession with the [typhus] disease, German doctors used public health to justify mass killings of Jews*" because they "*. . . posed a lethal threat to the German nation.*"[201] Another vector was the Reich Citizenship Law and the Law for the Protection of German Blood and German Honor (Nuremberg laws) were part of a mechanism that positively defined "*Aryan*" identity while negatively targeting those deemed racially undesirable, thus, establishing ethnicity based on bloodlines rather than religious identity.[202]

Another layer of deception that operated within both the ghettos and, even more starkly, the death camps was a façade of goodness—calculated displays of politeness, courtesy, and outward civility to mask acts of barbarity, as vividly portrayed in Borowski's *This Way for the Gas, Ladies and Gentlemen* (1976).[203] The ghettos themselves were products of profound hatred; antisemitic rhetoric and legislation formed the central philosophy for these pretexts in the first place. See Historical Anti-Semitism, 180. Weinreich (1999), for example, records Justice Robert H. Jackson's November 1945 address to the Nuremberg Tribunal, in which Jackson characterised the Reich government as a conspiracy of—

> [C]alculated cruelty . . . . Determination to destroy the Jews was a binding force which at all times cemented the elements of this conspiracy. Whose conspiracy? [. . .] This murder of a whole people was not perpetrated solely by a comparative small gang of the Elite Guard, or by the Gestapo [Secret State Police] Hitler's threats, the literature of the Nazi Party, Reich government, the Wehrmacht shows, German scholarship and the whole ruling class of Germany was committed to the execution of this crime.[204] (Alteration in original)

This enabled them to conceal the true intent of their racial hygiene policies, ultimately culminating in the Holocaust. This system was built upon falsehoods at multiple, often granular levels across society. Small wonder that Goldhagen (1996) highlights how pervasive the Nazi apparatus became, weaving itself into virtually every aspect of German life—

> No significant aspect of German society was untouched by anti-Jewish policy; from the economy, to society, to politics, to culture, from cattle farmers, to merchants, to the organization of small towns, to lawyers, doctors, physicists, and professors.[205]

The ghettos and the death camps were a positive, viewing them as financially lucrative and scientifically "*useful*" as "*. . . a vast reservoir experimental material*" for pharmaceutical companies drug production, such as I. G. Farben.[206] This involved culturing organisms and testing the efficacy of unnamed and untested drugs and vaccine

[201] Baumslag & Shmookler (2014), 39; Weindling, P. J. (2014). Delousing, 51.
[202] Friedlander (1995), 24; Goldhagen (1996), 138; Wenreich, 43.
[203] Borowski, T. (1976). This Way for the Gas; See also review Lawrence, A. (2015). [This Way for the Gas . .], 107–10.
[204] Weinreich, *6, 9.
[205] Goldhagen (1996), 8.
[206] Baumslag, *127, 129, *150.

serums on prisoners without their knowledge or consent.[207] Dr. Eugen Kogan (1980) writing as a political prisoner, described firsthand witnessing one "*Practice of Hell*" scene in Buchenwald where experimental subjects in typhus vaccines trials had a mortality rate from 30% to 40%, while other experiments resulted in 100% mortality, depending on the strength of the victim's immunity.[208] In other words, scientific progress was prioritised over and above the rights of the individual.[209] Accordingly, Baumslag (2005) observed that—

> Behring, Bayar, and Hoechst were the driving forces in the experimental drug sand vaccine research. . . . [Those that were blamed for, and convicted of the experiments,] [o]nly a few of their officials received short sentences and of those who did, upon their release they were given high positions with Hoechst and Bayer. Their sentences were light enough to please a chicken thief, and they showed no remorse.[210]

Animal tests were expensive, and rat procurements were more expensive than camp prisoners. Human experiments, therefore, made experiments more feasible.[211] Indeed, Baumslag (2005) noted that since 1.5 million prisoners died of Typhus, to act expeditiously—

> [T]he military were the driving force for measures to eradicate the disease, and the pharmaceutical industry derived profits from their drugs and vaccines. The SS doctors also experimented with German pharmaceutical company drugs for which they were paid. . . . Science was put ahead of humanity.[212]

Furthermore, the grandeur of the science provided opportunities for promotion, remuneration, prestige and higher standards of living.[213] Goldhagen (1996) shared this sobering thought that if Germany had not been defeated, they would have killed millions more.[214]

In accordance with Eugen Stähle's (1934) research results, he argued that certain blood diseases were specific to particular races, and that there existed an unspecified accumulation of diseases among Jews. Stähle further believed that it was an irony of biology that such conditions only afflict Jews.[215] Bergen (1998) emphasises that the underlying philosophy within the Nazi mindset was not that Jews had a disease, but that by virtue of their ethnic identity, they were the disease—

> The Nazi language depicting Jews as disease was not meant to be the hyperbole of invective but a literal identification. The disease was not something that was carried by the Jews-they were the disease. The disease was not specific but all the more terrifying because it was indefinite; it was everywhere, concealed in

[207] Ibid., 143–44.
[208] Kogon, E. (1980). The Theory and Practice of Hell, 94, quoted in Baumslag, 143.
[209] Baumslag, 130.
[210] Ibid., xxvii.
[211] Ibid., 136.
[212] Baumslag, xxv, citing Sofsky, W. (1997). The Order of Terror, 214-58.
[213] Baumslag, 127; Kogon, E. (1998). The Theory and Practice of Hell, 156–57.
[214] Goldhagen (1996), 4.
[215] Proctor (1988) 78–79, citing Stähle, E. (1934). Blut und Rasse.

> every Jewish act, constantly changing form.[216]

Himmler (1974) himself said as much when he remarked that: "*[w]e almost deloused, we have only some 22,000 lice left . . . .*"[217] Bergen's concluding remarks regarding "*constantly changing form*" aptly characterises how this constructed 'disease' narrative evolved into repugnant, dehumanising jokes. Dundes et al. (1983, 1988) catalogued these bleak jokes for educational purposes, preserving them as evidence of the ways in which hatred was circulated and normalised—

> Do you know why there is a shortage of soap?
> Because the authorities are trying to turn the soap back into Jews[218]

Thus, the modern appellation of 'dirty Jews' originated among Nazi Germans.[219] The implication of such language was the elimination of the Jews through a metaphorical "*cleansing*" of perceived impurities from the social and political body.[220] For example, Beinfeld (2014) documented in relation to the Vilna ghetto quarantine, a notice hung on the gate reading, "*Achtung! Seuchengefahr,*" warning of infection if any Aryan entered.[221] Hence, the public message was disseminated urging avoidance of the supposed 'carriers.'[222] The disease conditions in the ghettos reinforced the false paradigm in which containment and contagion were used to justify the imperative to "*kill the disease*" itself.

The next step was the death camps, framed as the final measure to address the alleged pathogenic threat. They were driven out of the ghettos and deported to the death camps of Auschwitz, Belzec, Birkenau, Sobibor, and Treblinka, all "*. . . staffed and operated by T4 personnel.*"[223] Seidelman (1989) explains the rationale how the death camps were equated with a medical cure: "*. . . a healing process, with mass extermination described by the Auschwitz physicians as the 'Great Therapy of Auschwitz,' referred to as 'The Central Hospital.'*"[224] This parallels the Black Death of the Middle Ages, which was actually believed to be typhus "*. . . caused by a small organisms called Rickettsia prowazekii which are spread from person to person by the human body louse.*" This instigated pogroms in Germany, and other European countries, followed by mass execution by burning.[225] See also heading Hatred transmission, 181. Thus, the idea became ingrained that Jews should be avoided, driven by the harrowing fear of contracting a plague-like disease.

Accordingly, killing sites were disguised as disinfection installations—shower facilities originally presented as measures to address typhus. Inmates were subjected to head-shaving and delousing, confronted with signage directing them to "*disinfecting baths,*" and instructed to inhale deeply under the pretence of infection prevention.[226]

---

216 Bergen (1998), 24. See also Proctor (1988), 198.
217 Smith, B. F. & Peterson, A. F. (Eds.). (1974). Heinrich Himmler, 200f, quoted in: Baumslag, 47; Glass, J. M. (1997). Life Unworthy of Life, 83; Hillgruber, A. (1984). War in the East, 129; Hillgruber, A. (1989). War in the East, 111.
218 Linke, U. & Dundes, A. (1988). More on Auschwitz, 6.
219 See also heading Historical Anti-Semitism, 180 for the jurisprudence side of the issue.
220 Linke & Dundes, 7. See also Dundes, A. & Hauschild, T. (1983). Auschwitz Jokes.
221 Beinfeld, 106. See also Baumslag, 51, 116. Cf. Seidelman, W. E. (1989). Medical selection, 436–37.
222 Baumslag, xxvii; Schafft, 18, 20, 31.
223 Friedlander (2002), 64–73.
224 Seidelman (1989), 436. See also Baumslag, 52, 67.
225 Baumslag, 2; Baumslag, N. & Shmookler, B. M. (2014). Typhus, 39–40; Cohn, S. K. (2007). The Black Death.
226 Baumslag, 22–23, 28, 62, quoting Tregenza, M. (1977). Belzec Death Camp, 8; Weindling (2014), 52.

Given that the Jewish population in Germany exceeded six million, Rosenberg unveiled the plan to use Zyklon B gas for mass extermination. It comes as no surprise that his ideas were shaped by Paul de Lagarde (1827–1891), whose ideology Rosenberg frequently acknowledged as a major influence.[227] Baumslag (2005) expressed that the rationalisation of genocidal medical evil were reconciled as "*. . . killing [for the] good [of] German society, and even an act of beneficence for the victim.*"[228] Dr. Jost Walbaum, chief health official of the General Government, claimed that the Jews posed a serious public-health risk.[229] The axiomatic truth, however, was that the Nazis favoured the use of Zyklon B, which Goldhagen (1996) explains as 'assembly-line killing,' a stock phrase commonly used in discussions of the event, because it was financially expedient and a more efficient means of murder.[230] By autumn 1941, Auschwitz was already fully functioning and remained so until its liberation on 27 January 1945.[231] Zyklon B was used to kill those labelled "*. . . both 'life unworthy of life' and of Jews, [a process] termed 'disinfection' . . . ,*" a euphemism for murder cynically presented as an effective sterilant intended to protect perpetrators and prevent the spread of pathogens.[232]

Once the typhus narrative had been established, it was then manipulated as a weapon of war, solidifying the recurring false claim that they were the source of disease.[233] First the Nazis created the ideal death environment: conditions in which disease and mortality flourished, producing a mortality rate of approximately sixty per cent.[234] Second, the squalor and deprivation to which inmates were subjected, included starvation and malnutrition, starvation-induced work exhaustion, drastic reductions in living space, and the removal of sanitary facilities were all known vectors for the spread of typhus. See Figure I-7, 168.

Third, chemical measures were employed to destroy the alleged bacterial threat, and fourth, bodies were burned to eliminate the source of infection. Baumslag demonstrated that the benefit-cost ratio of assigning large groups with small quantities was comparatively cheaper than using substantial amounts of disinfectant chemicals to delouse an entire camp.[235] The Nazis drew upon Dr Charles Nicolle's 1909 discovery that the typhus pathogen could be destroyed by heat; yet, according to Weindling (2014) "*. . . Zyklon might kill lice, but it did not kill bacteria, and the gas left clothing filthy.*"[236] Thus, the notion of "*killing the disease*" by burning Jewish bodies rested on a transient pretence, cloaked in the language of public health.

iii. Nazi Euthanasia: Targeting the Genetically Unfit and Socially Degenerate Classes

ELIMINATION WAS THE fate that awaited non-Jewish genetically impaired individuals and social reprobates. In fact, these concepts are arguably conjoined. According to Proctor (1988, 1992), while these two programs

227 Bausmlag, 204; Müller-Hill, 49, 50–51n.72, 52. See Müller-Hill's answer to question 4: "Why was the extermination of the Jews and of mental patients a secret?", 103–05; Neumann, 141.
228 Baumslag, xiii.
229 Bausmlag, 51–52, citing Walbaum, J. (1941). Kampf den Seuchen, 34.
230 Goldhagen (1996), 10.
231 Schafft, 118.
232 Baumslag, 2, 23–24, 203; Savage, 427, et seq.
233 Baumslag, xxiv, xxviii, 2, 17, 202–*03; Kershaw (2000), 256; Pellegrino, xv; Proctor (1988), 78–79, quoting Stähle, 4–6, and The New York Tribune, September 6, 1933. Cf. Beinfeld, 106.
234 Goldhagen (1996), 294; Latour, B. (1988). The Pasteurization, 20, 23.
235 Baumslag, 203.
236 Weindling, P. J. (2014). Delousing, 52. See also Baumslag, 11–13, citing Gross, L. (1996). How Charles Nicolle, Kiple, K. F. (1993). The Cambridge world history, 1088 and Nicolle, C. (1995). "Investigations on Typhus, 94.

initially seemed separate, they were in fact deeply connected. He explains—

> Historians exploring the origins of the Nazi destruction of lives not worth living have only in recent years begun to stress the links between the destruction of the handicapped and mentally ill on the one hand, and the Jews on the other. And yet the two programs were linked in both theory and practice. One of the key ideological elements was the "medicalization of antisemitism"—the view developed by Nazi physicians . . . .[237]

In May 1939, sterilisation was abandoned and was replaced with euthanasia centres.[238] Accordingly, euthanasia was a euphemistic, unassuming label for which Friedlander (2002) identifies as programme designed to conceal the blunt reality, officially termed *Vernichtung lebensunwerten Lebens* (trans. the destruction of life unworthy of life).[239] Professor Kleist, a psychiatrist, concluded his report of 24 March 1938 on the Herborn mental hospital stating—

> As long as there is no law 'for the destruction of lives unworthy to be lived', those who are beyond cure have the right to humane treatment which assures their continued existence.[240]

The writing was on the wall that they were intending to repeal the 1933 law, allowing them to bypass sterilisation altogether and proceed directly to destruction. To carry out this agenda, front organisations were created, the most prominent being T4. In many cases, the T4 euthanasia programme for adults operated as a secret killing initiative run by "*. . . Physicians and psychiatrists, mostly professors, hospital directors and bureaucrats,*" who were also medical experts.[241] Should this come as a surprise, given that this had been the original intent—the *sine qua non* of racial hygiene policy—to remove hospitalisation and humane treatment afforded to the genetically and socially inferior. See Table I-9, 150. Tenenbaum (1956) pointed out that in order to keep the true nature of the euthanasia programme secret, it was deliberately shielded from public scrutiny—

> [T]hree dummy corporations were set up: (1) The Reich Association of Mental Institutions; (2) The Charitable Foundation for Institutional Care, called the Stiftung; and (3) The General Sick Transport Company, equipped with special vehicles (big, black buses) for transportation of the victims from the institutions of care to the institutes of death.[242]

237 Proctor (1988), 194–95. Proctor (1992), 23–26; See also Lindenthal, 16, and generally, Burleigh, M. (1994). Death and Deliverance : 'Euthanasia' in Germany.
238 Kaminsky, 237–74; Mosse (1966), 60; Proctor (1988), 188–94; Proctor (1990), 35–39.
239 Friedlander, H. (2002). Physicians as Killers, 60.
240 Alexander, L. (1945). CIOS report, Item No. 24, quoted in Müller-Hill, B. (1998). Murderous Science, 12, 42. See also Zeidman, 352. Kleist in an interview with Alexander in 1933, expresses his opinion on the types conditions that meet the grounds for sterilization.
241 Friedlander (2002), 60.
242 Tenenbaum, 91n.5, citing the Testimony of Dr. Walter E. Schmidt. See also Gallagher, H. G. (1990). By Trust Betrayed, 11–13, for commentary, and 57–58.

Wertham (1966), Gallagher (1990) and Weiss (2010) expressed a colourful point that these mass killings were not linked to euthanasia whatsoever. The term implies a "*good death*" (eu='well;' thanatos'=death), denoting a merciful and compassionate act. Here, it was a state-driven action that set in motion a systematic and covert, meticulously planned "*euthanasia*" operation carried out under the cover of war.[243] They were not merciful deaths but merciless murders. As aforementioned, the unmanageable numbers housed within the segregation centres laid the groundwork for what the regime framed as a more "*efficient*" method of mass murder: they were cast as a financial burden on the state, a drain on resources, unfit for productive labour, and occupied hospital beds urgently needed for wounded soldiers—an exigent need that demanded the freeing of space and personnel.[244] Weiss (1987a) recorded a similar sounding statement, citing a statement the preface to the journal *Eugenik* in which she quoted—

> [U]seless individuals unworthy of life are maintained and taken care of in institutions at the expense of the healthy—of whom a hundred thousand are today without their own place to live and millions of whom starve from lack of work.[245]

It began with the preparation of questionnaires to be completed by psychiatric experts of recognised scientific standing. This formed the selection process for both adults and children patients destined for the T4 euthanasia programme at six designated extermination centres. On the questionnaire (also referred to as the registration form) they marked each case with a red plus or a blue minus to indicate whether the patient was to live or die. Selected patients were transferred to euthanasia centres in Brandenburg, Chelmno, Hadamar, Grafeneck, Bernburg, Hartheim and Sonnenstein, where they were gassed to death with carbon monoxide.[246] Friedlander (2002) reports that these institutions were created to murder with chilling efficiency, operating as an assembly line of death, later described as "*human abattoir[s]*" by an eminent judge of the United States Supreme Court.[247]

The gas chamber at the psychiatric institution in Hadamar, for instance, was located in the basement and deliberately constructed to resemble a shower facility. It was fitted with chairs and benches, yet emitted carbon monoxide under the deceptive label 'disinfection.' Those classified under Division II. in Table I-9 included, but not limited to individuals, institutionalised patients, disabled persons, individuals with intellectual impairments, healthy children labelled by teachers troublemakers or nuisances, and residents of children's homes.[248]

Indeed, they fabricated medical procedures under internal code names and euphemisms such as "*disinfection,*" "*special treatment,*" and "*cleansing the German body politic,*" all operating under the covert designation "*Community Foundation for Institutional Care.*" Gallagher (1990) writes that this is a professional practice used to make unacceptable taboo practices more acceptable by rebranding them with a new name, what psychiatrists call 'detoxification.' Thus, "*[t]he euphemism is used to detoxify*

243 Gallagher, 59, note at asterisk; Weiss (2010), 172; Wertham, 155.
244 Weiss (2010), 173. See also Bruns (2014), 222; Proctor (1988), 182; Schafft, 152-54, 158, *161, 167.
245 "Geleitwort," Eugenik, 1930, I, unpaginated, in Weiss (1987a) 224.
246 Friedlander (1995), 46, 83; Rotzoll, M. et al. (2006). The First National Socialist Extermination, 20; Schafft, 159-63; Weiss (2010), 173-74; Wertham, 169-*171.
247 Friedlander (2002), 62; Fedorenko v. United States, 449 U.S. 490, 494 (1981).
248 Gallagher, 14-19; Proctor (19988), 185-91.

the poison."[249] In physician committee meetings, mass extermination was described as "*negative population policies,*" and the verbal code for the sites where children were murdered was "*Specialist Children's Wards.*"[250] The killing hospitals (euthanasia centres) themselves operated under the internal cover name "*Reich Association of Medical and Care Institutions.*"[251]

Lifton (1986), Weindling (1993), and Weyers (1998), argue that this elaborate subterfuge, combined with Hitler's strategic issue of secret oral orders, was meticulously crafted to ensure that no directive could be traced to a specific document. These oral instructions were deliberate manoeuvres, intended to evade public scrutiny and to shield both the Führer's Chancellery and Hitler personally from direct implication in the procedures he had authorised.[252] More than this, the strategy served two underlying objectives: (i) the sanitisation or softening of language rendered morally and socially repugnant actions deceptively neutral, and (ii) it concealed the horrific nature of these actions, making it psychologically easier for perpetrators to discuss, justify, and implement such policies. For a more expansive discussion of the psychological dimensions, see The Role of Social Psychology, 256.

In instances where individuals escaped the initial culling, Hitler issued a secret order, backdated to 1 September 1939. Proctor (1992) and Rotzoll et al. (2006) reported that between 1940 and 1941, between 70,000 and 70,273 patients had been seized and subjected to *Gnadentod* (trans. "*mercy killings*") in the T4 extermination centres. Rotzoll et al. (2006) stated that these figures had come to light in 1941.¶[253] The consequent protests, and pressure from churches, Hitler created the illusion of dismantling the euthanasia programme. Moreover, military morale suffered tremendously due to pervasive rumours that wounded soldiers, having learned of their fate upon returning home, would be sent to the gas chambers. Despite the protests, Grodin (2010) explains that fervent nationalism and ironclad loyalty to the Third Reich made it nearly impossible to lodge rational objections.[254] Wertham (1966) maintains that, despite the outcries, operations were not abandoned; rather, they continued by altering their external mode of operation. Hence, the killings proceeded more cautiously, at a slower pace, and in different locations. Instead of gas, other methods were employed such as planned moderate starvation, poisoning, medication, drug overdoses of, and complications arising from these procedures[255]

Of course, such methods were represented as simple mistakes and therefore, counted as natural deaths. For example, Proctor (1988), Friedlander (2002), Schafft (2004) argue that under the auspices of 'lives not worth living,' Nazi doctors established a façade for medically killing genetically impaired children, by slowly administering poison by injection over days or a period of weeks, in order to mask the true cause of death.[256] Schafft further details how pervasive this lie became, to the extent that the entire process operated under guise of a medicine care—within a hospital wards, official

---

249 Gallagher, 59.
250 Ibid.
251 Ibid.
252 Lifton, 489; Weindling (1993), 549; Waller, 112–13; Weyers, 173.
253 Bruns (2014), 222; Proctor (1992), 23; Rotzoll et al., 22; Tenenbaum, 90.
¶Rotzoll et al. (2006) reports that 70,273 patients were killed by August 1941, Proctor (1992) cites a slightly lower figure, estimating around 70,000 victims within the same timeframe. This small discrepancy may stem from differences in source data or rounding, but both figures Emphasise the scale of the atrocities committed under the T4 Program.
254 Grodin (2010), 61.
255 Bauman, Z. (2000). Modernity, 217; Bruns (2014), 222; Friedlander (2002), 60–61; Rotzoll et al., 22; Weiss (2010), 173; Wertham, 183–84, 187–*88.
256 Friedlander (2002), 60–62; Proctor (1988), 115, 187; generally, chap 7, 177–222; Schafft, 159.

transport vehicle, mobile ambulances, or other healthcare providers—presented as routine medical activity. Therefore, it was increasingly difficult to distinguish euthanasia from a natural death—

> In order to keep the victims quiet once they arrived in these centers, a false sense of security was established through a doctor's supposed examination of the naked patient. The doctor at this time established a false reason for the death that was about to occur. Pushed into the gas chamber, the victims suffered a slow and excruciating death from carbon monoxide that lasted as long as twenty minutes. A death certificate was prepared with the previously established fictitious cause of death and also a fictitious place and date.[257]

In sum, it was precisely the same for Jewish people who died in the ghettos and death camps. German physicians concealed the true causes of death in typhus cases because camp authorities refused to take any action to control the epidemic. As a result, they erased evidence of outbreaks through false diagnoses or false negatives in death records—attributing deaths to "*natural causes*" such as cancer, heart disease, or other so-called natural conditions.[258] One account records: "*Patients without family connections or had incurable conditions were simply killed.*"[259] Essentially, even a generally healthy individual who happened to contract a cold or influenza during the winter months, and did not recover quickly, one's fate would be hanging in the balance.[260]

257 Schafft, 161. See also Baumslag, 42, and Müller-Hill, 69.
258 Alexander, L. (1949). Medical Science, 39–40; Baumslag, 60, 71–72.
259 Schafft, 162.
260 Ibid., 161.

PART II. CHRONICLES OF ANTI-SEMITISM: A JEWISH PERSPECTIVE

"Hatred of the Jew is not the result of a of a rational process. If it were, the absurdity of antisemitic accusations would stifle it still-born. . . . The attempt to prove logically and statistically that [antisemitism is rational] can quite apparently make no impression upon minds that are blind ab initio to the all-too-evident truth of the matter."

—Trachtenberg, The Devil and the Jews

CHAPTER 6. THE BEGINNINGS OF GERMAN JEWISH HATRED : ORIGINS, DEVELOPMENT AND CONTINUITY

A. Historical Anti-Semitism: Origins and Evolution

i. The Jurisprudential Principles of Jewish Hatred

HATRED COMES IN many guises, from indirect antisemitic rhetoric that never needs to mention Jews or Judaism. This may operate in a covert mode in what may be called "*discourses of silence*" and expressed as "*. . . the hate that dare not speak its name.*"[1] Another type clothes itself in politeness and counterfeit sincerity, falsely attributing crimes or conspiracies to Jews by portraying them as manipulating institutions such as the UN, left-wing political regimes, world leaders, international media outlets like the BBC, and global corporations. The manifestations of hatred may be understood in three broad phases, though these are not intended to be exhaustive:[2] (i) denial, (ii) propagation of hatred, and (iii) normalisation. Beneath them lies a more elemental form: plain, unadorned hostility requiring no narrative at all.

The first phase, denial, targets the past through Holocaust denial and its distortions (secondary antiemitism).[3] This consists of employing antisemitic tropes that relativise or downplay the memory of the Holocaust.[4] The second, propagation, accelerates hatred through deliberate promotion and propaganda. This involves the repeated use across everyday speech and public discourse of both ancient and modern antisemitic

[1] Grobman, A. (2011). License to Murder, 6–9 (foreword); Wodak, R. (2007). Pragmatics, 207, 208, respectively; Jikeli, G. (2009). "Jew" as a Slur in German and French Today, 210.

[2] One fundamental dimension of antisemitism was the academic and ecclesiastical practice of maligning Judaism. For an in-depth analysis, see Wiese, chap. 3:110–58. Theological antisemitism falls outside the scope of this paper. Wiese argues that, during the 1890s, covert and misleading scholarly critiques of rabbinical literature and Jewish scholarship contributed to intensifying hostility toward Jews and heightened social tensions. This erosion of Jewish apologetics was further exacerbated by libellous writings, "were exploited for antisemitic purposes," primarily those associated with Fritsch, Rudolf Kittel, and their followers. (Wiese, 254; Shavit, Y. & Eran, M. (2007). The Hebrew Bible Reborn). Borrowing Wiese's phrasing, such works "fanned the flames of hatred and distrust against the Jewish population and aggravated the social conflict" (Wiese, 110, 155). Consequently, Jewish religious law was frequently invoked in polemical discourse as evidence for portraying Jews as criminal. These critiques formed part of a wider network of social, theological, and political hostilities—a complex web of factors, with Liberal Judaism being a contributory factor (Wiese, 291). Efforts to normalise public disdain toward Jews further intensified these tensions (Cohen, N. W. (1984). The Challenges of Darwinism). See also Black, E. (2003). War Against the Weak, 139; Cohen, N. W. (1984). The Challenges of Darwinism; Ericksen, R. P. (1985). Theologians Under Hitler; Gerdmar, A. (2009). Roots of Theological Anti-Semitism, 608–610; Gotzmann, A. & Wiese, C. (2007). Modern Judaism; Greene, J. C. (1959). Darwin and Religion; Montgomery, J. A. (1914). 'Present Tendencies; Wiese (2005), ch. 4:160–215 and chap. 5:217–85.

[3] Gerstenfeld, M. (2006). Anti-Semitic Conspiracy, 173; Rensmann, L. (2017). Guilt, Resentment, and Post-Holocaust Democracy, 6; Rensmann, L. (2020). The Contemporary Globalization.

[4] Rensmann, L. (2017). Guilt, Resentment, 6, 13–15; Rensmann (2020).

stereotypes,[5] often regarded as innocent or harmless, such as "*wandering Jew,*" "*Jewish power,*" or "*Jewish money.*"[6] Wodak (2018) linguist and distinguished Professor and Chair in Discourse Studies, summarises these discursive approaches—

> Moreover, it is important to emphasize that there are several antisemitisms—*(i)* racist, capitalist, cultural, religious, or syncretic; *(ii)* Muslim or Christian; *(iii)* left- or right-wing; *(iv)* "old" or "new"; *(v)* traditional, structural, or secondary; *(vi)* hard-core or latent; *(vii)* explicit or coded; *(viii)* and soft or violent, *(ix)* the latter resemiotized in physical acts of hatred (references in text omitted).
>
> . . . .
>
> *(x)* . . . political ("Jewish world conspiracy"), *(xi)* secular (usury), *(xii)* religious ("responsible for the death of Jesus") and *(xiii)* racist ("Jewish character").[7] (Emphasis added)

In addition, myths and falsehoods are propagated through literary fabrications such as the *Protocols of the Elders of Zion* (the "Protocols"), all with the aim of devaluing and dehumanizing the Jews, their portrayal as social outcasts, and the projection of blame onto the victims themselves. This is a recurrent characteristic of genocidal propaganda and functions to incite hatred and violence.[8] A stark example is the antisemitic propaganda film Der Ewige Jude (The Eternal Jew, 1940), produced under Joseph Goebbels and directed by Fritz Hippler, which depicted Jews as "*dirty Talmud Jews,*" "*. . . morally bankrupt, dirty, lazy, ugly conniving, and disease-bearing rats,*" "*. . . subversive betrayers and corrupters of Christian nations . . . .*"[9]

The third phase, normalisation, embeds it within the fabric of society. Taken together, these dynamics form the preamble to corporate antiemitism. This is a skilful process designed, on one hand, to consolidate and activate latent antisemitic stereotypes in the public mind, while on the other, (i) turning neutral individuals into anti-Semites, and (ii) weakening strong pro-Israel or pro-Jewish allegiances until they capitulate. The goal is to inoculate minds with hatred.

ii. Hatred transmission to the Medieval Era, 486 BCE–1500 AD

HATRED FOR JEWISH people has always existed throughout human history, and in particular since the third century BCE onward.[10] In fact the Book of Esther, in the Bible (c.500 BCE to 201 BCE), reflects how the Persian government (modern-day Iran), under the command of the royal official Haman, sought to annihilate the Jews through legislation calling for their global destruction. Around the same period, the Hellenist dynasty sought their destruction. Then in the

5 Bledin, K. (2021). Further thoughts, 535; Lockwood, E. (2021). The Anti-Semitic Backlash, 265; Rensmann, L. (2017). Guilt, Resentment, and Post-Holocaust Democracy; Wodak, R. (2018). The Radical Right, 65–67.
6 Grobman, 18–19; Rensmann (2017).
7 Wodak (2007), 62, 63, respectively. See also Lockwood, 265.
8 Baumgarten, M. et al., (Eds.). (2009). Varieties of Anti-Semitism, 293; Bering, D. (2002). Gutachten über den antisemitischen, 173–86; Grobman, A. (2011). License to Murder, 13; Huggler, J. & Ensor, J. (2014) [online]. Anti-Semitism on the March; Jikeli, 211, 215–17, 226; Michael, R. (2006). Holy Hatred, 118, 125, respectively; Savage, 405; Wodak, R. (2007), 214, 217, 221.
9 Drawn from these sources: Rosenbaum, R. A. (2010). Waking to Danger, 166; Michael, R., 125, 145, respectively; Sutherland, A. (2006). I Was A Stranger, 35.
10 Ettinger, 208; See generally Feldman, L. H. (1997). Anti-Semitism in the Ancient World.

second century BCE, under Seleucid rule, King Antiochus Epiphanes IV desecrated the Jewish Temple by sacrificing a pig on the altar and transforming it into a site of pagan worship.[11] Later, the Romans, likewise targeted the Jewish people and their sacred spaces, culminating in repeated efforts to suppress Jewish religious life and national identity.[12]

The pivotal period of anti-Jewish activity in Europe spans from the reign of the Roman Emperor Constantine the Great (306–337 AD) to the time of the Crusades (c.1095–1291). Constantine issued a series of imperial edicts and laws, including prohibitions on intermarriage, restrictions on the construction of new synagogues, and limitations on Jewish participation in certain forms of employment. These measures were rapidly followed by a widening theological divide and the severance of Jewish ties with Church bodies, shaped in part by Augustine's (354–430) perspectives on Jews as articulated in his corpus of writings and sermons.[13] Anti-Jewish stereotypes persisted throughout the Middle Ages/Medieval Era (c.1000–1500 AD), a period marked by significant violence and persecution, crucial "*. . . for the overall study of the history of anti-Semitism.*" Langmuir (1990) argued that "*[s]ocially significant antisemitism first emerged in medieval Christendom, and it became ever more deeply rooted as the Middle Ages drew to an end.*"[14]

The legacy of animosity from earlier centuries was consolidated with the establishment of Christendom in the first centuries AD. Chazan (1997a) elucidates Christians of the Middle Ages were heirs or beneficiaries to a complex legacy[15] marked by "*. . . discrimination and persecution, . . . on social, cultural, or religious grounds.*"[16] This legacy intensified starkly between c.1182–1189, when 15-year-old Philip Augustus expelled Jews who had been settled in France for over a thousand years.[17] A second expulsion occurred again in 1306 by King Philip IV (Philip the Fair) expelled the Jews from France as a result of blood libel accusations. A third expulsion followed in 1394, after which no Jews remained in France.[18]

In the 1200s, the Fourth Lateran Council of 1215, convened by Pope Innocent III, initiated a persecution campaign against the Jews, portraying them as symbols of stubbornness and rebellion against the Church. Christians were urged not to engage in debate but to kill them. Additionally, Innocent forced Jews to wear a distinctive patch with an image of an ape on it, reminiscent of the identification measures imposed by the Nazis.[19] Between the 1100s and 1200s, anti-Semitism broke out, with pogroms in Worms, Cologne, and in Mainz, killing between 900 to 1100 Jews, without sparing women or children.[20]

In 1275, residents were prohibited from residing in any city without formal registration, enabling authorities to monitor and regulate daily life. Permissions and

[11] Feldman, 16; Goodman, M. (2008). Rome and Jerusalem 366–78; Grobman, 15; Schäfer, (1998). Judeophobia, 180–95; Webster, N. H. (1924). Secret Societies, 13, 17–18, 370–74.
[12] Goodman, 366–78.
[13] Bledin, 535; Marcus, J. R. & Saperstein, M. (1999). The Jew in the Medieval World, 3–8; Thomas, J. M. (2010). The racial formation, 1741–43.
[14] Langmuir, G. I. (1990). History, 305.
[15] Chazan (1997a), 54–55; Berger, 5.
[16] Chazan, R. (1980). Church, State, and Jew, 197–98.
[17] Marcus & Saperstein, 27–30; Thomas, 1745.
[18] Marcus & Saperstein, 27; Netanyahu, B. (1995). The Origins, 833; Poliakov (1965), 173, 176.
[19] Bethencourt, F. (2013). Racisms, 30–32; Berger, 4–6; Cohen, 35–38; Heng, 335–36; Le Goff, J. (1988). Medieval Civilization, 315, 317; Woolf, J. R. (2012). The Devil's Hoofs, 53; Weinberg, R. (2012). The Blood Libel, 275. See also Marcus & Saperstein, 153–58, for other Pope Innocent point.
[20] Le Goff, 318; Male, E. (1958). The Gothic Image, 193; Chazan (1980), 117–18; Chazan, R. (1997b). Medieval Stereotypes, 7–8; Woolf, 52.

licences became compulsory for establishing or changing one's place of residence, facilitating expulsions and the formation of segregated ghettos.[21] The fundamental principles of enforced separation, a recurring feature of later segregation policies, reflects what Nirenberg (2002) characterises as "*. . . a crisis of classification and identity . . .*" during the Aragonese dynasty in 1393. The difficulty of distinguishing converts to Christianity from practicing Jews led Ferdinand II, King of Aragon (1452–1516), to foster segregation and visible marking as the solution. These measures resurfaced between 1412 and 1415, propelled by St. Vincent Ferrer in collusion with the papal court in Avignon.[22] This policies sought to eliminate ambiguity in religious identity and, in doing so, consolidate social order and political control.[23]

Cohen (1981) argued that profound ideological changes in the anti-Jewish polemic, "*. . . came only in the thirteenth century [a crucial time in ecclesiastical history] with the inquisitorial and missionary efforts of Dominican and Franciscan friars,*" accompanied by pogroms on the road to the First Crusade.[24] During the 1300s, even Christians whose beliefs and values diverged from the dominant view supporting Jewish rights faced persecution and social ostracism.[25] In Spain, throughout the 1400s, Jews were forcibly converted to Catholicism. Yet purity-of-blood statutes stipulated that anyone with even a single Jewish ancestor was not regarded as a genuine convert.[26] By 1492, King Ferdinand and Queen Isabella expelled Jews from Spain.[27] In the words of Langmuir—

> By the end of the fifteenth century, Jews had been expelled from most of western Europe; where they were not, they were isolated in ghettos to protect Christians from them. Those expulsions were accompanied by the new wave of massacres that began at the end of the thirteenth century. [28]

The form of racism that emerged served several functions: (i) to regulate participation in, and segregation from, mainstream society, (ii) to distinguish converts from non-converts, and (iii) to enforce the separation based on Jewish ancestry through the requirement certification of "*purity of blood.*"[29] See Table I-8, 143 and discussion at heading Legislation and Genetic Health Courts, 153. The historical echoes of this concept reverberated through the Nazi ghettos. The notion of "*purity of blood*" stemmed from a rumour dating to the 1200s claiming that Spanish noble lineage could be traced to eighth-century heroes. This belief required individuals to prove the "*. . . purity of his pedigree on the evidence of blood.*"[30] As these misconceptions proliferated, older slurs were reworked and new ones introduced. Scammell (1991) cited a chronicle positing

---

21 Heng, G. (2011). The Invention, *334–35.

22 Nirenberg, D. (2002). Mass Conversion, 10–11.

23 Chazan, R. (1997a). Medieval Anti-Semitism, 49, 54–55; Katz, J. (1980). From Prejudice, 13; Poliakov, L. (1965). The History, 41–72, 127. See also Feldman, L. H. (1997). Anti-Semitism in the Ancient World.

24 Berger, D. (1997). Anti-Semitism, 8; Browning, 3; Cohen, J. (1982). The Friars, 32–33; Cohn (1981), 21–22; Poliakov (1965), 41–72.

25 Cohen, J., 261.

26 Friedman, 3–8, 16.

27 Browning, 3; Chazan (1980), 79; Cohn (1981), 22; Netanyahu, B. (1995). The Origins of the Inquisition.

28 Langmuir, 304. See also Browning, 4.

29 Popkin (1974a), 126–27; Hannaford, 100–01. See also Bejarano, L. G. (2012). Anti-Semitic Discourse; Bildhauer, B. (2003, 2020). Blood, Jews, and Monsters; Friedman, J. (1987). Jewish Conversion, the Spanish Pure Blood Laws; Kaplan, G. B. (2012). The Inception of Limpieza de Sangre (Purity of Blood); Nirenberg, D. (2002). Mass Conversion; Nirenberg, D. (2008). Race and the Middle Ages; *Thomas, J. M. (2010). The Racial Formation.

30 Friedman, J. (1987). Jewish Conversion; Hannaford, Race, 123.

that "*[b]y the mid-1500s, anti-Semitism had reached an unprecedented and hysterical intensity. . . .*"[31] Hints of "*scientific racism*" emerged with the claim "*. . . that the defects of Jews were biological, 'hereditary in them', and transmissible,*" expressed in the writings of Diego de Simancas in 1575. Scammel further argued that a child of pure blood suckled by a Jewess would be irrevocably tainted.[32] As these ideas circulated, Jews were accused of every imaginable offence linked to blood.

During the Renaissance, venomous stereotypes and accusations were levelled at Jews. They were portrayed as agents of the devil, assigned phenotypical traits associated with demonic figures, branded as devil-worshippers, and demonised as the devil incarnate in human form. Jews were further maligned as "*Christ-killers,*" accused of practising witchcraft and sorcery,[33] and falsely linked to ritual murder, notoriously through the "*blood libel*" myth—a subset of ritual-murder allegations. Nirenberg (2008) argued that such beliefs laid the groundwork for the doctrine of limpieza de sangre (trans. purity of blood), which held that cultural traits were inseparable from physical lineage. In this polemical tradition, he wrote, Jews were cast as "*. . . the lineage of demons and of monsters, the mule and the sow their adoptive mother.*"[34]

According to the blood-libel myth, Jews were accused of murdering Christian children to use their blood in Passover rituals. Notably, in 1144, Norwich, UK, witnessed "*. . . the first known case of the ritual-murder . . . [w]hen a mutilated corpse of a Christian youngster was discovered outside the town . . . .*" This led to baseless assumptions that local Jews were culpable.[35] The alleged motive was attributed to Kabbalistic teachings, and the attendant belief that Jews could obtain sacred or magical powers through the use of a child's blood.[36] Matteoni (2008) found that—

> [B]lood [was considered] as a magic strengthening drink, capable of giving back health and youth . . . therefore, [it] has a double function—that of consuming a living being to empower another.[37]

As a sidenote, Trachtenberg (1983) observed that the inversion of the blood-libel myth appeared in 1784, when two women, believed to be witches, were said to have murdered a Jew to use his blood for magical purposes.[38] The first recorded instance of Jews being falsely accused of blood libel in Europe occurred in Blois, France, in 1171, following the discovery of a Christian child allegedly murdered during Passover.[39] Accusations of cannibalistic practices soon proliferated, including assertions that Jews used Christian children's blood in the baking of unleavened bread, drank it as a universal remedy for ailments, diseases, and even male menstruation.

---

31 Scammell, G. V. (1991). Essay and Reflection. 505.
32 Ibid.
33 Bildhauer, B. (2003, 2020). Blood, Jews, and Monsters; Cohn (1981), 41; Ettinger, 214; Reichmann, 56; Thomas, 1747; Trachtenberg, 11-31, 44–53, 196–216. See also the above and below footnotes. See generally Poliakov (1965), 137–45, for demonology.
34 Nirenberg, D. (2008). Race, 82.
35 Chazan (1997a), 57 et seq, 61; Gilman, S. L. (1987). The Struggle, 303; Johnson, W. (2012). The myth, 279–880; Marcus & Saperstein, 135–141; Matteoni, F. (2008). The Jew, 185; Poliakov (1965), 58, full discourse, 56–64; Weinberg, 275–85.
36 Cohn (1981), 42; Fredrickson & Camarillo (2002, 2015). Racism, 53.
37 Matteoni, 184. See also Trachtenberg,134–35.
38 Trachtenberg, 140–41. See also Matteoni, F. (2008). The Jew, 182.
39 Marcus & Saperstein, 142–46; Poliakov (1965), 60; Matteoni, 182–97; Trachtenberg, 124–155.

The "*. . . first recorded allegation of Jews requiring Christian blood as a cure was raised in the German town of Fulda in 1235.*"[40] The lore of blood was imagined as a cure for leprosy, various congenital conditions, rejuvenation, and even as an aphrodisiac or a source of vitality.[41] Accordingly, blood played a duplicitous role, simultaneously viewed as both cure and disease, as Matteoni (2008) observed.[42] The blood libel gained prominence in the 1100s, spurred by incidents such as the discovery of a child's corpse in England. Further accusations followed in France and Germany, and with the advent of the printing press, the myth spread widely across Europe and beyond. Between 1473 and 1475, the death of a toddler named Simon of Trent triggered a sensational trial against Jews in Italy, leaving enduring repercussions.[43]

Between the late 1400s and early 1500s, antisemites appropriated the concept of purity of blood and revitalised anti-Jewish hostility by intertwining misogynistic beliefs about women's menstrual blood with the notion of male menstruation. This fusion portrayed Jewish men as diseased, perpetually menstruating or haemorrhaging, centred on the idea of purging impure, polluted blood.[44] This concept persisted into the 1800s, expanded into German and Austrian folklore, linking the stereotyped elongated Jewish nose with sexual difference, in particular, male genitalia. Wilhelm Fliess (1858–1928) and Sigmund Freud later advanced the idea that the pathology of the circumcised penis could be altered through surgery on the nose.[45]

More still, according to antisemitic thinking, the nose was not merely a sign of difference; it was a feature that could not be altered or reformed—"*. . . [t]he Jewish type is indestructible,*" Gilman (1998) wrote, because it was believed to embody an immutable, pathological nature.[46] Expanding on these antecedents of blood impurity and their cumulative thematic development, a new form of antiemitism emerged, forging a nexus between stereotypes of dirtiness with the depiction of Jews as carriers of disease and contagion.[47]

Lindenthal (2014) maintains that the next link in this chain was the claim that their "*. . . eradication [which] would serve to improve the health status of the rest of the world.*"[48] See the Nazi period discussion at Racial Hygiene at its Zenith, 160. These accusations had earlier roots between the 1200s and 1500s. Fishberg (1911) and Poliakov (1965) found that Jews were accused of deliberately spreading the "*Black Death*" in Europe, including Germany, during the years 1347–49. They were portrayed as emissaries of Satan, allegedly using themselves as biological weapons to poison wells and springs,

40 Bildhauer, B. (2020). Blood, 195–96, 197; Cohn, N. (1993). Europe's Inner Demons, xi, 35–41; Kruger, S. F. (1992). The bodies, 303, cited in Thomas, 1745; Gilman (1992), 235; Heng, 335; Johnson (2012), 293; Matteoni, 185, 191–92; Poliakov (1965), 57, 142–43.
41 Gow, A. C. (2003). 'Sanguis Naturalis,' 129–58; Matteoni, 192; Newall, V. (2004). "The Jew as a Witch-Figure," 95–129; Strack, H. L. (1909). The Jew, 30–42, 50–65, 132–146.
42 Matteoni, 192.
43 Browning, C. (2004). The Origins, 5; Chazan (1997a), 56, 62–33; Grobman, 15–19; Cohn (1981), 25; Gerstenfeld, 172; Hannaford, 116, 123; Montagu, 21; Mosse, G. L. (2000). The Jews, 195–96; Matteoni, 186; Nirenberg, D. (2014).170–71; Poliakov (1965), 123–69; Resnick, I. M. (2000). Medieval Roots, 242–43n.4. Resnick traces the origins of the brutal murders of children for Pesach, to Thomas of Monmouth, in the year 1171. Resnick noted that he helped to popularize Pesach accusation in his book, '*Life and Passion of St. William, the Martyr of Norwich*;' Voegelin, 308–10; Weinberg, R. (2012),3. See also Jessopp, A. & James, M. R. (1896). The Life and Miracles.
44 Beusterien, J. L. (1999). Jewish Male; Gilman (1991), 58; Gilman (1992), 235; Heng, 335. Johnson (2012), 273–78, 286–95; Matteoni, 190–96.
45 Gilman, S. L. (1987), 299–302; Gilman, S. L. (1998). Creating Beauty, 84–91. See also Lindenthal, J. J. (2014). The Epidemiological Status, 15.
46 Gilman (1998) 72, 74, 82, generally, 72–83.
47 Friedman, 17–18; Savage, 407.
48 Lindenthal, 15.

thereby causing the epidemic.[49] Metcalfe (2002) quotes a well-known documented statement associated with this legend—

> Even though bacteria and viruses were unknown, the practice of poisoning enemy wells with bodies of dead soldiers to spread diseases is as old as war itself.[50]

In sum, this thematic development persisted into the 1800s, linking the Jew's distinctive foot to the cloven-footed devil and disease, reflecting early modern European beliefs connecting physical traits to demonic influence and illness[51] Within a broader matrix of limited toleration, which granted Jews only a conditional and precarious existence, stereotypes conflated accumulated myths, wilful ignorance, and the perception of Jews as perpetual threats and menaces, all of which pervaded European culture.[52] During plague outbreaks, fearful citizens resorted to burning Jews en masse, a precursor to later genocidal logics that resurfaced in Nazi Germany. [53] This lie also fuelled the further accusation that Jews were manufacturers of biological weapons.

These caricatures formed the building blocks of a worldview that ultimately enabled genocidal thinking, or better put in Nirenberg's (2008) words, "*. . . medieval ideologies about Jews [were] early coordinates in a trajectory clearly spiralling toward destruction.*"[54] In Germany, these revived medieval tropes were mobilised directly by Nazi leaders, forming a preparatory step for the Protocols and then for the Holocaust. Weinreich (1999) reported that demonology doctrines, the blood libel and ritual murder, ideologies of Jewish world domination, notions of hereditary Jewish criminality, and portrayals of Jews as disease carriers were all revived in Nazi Germany by its leaders.[55] These revivals informed the rhetoric of the "*Jewish Question,*" the "*Jewish Problem,*" and ultimately the "*Final Solution.*"

---

49 Fishberg, 276–77; Gilman (1991), 221; Gilman (1992), 224; Marcus & Saperstein, 49–55; Poliakov (1965), 104–05, 112–13. Trachtenberg, 97–108. See Poliakov (1965) for Jewish discourse on the 1300s, 101–122. For a broader discussion with emphasis on Germany, see Staub (1989) 100–04.
50 Cohn, S. K. (2007). The Black Death; Metcalfe, N. (2002). A Short History, 272, quoted in Baumslag, 204; Nikiforuk, A. (1993). The Fourth Horseman, 47 et seq, 160.
51 Gilman (1991), 39–59; Gilman, S. L. (1992). The Jewish Body : A foot-note.
52 Browning, 3; Chazan (1997a), 54, 63–64; Review of the book [*The History of Anti-Semitism*,1979].
53 Nikiforuk, 47 et seq; Savage, 407.
54 Nirenberg (2008) 73.
55 Weinreich, 202–07.

CHAPTER 7. THE PROTOCOLS OF THE ELDERS OF ZION AUDIT TRAIL : CENTURIES OF ACCUSATIONS

A. German Racial Hygiene subordination of Jewish hatred—preparatory content for the Protocols, 1600s–1900s

Following the expulsions of 1200–1500, European societies underwent major transformations from the 1500s to the *fin-de-siècle*. The Enlightenment shaped in part by figures such as Voltaire, ushered in a multi-phased intellectual revolution.[1] Indeed, Mosse (2020) argues that when looking at the history of European racism in particular, it is generally acknowledged that its origins lay in the intellectual shifts of the Enlightenment in the 1700s—

> [The study period] must be seen as originating in the eighteenth century . . . It was in the eighteenth century that the structure of racial thought was consolidated and determined for the next one and three-quarter centuries.[2]

Mosse's (2020) concurs with Berger's (1997), emphasising continuity: "*[a]s the Middle Ages draw to a close a new spectre began haunting the Jews of Europe—the spectre of demonology.*"[3] Langmuir's analysis similarly highlighted that "*[s]ocially significant antisemitism first emerged in medieval Christendom . . . became ever more deeply rooted as the Middle Ages drew to an end.*"[4] In other words, the pogroms evolved, transitioning into a more intellectualised, literary form during the 1600s, reflected in early racial classification theories, most notably those of François Bernier in 1684.[5] Thomas (2010) found that as early as 1671, in Brandenburg, Germany, Jews were prohibited from constructing synagogues and were berated as daily abusers of Christ in their temples,[6] echoing the Renaissance-era diatribes of Christ-killers.

After the close of the medieval period, Langmuir (1990) argues that hostility toward Jews declined between roughly 1500 and 1800.[7] It is not entirely clear, however, whether

1 Langmuir, 318; Nordenskiöld, 130–31. For a general discussion about the Enlightenment, see Stuurman, S. (2017). The Invention of Humanity.
2 Mosse (2020), xxiv, xxvii, xxxix, 3, respectively.
3 Berger, 7–8; See also Chazan (1997a), 61; See also Cohn (1981), 22–23.
4 Langmuir, 305.
5 Arendt (1944), 43n.12.
Arendt noted that the finding of Ernest Seilliere, that Francois Hotman, French 16th century author of *Franco-Gallia*, was the forerunner of 18th century racial doctrines, a misconception, corrected by Theophile Simar in *'Etude Critiaue sur la Formation de la doctrine des Races au 18e et son expansion au 19e siecle (Bruxelles, 1922)'* 20.;
Hannaford, 158, 409n.11; Stuurman (2000), 2. Cf. Schafft, 205, re her point about the fictitious schema of racial classification.
6 Thomas, 1750.
7 Langmuir, 318; Vartanian, A. (2015). Diderot and Descartes, 6.

he meant a reduction in overt violence or a shift toward more covert, subdued forms of antiemitism. The persistence of inherited prejudices across successive epochs is crucial to recognise.[8] Browning (2004) suggests that the temporary pause in anti-Jewish virulence noted by Langmuir stemmed from several developments: (i) the Renaissance, (ii) humanism, (iii) the gradual secularisation of early modern Europe, (iv) the fracturing of religious unity during the Reformation, (v) the scientific advances of Galileo and Newton, and (vi) the Enlightenment. Browning added that "*[t]he French Revolution signalled the emergence of liberalism and nationalism, and the Industrial Revolution [inaugurated] economic and social transformation.*"[9] He applied the term "*Dual Revolution*" to these combined forces of the late 1700s and early 1800s, which proved to be a significant boon for European Jewry.[10]

How did German Jewry move so dramatically from the promise of liberty to the emergence of national hatred? Hannaford (1996) postulates that the key turning points "*. . . were the defeat of Napoleon in 1805, the influences of Darwin from 1859 onward, and the defeat of France in 1871.*"[11] Although the Enlightenment marked an important historical shift, it was not the decisive one. It provided the philosophical basis for emancipation, while the French Revolution (1789–1799) translated these ideals into laws promoting liberty and equality across Europe. For Jews in Germany, the French Revolution, an outgrowth of Enlightenment thought, initially opened the door to equal citizenship in various European states.[12] Yet this progress was met with resistance within Germany, where anti-Jewish hatred metamorphosed into an unusual alliance of disparate groups united by a shared hostility toward Jews.[13]

The medieval antiemitism in Germany stemmed in part from Germany's spiritual father, Martin Luther (1483–1546). His pernicious 1543 treatise, '*Against the Jews and heir Lies*', endured across the centuries. In it, he described the Jews as a "*. . . heavy burden, 'like a plague, pestilence, pure misfortune in our country.*'"[14] Rather than dissipating, his polemic became a prototype for the incendiary pamphlets that proliferated throughout the 1600s and 1700s.[15] According to Tal (1975) this was a continuation of incitement to hate and violence. While it is admittedly something of a *post hoc ergo propter hoc* leap to claim that Luther's pamphlet directly precipitated the expulsion of German Jewry from Frankfurt in 1616, it is no stretch to argue that the barrage of pamphlets that followed had an inflammatory and inculcative effect that contributed to the expulsion.[16] This climate of agitation clearly fostered a culture of intolerance and aggression that paved the way for such actions. Thus, these pamphlets, building upon Luther's initial incitements, exerted an enduring and pernicious influence on German society.

Between 1816 and 1848, antiemitism had dissipated in Germany, though only marginally. During this period, Germany repudiated the civic-national ideal shaped by the Enlightenment in favour of a conception of national belonging grounded in ethnicity. According to Massing (1949) and Bracher (1970), the chief disseminators

---

8 Katz (1980), 13, 664; Langmuir, 305.
9 Browning, 4.
10 Ibid., 4–5; Fredrickson & Camarillo (2002, 2015), 64–65.
Although this appeared to be an extraneous respite, it was actually a pivotal time to found a completely new pillar of Jewish hatred. This was rationalism. This is mentioned for chronology purposes, but we will return to this discussion later.
11 Hannaford, 316, citing Mosse, but no reference to the source.
12 Gordon (1984), 7; Israel, J. I. (2006). Enlightenment, 3; Schleunes, 48.
13 Cohn (1981), 25; Katz (1980), 52–54; Pipes, D. (1997). Conspiracy, 27, *59–60; Massing, W. (1949). Rehearsal For Destruction, xv–xvii, 4–6; Pulzer, 25; Stuurman (2017), 347, 353–54.
14 Hilberg (1985), 14. See also Kressel, N. J. (1996). Mass Hate, 138.
15 Poliakov (1965), 216, 240–41nn.14, 17.
16 Ibid., 244; Tal, 224–25.

of the most virulent antiemitism were not the rural, blue-collar classes, as one might assume, but rather the urban, higher-educated milieu.[17] Jews were faced with a stark choice: "*[e]ither they . . . surrendered their Jewishness and become good Germans or there would be no place for them.*"[18]

However, the Enlightenment mitigated the latent antisemitic accusations of the medieval period—granting Jews partial emancipation in the form of economic, social, and cultural advantages—a deep-seated intolerance "*. . . always lurk[ed] in the background of German social structure, it could be conjured forth to serve those who understood its political potentialities.*"[19] Reichmann (1951) furthers this notion with the interesting point that such superstitions are "*. . . reawakened by an intense propaganda . . . driven back into the subconscious [to] reassert[] itself.*"[20] The benefits of emancipation and liberalism granted by the "*North German Federation of July 3, 1869, valid in 1871,*" were conditional. Jews were compelled to promise to assimilate into German society[21] and demonstrate a "*heartfelt*" conversion to Christianity. Consequently, their legal rights were "*. . . curtailed in practically all German states,*" owing to what the Prussian aristocracy interpreted as a lack of "*gratitude*" for what it regarded as an act of generosity.[22]

From the 1870s onwards, growing demands for the revocation of Jewish rights intensified, accompanied by a surge in virulent racial antisemiticism. Between 1890 and 1895, an entirely new a mass basis for antisemitic thought, as Wiese (2005) observed.[23] Little wonder that from the 1870s and 1880s—the latter half of the nineteenth century—the benefits of emancipation, including civic equality, began to decline disastrously. This was driven partly by the rise of "*respectable*" and "*intellectual*" antiemitism, which targeted mythical rather than "*real*" Jews, and thereby rendered such hostility socially and culturally acceptable.[24] Tal (1975) elaborates on this that antiemitism had become an intrinsic component of Aryan race propaganda. By February 1887, racial antisemitism was proclaimed to be "*. . . a pioneer in the creation of a new religion.*'"[25] See also heading, Broadening the Scope of Racial Hygiene, 113. These developments, influenced by scientific racism—cloaked in academic respectability, cemented antiemitism's growing acceptance among the intelligentsia—reawakened latent medieval German hatred.[26]

Thus, antisemitism resurfaced in c.1848 and later merged with Darwinist ideology and Aryan-supremacist ideals in the 1890s. (See Weimar Era Refinements, 131. Interestingly, Cohn (1981) emphasised the magnitude of this expansion, remarked that "*[w]hen, in the 1880s, antisemitism became an important political movement, Germany*

17 Bracher, 37–45; Landheer, B. & Massing, P. (1950). Rehearsal for Destruction; Massing, 47, 48, 77, 97. See Maymind, I. (2016). On the Concept of Self-Hatred.

18 Fredrickson & Camarillo (2002, 2015), 69-*71, 72; Hertz, 53; Pulzer, 7.

19 Massing, xiii. See also Fredrickson & Camarillo (2002, 2015), 71, 77–78; Pulzer, 28–29, 31–32; Reichmann, 56; Craig, G. A. (1978). Germany, 83–85, 204; Wiese, 52.

20 Reichmann, 56.an

21 For an elaborate discussion on the importance of understanding assimilation and its cultural implications in the gradual disappearance and dissolution of German Jewry, see C. Wiese, 61n.86, 62, 74, 81–82. Wiese wrote that "*. . . members of the society continuously insisted that the price of emancipation was a complete merger with the general culture, the gradual renunciation of all distinct features.*" Wiese, 68n.117, citing, B. Suchy, loc. cit., 228; references 225ff.; see I. Schorsch, loc. cit., 99. As already established in the previous section, in the Third Reich era, the concept of assimilation became irrelevant. The regime systematically pursued elimination based solely on ethnicity, determined through blood tests.

22 Krausnick (1970), 22–23; Levy, R. S. (1975). The Downfall, 22, 26; Massing, 3–5, 106–09; Wiese, 43.

23 Wiese, 46.

24 Endelman, 97, 101–02; Gordon (1984), 24, 45–46; Gutteridge, 3–4; Krausnick (1970), 19, 22–23; Levy, 11–16, 28; Pulzer, 69.

25 Tal, 266–67.

26 Kershaw (2000), xliii, n.19, citing Krüger, A. (1972). Die Olympischen Spiele 1936, 200; Marten, 35–36; Schafft, 206; Pulzer, 240.

emerged as the leading producer of antisemitic propaganda of every kind."[27] Weikart (2004) similarly maintained that many race theorists became antisemitic largely through their embrace of racial hygiene and biological racism. Although, "*[n]ot all of this was directly tied to Darwinism, but much of it received some stimulation from Darwinian social thought.*"[28] According to Weikart, antiemitism found a ready ally in the writings of numerous avid Darwinists at the *fin-de-siècle*, particularly among antisemites and the emerging bacteriologist scientists.[29] See Racial Hygiene at its Zenith, 160. Moreover, Fritsch connected Darwinian ethics and eugenics seamlessly in a published article, an association documented by Weikart—

> Morality and ethics arise from the law of preservation of the species, of the race. Whatever ensures the future of the species, whatever is suited to raise the species to an ever higher level of physical and mental perfection, that is moral. For Fritsch this was not just a minor side issue, but rather a central doctrine of his antisemitic worldview.[30]

Post-1880, antisemitism took root. These included revolutionaries, who yearned, first, for "*. . . antisemitism to catch on as a mass movement*" and, second, for political power to reorder the world in a radical way, "*. . . and only then would the Jewish race be dealt with in a final way.*"[31] This was a coalition that consisted of teachers, students, industrial and commercial employees, professionals, nature-worshippers, followers of cults, members of the "*life reform*" movements, the agrarian community, the land-owning aristocracy, and reactionary clergy.[32] The chief organiser of "*the Anti-Semites Petition,*" the revolutionary anti-Semite Bernhard Förster, succeeded in gathering a quarter of a million signatures for the petition—

i. seeking emancipation from the Jews;
ii. demanding that their civil rights be curbed;
iii. exclusion of Jews from all sectors of employment and professions (in addition to the Nazi era);
iv. a special census of the Jewish population.[33]

Even though the petition failed and without lasting influence, the revolutionary outlook it set in motion nonetheless flourished after 1900.[34] Wilhelm Marr, one of the first agitators and the "*founding father*" and "*patriarch*" of German racial antisemitism in the 1870s[35] was later followed by Fritsch, whose antisemitic writings came after Marr. In fact, Marr is widely regarded as the first to coin and popularise the term "*antisemitism*" which, "*. . . later became the sine qua non of international vocabulary.*"[36] Although Marr's 1862 pamphlet *Der Judenspiegel*, was perceived as anachronistic and attracted little attention, it did, however, plant the seed for future developments and lay the groundwork for Marr in 1873, in which he transformed anti-Jewish sentiment into racial categories.

27 Cohn (1981), 39.
28 Weikart (2004), 122.
29 Gilman (1991), 62; Weikart (2004), 123.
30 Weikart (2004), *55, 69; Weikart (2006), 107–08.
31 Levy, 29, 111, 119.
32 Bracher, 39; Massing, xiii, 75–76, 102.
33 Levy, 21, 29; Massing, 39–40.
34 Levy, 30.
35 Ibid., 16; Massing, 6, 18, 85; Bruns, C. (2011). Toward a Transnational History, 124, 125, respectively.
36 Bruns, 125; Flowers, 38; Volkov, 38–39.

Between 1873 and 1879, he published numerous pamphlets, "*. . . whose main work reached a mass audience,*" one of which was titled, *Der Sieg des Judenthums aber das Germanenthum* (trans. The Triumph of the Jews over the Germans).[37] Marr "*. . . invoked social Darwinian principles to justify a struggle against the Jews.*"[38] Anti-Semitic agitation was heavily promoted and amplified by the press, especially in Berlin.[39] Simply put, the Enlightenment ushered in the romanisation of intellectual racism—classifications articulated with philosophical eloquence and beauty. Indeed, Mosse observed (2020) that "*[t]he Enlightenment was a revolution in aesthetic and intellectual tastes and conventions.*"[40] This produced a paradox: on one hand, the Enlightenment celebrated intellectualism, liberty, scientific discovery, and romanticism; on the other, "*. . . the not-so-enlightened theories of the inferiority of non-Europeans sprung forth.*"[41] Popkin (1974b) further elaborated on this paradox. Race was posited in elevated philosophical terms defining the human race as a comprehensive classification of humanity—

> They all define man in terms of mental and psychological characteristics. Size, skin color, religious beliefs, etc. do not enter into the question of whether a given individual is to be considered human, and whether he is to be treated in certain ways which differentiate him from animals or machines. However, the same people in the Enlightenment . . . provide the bases for theories claiming that [millions of individuals] were less than men [or] less than human in mental life, because they were dark, or accepted the wrong religion.[42]

Massing (1949) made a similar point, and a somewhat curious one, expressing that, "*'[r]owdy' antisemitism had given way to 'scientific' antisemitism, which remained as yet the ideology of a rather esoteric cult of intellectual and artistic endeavor.*"[43] It is not entirely clear whether he was referring to participants in the esoteric occult fraternities, but the implication is that antiemitism had migrated into circles that prided themselves on intellectual sophistication. Once the Enlightenment had reached intellectual maturity, it influenced the formulation of two primary treatises.

The first is found in a passage from the 1967 edition of Warrant for Genocide, where Cohn (1981) emphasises that "*. . . the world-conspiracy grew straight out of these age-old superstitions . . . .*" Ettinger (2011) succinctly summarises this continuity that "*[m]odern-day antisemitism is a new version of an ancient hatred.*"[44] The second treatise traces the continuous transmission of this idea across centuries. The purpose for re-engaging the past, according to Cecil's (1972) assessment, was "*. . . to make the bald accusations more plausible by adding a wealth of historical data.*"[45] Thus, this was no new hatred or movement. The truth, however, is profoundly more complex. Just as the venomous ancient hatred was transmitted from the first century to subsequent centuries,

---

37 Levy, 16–17, 20; Marten, 37–38; Massing, 6, 18, 85 (8, the press considered Marr to be a frustrated journalist); Mosse (2000), 199; Schleunes, 24–25. Massing, 80, states that Wilhelm Marr's pampflet was usually referred to as a "landmark" in the evolution of racial antisemitism.
38 Weikart (2006), 102. See generally Mosse (1964), 130.
39 Massing, 40.
40 Mosse (2020), 3.
41 Popkin (1974b), 246; Vartija, D. (2021). Revisiting Enlightenment, 603.
42 Popkin (1974b), 246, 251.
43 Massing, 148. See 163, for the psychological and sociological aspects of antisemitism that were in the foreground of scientific interest.
44 Cohn (1967), 254; Pipes, 54, citing Cohn (1967), 254; Ettinger, 208, 213; Friedman, J. (1987). Jewish Conversion; Gordon (1984), 25.
45 Cecil, 74-75.

culminating in the First Crusade (1096–1099), so too did it perpetuate into later epochs, forming the sinister foundation to the Protocols of the Elders of Zion ("the Protocols") myths in the late 1800s.

The second treatise development is found in the medieval hatreds, which formed the basis for the final act: the conceptual development of Germany's racial-science policy in the late 1800s, ultimately directed against the Jews. The 1700s played a crucial role in shaping the intellectual foundations of scientific racism in the 1800s, and the cultural trends of that earlier century laid the groundwork for later racist thought.[46] The impact of these proto-ideological protocols on Jewish racial anthropology during that period was far from negligible. Katz (1980) for example, found that Johann Andreas *Eisenmenger's Entdecktes Judenthum* (trans. The Traditions of the Jews), published in 1710, recycled ideas indirectly nourished the antisemitic movement despite repeated attempts to ban the book.[47]

In fact, interbreeding first emerged at the turn of the 1800s, representing the primitive stages of racial hygiene policy formation, but *in extremis*, promulgated the accusation of a "*. . . Jewish racial conspiracy, which exploited racial mixing as a strategy of domination.*"[48] Within this period, the devil stereotypes from the Renaissance were exhumed, and artfully reworked and synthesised into a harsher form of racism to justify Jewish hatred.[49] See discussion earlier Hatred transmission, 181.

Demonology and devil-worship polemics, first conceived between the first and second centuries, re-emerged in the 1200s, when Cohn (1967, 1981) asserts that they revived and "*. . . developed into a coherent and terrifying demonology.*"[50] These were the old modes of antiemitism, evident in antiquity, the medieval era, and the early periods, that "*. . . never vanished from the European consciousness even during the enlightenment.*"[51] Medieval anti-Semitism dwindled but continued to be nurtured through successive stages of hatred, awaiting revitalisation with renewed force.[52] For instance, Bein (1964) writes that the concept of the Jew as a parasite was used synonymously with that of a vampire—

> The concept "parasite" merges by open or hidden association with the age-old image of the Jew, an association from former epochs of history forming . . . The devil, the blood-sucking usurer, the exploiter, the international conspirator as depicted in The Protocols of the Elders of Zion, and the mythological figure of the vampire are synonymously used with parasite, despite the fact that the original term "parasite" had nothing at all to do with those designations.[53] (Footnote call numbers omitted)

Vampirism, leeches, and parasites, of course, carry connotations of the old medieval adage of blood-libel, blood-drinking, and devil worship. Bein extends this list, drawing on Rosenberg and other depictions, that they are "*. . . pest and plagues, as locusts and leeches , as spiders and vampires, with the notion of the parasite and its connotations of*

46 Mosse (2020), 3.
47 Katz (1980), 14, 23, respectively.
48 Hutton, 66, citing Chamberlain, H. S. (1899) Die Grundlagen des neunzehnten, 2:383.
49 Resnick, I. M. (2000). Medieval Roots, 241; Snyder, L. L., 80–81; Stuurman (2017), 351, 556.
50 Cohn (1967), 258; Cohn (1981), 22.
51 Davies, 24; Heng, G. (2018). The Invention of Race, cited in Vartija, D. (2021). Revisiting Enlightenment, 603; Mosse (2000), 196.
52 Krausnick (1970), 20; Langmuir, 306–18. Mosse (2000), 195, 198. See for also Katz (1980) 23–91.
53 Bein, 18. See also Kurlander (2017a), 252–58, 283.

bacteria . . ."[54] Cassirer's (1874-1945) detailed philosophical analysis of language and its connection with myth, provided an invaluable insight into how old sayings were resurrected and imbued new force as "*magic*" or power words—

> New words have been coined; and even the old ones are used in a new sense; they have undergone a deep change of meaning. This change of meaning depends upon the fact that those words which formerly were used in a descriptive, logical, or semantic sense, are now used as magic words that are destined to produce certain effects and to stir up certain emotions. Our ordinary words are charged with meanings; but these new-fangled words are charged with feelings and violent passions.[55]

The Nazis embraced the world-domination conspiracy—the belief that Jews secretly ruled global affairs. The rise of the new Germany was marked by efforts to dismantle existing structures through elimination and replace them with a new leadership.[56] From its origins through subsequent conveyors and its development in the late 1800s, historical antisemitic ideologies left a sporadic trail across multiple eras. These antisemitic demonology preparations, as one scholar notes, "*. . . establish[ed] their core status in Nazi philosophy during the early 1920s.*"[57]

Indeed, the medieval charges, in addition to ritual murder "*. . . that had long been considered outmoded,*" revived in 1892, acting as an effective vector of agitation for the antisemitic movement.[58] This resurgence was also evident in 1891 in the Lower Rhenish region of Germany.[59] Coupled with other antisemitic forgeries and fabrications—the repeated recycling of the same medieval hatred refrain—the fusion of the medieval with the modern culminated in the birth of the Protocols. Additional conspiracies that bore no real relation to these latent medieval motifs, such as the assassination attempts on the Kaiser's life in 1878, simply made good propaganda to entrench these antecedents.[60]

It was a prelude, another layer in a repertoire of preparatory narratives that paved the way for the ideological construction of the Holocaust.[61] Arendt (1973) said as much, inasmuch as the Protocols was a patent forgery, it was nevertheless believed by vast numbers of people. Once it became "*. . . the text of [the] whole political movement, the task of the historian is no longer to discover a forgery . . . ,*" nor imperative to establish its validity.[62] In short, they framed or falsely incriminated the Jews for destruction by intellectual poise. Thus, it was a structural process.

For instance, this enabled Rosenberg to employ pseudo-historical methods, projecting backwards into history, to reignite a fresh wave of antisemitism, by invoking latent Jewish hatreds that had seemingly been resigned to history. Indeed, Gilman (1990) captures this disparagement mechanism with precision. Initially discussed in the context of Jewish self-hatred,[63] where the individual is cast as a "*traitor.*" The device functions by reactivating historical hatreds and repurposing them to produce contemporary

54 Bein, 23.
55 Cassirer, E. (1955). The Myth, 356. See also Bein, 5.
56 Cohn (1981), 22–23; Grobman, 34; Katz (1970), 1; Proctor (1988), 70–71, 75; Weinreich, 80, 118.
57 Goodrick-Clarke (2002), *1, 3.
58 Wiese, 110.
59 Ibid., 111–12.
60 Massing, 92.
61 Cohn (1981), 14; Davies (1980).
62 Arendt (1973), 7.
63 See Maymind, I. (2016). On the Concept of Self-Hatred, for self-hatred discourse.

anti-Jewish narratives, which are then cited as evidence of those past prejudices.[64] Its relevance here lies in how such rhetorical vectors continue to perpetuate modern-day biases. These include jokes, debates, wordplay, newspaper cartoons, and the theatre stage. Endelman (1997) aptly noted the subtlety with which this construct reinforced new caricatures built upon long-standing popular tropes—

> Charles Dickens, whose villainous Fagin is one of the best known Jewish characters in English literature, naively claimed in 1854, "I know of no reason the Jews can have for regarding me as 'inimical' to them." Dickens's Fagin and thousands of other less skilfully drawn literary and journalistic portraits reinforced the popular notion that Jews were dangerous creatures, up to no good, untrustworthy and unscrupulous. Jews everywhere came in frequent contact with these popular images, and in the last resort the discomfort they inflicted may have done more to encourage the flight from Jewishness than social exclusion and occupational discrimination.[65]

[64] Gilman, S. L. (1990). Jewish Self-Hatred, 67. Cf. Gerstenfeld, M. (2006). Anti-Semitic Conspiracy.
[65] Endelman, 103; Poliakov (1965), 79. For wordplay and jokes specifically, see Dundes, A. & Hauschild, T. (1983). Auschwitz Jokes; Hughes, E. C. (1962). Good People and Dirty Work; Linke, U. & Dundes, A. (1988). More on Auschwitz Jokes.

B. Constructing the Protocols: Influences from the Occult

i. The occult influences and migration to Germany

THE DEVELOPMENT OF the Protocols can be situated within a broader historical narrative.[66] Cohn (1967, 1981) dates the final iteration of the forgery between 1897 and 1898, a period that coincided with Germany's entrenched antisemitic cultural climate.[67] The text itself comprises 24 protocols and draws on lectures, notes for lectures, speeches, and alleged correspondence spanning back to the Constantinian dynasty onwards, all presented as spurious documentary evidence.[68]

Webb (1976) catalogues the range of historical accusations levelled against Jews for world domination, originally attributed to the Jesuits, "*... the basis of which was a tract entitled Monte Secreta, published in 1612 ....*" This involved overthrow of governments, the construction of a subterranean network to blow up the capitals of Europe, domination of trade, the seizure of financial and commercial activity, the acquisition of control over all money, and the manipulation of law-enforcement agencies to advance Jewish interests. Collectively, they reveal a coherent conspiratorial pattern that underpins the Protocols' structure.[69] Accordingly, the recurring themes structuring this schema—contribute to the Protocols' enduring characterisation as a successful forgery—may be summarised as a systematic repertoire of conspiratorial tropes—

i. the portrayal of Jews as perpetually engaged in clandestine plotting;
ii. the assertion that Jews possess an omnipresent, global reach;
iii. the claim that Jews operate behind, and exert influence over, all major institutions;
iv. the depiction of Jews as subordinated to a centralised, hidden authority—the "Elders";
v. the association of Jews with exceptional wealth, luxury, and disproportionate success.[70]

The Protocols are a nefarious spider's web, an intricately woven network implicating a wide array of actors—including German Social Darwinist ideologists, French clerics, Russian aristocrats, journalists, occultist writers, Medieval Templars, the Bavarian Illuminati, and Hitler himself—rather than Jewish people. Discussion follows below.[71] Moreover, the earliest edition of the Protocols, published in 1903 in the Russian newspaper 'Znamya', is fragmentary and internally inconsistent. The translator claims that the document originated from the Central Chancellery of Zion in France, yet the editor openly concedes uncertainty—

> [W]e do not know how, where or by what means the minutes of these meetings, which took place in France, could be copied down, nor above all who copied them ...."[72] (Alteration in original)

66 See generally Cohn, N. (1967, 1981). Warrant; Webb (1976), 213–273.
67 Cohn (1967, 1981), 113. Webb (1976), 217.
68 Cohn (1981), 60–61.
69 Webb (1976), 214, 227, respectively.
70 Pipes, D. (1997). Conspiracy, 86–87.
71 Cohn (1981), 45–47, 51.
72 Cohn (1981), 68, citing Znamya, St Petersburg, issue of 26 August 1903.

> The editors of various later translations [(French, German, Polish)] of the Protocols [built upon the originals, adding more fabricated stories].[73]

By overlaying new layers upon earlier prejudices, "*. . . they could be adapted to new circumstances, both material and spiritual,*" and "*. . . combin[ing] the worst of both its ancient and Christian antecedents.*"[74] A succession of fabrications, myths, and deranged ideological constructs was collated, organised, and consolidated within the Protocols, each element anchored in earlier traditions to lend the forgery an air of historical depth. Put differently, the Protocols emerged from the pilfering and recombination of disparate sources, produced by numerous authors across different periods, in multiple languages, dating from 1763 to 1903/4.[75]

These ideas were unreservedly embraced by many Germans and, crucially, by Hitler himself in Mein Kampf. In time, Hitler and his Nazi acolytes acquired the capacity to act as though these delusions were factual, thereby transforming ideological fantasy into state policy. Put another way, the belief that Jews were a malevolent force in the world became the formative pretext, adding a veneer of validation to his genocidal policies.[76] As Arendt (1973) wrote, this delusion was rooted in an "*. . . already existing Jewish world domination formed the basis for the illusion of future German world domination.*"[77] Germany then behaved as though the Jews were indeed dominating the world and "*. . . needed a counterconspiracy to defend itself.*"[78] Unsurprisingly, Redles (2012) found that Hitler utilised the Protocols as a guide.[79] He records Hitler's 20 April 1923 speech, entitled "*Why are We Anti–Semites,*" in which Hitler parallels, describing Jews as—

> '[D]estroyers of peoples' who attempt to use democracy and Marxism to destroy morality . . . . We may be inhuman! But if we save Germany, we have achieved the greatest deed in the world. We may do an injustice! But if we save Germany, then we have removed the greatest injustice in the world. We may be immoral! But if our people is saved, then we have once again broken a path to morality.[80]

Hence, the Protocols accomplished their singular objective—their transmission as an instructive, quasi-pedagogical manual for the masses, the foundation for German national policy and even becoming compulsory material in schools.[81] Regarding the world-conspiracy aspects, its intellectual pathology can be traced to the French Revolution, specifically to the work of the French Jesuit Abbé Augustin Barruel (1741–1820).[82] Between 1797 and 1798, Barruel produced his five-volume *Mémoire pour servir à l'histoire du Jacobinisme*, a text that supplied one of the earliest systematised conspiracy frameworks later absorbed into the Protocols' ideological architecture.

In this comprehensive study, Barruel argued that the French Revolution was the culmination of an ancient conspiracy allegedly orchestrated by secret societies, through which he accused Jews of manipulating these groups. He claimed that the conspiracy

73 Cohn (1981), 68, citing Znamya, St Petersburg, issue of 26 August 1903.
74 Chazan (1997a), 64; Katz (1980), 324. See also Davies, 20.
75 Cohn, 25–38, 56; Goldstein, 68; Grobman, 33.
76 Cecil, 77–79; Hitler (1971), 307–08; Mosse (2000), 200; Redles, D. (2012). The Turning Point, 118.
77 Arendt (1973), 360.
78 Arendt (1973), 362; Pipes, 177.
79 Redles, 118–19, 123. See also Pipes, 95, 132, 175–77.
80 Redles, 120–21.
81 Arendt (1973), 358n.41; Cohn (1981) 40; Mosse (2020), xxxvii. See also Mosse (1961), 92.
82 Cecil, 73; Cohn (1981), 23, 25.

originated with the Medieval Order of the Templars in 1314, extending his narrative back to ancient Iran, formerly Persia.[83] According to Barruel, the Order did not dissolve but instead survived, reorganised, and sought to establish a world-republic, while Jews purportedly captured the freemason order. The reverse is true. It was alleged that Jews created a literary society comprising Voltaire and other figures, yet such claims collapse under scrutiny. In reality, freemasonry has existed in diverse forms and practices since antiquity, long before any supposed Jewish collectivist influence, operating under different names across periods. If anything, freemasonry appropriated Jewish symbols and traditions, distorting them into contrived and menacing ideological constructs.

Barruel maintained that Jews had to be stopped because, in his view, this small group would ultimately dominate the world. In 1806, Barruel received one of the earliest forged documents later incorporated into the infamous Protocols—the Simonini letter. Attributed to an army officer named Jean Baptiste Simonini, whose identity was largely unknown, the letter claims that Simonini obtained information from the Jewish community through a ruse. Among its allegations, the letter asserts that the freemasons and the Illuminati were both founded by Jews, and that these orders were the driving force behind revolutionary upheaval.[84]

In 1942, Louis Darquier de Pellepoix wrote the introduction to the Protocols, reportedly remarking that he did not care whether they were true; what mattered was the vision they projected.[85] The primary authorship of the Protocols was generally the work of Russian agents c.1900, responsible for both its formation and publication. Hagemeister (2008) observed that the malevolent world-domination myth gained renewed force, noting—

> Mikhail Lepekhin had discovered who was behind the Protocols of the Elders of Zion. After five years of research in formerly inaccessible and secret Russian archives, he named Matvei Golovinskii, a reactionary journalist and writer, author of the notorious "document" that pretends to describe the secret plan of a Jewish conspiracy . . . .[86]

That said, Herman Goedsche (1815–1878), writing under the pen name Sir John Radcliffe (also spelt Retcliffe), had earlier popularised the idea of a Jewish conspiracy in his novel Biarritz (1867).[87] According to Cohn (1981), Goedsche appears to have authored a text that was the prototype for the Protocols and became the basis for an pivotal antisemitic forgery. The novel is set in the Jewish cemetery in Prague, a location that provided an ideal platform for Goedsche and other writers to embellish and sensationalise a shadowy graveyard scene filled with supposed Jewish mysteries and secret dealings. They drew in particular from Kabbalah legends, linking them to a fictional meeting of thirteen Jewish elders in this cemetery.[88] Mosse (1961) adds an interesting annotation that Kabbalah was wrongly regarded as Jewish; consistent with earlier analysis, its deeper origins lie with the ancient Sanskrit-speaking Germans, who compiled archaic Sanskrit wisdom attributed to Dionysus, later fused with Platonic

83 Cohn (1981), 25–32; Outram, D. (2013). Enlightenment, 131–33; Pipes, 67–75, *70, *74.
84 Cohn (1981), 27–28, 32; Hannaford, 318; Webb (1976), 237.
85 Mosse (2020), 206–07. See also Davies, 21.
86 Hagemeister, M. (2008). The Protocols, 83, 90.
87 Arendt (1973), 241, 358n.42; Goodrick-Clarke (2002), 186; Grobman, 33; Mosse (2000), 197–98; Redles, 114–15; Hannaford, 318; Mosse (1964), 128.
88 Cohn (1981), 33–37, 41; Mosse (2000), 97; Webb (1976), 217.

doctrines in Alexandria.[89] Cohn (1981) introduces a compelling twist regarding Goedsche's pseudonym Radcliffe and the origins of a speech attributed to him, known as "*The Rabbi's Speech.*" The story goes that in 1896, François Bournand published a "*speech*" in *Les Juifs et nos contemporains*, presented a startling falsity—

> We find the programme of Jewry, the real programme of the Jews expressed by . . . the Chief Rabbi John Readclif. . . .[90] (Alteration in original)

The speech purportedly expressed the true programme of Jewry. Consequently, its history had become intermeshed with the Protocols. Cohn concluded with these final remarks—

> It was a sad end for a man who, if he had been a German novelist, had also been an English diplomat and historian, and who if he had been a heroic anti-Semite had also been a chief rabbi.[91]

Another forgery narrative claims that the Protocols originated at the first congress of the Jewish Zionist Movement—a clandestine gathering comprising 300 Jewish elders, high-ranking French Masons, and others—said to have taken place in Basel in 1897, at the *fin-de-siècle.* The Protocols were presented as the purported minutes of this session, in which participants were allegedly plotting with masonic lodges to seize control of the world and establish a Jewish world-state. While there is a modicum of truth in the idea that freemasonry plays a role in understanding notions of clandestine forces shaping world affairs, the agenda, minus Jewish involvement, would have continued unchanged. Katz (1970) rightly points out, the pairing of Jews and freemasonry became a popular propaganda slogan in antisemitic circles across Europe, particularly in Germany.[92]

Katz comments that Jews were largely excluded from German masonic lodges during the 1860s, 1870s and especially the 1890s, primarily due to pervasive antiemitism. Those who had been members prior to these periods were typically expelled. In many German lodges, Jewish membership was contingent on adherence to Christian beliefs, a requirement that undermines the claim of Jewish–masonic collaboration. This stands in stark contrast to the notion that Jews were working with masonic lodges to exert global control, whether in Europe generally or within German contexts in particular.[93]

Hitler, along with Fritsch and Rosenberg, was especially captivated by the Protocols. He claimed to have learned how to wield power from them and to have derived his

---

89 Mosse (1961), 86; Blavatsky, H. P. (1877). Isis Unveiled, 38, 639; Goodrick-Clarke (2002), 81; Kuhn, A. B. (1930). Theosophy, 144; Pulzer, 238.

While Blavatsky's work endeavours to trace the origins of occult wisdom, her conclusions must be approached with caution, as they are highly speculative and frequently contradict established empirical evidence. For a critical perspective, see Gertrude L. Williams, Priestess of the Occult (New York: Alfred A. Knopf, 1946). Blavatsky notably rejected the notion that Kabbalah was of Jewish origin. A. B. Kuhn notes in his commentary in Theosophy (1930), 116–17, that Blavatsky's intent, "*. . . was to reveal the traces, in ancient and medieval history and literature, if a secret science whose principles had been lost to view.*"

90 Cohn (1981), 37.

91 Ibid.

92 Katz, J. (1970). Jews and freemasons, *159, 160–73.

93 Grobman, 33–34; Hagemeister, M. (1998). Die Protokolle der Weisen, 250–273; Hagemeister (2008), 89; Katz (1970), 1–3, *148–59 passim; Redles, 114, 128n.8; Webb, J. (1976). The Occult Establishment, 217.

strategy for world domination from their contents.[94] Redles (2012) explains that the Protocols drew on two earlier sources: the anti-Napoleon III satire *Dialogue aux Enfers entre Machiavel et Montesquieu* (1864) by Maurice Joly, and Goedsche's "*The Rabbi's Speech.*" Redles further emphasises that the forgers adopted Joly's structure and language, extrapolated its themes of cynical manipulation of democracy and the press, and drawing on "*The Rabbi's Speech,*" placed the "*. . . blame for the horrors of modernity onto Jews.*"[95]

Cohn (1981) argues that the Protocols were first composed in Paris, France, rather than in Russia. The forgery was later conveyed into Russia by Yuliana Dmitrievna Glinka (1844–1918), the daughter of a Russian diplomat and a contemporary of Hartmann and Hübbe-Schleiden. See The Superman Doctrine, 72. During this period, Glinka divided her time between Paris and St. Petersburg. This chain of succession eventually led to their first publication in Russia by P. A. Krushevan.[96] Cohn traces the connection by reconstructing the chain of custody through which the text entered Russia.

First, Glinka produced a hand-written Russian version of the material and passed it to Alexy Sukhotin, her personal protector. Sukhotin then transferred it to Filip Petrovich Stepanov, who, in a sworn affidavit to the Ecclesiastical Synod of Moscow, stated that he had received the Protocols from Sukhotin. He had obtained them from an unnamed lady in Paris, a woman of his acquaintance, who translated the material, left Paris with it, brought it to Russia, and then passed it on to him.[97] This chain of custody eventually led to their first publication in Russia by P. A. Krushevan. From Russia, the Protocols migrated to Germany, their final resting place, where they cemented the foundations of German scientific racism in the late 1920s and early 1930s. Webb (1976) documented the subsequent transmission and reception within Germany—

> The Protocols arrived in Germany under the wing of Russian nationals P. N. Shabelsky-Bork and Colonel F. V. Vinberg . . . From Vinberg they passed to the Protocols were passed to the völkisch anti-Semite Ludwig Müller von Hausen, who published six editions of his German translation in 1920.[98]

In the section titled The Superman Doctrine, we explored the occult's interlacing with the Nazi regime; however, we have yet to examine its intersection with antiemitism. There is some overlap. First, after Blavatsky visited Germany and eventually settled there, she founded the inaugural Theosophical lodge in 1884, from which the movement expanded rapidly.[99] By contrast, freemasonry lodges had already spread across Europe and were established in Germany by 1758, with several operating in Berlin by that time.[100]

Second, Germany adopted Blavatsky's racial schema from The Secret Doctrine (1888), which emphasised a hierarchical progression of human evolution. The German translation of her treatise by Hartmann, completed between 1897 and 1903, greatly facilitated the dissemination of these ideas.[101] Within this framework, she identified the "*Fifth Root-Race*" as a stage of evolution generally associated with the Aryan race.

94 Katz, 185–187, 277n.71.
95 Redles, 114.
96 Goldstein, 68; Cohn (1981), 56, 100–01; Levenda, 47–49; Webb (1976), 217, 219, 222.
97 Cohn (1981), 100–07; Goldstein, 68; Webb (1976), 217, 239, 242. See also Toczek, N. (2015). Haters, Baiters.
98 Webb (1976). 295.
99 Campbell, 40, 97; Spielvogel & Redles, 85.
100 Goodrick-Clarke (2008), 132–51, 168.
101 Flowers (2022), 52; Webb (1976), 281.

Third, Theosophy was a bridge between the occult worldview and traditional antisemitic imagery, thereby buttressing and expanding doctrines of racial supremacy. It created an even more volatile mix by synthesising theosophical ideas with German nationalism.[102] As discussed in The Superman Doctrine, 72, Hess and the SS leadership had a marked taste for occultism, astrology, and alternative medicines. Hess also had a strong bond with Himmler, who sent an expedition to Tibet to explore supposed Aryan origins, ideas ultimately derived from Blavatsky. Additionally, Himmler maintained a deep interest in occultism, the mystique of blood runes, neo-paganism, religious symbolism, and reincarnation, and sought to assimilate the wider Occult Revival (1880–1910) into a form of illuminated nationalism.[103]

In sum, both Glinka and Blavatsky were significantly immersed in the realm of the occult. Glinka, a staunch anti-Semite and devoted follower of Blavatsky who subscribed to her Theosophical doctrines, was the first person to bring the Protocols from Paris to Russia.[104] German accusers exhibited duplicity, alleging that Jews were prominent freemasons while conveniently overlooking the participation of its own German citizens and the early establishment of masonic lodges within Germany. Blavatsky's influence extended to Guido von List, whose extensive Theosophical writings and interpretations of hieroglyphs, which includes the swastika, were pivotal in The Occult Revival (1880–1910) in Germany. His mystical concepts were appropriated and assimilated by the elite establishment. As previously substantiated, ultimately, the malevolent occult development of Nazism all point back to Blavatsky.

ii. Theodor Fritsch—the chief promoter of the Protocols

Theodor Fritsch was a vehement and indefatigable antisemitic publicist, and, probably, the most significant individual in German antisemitism, whose efforts surpassed those of many contemporaries. Through his extensive publishing empire, the Hammer Publishing House, he became the central architect in shaping the ideological environment that enabled the early emergence of the Nazi regime. He was hailed as the doyen and progenitor of "*practical antiemitism,*" commonly referred to as "*the Father of Anti-Semitism.*" His influence was so pronounced that Nazi ideologues, apart from Hitler, revered him as a supreme authority.[105] For Fritsch, the polemical denigration of Jews, combined with Social Darwinism and hereditary biology, formed a tightly conjoined triad. As early as 1904, Ploetz lavished praise upon Fritsch in his journal, and Poliakov (1974) went on to mention that Fritsch was an intellectual mentor to Hitler—though Hitler would have been only thirteen at the time of Ploetz's writing.[106]

Furthermore, Willibald Hentschel, a prominent antisemitic theorist who studied under Haeckel, was also associated with Fritsch. Together, Hentschel and Fritsch were connected by a continuous thread to Ploetz and his cofounder Schallmayer, the founders of racial hygiene. As discussed in The Superman Doctrine, Webb (1976) noted that—

102 Goldstein, 71.
103 François, 21–22, 25, 37; Kurlander (2017b), 82; Webb (1976), 319; generally, 318–22.
104 Cohn (1981), 99–101; Goldstein, 68n.25; Webb (1976), 217, 222–23, 233.
105 Flowers, 120–22; Gutteridge, 341; Phelps (1963), 247–49; Pulzer, 51; Webb (1976), 278.
106 Ploetz, A. J., & Nordenholz, A. (1904). Archiv, I:885, 888; Poliakov (1974), 296. Poliakov's reference to Fritsch being one of Hitler's mentors, whilst most probably true given their common ties of hatred for Jews, though it cannot be verified.

> The Germanen Order was founded in 1912 as an 'Aryan' antisemitic group composed of members of the already-existing Hammerbund of Theodore Fritsch.[107]

These circles actively integrated Darwinism into latent antisemitic ideologies, rendering them newly relevant to their time. Fritsch believed that antiemitism should be elevated to the status of a comprehensive worldview and, as such, Serge Tabary observed, this is precisely what he achieved.[108] Fritsch cited Hentschel's book, Varuna (1901), which proposed an "*. . . Aryan racial utopia named Mittgart . . . .*" He further asserted that Hentschel had raised antiemitismto scientific status by grounding it in Darwinian principles.[109]

The finale, following the completion of the Protocols, now, all that was needed was a marketeer. Fritsch assumed that role. He was regarded as one of the most important early racist and antisemitic publicists.[110] Katz (1970) stated that Fritsch initially doubted the authenticity of the Protocols and recognised them as forgeries. Yet Fritsch's conclusion was that, forgery or not, he did not matter, as he believed that they accurately portrayed the alleged Jewish mindset and conspiratorial designs.[111] He also maintained that freemasons were agents of the Jews, mirroring the same indifference toward factual accuracy displayed by Pellepoix, author of the introduction to the Protocols.[112] With this, Fritsch's publicist reign of ideological terror began: he was the darling of the German pre–Holocaust apparatus. Indeed, Cohn (1981) informs us that he was "*. . . the Nester of German antisemitism.*"[113] Goodrick–Clarke (2004), specialist on Nazi ideology, corroborates this, stating—

> Both the Reichshammerbund and the *Germanenorden* were virulently antisemitic groups, the origins of which lie in the organizing ability of Theodor Fritsch, a major figure of pre-war German antisemitism, and in the politics of Germany between 1900 and 1914.[114]

In addition, Fritsch's sphere of influence extended from 1881 to 1920, and his conduct throughout that period was anything but normal. In fact, Fritsch's work *Antisemiten-Katechismus or catechism* (trans. *Handbook of Anti–Semitism*) (1881–93)) was of seminal importance. Levy (1975) and Gilman (1987) described it as a compendium of misinformation and Jewish quotations from the classics of world literature—an encyclopaedia of German antiemitism. Small wonder that Hitler, as cited by Luhrssen (2012), stated that the Antisemiten-Katechismus "*. . . helped prepare the ground in a very special way for the national socialistic anti–Semitic movement.*"[115] It also amalgamated nationalist antisemitism, fine-tuned it in ways Fritsch was eager to promote, and facilitated its dissemination "*. . . in all areas of German life, especially the*

107 Webb (1976), 296,
108 François, 10, 16n13.
109 Flowers, 140–42; Webb (1976), 278; Weikart (2004), 122–23; Weikart (2006), 108–09; Weikart (2013), 540; Weindling (1993), 136.
110 Cohn (1981), 70, 133, 138, respectively; Goodrick–Clarke, (2004), 123; Katz (1980) 304; Levenda, 73, 101; Luhrssen, D. (2012). Hammer, 63; Mosse (1964), *112, 142–43; Webb (1976), 130.
111 Katz, (1970), 185.
112 Hagemeister, 87, citing the Chicago Sun–Times; Mosse (2020), 206–07.
113 Cohn (1981), 70, 182.
114 Goodrick-Clarke (2004), 123. See also Goldstein, 72–73.
115 Luhrssen, 63, 238n.86. See also Gilman (1987), 304; Levy, *37–38.

schools."[116] Even by April 1880–81, the national mood in Germany was already defined by racially antisemitic.[117] See also Education of Society Matrix, 55.

He worked tirelessly publishing antisemitic works, to which it could be said he possessed the hatred of a thousand men, although such intensity is impossible to quantify. He had taken up the baton of racial hatred from his predecessors and "*. . . became a major propaganda center for racial antisemitism,*" He inaugurated virulent anti-Jewish agitation in Germany to indoctrinate children[118] who, thirty to forty years later, would become the merciless killers of the Jews. Wiese (2005) discloses that Fritsch employed what appears to have been a systematic strategy for people to adopt antisemitic mind-sets, carte blanche—

> [O]ne of the first influential representatives of the theory of an "Aryan Jesus" and the demand for a national-racist renewal of Christianity. Fritsch used . . . [h]istorical aspects in order to allow Christians who wanted . . . to embrace anti-Semitic thinking.[119]

Luhrssen (2012) elaborates on this notion, arguing that Fritsch, "*. . . was more concerned with changing the cultural paradigm than with changing the government.*"[120] Luhrssen emphasised that Fritsch's intent was not merely to disseminate propaganda in the hope of democratic change. It was an overt, calculated strategy—a pervasive, radical plan to instil and indoctrinate all social classes with darkening hatred, ultimately culminating in pogroms of extreme violence reminiscent of the medieval period.[121] To this end, Katz (1980) reiterates that Fritsch did not envisage a "*cultural paradigm shift*" that consisted of a change of system that granted Jews legal protection. Rather, he anticipated the eruption of uncontrolled violence against them.[122] Through ownership of his own publishing house, he was able to produce innumerable pamphlets and periodicals, with a particular emphasis on "*scientific*" studies of race in which—

> [Fritisch] proclaimed that by proving 'Scientifically' the depravity of the Jewish and the sublimity of the German 'race', German racists were inaugurating not only a prodigious advance in human knowledge but a new epoch of human history.[123]

He produced an edition of the Protocols, serialised in a number of newspapers, which went on to become a best-seller in 1920.[124] He later renamed it The Zionist Protocols, a move that proved highly profitable: by 1933, more than 100,000 copies had circulated across thirteen editions.[125] A selection of what he wrote is as follows, but not all—

---

[116] Levy, 37; Mosse (1964), 135, 143.
[117] Massing, 40, 80.
[118] Goodrick-Clarke (2004), 123–26; Massing, 91.
[119] Wiese, 252. See also Goodrick-Clarke (2004), 126.
[120] Goodrick-Clarke (2004), 124; Luhrssen, 64.
[121] Katz (1970), 304–05; Luhrssen, 84.
[122] Katz (1970), 311.
[123] Cohn (1981), 171; Goodrick-Clarke, 124.
[124] Luhrssen, 161.
[125] Cohn (1981), 132–33, *138, 221, respectively; Goodrick-Clarke, (2004), 123.

i. 1881: *Huchtkugeln* (trans. Fire-Balls), a collection of pan-German and antisemitic sayings;
ii. 1884–85: His first antisemitic organisation, the *Leipziger Reformverein* (trans. Leipzig Reform Association), for which he issued the periodical *Antisemitische Correspondenz* from 1885;[126]
iii. 1881–93: *Antisemiten-Katechismus* (trans. Handbook of Anti–Semitism or Handbook of the Jewish Question).[127] Fritsch later published forty editions by 1936, alongside the pamphlet series Brennende Fragen (trans. Burning Questions);
iv. 1886: *Zur Bekämpfung zweitausendjähriger Irrtümer* (trans. Struggle Against Two Thousand Years of Error)

In sum, Fritsch's polemic became en vogue in scholarly antisemitic thought, but with a twist: it was couched in the language of racial anthropology and cultural differentiation[128] first developed during the Enlightenment. In line with his radical racist views, Fritsch declared in his book, *Struggle Against Two Thousand Years*, that he "*. . . demanded a complete restructuring of German society to the exclusion of the Jews.*"[129] Significant anti-Judaic events in the 1800s produced a volatile fusion of Darwinism and German rationalism, which formed the pillar of German racial policy.[130] The Protocols, in this context, was the icing on the cake—a contrived rationale used to legitimise exclusionary and eliminationist policies. Germans saw in Jews a symbol of everything they despised.[131]

[126] Cohn (1981), 37, 132; Goodrick–Clarke, (2004), 123–26; Katz, 304–05; Luhrssen, 64; Massing, 91; Mosse (1964), 142.
[127] Levy, 37; Gutteridge, 342; Steinweis, 9. See also Massing's list of Fritsch's charges', 77–79.
[128] Mosse (1964), 134; Katz (1970), 307.
[129] Luhrssen, 63.
[130] Cohen, N. W. (1984). The Challenges of Darwinism, 125, 139, 143–144,150; Habib, M. A. R. (2005). A History, 470; Latham, A. (2005). The Naked Emperor: Darwinism Exposed, 195.
[131] Cohn (1981), 24; Langmuir, 321. Langmuir does not expressly mention the Protocols, but it is implied, since he mentions the content of the protocols.

C. Contemporary Anti-Semitism: The New Wave Revival of 'Protocols'—Inspired Ideologies

A RENEWED WAVE OF antiemitism is emerging globally, heightened by contemporary media ecosystems, escalating social-platform circulation, and the political fallout of Middle East conflicts and Palestinian wars. These forces have sharpened existing tensions and triggered hostility on an unprecedented scale. The Protocols, long exposed as a fabricated antisemitic tract, have nevertheless resurfaced over the past half-century, gaining transnational reach. Their revival is sustained by digital dissemination and political movements skilled in mobilising conspiratorial narratives.

The current resurgence is driven chiefly by Islamist networks operating internationally, recycling tropes rooted in, and derived from, the Protocols across parts of the Arab world. The propaganda once deployed in Nazi Germany to mobilise Europe against Jewry now serves as a strategic template for attempts to turn global opinion against world Jewry. Taguieff (2004) frames this 'new antiemitism' through the revival of the Nazi metaphor of eliminating cancer and—

> The spread of so-called "anti-Zionist" rhetoric now affects all Western countries, not only France . . . the term refers to both propaganda and an ideological project . . . [aimed at destroying] the State of Israel or . . . the "removal of the Zionist entity." In Israelophobic polemics, metaphors from pathology, such as 'cancer' to 'chance,' [are] used as a strategy to eliminate the State of Israel.[132]

Other examples include ritual murder, world domination, media and market control, and the portrayal of Jews as global conspirators. These motifs informed the ideological climate preceding September 11, 2001 ("9/11"). They were also cited by Islamist movements as part of the justificatory framework for the attacks. The aftermath triggered a fresh wave of Protocols rhetoric, claiming that the attacks had been engineered by a Zionist conspiracy.[133] Since Nazi Germany, the Protocols has circulated in multiple linguistic and regional editions: a French version (the original), English, German, Russian, an Iranian edition published by the Islamic Propagation Organisation, as well as Japanese, Pakistani, and numerous Indonesian versions.

Ma'oz (2011) underlines the widespread reach of the Protocols, asserting that the Indonesian versions alone have been reprinted six times. These editions are not indigenous texts but translations from Arabic sources. Kailany & Baraja's (1988) Indonesian version derives from a Saudi copy; Sulaiman's et al. (1990) translation includes an introduction by a Kuwaiti author; and Darouza's (1992) edition is based on a Syrian copy first issued in 1970 and reprinted in 2005.[134] The Palestinian version is treated as an authentic document and is deeply embedded in Palestinian Authority ideology. It is incorporated into the official Hamas charter, specifically, Article 32. Within this indoctrinatory framework, Palestinian households are expected to keep a copy of the Protocols to instruct children in its antisemitic content.[135]

[132] Taguieff, P.-A. (2004). Rising, 17. See also Glazer, 116.
[133] Glazer, S. (2010). Issues in Peace, 109; Küntzel, M. (2007). Jihad and Jew-hatred, 70, 95, *123–161.
[134] Ma'oz, M. (2011). Muslim Attitudes, 231–32. See also A'lam, J. et al. (1987). Dendam Barat; Darouza, M. I. (1992). Mengungkap tentang Yahudi; Kailany, M. & Baraja, A. (1988). Bahaya zionisme; Sulaiman et al. (1990). Ayat-ayat setan Yahudi.
[135] See Marcus, I. & Crook, B. (2011). The Protocols of the Elders of Zion : An Authentic Document 152-160.

Moreover, the spread of the Protocols extends beyond Western and Middle Eastern contexts, reaching the Far East. Goodman's (2011) research shows that the text entered Japan between 1921 and 1945, becoming part of local intellectual discourse. In 1958, Japanese scholar Matsumoto Fumi reprinted it, and by the 1980s it resurfaced again, by exploiting the idea of a 'global Jewish conspiracy' to destroy Japan. Goodman also records that the 10 July 1992 issue of the Shūkan Post (週刊ポスト), claimed Jews had orchestrated a drop in the Japanese stock market to wreck the nation's economy. Given this escalating hostility trend, Staub (1989) offers a theoretical view in which genocide may be organised by those in power—

> Genocide is usually organized and executed by those in power, by a government or ruling elite. Governments will commit genocide if the way of thinking and motivations out of which genocide evolves are already consistent with the culture or if they become so under the influence of the government.[136]

Applying Staub's (1989) proposition sharpens the argument: governments are more likely to commit genocide when public mindset, social mood, and cultural norms correspond with elite intent, with the critical "*. . . one precondition for genocide [being] widespread antiemitism,*" Staub states.[137] Between late 2023 and spring 2024, more disturbingly, even prestigious institutions of Harvard and the University of Pennsylvania have permitted students to engage in collective acts of aggression against Jews—a sign of normalised aggression. Prominent film actors, including George Clooney and Whoopi Goldberg, Mel Gibson, and most noabaly Robert De Niro and others, have also publicly expressed disdain for Jews, helping to generate a wider cascade effect across global audiences.

Kelman & Hamilton's (1989) account of genocidal societies shifts the focus from individual psychology to the structural conditions that erode moral restraint. They argue that it is more instructive to examine the situational pressures and cognitive biases under which ordinary moral limits weaken.[138] In this sense, global antiemitism emerges and intensifies only when socio-political conditions are conducive; otherwise, it remains latent until specific events activate it. Notably, the aftermath of the 7 October 2023 Hamas attacks on Israel, and subsequent retaliatory actions, activated international riots, amplified anti-Israel sentiment on a global scale, and triggered widespread violence against Jewish communities, reflecting patterns embedded in Islamic ideological narratives.

The agency of media editorial biases—particularly in bystander outlets such as CNN, NBC, the New York Post, and the BBC—are a perpetual conduit for anti-Israel narratives. Through curated portrayals of events, these platforms have developed into a sophisticated propaganda apparatus, mobilising global opinion in ways reminiscent of Nazi Germany's strategic use of propaganda. Their coverage has seductively shaped public perception, steering social attitudes and cultural narratives against Jewish affairs and conduct. Selective imagery, event framing, and editorial choices can either disrupt or reinforce harmful stereotypes. In this case, many outlets are not curbing the spread of antisemitic sentiment but are instead amplifying prejudicial narratives through conformity, omission, or tacit collusion.

136 Staub, E. (1989). The roots of evil, 24; Grodin & Annas (2007), 651.
137 Staub (1989), 100.
138 Kelman & Hamilton, 15. See also Kressel, 198, 214.

In the UK and Europe, Muslim animosity and state-sanctioned police violence are rising sharply, epitomising Stab's (1989, 1990, 1995, 1999) and Browder's (1996) concept of state authorised societal violence. This surge is fuelled by narcissistic, ego-driven Muslim dehumanisation propaganda,[139] amplified by governments and backed by established institutions. Jews already feel forced to conceal their cultural and religious identities amid growing societal hostility.[140] Group violence and bystander complicity are not limited to the occult; society can assume these roles. "*Bystanders also learn and change through passive or semi-active participation,*" and even partial conformity or tacit approval of perpetrators signifies movement along this continuum.[141] Moral exclusion and diffusion of responsibility—seeing certain groups as unworthy of moral concern or feeling diminished personal accountability—can escalate to genocide, framed as a service to humanity, as in Nazi Germany if unchecked.[142]

Staub's (1989) framework is evident in Europe and the UK, where Jews have faced attacks and, at times, police indifference, making daily life perilous. Should openly antisemitic governments gain power in Western nations, the risk of state-sponsored violence would rise, as their ideological commitments already resonate with broader shifts in global political culture. Historical analysis of the 1890s and the Weimar era shows that aggressive antiemitism was then Europe-wide, not global as it is today. That earlier trajectory culminated in the atrocities of 1933, and elements of it appear to be re-emerging in the contemporary geopolitical landscape. Within this framework, the occult plays a central role, subtly reinforcing societal norms that indirectly sanction violence against Jewish communities. Poliakov (1965) observed, the impulse to assign collective culpability to Jews is deeply rooted in historical precedent—a pattern that carries a significant probability of recurrence—

> The countless anti-Jewish excesses of the preceding centuries, even though sporadic, had already prepared the ground sufficiently so that in a grave crisis, a collective disaster, the Jews were immediately designated as responsible.[143]

This dynamic is illustrated by Bledin (2021), who reports that antisemitism surged sharply, driven by conspiracy theories promoted by official government sources falsely claiming that Jews and Israel were responsible for the global spread of the Covid pandemic.[144]

---

139 Baumeister, R. F. et al. (2000). Self-Esteem, Narcissism; Baumeister (2009), 158; Waller, 193.
140 Browder, G. C. (1996). Hitler's Enforcers, 33–34, 40, 157, 171.
141 Bettelheim, B. (1952). Remarks on the Psychological Appeal of Totalitarianism; Staub (1989), 82, 86–88.
142 Staub (1989), 83, 132.
143 Poliakov (1965), 102.
144 Bledin, K. (2021). Further thoughts, 534.

PART III. COVID-19 AND PUBLIC HEALTH CONTINUITIES

"Until bioterrorism's true nature as an epidemic disease event is fully recognised, our nation's preparedness programs will continue to be inadequately designed: the wrong first responders will be trained and equipped; we will fail to fully build the critical infrastructure we need to detect and respond; the wrong research agendas will be developed; and we will never effectively grapple with the long-term consequence management needs that such an event would entail."

—Dr. Margaret Hamburg, in testimony to the US House Committee on Government Reform, 23 July 2001 quoted in Noah, D. L. et al. (2002). The history and threat of biological warfare and terrorism

CHAPTER 8. FROM CRISIS TO PREPAREDNESS : STRENGTHENING CRITICAL INFRASTRUCTURE THROUGH GLOBAL PANDEMIC RESPONSE

Introduction

IT IS THIS author's personal, analytical interpretation that COVID-19 ("Covid") represented an unprecedented global framework—a preliminary testing protocol for a prospective transnational authoritarian regime structured around virological safeguards. The reality is that this phenomenon should be perceived as a dress rehearsal, or an introduction to a new totalitarian Global Order, and a preparatory exercise for it. Professor Klaus Schwab, Founder and Executive Chairman of the World Economic Forum, restricts himself to explicitly arguing that Covid is paving the way for a transnational authoritarian regime. However, Schwab & Malleret's (2020) *The Great Reset* maintain that, in the post-Covid context, there will be future "*momentous changes*" in the reconfiguration of global economic and governance systems.[1]

This section will demonstrate how the continuity of emerging developments is steering towards the complete realisation of the *New Atlantis* narrative, previously analysed in this paper, with the scientific and medical community at the centre. While the Nazi regime constituted a partial realisation of authoritarian ideals localised to Europe and ultimately failed, the current trajectory suggests a more comprehensive and worldwide iteration. Indeed, the coordinated global response to Covid has fuelled widespread speculation about an authoritarian "*New International Scientific Order*" or "*Global Reset.*" This perception arises from the unprecedented coordination of global response strategies to Covid.

If the devastating Ebola or the Black Death virus were to re-emerge on a global scale, an elaborate informed response would not just spontaneously materialise into a well-coordinated and preemptive strategy. The global transmission of a dangerous pathogen via international transport networks heightens the existential risk of spreading contagion.[2] This necessitates the imperative to meticulously craft contingency plans to effectively address to such global hypotheticals.

The primary focus of Covid was devising new systems and updating and refining existing preparedness strategies. Generally speaking, these strategies demand the collaborative efforts of international, national, and regional focus groups and think tanks. This requires extensive strategic planning over several years. This is precisely what the Covid pandemic accomplished—the revision and finalisation of existing protocols by the global alliance, culminating in the establishment of robust strategic frameworks. The Covid pandemic involved both the creation of new systems and the refining of existing ones.

1 Schwab, K. & Malleret, T. (2020). COVID-19 : The Great Reset, 16.
2 Berenbaum, 5; Garrett, L. & Fidler, D. P. (2007). Sharing H5N1.

On one hand, the development of new systems involved establishing robust infrastructures such as advanced communication networks, expanded supply chains, and new healthcare facilities. These systems were critical in managing the Covid crisis and ensuring rapid response capabilities. The swift mobilisation into a coalition by March 2020 demonstrated that a global alliance approach had already been in place prior to the start of Covid in January 2020. On the other hand, updating and refining existing preparedness strategies included revising protocols, enhancing training programmes, and improving resource allocation—ensured that emergency planning was comprehensive, adaptable, and transnational in scope. First, the following discussion in no way denies the reality of the Covid virus, its prognosis, or its symptoms. Second, the post-Covid analyses indicate that the alliance constructed a narrative that far exaggerated the pathogenic threat of Covid from a flu-like illness to one of extraordinary mortality "*. . . for political and economic purposes.*"[3]

A. Shaping Public Consciousness: Influencing Perceptions and Collective Thought

There are many facets to the Covid debate. The primary overarching head, when it spills over into sociopolitical contexts, is institutional invalidation and subsequent polarisation. This will be the first point of discussion. The second strand involves a critical evaluation of opposing arguments. Many maintain that "*scepticism and fear of vaccinations*" were the driving metrics behind vaccine hesitancy. Another contention asserts that misinformation and vaccine myths were key drivers of vaccine hesitancy.[4] Collectively, these dynamics created an atmosphere of fear, paranoia, and suspicion. The third area of analysis focuses on the Covid virus and vaccine outcomes.

i. Managing Dissent and Deployment of Control Measures

WITH COVID NOW buried in the past, discussions on governance and policy responses still continue to unfold. Systemic delegitimisation created a polarised society. Hillary Clinton's reckless assertion that critical and vital debate should be silenced underlines how the crisis of authority unfolded. She leveraged Covid-era censorship to advocate for a broader policy aimed at curtailing free speech. In the 2024 United States Congressional Record, Clinton framed social media regulation under the pretext of combating "*misinformation,*" warning that "*. . . social media companies . . . must moderate content on their platforms or we will lose total control.*"[5]

This is not a new perspective. Cited earlier in Education of Society Matrix, Pearson (1897) argued that the regimented side of socialism aimed to utilise state power to organise segments of society by interfering with individual freedoms, requiring "*. . . implicit obedience of the individual to a governmental commander-in-chief. . . .*"[6] Regardless of intent, her position reflects dystopian governance underpinned by inequality. Indeed, the US federal government and European Parliament either pressured social media to suppress dissenting Covid narratives or summarily dismissed them as "*misinformation*" or "*disinformation.*"[7] This approach was only exacerbated by what Berenbaum (2022) describes—

3 Anisin, 2.
4 Coustasse, A. et al. (2021). COVID-19 and Vaccine Hesitancy; Naeem, S. B. et al. (2021). An exploration; Piccaluga, P. P. et al. (2022). COVID-19 Vaccine.
5 Congressional Record. (2024). Concepts that are inconvenient to the country, 170(173), p.H6179.
6 Pearson (1897), 112, accessed, March 30, 2026.
7 Anisin, 34–*36.

> [P]oliticization of medical information, the falsification of medical data to advance a political agenda . . . [and] the promotion of false information to satisfy the whims of political leaders, national and local.[8]

This phenomenon is not without historical precedent. Alexander (1949) specifies that, "*Nazi propaganda was highly effective in perverting public opinion and public conscience, in a remarkably short time,*"[9] from its inception in 1933. We, the people, have forgotten the lessons of history and the manner in which the Nazi regime's orators succeeded in captivating the hearts and minds of the majority within a matter of months. The underlying reason can be discerned in the words of Pellegrino (2005) who posited that—

> When survival of any society is threatened, there is an understandable tendency to lay ethics aside. This is just when ethics meets its severest test.[10]

Today's world leaders have shown they could replicate similar outcomes. What about the dissidents? As noted earlier, Covid responses loosely mirror the Third Reich's actions against its citizens, with contemporary governments and elites allegedly conspiring to harm the populace. Once the Nazi utopian superstate took hold, the rank and file fell in line on a national scale, minimising dissidents and the potential for civil unrest. This is precisely what transpired in the Covid pandemic. It ushered in a new provisional police state authority. However, it is important to recognise that reprisals for expressing dissenting opinions during Covid did not, to our knowledge, involve "*boot-smashing*" operations or Hitler's mobile death squads to identify and eliminate non-conformists.

Furthermore, Schafft (2004) makes a notable observation regarding new Nazi legislation enacted under the guise of eradicating "*communism,*" which, in reality, targeted activists and regime opponents. This law, deceptively titled the "*Order for the Protection of People and State,*" mandated the dismantling and erosion of the constitution, similar to the Covid outcomes, which read as follows—

> [A]s a defensive measure against communist acts of violence endangering the state . . . [this act imposes] restrictions on personal liberty, the right of free expression of opinion, including freedom of the press, on the right on assembly and association. And violations of privacy of postal, telegraph, and telephone communications, and warrants for house searches, orders for confiscation as well as restrictions on property rights are permissible beyond the legal limits otherwise prescribed.[11] (Alterations in original) (Omission in original)

In 2026, the political landscape continues to encompass far-left socialist and communist movement, juxtaposed with far-right fascists. The critical variable lies in he endorsement of particular ideologies among elites and broader society, fostering a paradigm that champions the majority perspective. Reflecting on the Nazi regime, attention rightly centres on the Holocaust, yet Nazi science and medicine ominously

8 Berenbaum, *4–5. See also Blaylock, R. L. (2022). COVID UPDATE; Schwab, K. & Malleret, T. (2020). COVID-19 : The Great Reset.
9 Alexander (1949), 39.
10 Pellegrino, E. D. (2005). Foreword, xviii.
11 Schafft, *66, 67, citing Noakes, J. (1984b). Nazism, 1:142.

lurked in the background, inadvertently overlooked. As demonstrated earlier in, Racial Hygiene at its Zenith, 160, German science and medicine first legitimised isolation through the belief that Jews were dehumanised through biological metaphors of disease and contamination requiring containment. This then facilitated the Holocaust itself, where the constructed "*pathogen*" threat was eliminated through disinfection agent Zyklon gas.

Analogously, during the Covid pandemic, the dominant view of democratic accountability crystallised, making the transition to an unprecedented technological, global surveillance state almost seamless.[12] This transformation saw the scientific community assuming the mantle of governance, a quasi-political-industrial governmental medical milieu was formed, spearheaded by de facto leader, Dr. Fauci, who exerted control top-down. Blaylock (2022) explains the relevance of this shift, highlighting how the pandemic response led to unprecedented levels of state intervention and control over individual liberties—

> For the first time in history medical treatment, protocols are not being formulated based on the experience of the physicians treating the largest number of patients successfully, but rather individuals and bureaucracies that have never treated a single patient—including Anthony Fauci, Bill Gates, EcoHealth Alliance, the CDC, WHO, state public health officers and hospital administrators.[13]

This development signifies the partial resurgence of practices observed in Nazi science and medicine within imperial Germany, foreshadowing their full realisation in a future totalitarian New International Order. White (2015) suggests that in order to weigh the true value of democracy, one must not focus on sovereignty, political power, the state, or the law, but rather on moments when the rulebook is set aside. In this context, the transformation of power and control dynamics during the Covid pandemic demonstrated that the rule of law and democratic freedoms can be easily overridden, suspended, and systematically stripped away as the situation dictates, when global health security issues, such as vaccine diplomacy, are at stake.

This enabled a limited freedom window "*to do things you could not do before.*"[14] Anisin (2024) noted Lewkowicz's et al. (2022) methodical analysis of 'democracy backsliding' (institutional corrosion) which found that the stronger the rule of law is and the higher the levels of democracy are, the lower the risk of direct violations of democracy during a global pandemic.[15] This principle is evident when comparing less-developed countries with those possessing strong democratic institutions in the West. However, during Covid, the global alliance railroaded the rule of law. For instance, emergency approval for the use and distribution of vaccines, bypassing FDA clinical trial protocols, marked a clear departure from established legal and ethical standards.

Indeed, Wilkinson (2021) refers to the Nazi era when he asserts that Nazism arose from an "*. . . excess of liberalism . . . pursued by political elites ruling in emergency mode.*"[16] A fuller definition of 'authoritarian liberalism,' drawn and adapted from Wilkinson's articles is—

12 Thomson, S. & Ip, E. C. (2020). COVID-19 emergency measures, 20.
13 Blaylock, 1, citing Kennedy, R. F. (2021). The Real Anthony Fauci, 135 and online websites.
14 Friedman, S. D. (November 25, 2008). Recessions.
15 Anisin, 7, 14; Lewkowicz, J. et al. (2022). COVID-19 and erosion, 11.
16 Wilkinson, M. A. (2021). Authoritarian Liberalism, 25.

> [First] an avowed antipathy towards representative democracy in favour of autocracy and dictatorship, using emergency powers . . . ; and [second, while undermining the welfare state, by fostering] an ideological defence of free-market liberalism and the protection of the interests of capital [by subsidising] major industrialists . . . , [creating the illusion of the separation of the political and economic realms] [to ease] the transition to National Socialism.[17]

Arguably, Covid seems to match the criteria of Wilkinson's quote. Covid was a period marked by leaders ruling through "*. . . emergency decrees or during states of emergencies.*"[18] It was a façade of separation between political and economic spheres; in reality, it was a biopolitical authority operating in emergency mode, appropriating the state's powers to govern both life and the populations. Counterarguments might assert that extraordinary times call for extraordinary measures. However, the balance between practicality and adherence to democratic principles does not justify the creation of a medical-science police state.

A case in point, Gaudillière (2000) documents a parallel in vaccine development between 1943 and 1945 in the United States. Wendell Stanley, a biochemist and advocate of ultracentrifugation, secured a contract with the Office of Scientific Research and Development ("OSRD") for influenza ("flu") studies. He developed a flu vaccine for the U.S. Army using untested "*. . . ultracentrifugation as a means of concentrating and purifying large amounts of influenza viruses.*"[19] Despite the contested centrifuged vaccine laboratory data, and incomplete trials, the vaccine was approved by the Army Influenza Commission in 1945. Although its results were later challenged and temporarily halted, it had already been mass-produced by commercial pharmaceutical firms and deployed through military networks.[20]

A more recent example is the 2008 Euro crisis. White (2015), Wilkinson (2021), and Anison (2024) document how elected heads of government in Italy and Greece were replaced with technocratic or expert-led administrations. This was justified on the grounds of necessity and emergency, frequently expressed through sensationalised rhetoric.[21] Capra (1982) maintains that the elevation of scientific knowledge as the sole epistemic authority, coupled with the summary rejection of other valid forms of understanding (e.g. intuitive knowledge or personal life experiences), bolsters the supremacy of "*scientific progress*" and "*technological efficiency*" as societal ideals. This, in turn, constrains the exercise of human will and freedom. Accordingly, scientific communications often carries automatic credibility, regardless of whether the ideas are scientific in character. Gross (1990) aptly refers to this as, "*. . . a built-in ethos of especial intensity,*" quoted by Hammer (2003) and elaborated upon by Barbour (1997).[22] Yet, scientific conclusions themselves are based on experimental tests that are indicative of uncertainty and instability, as Barbour (1991) rightly points out—

> [A]re always incomplete, tentative, and subject to revision. Theories change in time, and we should expect current theories to be modified or overthrown, as previous ones have been. But science does offer reliable procedures for testing

17 Wilkinson (2021), 25–43; Wilkinson, M. A. (2022). Authoritarian Liberalism, 194.
18 Anisin, 5.
19 Gaudillière, J.-P. (2000). Rockefeller Strategies, 504.
20 Ibid., 505.
21 Anisin, 32; Wilkinson (2021), 206; White (2015), 303, 305.
22 Capra, F. (1982). The Turning Point, 22, 42; Hammer, 234.

> and evaluating theories by a complex set of criteria.[23]

Todorov (1993) was on point when he argued that science is not to blame, but scientism is. In his nuanced analysis, he attempts to reconcile the intersections between totalitarian systems, politics, and ethics, differentiating humanism from scientism. His analysis can be summarised as: reductive, unbridled scientific advancement, when appropriated for political ends, can erode human freedom. Todorov warns that although scientism is not intrinsically totalitarian, it can facilitate the emergence of a totalitarian society when science is in submission to political and ethical dictates, resulting in its annexation by the government,[24] as it did with Fauci under the Biden administration.

More to the point, and quite surprisingly, Thomson & Ip (2020) reiterate that the Covid regime inaugurated a suspension of civil liberties and fundamental freedoms *salus populi suprema lex esto* (Cicero), underpinned by the deployment of emergency powers, which "*. . . served as facilitators of authoritarianization . . . .*"[25] This created, in the blink of an eye, a surveillance police state, which comprised the imposition of curfews, nationwide lockdowns, travel bans, and the escalation of citizen surveillance. Enforcement was generally carried out using biometric facial recognition cameras, police patrols, roadblocks with force applied when necessary to ensure compliance with stay-at-home orders. Drones were deployed in France, Spain, Malaysia, and mainland China to enforce lockdowns.[26] Mobile phone location monitoring became commonplace in the majority of countries.[27] In Poland, smartphone holders were "*. . . mandated to send a photograph of themselves within 'twenty minutes of receiving a prompt from government officials.'*"[28]

Penalties for non-compliance varied: on-the-spot fines were imposed in the UK and Florida, while more stringent measures were enacted in France and the Cayman Islands, where violations resulted of up to six months' imprisonment.[29] In Cambodia, failure to adhere to emergency Covid measures carried sentences of up to ten years' imprisonment. Bosnia and Herzegovina introduced indefinite curfews and substantial fines for the dissemination of false news or claims causing panic.[30] In India, quarantine notices were affixed to homes, causing socio-psychological turmoil, and under the Indian Penal Code, acts likely to spread infection carried sentences of up to six months' imprisonment.[31] In Spain, individuals breaching lockdown regulations were fined up to 600,000 euros or faced 12 months in prison.[32]

Additionally, the new authorities assumed a role akin to "*doctors of the world.*" This phenomenon is not unprecedented. As discussed in the section Racial Hygiene at its

---

23 Barbour, 110. See also Jones's, 36–40, for critical analysis.

24 Barbour, 81; Gross, A. G. (1990). Rhetoric of Science, 21, quoted in Hammer, O. (2003). Claiming Knowledge, 235; Todorov, 169–70.

See generally Hammer (2003), chap. 5:201–330 for analysis on scientism; see specifically, 205–*06, for his definition of definition of "scientism" as the ". . . active positioning of one's own claims in relation to the manifestations of any academic scientific discipline, including, but not limited to, the use of technical devices, scientific terminology, mathematical calculations, theories, references and stylistic features—without, however, the use of methods generally approved within the scientific community, and without subsequent social acceptance of these manifestations by the mainstream of the scientific community through e.g. peer reviewed publication in academic journals."

25 Thomson & Ip, 4.

26 Miyamoto, I. (2020). Mass Surveillance; Thomson & Ip, 29.

27 Anisin, 54–55; Boudreaux, B., et al. (2020). Strengthening Privacy; Thomson & Ip, 29.

28 Garrett, P. M. (2021). 'Surveillance Capitalism, 1758. See also Anisin, 27.

29 Thomson & Ip, 4, 8–10.

30 Ibid., 16–17.

31 Ibid., 17, 21.

32 Anisin, 52.

Zenith, 160, doctors transitioned from "*physicians of the individual*" to "*physicians of the nation.*" They too decided the fate of the patient in Nazi Germany—who lived and who died. Berenbaum (2022) noted, widespread violations of medical ethics by doctors and medical personnel had becoin which they determined who lived, who died, and "*. . . what priorities to give to treatment and who should be allowed to die by neglect.*"[33]

ii. Philosophy of Rule and the Crisis of Legitimacy

MANY COVID VACCINE advocates have scrutinised decisions to refuse the vaccine, generally attributed them to fear, scepticism, hesitancy, and sectarianism. But why? While this author is not a behavioural psychologist, it is clear that these reactions are corollaries of a deeper, broader crisis of authority. This is not merely an anti-vaccine response from a specific separatist group, city, or nation. It was a global phenomenon, in which populations worldwide have expressed vaccine refusal, transcending religion, ethnicity, cultural norms, and scientific and political divides. This widespread response was not simply driven by fear, scepticism, hesitancy, or sectarianism; it was indicative of a deeper crisis of trust in authoritative institutions.

A plethora of academic articles on vaccine hesitancy, predating 2019, addressed these reactions but failed to examine the macro-level structural failures underpinning them. Three such articles by Coustasse et al. (2021), Naeem et al. (2021), and Piccaluga et al. (2022)—highlighted these reactions but overlook their underlying causes, rendering their conclusions incomplete. Piccaluga et al. (2022), drawing on the World Health Organization's ("WHO") definition, described vaccine hesitancy as—

> '[T]he reluctance or refusal to vaccinate despite the availability of vaccines.' Hesitancy towards vaccinations is a phenomenon that has existed since the first vaccine became available.[34] [Footnote call numbers omitted]

The WHO definition, while widely cited, was not fully applicable to the Covid context. There were extraneous concerns about the efficacy and rapid development of mRNA vaccines. The quote omits a further critical dimension. Vaccine hesitancy did not enter the global public consciousness until the Covid pandemic in 2020. Prior to this, while there was ongoing academic debate, but it did not command the same level of attention. The hesitancy that emerged gained international prominence, exacerbated by the unprecedented exposure Covid received. Even then, as illustrated in Table III-1 below, there were reasonable grounds for hesitancy when refracted through Level 1 structural driver logic fractures—

33 Berenbaum, 5.

34 Piccaluga, P. P. et al. (2022). COVID-19 Vaccine, 3; Naeem, S. B. et al. (2021). An exploration; Coustasse, A. et al. (2021). COVID-19.

Table III-1. Crisis of Authority Social Fragmentation

Level 1: Systemic Conditions (Structural Drivers)

↓

Opaque Governance and Lack of Transparency

↓

Misinformation / Disinformation

↓

Crisis of Expertise and Trust in Expert Opinions

↓

Erosion of Institutional Legitimacy

↓

Economic Precarity and Inequality

Level 2: Incidental Behavioural Reactions

↓

Scepticism & Suspicion (initial reaction)

↓

Hesitancy About Official Narratives

↓

Distrust and Disengagement

↓

Desire for Alternative Explanations

Level 3: Epistemic Break (Logic fractures)

↓

Embrace of Conspiratorial Frameworks

↓

Rejection of Consensus Science

↓

Distrust Evolves into Counter-Belief

↓

Parallel truth economies (QAnon, antivax, etc.)

Level 4: Sociopolitical Consequences

↓

Polarization of Beliefs and Sectarianism

↓

Political Radicalisation and Extremist Leanings

↓

Breakdown of Civil Discourse and Democratic Norms

↓

Violence and Collective Mobilisation

Legend

Level 1 Macro-level structural failures, systemic weaknesses, and a loss of trust shape individuals' information environments, reflecting a Crisis of Authority and systemic delegitimisation.

Level 2 Rejection of vaccines, public health advice, media coverage, or government messaging is not driven by ignorance. Rather, it's because institutions issuing guidance are no longer seen as legitimate or credible. Scepticism and hesitancy are effects, not causes—they are symptoms of a crisis of authority, transitional states that accelerate social fragmentation.

Level 3 Distrust evolves into a new worldview. People embrace conspiracies, reject consensus science, and seek hidden truths. Parallel truth economies, like QAnon and anti-vax movements, emerge, marking a shift from scepticism to counter-belief.

Level 4 Society fractures. Polarisation deepens, political views radicalise, and civil discourse breaks down. Violence and collective mobilisation rise as common ground erodes. Sectarianism entrenches the divide, reinforcing misinformation and delegitimisation, creating a self-sustaining cycle of polarised beliefs.

The prevalent themes in pro-Covid author arguments, with few exceptions, share common structural elements: (i) vaccine-acquired immunity, (ii) pharmaceutical profits, (iii) misinformation driving vaccine hesitancy, (iv) public concerns over vaccine safety and rollout speed, (v) mRNA technology fears, (vi) viral mutations and variants, (vii) conspiracy theories (see Table III-1 Level 3), and (viii) herd immunity and public allegiance. These ideas are recycled, referencing and reinforcing one another, constructing self-reinforcing, circular logic.[35] These principles are aptly labelled Epistemic Breaks (Logic Fractures, Level 3), while vaccine hesitancy and scepticism fall under Incidental Behavioural Reactions (Level 2). The WHO officially listed "*vaccine hesitancy*" and "*infodemic*" as top-ten global health threat by the WHO in 2019, these reactions emerged alongside the Covid pandemic.[36] The rejection of vaccines, public health guidance, and official messaging reflects broader societal fractures, not isolated phenomena. Piccaluga et al. (2022) built on Dubé et al. (2014) and Naeem et al. (2021), stating—

> Vaccine hesitancy is a complex problem closely linked to the individuals social context, income, and level of education, posing dangers to both the individual and their community and increasing the risk of vaccine-preventable disease outbreaks and epidemics.[37]

Danchin & Buttery (2021) employed an Appeal to Authority fallacy, by citing a similarly worded statement from the WHO's (2021) *Behavioural and Social Drivers of Vaccine Acceptance guidance*. However, these vaccine hesitancy framings appear largely irrelevant in the context of the Covid pandemic in this case.[38] The central issue was not social context but the perceived efficacy of the mRNA vaccines. Dodd et al. (2021) and Machado et al. (2022) succinctly capture the link between the rapid rise of vaccines, the loss of trust in public health professionals, and the hysteria that followed—

35 For opposing arguments see Doerfler, W. (2021). Adenoviral Vector DNA- and SARS-CoV-2 mRNA-Based Covid-19 Vaccines.
36 Coustasse, A. et al. (2021). COVID-19; Löffler, P. (2021). Review : Vaccine Myth-Buster, 2; Naeem et al., 143–44; Piccaluga et al. (2022), 4; Sisco, H. K. F. & Brummette, J. (2024). mRNA Vaccine.
37 Piccaluga et al. 3; Dubé, E. et al. (2014). Vaccine hesitancy.
38 Danchin, M. & Buttery, J. (2021). COVID-19 vaccine hesitancy, 1987.

> [The] rapid development associated with the limited follow-up time post-vaccination and lack of information about long-term side effects of the vaccines aroused great public concern about the safety profile of the available vaccines.[39]

On a different note, evidence shows, despite the False Dilemma construction, Piccaluga et al. (2022) introduce the counter-narrative of health over profit as the only possibility.[40] This narrative overlooks the fact that pharmaceutical companies did not just financially benefit; the inner investors such as Warren Buffet, Bill Gates and other contemporaries had significant stakes in vaccine development, which coincided with the sudden development of the Covid vaccine, generating substantial profit margins. See heading Constructing Robust Preparedness, 232.

The financial aspects and herd immunity debates deepen public distrust, intensifying Level 2 scepticism, disengagement, and broader concerns about corporate profiteering, economic incentives, and the influence of powerful actors in shaping health policy. Löffler (2021) employs a classic False Dilemma (or "*black-and-white*" fallacy) to reinforce the narrative that greater vaccination coverage equals greater immunity. However, this claim is highly questionable. Morens et al. (2022) argue that unlike polio or measles, Covid lacks phenotypic stability: "*. . . increasing herd immunity may be an elusive goal.*" Other researchers agree, noting how the persistence and evolution of new variants undermine long-term vaccine efficacy.[41]

Scepticism acted as an accelerant, signalling a shift toward counter-belief systems and eventual epistemic breaks, characterised by conspiratorial thinking, parallel truths, and Level 4: Sociopolitical Consequences, which stopped short of violence and collective mobilisation. These factors extend beyond the mere locus of vaccine hesitancy and scepticism, which should instead be considered within the sub-divisions of the Level 1 crisis of authority. The lack of biopolitical transparency within the de facto quasi-governmental medical-scientific institutions, combined with visible institutional hypocrisy, contributed to a form of opaque dictatorship. Piccaluga et al. (2022) maintain that—

> [T]he anti-vaccine community argues that vaccines, contrary to the claims of most of the scientific community do not have a sufficient or adequate safety profile.[42]

This argument, along with other straw man claims, overlook the real underlying issues. Ironically, misinformation did not solely emanate from polarised sectors (Levels 3 and 4); it also came from media conglomerates and Western governments themselves or were actively endorsed by them. [43] The truth is, the entire pandemic era was marked by "*misinformation*" propaganda from authorities. But what triggered this dystopia? It was not exclusively the social media platforms; it has been conclusively shown that official authorities and various agencies manipulated the pandemic narrative—

39 Dodd, R. H. et al. (2021). Concerns; Machado, B. A. S. et al. (2022). The Importance, 8; See also Hampton, L. M. et al. (2021). General determination of causation.
40 Piccaluga et al., 4.
41 Ferranna, M. (2023). Causes and costs, 471; Glette-Iversen, I. et al. (2023). A risk science, 2784, 2790; Italia, M. et al. (2023). Model-informed, 11; Koh, K. P. et al. (2024). Understanding the determinants, 2; Löffler, 2; Morens, D. M. et al. (2022). The Concept, 1; Obeng, E. M. et al. (2022). Multivalent ACE2, 2.
42 Piccaluga et al. (2022), 3.
43 Calisher, C. et al. (2020). Statement in support of the scientists; Naeem et al., 145.

> The media (TV, newspapers, magazines, etc), medical societies, state medical boards and the owners of social media have appointed themselves to be the sole source of information . . . Websites have been removed, highly credentialed and experienced clinical doctors and scientific experts in the field of infectious diseases have been demonized, careers have been destroyed and all dissenting information has been labeled "misinformation" and "dangerous lies", even when sourced from top experts in the fields of virology, infectious diseases, pulmonary critical care, and epidemiology.[44]

Blaylock's (2022) makes a similar point—

> [Covid] is one of the most manipulated infectious disease events in history, characterised by official lies in an unending stream lead by government bureaucracies, medical associations, medical boards, the media, and international agencies.[45]

Dubé et al. (2014) pre-covid, posit an interesting angle on hesitancy, relying on Mnookin (2011), who states, "*. . . vaccination has become a source of fear and a target for misinformation.*" His point was intended to signify that vaccination has become a subject of fear, largely driven by misinformation. However, in doing so, it inadvertently illuminates the greater truth: that *bona fide* reasons are now themselves mischaracterised by vaccine advocates as misinformation, or attributed to misinformation, or simply dismissed.[46] Furthermore, even within the scientific community, there is ongoing debate over mRNA vaccines. While these disagreements are part of legitimate discourse, pro-vaccine messaging is not exempt from contributing to the further blurring of the lines. The misinformation delirium can be dissected into two core principles—

i. **Negative Principle:** This involves the systematic suppression or marginalisation of alternative scientific perspectives, including the exclusion and silencing of dissenting peer-reviewed literature. It also entails the dismissal of scientists and medical professionals who challenge prevailing consensus positions, undermining intellectual diversity and stifling critical discourse within the scientific community.

ii. **Active Principle:** This principle focuses on the promotion of pharmaceutical products through industry-funded research, authorship practices, and the selective presentation of data in published studies. It is perpetuated by industry, government agencies, medical institutions, and media, collectively advancing dominant narratives around vaccine safety. This coordinated effort narrows genuine scientific inquiry and ethical considerations, consolidating a single, monolithic perspective.[47]

Indeed, in both constructs, Kennedy (2021) argues that various internet agencies "*. . . propped up all official pronouncements while scrubbing all dissent.*" He highlighted how conglomerates were incentivised with the strategic benefits for the larger enterprise

---

44 Blaylock, 1.
45 Ibid.
46 Dubé, E. et al., 1765; Mnookin, S. (2011). The Panic Virus, 20.
47 Blaylock, 2.

of encouraging public compliance . . . ,"[48] while simultaneously rebutting and censoring any perspectives that deviated from the official narrative regarding Covid vaccines and pharmaceutical treatments. Unsurprisingly, potentially credible postulations were either lost in a flood of conspiracy theories or were immediately labelled as such—blurring the boundary between truth and conspiracy. In other words, any potential truth outside the dominant narrative was broad-brushed as conspiracy, leaving the established position unchallenged. Caulfield (2020) articulates this issue quite acutely—

> The fight against pseudoscience is weakened if trusted medical institutions condemn an evidence-free practice in one context and legitimise it in another. We need good science all the time, but particularly during disasters.[49]

By implication, this stresses the necessity of a reciprocal framework. There are always multiple professional medical bodies with differentiated opinions, diverse political parties to debate and evaluate key issues, legal actors upholding the due process of law, and three branches of government designed to prevent the corruption or tyranny of power. Essentially, this is the structural significance of how Caulfield's (2020) point operates.[50] Llewellyn (2020) and Naeem et al. (2021) cite Wardle (2020), a disinformation expert at Harvard University, as saying—

> The best way to fight misinformation is to swamp the landscape with accurate information that is easy to digest, engaging, and easy to share on mobile devices. It should also answer people's questions and, ultimately, fears.[51]

In normal circumstances, this might constitute sound advice. However, Wardle's solution is an '*Oversimplification*' construction, which fails to account for the underlying complexities and vectors of misinformation. In short, the variable presented does not address the underlying dilemma. Indeed, government officials themselves were issuing misinformation and violating their own guidelines (e.g. Obama's 60th birthday event, Rishi Sunak and Boris Johnson party scandal).[52] These instances, while significant, are part of a broader pattern. Misinformation was deeply entrenched at every level. The saturation of curated falsehoods had reached such an extent that it prompted widespread distrust (Level 2: Incidental Behavioural Reactions in Table III-1). The flooding of accurate information with misinformation contributed to its proliferation, creating ostensibly credible versions of the truth, until truth and falsehood became indistinguishable.

Indeed, both Vraga & Bode (2017), Islam et al. (2021), and Sisco & Brummette (2024) along with others cited in this section, present an '*Appeal to Authority*' construction, by selectively focusing on misinformation from social platforms. While relevant, their analyses rely heavily on institutional credibility and fail to critically examine misinformation originating from scientific, medical, and governmental authorities. Although Sisco & Brummette categorise critiques of Dr. Fauci as misinformation,

48 Kennedy Jr, R. (2021). The Real Anthony Fauci, unpaginated.
49 Caulfield, T. (2020). Pseudoscience, 2.
50 Naeem et al., 146.
51 Llewellyn, S. (2020). Covid-19; Naeem et al., 148.
52 Additionally, Silvio Berlusconi (Italy), Emanuele Pozzolo (Italy), Matt Hancock (UK), Gavin Williamson (UK).

they offer no defence of his actions, nor do they interrogate the institutional dynamics that contribute to public distrust. This is a '*Tu Quoque*' argument in which Sisco and Brummette implicitly align themselves with Dr. Fauci by exempting him from scrutiny, while reserving extreme censure for non-institutional actors. They criticise misinformation on social platforms but fail to apply the same standards of scrutiny to institutional representatives such as Fauci.[53] This is not neutrality; it is selective framing. It replaces balanced inquiry with the one-sided protection of institutional voices. In short, the discourse reveals a fundamental lack of symmetry.

This rhetorical structure borders on a Bandwagon Fallacy, in which widespread acceptance is mistakenly equated with inherent validity. While scientific consensus is vital for evidence-based policy, invoking the popularity of expert opinion to silence or delegitimise dissent risks breaching epistemic integrity. The central concept here is "*mass suggestion.*" The critique does not conflate consensus with popularity per se; rather, it interrogates how apparent unanimity can reflect evidential bias by privileging majority viewpoints and excluding controversial or alternative analyses. This framing incorporates two additional argumentative constructs: Epistemic Inversion and the Virality Fallacy. Importantly, the critique does not reject expertise; it questions how expertise is framed, amplified, and institutionally protected—

I. Dissent in this instance is erased through *epistemic conformity*, where truth becomes indistinguishable from power. In such environments, truth-seeking demands friction—not consensus. As seen during the Covid pandemic, some conglomerate institutions were the epistemic gatekeepers, defining what counted as valid knowledge and marginalising dissenting or emerging perspectives, even when methodologically sound. The contention that this is an overgeneralisation of institutional behaviour—arguing that dissenting views are not universally suppressed—is recognised in this critique. Indeed, numerous journals participating in global surveillance projects reported vaccine-related adverse events, demonstrating that not all institutions fostered *epistemic closure.*

   To reiterate, scholarly practices that reinforce one another's positions through citation loops—rehashing the same circular arguments—create a feedback mechanism: credibility is granted to those who consolidate institutional norms, not those who challenge or refine them. As a result, this creates thousands of high-impact journal articles, all "*beating the same drum,*" further entrenching a narrow range of ideas and creating an illusion of consensus. This phenomenon is best described as an intellectual "*echo chamber,*" wherein the consolidation of a particular worldview takes precedence over the advancement of knowledge. Covid was not the first instance.

   In fact, this pattern has occurred before. Leiss and Chociolko (1994) cite an instance where scientific opinions on the health effects of power frequency and electric and magnetic fields ("PF E/MF") from power lines were sharply polarised. In this case, researchers undermined legitimate scientific inquiry by producing biased studies designed to support opposing positions, thereby distorting the evidence base. These studies were systematically manipulated to adhere to specific, entrenched viewpoints. Regarding PF E/MF they stated—

[53] Sisco & Brummette, 542–43.

> [T]echnical expertise was . . . exploited by all parties . . . to justify their views, to create legitimacy, and to control . . . the debate. Influence depended on the ability to manipulate knowledge or challenge evidence presented to support a particular position. Because of the state of scientific understanding . . . it did appear that studies were lined up on either side of the public debate about regulations, or, at least, that evidence could be found to support any value position in the regulatory debate. The credibility of the experts was questioned, in both their roles as researchers (because studies were "rigged") and as expert witnesses or advisers (because of "selective" use of studies). . . . It was often observed that 'science could be bought.'[54] (Footnote call numbers omitted)

The key distinction between the PF E/MF issue and the debates surrounding Covid, however, is the nature of the opposing views: in the case of the former, multiple, competing bodies of opinion existed. The irony, however, is that Leiss and Chociolko's paper was specifically dedicated to addressing how to overcome "*obstacles to consensus*" in scientific discourse. In contrast, the scientific community's response to Covid largely presented a unified, consistent consensus. Yet, despite the diversity of opinion on the PF E/MF issue, the final conclusion reached, Leiss and Chociolko (1994) reported that "*[t]he scientific evidence was 'insufficient,' 'inconclusive,' and 'unreliable.'*"[55]

Several prominent publications, representative of a broader trend in the literature,[56] gave no meaningful deference to vaccine concerns. Instead, they buttressed a narrative that sidelined dissent as "*misinformation*" rather than interrogating whether such vaccine safety claims held any validity. While there was enormous misinformation in the global public domain, it bears on my earlier point that misinformation was widespread. It is difficult to understand how anyone could maintain that the general public's "*safety concerns*" were speculative when official global agencies, formally mandated to oversee safety, were simultaneously publishing findings on verifiable adverse events. The facts were readily available.

Their omission may not rise to the level of epistemic bad faith, but it invites adverse inferences. These condemnations were not only premature, but fundamentally at odds with the evidence emerging from the official global surveillance efforts. This omission is especially striking in Sisco & Brummette's publication, given that adverse event analyses from biomedical researchers were available as early as 2021–2022. Three years was ample time to engage with these findings. What makes this instance of *Epistemic Inversion* more compelling is that it would be understandable if these articles were only recently published, with research still emerging; in that case, it could be construed as premature to

---

[54] Leiss, W. & Chociolko, C. (1994). Risk, 138. See also Fischer, F. (1990). Technocracy, 35–36.
[55] Leiss & Chociolko, 139.
[56] See Sisco & Brummette (2024), Piccaluga et al. (2022), Naeem et al. (2021), Löffler (2021), Islam et al. (2021), Danchin & Buttery (2021), and Coustasse, A. et al. (2021). These authors consistently framed vaccine safety concerns as "misinformation," without engaging substantively with adverse event data published by biomedical researchers between 2021 and 2022. This pattern of omission is especially notable given the availability of global surveillance findings during that period.

draw definitive conclusions. However, the data had long been available, with initial publications dating back to 2021–2022.

By contrast, a representative portion of researchers, such as Mohammadi et al. (2025), Faksova et al. (2024), and Raethke et al. (2023) provided comprehensive data disclosures and classification methodologies for injuries associated with Covid vaccines—findings that the aforementioned institutional commentators labelled "*misinformation.*" Islam et al. (2021), for example, was an early study that closely mirrored Sisco & Brummette's (2024) thematic analysis of misinformation. Given the release of Islam's et al. in 2021, they could reasonably be excused for not incorporating reports on vaccine harms, yet it is quite plausible that they were aware of these early studies. Once more, no balanced discourse.

II. *Epistemic Inversion,* is where legitimate concerns are prematurely dismissed as dangerous misinformation, only to be later validated. Sisco & Brummette acknowledge that researchers identified problems with vaccine efficacy, but only in passing mention.[57] In their results section, concerns reappear as a 'tweet' categorised under Cluster Four misinformation—despite documented evidence to the contrary.[58] Indeed, a broad range of adverse events linked to Moderna, Pfizer, and AstraZeneca vaccines were reported worldwide.[59],[60]

In particular, AstraZeneca's vaccine was suspended in several countries — Denmark, Norway, Sweden, and Finland on March 11, 2021, followed shortly by Germany, France, and Italy—due to concerns over rare thrombotic events. In contrast, neither Moderna nor Pfizer vaccines were suspended by any government. The decision to suspend AstraZeneca's use in multiple countries suggests a level of credible concern. While both the United Kingdom and the United States recorded adverse events, the UK did not suspend AstraZeneca but instead recommended that individuals under 30, later extended to those under 40, be offered an alternative.[61]

---

57 Sisco & Brummette, 539.

58 Ibid., 541–42.

59 Aven, T. et al. (2024). Special issue; Abolmaali, M. et al. (2022). Guillain-Barré syndrome; Blumenthal, K. G. et al. (2021a). Safety Surveillance; Blumenthal, K. G. et al. (2021b). Acute Allergic Reactions; Dias, L. et al. (2021). Cerebral Venous Thrombosis; Diaz, G. A. et al. (2021). Myocarditis; Faksova, K. et al. (2024). COVID-19, 2201; Glette-Iversen et al., 2784; Hampton, L. M. et al. (2021). General determination; Kim, K. H. et al. (2022). Onset of various CNS inflammatory; Kayser, V. & Ramzan, I. (2021). Vaccines and vaccination; Lee, J. et al. (2023). Knife's edge, 1308–11; Mohammadi, S. et al. (2025). COVID-19 vaccine safety; Patone, M. et al. (2021). Neurological complications; Parhiz, H. et al. (2024). mRNA-based therapeutics; Paterlini, M. (2021). Covid-19; Shimabukuro, T. & Nair, N. (2021). Allergic Reactions; Yu, M. et al. (2023). Guillain-Barre syndrome.

60 **Here is a summarised list of all reported adverse events, compiled from the above journal vaccine safety literature and surveillance data:** Guillain-Barre Syndrome, Transverse Myelitis, Facial Palsy, Acute Disseminated Encephalomyelitis, Generalized Seizures, Febrile Seizures, Encephalitis, Meningitis, Peripheral Neuropathy, Multiple Sclerosis, Optic Neuritis, Small Fiber Neuropathy, Cerebral Venous Sinus Thrombosis, Splanchnic Vein Thrombosis, Pulmonary Embolism, Thrombocytopenia, Immune Thrombocytopenia, Deep Vein Thrombosis, Disseminated Intravascular Coagulation, Hemolytic Anemia, Thrombotic Thrombocytopenic Purpura, Lymphadenopathy, Myocarditis, Pericarditis, Acute Myocardial Infarction, Cardiac Arrhythmias, Hypertension, Stroke, Venous Thromboembolism, Takotsubo Cardiomyopathy, and Heart Failure.

61 EMA. (2021). EMA finds possible links; Glette-Iversen et al., 2784; Petersen, M. B. et al. (2022). Did the European suspension; Pottegård, A. et al. (2021). Arterial events; Pugh, J. et al. (2022). Vaccine suspension; Vogel, G. & Kupferschmidt, K. (2021). New problems; Xu, S. et al. (2024). Mortality risk.

In Hill's (1965) classic analysis on causality in illness, Hampton et al. (2021) distilled the key criteria for assessing vaccine-related injuries, including "*. . . temporality, effect size and its measurement uncertainty, coherence across multiple lines of evidence, and biological plausibility.*"[62] While causation is complex and multifactorial, it is statistically improbable that three distinct vaccines, administered across multiple nations, would consistently exhibit similar adverse effects without some underlying connection. Subsequent investigations identified a causal link; notably, Moderna and Pfizer were "*. . . associated with an excess risk of serious adverse events.*"[63] These findings emerged alongside global surveillance initiatives such as the CDC-funded Global COVID Vaccine Safety (GCoVS) Project. According to Faksova et al. (2024), the study period was generally December 14, 2020 to December 14, 2021—a timeframe during which biomedical researchers from around the world participated in a coordinated global collaboration—

> [F]rom 10 sites across eight countries participating in the unique Global COVID Vaccine Safety (GCoVS) Project . . . is a Center[] for Disease Control and Prevention (CDC) funded global collaboration of investigators and data sources from multiple nations for the purpose of COVID-19 vaccine safety monitoring.[64]

Raethke et al. (2023) illustrate the classification methodology employed by a regional sub-study of the broader GCoVS collaboration across seven European countries. Although the principal GCoVS report does not detail its methodology, Raethke's et al. sub-study offers a glimpse into the epistemic architecture behind the data; both studies pooled aggregated results across sites: adverse reactions were stratified as "*. . . solicited, non-solicited; serious/non-serious; and adverse events of special interest and stratified by vaccine brand.*"[65] The primary remit of the GCoVS protocol was to conduct a retrospective cohort safety surveillance study, which Faksova et al. (2024) pointed out as a large-scale surveillance initiative—

> [C]ompiled electronic healthcare data on AESI (Adverse Events of Special Interest) related to COVID-19 vaccines from participants across multiple sites within the GVDN (Global Vaccine Data Network) [from around the world] . . . .[66]

Referring back to my earlier point that not all conglomerate institutions adopted epistemic closure, these projects are demonstrably clear that candidate data was not only collected, but also actionable, even if selectively used. Given the scale and rigour of this global collaboration, it is difficult to reconcile how concerns about vaccine safety were so readily dismissed as 'misconception, myth, conspiracy,

62 Hampton et al. (2021), 1478; Hill, A. B. (1965). The Environment and Disease.
63 Black, S. & Evans, S. (2023). Serious adverse; Fraiman, J. et al. (2022). Serious adverse events.
64 Faksova, K. et al. (2024). COVID-19, 2201.
65 Ibid., 2203; Raethke, M. et al. (2023). Cohort Event Monitoring, 391.
66 Faksova et al. (2024), 2201.

and misinformation.' These characterisations emerged when global agencies were publicly reporting adverse event data in real time by a global agency tasked with identifying such risks. The point about stratification and aggregation of adverse reactions was central to a protocol designed specifically to monitor and report safety signals. These findings directly contradicted the tendency to dismiss safety concerns as baseless, yet the rhetoric persisted. Moreover, the infrastructure behind these efforts was misrepresented as a response to vaccine hesitancy—a narrative promoted not only in public health messaging but also by researchers themselves.

Scientific institutions risk falling into the *Virality Fallacy* argument when institutional commentators fail to critically examine the visibility and influence of figures like Fauci. The lack of transparency (Level 1 within Table III-1) in refusing to issue substantive censure for controversial decisions—while simultaneously labelling dissenting claims (such as those from Sisco & Brummette's Cluster Five group)[67] as misinformation is a rhetorical backdoor to virality-based validation. This created a proxy for truth that mimics the logic of Hitler's rallies: the more people cheer, the more persuasive the message appears. In White's (1956) review of Linebarger's Psychological Warfare, he asserts that manipulative information is most effective when it is—

> Developed 'scientific in spirit' and as a teachable skill, when its premises are clearly stated, its mission defined, its instruments put in systematic readiness, and its operations subject to at least partial check.[68]

Fauci's scientific omnipresence in the media and journals, regardless of the accuracy of his claims or the implausibility of some statements, elevated his credibility through constant repetition and reinforcement—commonly referred to as the "*illusory truth effect*." This phenomenon represents an old-school propaganda and psychological warfare technique, and a form of occult 'incantation.' See also Counterarguments, 95. Unkelbach & Greifeneder (2007, 2009, 2017, 2018, 2019), Fazio et al. (2019) and Hassan & Barber (2021) affirm that repeated exposure to statements or declarative information leads to the perception that these statements are, "*. . . more easily processed and understood, and this processing fluency is used as a signal for truth.*"[69] This parallel is especially relevant, as noted in Part IV. (Romanticism Dynamics, 250), where Hitler's herd mentality propaganda approach in Mein Kampf is laid bare—

---

[67] Sisco & Brummette, 542–43.

[68] White, H. (1956). [Psychological Warfare . . .], 132.

[69] Fazio, L. K. et al. (2019). Repetition, 1705; Hassan, A. & Barber, S. J. (2021). The effects, 1–3; Unkelbach, C. (2007). Reversing the truth effect; Unkelbach, C. & Stahl, C. (2009). A multinomial modeling approach; Unkelbach, C. & Rom, S. C. (2017). A referential theory; Unkelbach, C. & Greifeneder, R. (2018). Experiential fluency, 78–79; Unkelbach, C. et al. (2019). Truth by Repetition.

See also a small selection of articles, including Crossman, R. H. S. (1958). The Creed; Daugherty, W. E. & Janowits, M. (1958). A Psychological Warfare Casebook; Hargie, O. (2010). Skilled Interpersonal Communication; Institute for Propaganda Analysis. (1938). How to Detect Propaganda, 5–8; Jowett, G. & O'Donnell, V. (2006). Propaganda and Persuasion; Lasswell, H. D. (1927). Propaganda Technique; Lee, A. M. (1952). How to Understand Propaganda, 10, 88, 132, 180 passim; Perusse, R. I. (1958). Psychological Warfare; Pratkanis, A. R. et al. (2001). Age of Propaganda; Rankin, C. H. et al. (2009). Habituation revisited.

> If [an adherent of a movement] steps for the first time out of his small work shop or out of the big enterprise, in which he feels very small, into the mass meeting and is now surrounded by thousands and thousands of people with the same conviction . . . he himself succumbs to the magic influence of what we call 'mass suggestion.'[70]

In line with Sisco and Brummette's (2024) call to action, their final concluding remarks state that if individuals must scrutinize sources to make informed health decisions, ". . . *we can create a more informed and resilient society* . . . ,"[71] then according to the same reasoning, institutions must be also held to the same exacting standard. No one, not even the experts, is above the law or immune to scientific rebuttal. Truth demands friction, not blind consensus. To borrow Löffler's (2021) approach to the Covid outcomes: health experts should publicly report in a fair, unbiased, independent ". . . *balanced manner, which could strengthen the general public confidence in science.*"[72] It calls for epistemic humility, transparency, and pluralism, especially in times of crisis when the temptation toward conformity is strongest.

The overwhelming diversity of arguments promoting the Covid vaccines can be understood through the lens of classical propaganda devices. One such device that fits this criterion is known as the "*Card Stacking*" technique, or "*Hasty Generalization.*" Yourman (1939) defines this device as follows—

> 'Card stacking' is a device . . . . to win our support for himself, his group, nation, race, policy, practice, belief, or ideal. He stacks the cards against the truth. He uses underemphasis and overemphasis to dodge or evade facts.[73]

The killing of Charlie Kirk in 2025 further intensified concerns that freedom of speech—in particular, meaningful, honest, analytical debate in public forums—is no longer being tolerated. In any future crisis, not necessarily a pandemic, the lawful exercise of dissent and free speech, may be curtailed if current trends persist, justified by policy logic akin to "*herd immunity*" propositions. Sisco & Brummette cement this logic through a reference-driven 'Appeal to Consequences' argument, framing the issue as one of critical societal importance—

> [M]isinformation [is] one of the most critical issues we face as a society. It has worsened during the COVID-19 pandemic partly because the severity and novelty of the virus have caused high levels of uncertainty among the scientific community, and accuracy, or what is considered the truth, has become a moving target.[74]

[70] Mosse (1999), 92; Neumann, 439, quoting Hitler, Mein Kampf (1971), 715–16. Although, See Hitler, A. ([1925] (1971), 478–79.
[71] Sisco & Brummette, 545.
[72] Löffler, 6. See also Doerfler, 5; Limb, M. (2011). "Vaccine hesitancy."
[73] Yourman, J. (1939). Propaganda Techniques, *156–58.
[74] Sisco, H. K. F. & Brummette, J. (2024). mRNA Vaccine, 538; Vraga, E. K. & Bode, L. (2017). I do not believe you.

The implication is clear: legislative reforms are positioned as necessary to curtail certain freedoms. See also Class-Based Stratification, 283.

iii. Covid Virus Outcomes and Vaccine Outcomes

THIS HEADING AIMS to demonstrate that the new de facto leadership, by incentivising healthcare and nursing facilities under the umbrella of the 'medical body of opinion,' adopted a regressive approach to the Hippocratic Oath, expediting the demise of hospitalised patients. Blaylock (2022) and McCullough (2020, 2021) contend that the majority of hospitalisations and deaths might have been averted had physicians been permitted to employ treatments such as Ivermectin, hydroxychloroquine, azithromycin or Lopinavir and Ritonavir.[75] The form rebuttal response argues that the efficacy of hospital-administered drugs is consistent with acceptable treatment protocols within the scope of professional medical body of opinion. Nevertheless, alternative bodies of opinion have also challenged this view, asserting that these hospital treatments did not at all alter adverse Covid outcomes. Concurrently, other randomised control trials have reported that Ivermectin, hydroxychloroquine, and azithromycin were safe and effective.[76]

Thus, from a medical standpoint, the diversity of divergent valid approaches presents a troubling paradox: one can justify medical murder/misconduct if the treatment protocol adhered to, falls within the body of responsible peer professional opinion. The higher argument suggests that the CDC mandated hospitals, "*. . . to follow treatment protocol that resulted in the deaths of hundreds of thousands of patients,*"[77] irrespective of other viable treatments. Consequently, healthcare practitioners found their hands tied. Additionally, proponents of hospitals argue that healthcare facilities were overwhelmed; Covid exerted a significant strain on resources, exacerbated by inadequate staffing levels.

When Thomson & Ip (2020) expressed that during Covid, "*. . . dubious healthcare practices generat[ed] significant medical ethics controversies;*"[78] they were referring to incidents within the NHS, UK, where medical professionals advised individuals with chronic illnesses to sign, "*Do Not Attempt Cardiopulmonary Resuscitation*" ("DNACPR") forms. Moreover, Thomson & Ip reported that nursing homes similarly encouraged or pressured patients to sign DNACPR forms. In the United States, The New York Times reported that one-third of all U.S. Covid deaths, occurred in New York nursing homes and other long-term care facilities.[79] A report from France indicates that poor management of nursing homes led to overcrowding and congregate settings that significantly increased mortality rates. Similar issues have been reported in Hungary,

---

75 Blaylock, 3–4; McCullough, P. A. et al. (2020). Multifaceted; McCullough, P. A. et al. (2021). Pathophysiological Basis. See also Alexander, P. E. et al. (2021). Early multidrug; Choudhury, A. et al. (2021). Exploring the Binding Efficacy; Derwand, R. et al. (2020). COVID-19 outpatients; Heidary, F. & Gharebaghi, R. (2020). Ivermectin Reis, G. et al. (2021). Effect of Early Treatment; Santin, A. D. et al. (2021). Ivermectin. For a contrasting perspective that characterizes alternative treatments as misinformation, see Sisco & Brummette, pp. 542–43.

For an opposing finding see Ladapo, J. A. et al. (2020). Randomized Controlled Trials. Five randomized controlled clinical trials enrolling 5,577 patients were included. Hydroxychloroquine demonstrated only a 24% reduction in Covid.

76 Ahmed, S. et al. (2021). A five-day course of ivermectin; Beltran Gonzalez, J. L. et al. (2022). Efficacy and Safety; Cadegiani, F. A. et al. (2021). Early COVID-19 therapy; Choudhary, R. & Sharma, A. K. (2020). Potential use of hydroxychloroquine.

77 Blaylock, 3.

78 Thomson & Ip, 13.

79 Yourish, K. et al. (2020). One-Third; See also Kennedy, "Nursing Homes and Quarantine Facilities," unpaginated.

Italy, the Netherlands, and Singapore.[80] This is not without historical precedent either. As previously discussed, in section Nazi Euthanasia, 173, actions taken against those designated as weak was the executive decision of "*expert*" groups to which Weindling (2017) identifies it was the medical profession—

> German medical profession mobilised for mass murder of the mentally ill, disabled and the elderly infirm. . . . This is part of a revised narrative of the origins of the Holocaust, which itself gained in historical recognition, as well as of power structures and categories of persons deemed undesirable from a reconstructed new Germany. "Expert" groups such as physicians attempted to extend their power in the spheres of race and gender, and on the other hand, the categories of groups deemed to be an undesirable burden on the racially 'fit.'[81]

Post-Covid vaccinations, a prevalent belief emerged suggesting that the mortality rate had exponentially increased due to the Covid vaccines, surpassing deaths caused by the virus itself. This is not without precedent either. Baumslag (2005) reports that as early as 1936, large-scale experiments by Japanese scientists involving disease vectors such as anthrax, typhus, cholera, and the plague were conducted on human subjects. Harris (1994) adds that the same facility was producing up to twenty million doses of vaccine annually. Of course, while Covid is not in the same league of severity in terms of pathogenicity and mortality, the underlying principle under scrutiny is perennially the same.[82] The current establishment, however, maintains that high mortality rates were predominantly among the unvaccinated.

There is another historical parallel. Gallagher (1990) and Proctor (1988) remark that Nazi doctors as part of the broader medical community, passively participated in the euthanasia programme because they were aware of its existence. They argue that this awareness stemmed from the mandatory, detailed documentation issued by Dr. Leonardo Conti, the Reich Health Führer, which physicians were required to complete for each chronically ill patient.[83] With Covid, there is no credible published evidence of any ill-doing or clandestine programmes by governments or public health institutions. However, if the Covid-era world governments were eliminating dissidents and the genetically unfit, a cursory examination of the Nazi regime reveals that it was relatively easy to conceal such killings under the official cover of covert medical legislation titles and policy labels.

Readers will recall from the analysis of Nazi Euthanasia, 173, that during the Nazi era, rumours circulated about patients transported to killing hospitals. For example, the Nazi government, to avoid detection and limit public scrutiny, Hitler issued secret directives and deployed confidential killing initiatives by establishing dummy legislation and bureaucratic structures, such as the '*Community Foundation for Institutional Care*' (T4 euthanasia programme) or the '*Reich Association for Hospitals*' (euthanasia centres) or *The General Sick Transport Company* (transportation of the victims). Moreover, similar to the Nazi regime, fervent nationalistic pro-vaccine support and widespread compliance during the Covid era made it arduous to lodge lawful, rational protests or question establishment narratives about alleged rumours circulating on social media platforms.

80 Alexander et al. (2021), 5.
81 Weindling (2017), 13.
82 Baumslag, 206–07; Harris, S. H. (1994). Factories of death, 55, 59; Miller, J. et al. (2001). Germs.
83 Gallagher, *62–68, includes Dr. Conti's memorandum, registration form, instruction leaflet, and sample form; Proctor (1988), 94.

Regarding the falsification of death certificates, after the Nazi medical profession carried out the murders, they falsified the cause of death on official documents. These claims were, of course, stringently denied. Jewish individuals who died of typhus, as discussed in From Disease Legitimisation, 162, the medical establishment recorded false diagnoses or false negatives on death certificates, citing cancers, heart disease, or anything but typhus. It was only post-WWII that the truth came to light. With Covid, the inverse occurred. False diagnoses and false positives of Covid as the sole cause were recorded on death certificates. When these accusations were levelled against the medical authorities, the form response was an unequivocal denial, reminiscent of the tactics historically associated with the Nazi regime's ruling apparatus. To support and validate these denials, the standard rebuttal consisted of routine assertions that the accusations have been "*fact-checked*" and were inaccurate. Yet the reliability of the purported fact-checking bodies was dubious at best. Indeed, the agencies or individuals conducting the fact-checks, much like during the Nazi regime, were either subordinate to, or aligned with, a restructured, Nazi-controlled quasi-medical-political establishment. Yet, currently, it would be virtually impossible to verify within any established evidentiary frameworks. Simply put, any credible evidence that such calculated acts have occurred in modern society would be summarily and intellectually dismissed.

What was written over 100 years ago is relevant today. The adage "*Can a leopard change its spots?*" highlights the persistence of extreme political manifestos. Though these ideologies may come and go, adherents to these philosophies have never disappeared. Frank Walter Steinmeier, the current President of Germany, made a similar observation in his in an address at Yad Vashem, on January 27, 2020, at the commemoration of the 75th anniversary of the liberation of the Auschwitz death camp. While Steinmeir's statement was in context of the Holocaust, it still has relevant today—

> Of course, our age is a different age. The words are not the same. The perpetrators are not the same. But it is the same evil.[84]

Indeed, it is noteworthy that as of 2026, there appears to be an evident absence of unbiased, peer-reviewed journals and books critically examining the outcomes of Covid vaccines. While there is an impressive body of peer-reviewed research on the overarching Covid governance regime,[85] critical formal works on the Covid vaccines themselves are perennially fragmented. Much of the discourse is dominated by a mixed bag of conspiracy theories, blogs, and newspaper articles, rather than rigorous academic scrutiny. To validate the establishment's claims that Covid vaccines are safe, unbiased examination through peer-reviewed scholarly works is crucial.

White (2015) aptly notes, "*. . . public accountability and scrutiny are an appropriate expectation*" to offset arbitrary governance.[86] Honest, unbiased examination through peer-reviewed scholarly works is vital to build and restore trust.[87] This emphasises the need for balanced and transparent research to ensure that public health decisions are based on solid evidence, not skewed unilaterally in favour of the vaccines. Transparency without trust is to an architect what a witch doctor is to a medical physician; simply

---

84 Berenbaum, 2; Steinmeier, F.-W. (2020). Address by President of Germany to the Fifth World Holocaust Forum.
85 Anisin, 7; Lewkowicz, J. et al. (2022). COVID-19 and erosion.
86 White (2015), 303.
87 Boudreaux, B., et al. (2020). Strengthening Privacy.

presenting and endorsing a public health decision does not guarantee the safety of the medicine. Balancing scepticism with a commitment to seeking truth is the cornerstone of a well-informed society. During the inauguration of the Nazi regime, as already discussed, the stigma against Jews was specifically propagated by the medical science establishment. Baumslag (2005) commented that—

> It was not cultural propagandists who organized the infamous "special treatment" of the Jews; it was the public health officials, the scientific journals, the physicians, the administrators, and the lawyers who feared that the very presence of the Jews would endanger their families and ultimately their lives.[88]

Although Baumslag was addressing the threat of biological weapons, her observation draws a parallel to recent events. Indeed, during the Covid pandemic, individuals who declined the mRNA vaccines and questioned their efficacy were labelled with derogatory terms such as "*virus carriers*" or "*Covid spreaders.*"[89] This is congruent with Weiner's (1980, 1985, 2005) attribution theory, which explains how people assign responsibility for outcomes.[90] In the context of Covid, the attribution of responsibility led to defamatory abuse of the unvaccinated, as they were blamed for the spread of the virus, the resulting hospitalisations and the mortality rates, which were viewed as avoidable. Consequently, vaccinated individuals regularly exhibited no sympathy toward the unvaccinated, evoking anger and aggression instead. For example, Canadian Prime Minister Justin Trudeau publicly declared that "*. . . if you don't want to get vaccinated, that's your choice, but don't think that you can get on a plane or a train besides vaccinated people and put them at risk!*"[91]

These remarks were inflammatory at best, crafted to provoke attribution theory en mass against the unvaccinated. By shaping public perception and conduct, they influenced societal attitudes and reinforced measures linked to the subsequently reported failure of Covid herd immunity,[92] generated social tensions and aggression. The principal point to extrapolate from this context is that the propagation of such stigmatisation was not driven by overt aggression from executive institutions. Rather, it stemmed from pronouncements by prominent public figures, including Trudeau, Pfizer's CEO Albert Bourla, and sensationalised media op-eds.[93] It is apparent how easily these interventions incited peer pressure conformity, which became the primary mechanism driving these detrimental attribution responses. Consequently, objectors

---

88 Baumslag, 213. See also Baumslag & Shmookler (2014), 39–44.

89 Bor, A. et al. (2023). Discriminatory attitudes; Claudy, M. C. et al. (2022). Reckless spreader; Demirtaş-Madran, H. A. (2020). Exploring the Motivation; Drążkowski, D., & Trepanowski, R. (2024). "It's Your Fault!"; Filsinger, M. & Freitag, M. (2024). Asymmetric affective polarization; Kubo, Y., & Okada, I. (2022). COVID-19 health certification; Savulescu, J. (2021). Good reasons to vaccinate

90 Weiner, B. (1980). A cognitive (attribution)-emotion-action; Weiner, B. (1985). An Attributional Theory; Weiner, B. (2006). Social Motivation. Cf. Mynatt, C. & Sherman, S. J. (1975). Responsibility attribution, 1112–13, for discussion on self-attribution within a group.

91 Anisin, 66.

92 See Koyama, T. et al. (2020). Emergence of Drift Variants; Morens, D. M. et al. (2022). The Concept of Classical Herd Immunity; She, J. et al. (2022). Challenges of vaccination; Yewdell, J. W. (2021). Antigenic drift. Essentially, Covid is a subtype of influenza. Such viruses are unstable (unlike polio and measles) and undergo rapid mutation, resulting in continuous antigenic drift and other variables. These factors limit the durability of immunity, making it untenable to attain the desired effects of classical herd immunity.

See also Pilz, S., et al. (2022). SARS-CoV-2 reinfections, as authority for the premise that the efficacy and duration of natural immunity protect against reinfection, and that natural immunity "may offer equal or greater protection." See Anisin, 108, citing Pilz.

93 Anisin, 104, 106.

(the unvaccinated) constituted a minority, while adherents (the vaccinated) formed the majority. Staub (1989) was on point in articulating the principle identifying the effectiveness of social influence—

> If everyone seems to be thinking the same way, it may stifle doubt or resistance, even inner resistance. Many people report seeing two lines of clearly different length as equal when a number of other people before them report seeing the lines as equal.[94]

Thus, the point is that it demonstrates how individuals will conform to the majority view, even when it is clearly erroneous or suspected to be so. The unvaccinated were cast as social transgressors and reprobates, prompting calls for punitive measures, stricter regulations, and severe societal sanctions against them for refusing vaccinations. This contemporary stigmatisation echoes latent medieval and pre-Holocaust associations, in which Jewish people were labelled as "*disease carriers,*" a notion that surfaced in various adverse bacteriological metaphors in the 1890s. See discussion in From Disease Legitimisation, 162).

In summary, the definitive truth about Covid vaccines are elusive. Indeed, in the aftermath of the pandemic, it has become increasingly evident that official authorities manipulated the response. The lack of transparent oversight and the presence of censorship raise questions about the integrity of the global academic and scientific communities. The presence of impartial, peer-reviewed literature is critical for sustaining accountability and trust within the medical field.

94 Staub (1989), 65, citing Asch, S. E. (1951). Effects of Group Pressure, 178, 187.

B. Constructing Robust Preparedness Response Systems and Critical Infrastructure

i. Establishing the Narrative for Future Pandemics and Vaccine Distribution Networks

HISTORICALLY, GLOBAL HEALTH issues were the purview of foreign policy and diplomatic discussions aimed at resolving pressing foreign policy issues. Hotez (2014) and Feldbaum et al. (2010) assert that in 2001, vaccine diplomacy and vaccine science diplomacy emerged from the broader framework of global health diplomacy. This was demonstrated when the WHO exercised its first-ever use of its authority to create a global health treaty—the Framework Convention on Tobacco Control ("FCTC"), adopted in 2003.[95] The definition of global health, initially viewed as distinct and derivative of international and public health, was later revised as inherently part of public health.[96]

Vaccine diplomacy emerged as a paramount strategy during the socio-psychological upheavals, championed by scientists to bolster global health security and encourage international cooperation. Feldbaum et al. (2010) explain that diplomacy generally refers to a body of delegates implementing foreign policy to negotiate alliances, trade agreements between nations, and other treaties. Vaccine diplomacy encompasses a wide spectrum of efforts and strategies aimed at addressing global health challenges and fostering international collaboration. Its purpose is to establish a cooperative framework for the production of vaccines and ensure that health interventions are accessible and effective, not only managing current health crises but also preparing for future pandemics. While it spans numerous focal areas, the field is inherently dynamic, continually evolving and adapting. The principal remit is—

A. **Diplomatic Issues and Health Concerns:**
   1. **Health concerns:** Addressing global health issues through diplomatic channels.
   2. **Negotiations:** Engaging in bilateral and multilateral negotiations on health issues.
   3. **Dissent resolution:** Resolving dissent behaviour on vaccines and virus-and-benefit-sharing of viral samples.
   4. **Pandemic challenges:** Tackling social and psychological challenges posed by pandemics.

B. **Strategies and Partnerships:**
   1. **Strategy development:** Formulating strategies and partnerships within countries to improve health programs.
   2. **Health interventions:** Implementing targeted health initiatives and programs.

C. **International Cooperation and Global Health Security:**
   1. **Global cooperation:** Fostering international cooperation to ensure global health security.
   2. **Emergency response:** Coordinating responses to health emergencies.
   3. **Vaccine access:** Ensuring fair vaccine distribution.
   4. **Regulatory standards:** Harmonizing health regulations across countries.
   5. **Joint R&D:** Promoting collaborative research and development projects.

[95] Hotez, P. J. (2014). "Vaccine diplomacy; Feldbaum, H. et al. (2010). Global Health, 84.
[96] Fried, L. P. et al. (2010). Global health is public health; Koplan, J. P. P. et al. (2009). Towards a common definition. See also Kampmark, B. (2020). The pandemic surveillance state, 67n.5.

D. **Healthcare System Enhancement:**
   1. **System enhancement:** Strengthening national and local health infrastructures.
   2. **Vaccination campaigns:** Conducting mass immunization efforts.
   3. **Public education:** Informing and educating the public about health measures.
   4. **Vaccine monitoring:** Tracking the efficacy and impact of vaccines.

E. **Virus Sample Sharing and Safety Concerns:**
   1. **Negotiating virus sample sharing:** Facilitating the exchange of virus samples for research and vaccine development.
   2. **Vaccine Safety and IP:** Addressing vaccine safety and resolving intellectual property issues.

F. **Combating Misinformation and Managing Impacts:**
   1. **Combating misinformation:** Addressing and correcting false information about vaccines.
   2. **Managing societal and mental health impacts:** Dealing with the social and psychological effects of pandemics.

Globalisation is exposing countries to growing risks of newly emerging and re-emerging contagious diseases and bioterrorism that threaten international and national security. Interdependence, a key feature of globalisation, is prompting nations to collectively re-prioritise and rethink national health preparedness strategies. This is achieved by advancing a panoply of global health initiatives, that transcend conventional boundaries to reduce the threat and/or contain outbreaks of infectious diseases.[97] Farmer (1999) and Aginam (2005) eloquently capture this phenomenon, noting several factors. One is that globalisation has created a "*global village environment*" and has brought about what has been described as the "*death of physical and geographical distance.*"

This transformation has redefined geographic divisions, conceiving the world as a singular, interconnected entity. Goodrick-Clarke (2002) observes, "*[g]lobalisation is unleashing a massive flow of capital, information, skills and personnel across national borders,*" alongside heightened international travel and commerce, evolving demographic patterns, microbial adaptation across borders, and the deepening of shared experiences.[98] He & Chen (2021) remarks that "*. . . a single nation-state cannot solve the aforementioned problems individually.*"[99] In November 2001, the Global Health Security Initiative ("GHSI")[100] was launched in response to the 9/11 as the vector for greater integration. Kickbusch et al. (2007b) explained its mandate as an—

> [I]nternational partnership to strengthen health preparedness and response globally to biological, chemical, radiological, nuclear, and pandemic influenza threats. The founding members include Canada, the European Commission, France, Germany, Italy, Japan, Mexico, the United Kingdom, the United States of America, [and the] WHO . . . .[101]

97 Feldbaum et al., *83, 84, 87–88; Fidler, D. P. (2005). Health as foreign policy, 184; McInnes, C. & Lee, K. (2006). Health, Security; Michaud & Kates, 27; National Intelligence Council. (2000). National intelligence estimate.
98 Aginam, O. (2005). Bio-terrorism, 462; Farmer, P. (1999). Infections and inequalities, 39–40; Goodrick-Clarke (2002), 5.
99 He, Z. & Chen, Z. (2021). The Social Group, 70, citing Hutchinson, J. (2003). Nationalism.
100 GHSI established the actual foundational framework for international health cooperation.
101 Kickbusch, I. et al. (2007b). Global health, 971.

This coalition was further solidified when the United Nations Security Council recognised HIV/AIDS as an epidemic and a national security threat.[102] This acknowledgment enabled Feldbaum (2010) to assert that "*. . . foreign policy is now being driven substantially by health . . . ,*" and Kickbusch et al. (2007a, 2013) add that this is also linked to "*. . . protect[ing] national security, free trade, and economic advancement.*"[103] This is more significant than ever. The notion was the catalyst for the Oslo Ministerial Declaration on 20 March 2007. The Ministers of Foreign Affairs from Brazil, France, Indonesia, Norway, Senegal, South Africa, and Thailand issued a joint declaration, articulating their commitment to forge a strategic alliance and appealing to the global community to do the same—

> [T]here is an urgent need to broaden the scope of foreign policy. Together, we face a number of pressing challenges that require concerted responses and collaborative efforts. We must encourage new ideas, seek and develop new partnerships and mechanisms, and create new paradigms of cooperation [in response to global health challenges.][104]

This simple yet insightful statement illuminates a paradigm shift from the embryonic stages of vaccine diplomacy, following 9/11. The increasing number of international agreements within the context of the aforementioned treaties became the apparatus to identify new ground. This was accomplished primarily by International Health Regulations ("IHRs"). See The Role of Surveillance, 312. Kickbusch (2007a) phrased it that this is achieved "*. . . by mapping many activities in global health . . . to establish new mechanisms of coordination.*"[105] Notably, six months later, in September 2007, the International Health Partnership+ ("IHP+") initiative was launched. The Oslo Declaration anticipated IHP+,[106] focused, specifically, on better coordination and effectiveness in health aid, an area not fully developed in the Oslo Declaration.[107] See Figure III-1 below for the timeline of key global health initiatives.

While the burgeoning demand for greater union by nations in the Northern Hemisphere, it left the Islamic world standing aloof. This is where President Barack Obama stepped in. In an endeavour to improve foreign relations and promote peace, he was the apt person to launch a new Science Envoy programme to Islamic nations, given that "*. . . Obama's election was celebrated in many parts of the Muslim world . . . .*"[108] This initiative aimed to bridge the gap and foster collaboration between the United States and Islamic nations, leveraging scientific diplomacy as a tool for building mutual cooperation. In his Cairo speech on June 4, 2009, he promised a new beginning with the Muslim world. This programme fostered scientific collaboration and transformed into a meaningful bipartisan initiative in vaccine R&D diplomacy. It was designed generally "*. . . to connect with scientists in the Islamic world and produce a new generation of life-*

[102] Aldis, W. (2008). Health security, 372.
[103] Feldbaum et al., 82, *88; Kickbusch et al. (2007b), 971; See also Kickbusch et al. (2007a). Global health; Kickbusch, I. & Kökény, M. (2013). Global health; Labonté, R. & Gagnon, M. L. (2010). Framing health. See also Hotez, P. J. (2014). "Vaccine diplomacy."
[104] Amorim, C. et al. (2007). Oslo Ministerial Declaration. See also Enserink, M. (2007). Indonesia Earns Flu Accord; Feldbaum et al., 86–87; Labonté & Gagnon (2010). Framing health.
[105] Kickbusch et al. (2007a), 231.
[106] IHP+ is a global health partnership that includes countries, development partners, and civil society organisations. It also address funding alignment issues not adequately covered by earlier initiatives.
[107] Labonté & Gagnon, 6.
[108] Zewail, A. H. (2010). Science in Diplomacy.

saving products."[109] Fidler's (2005) insightful point on the conception of the "*science-politics dynamic,*" frequently discussed under the term "*scientopolitics,*" is indicative of how the landscape of governance is changing. Power is shifting from normative sovereign state governance towards a science-based politics, a *New Atlantis* utopia, reminiscent of the ideas it proposed 400 years ago. Only in this epoch is this paradigm shift evident in the increasing tendency of—

> [G]overnments, international organisations, and non-state actors are increasingly linking communicable disease problems to critical areas such as national and international security, international trade, and economic development.[110]

This intricate integration underpins the growing influence of scientific insights on political decision-making, reflecting a governance model that adeptly merges safeguarding national security, free trade, and economic advancement. Put simply, the pressing imperative was to develop innovative paradigms of global health governance. Thus, the spectre of a New International Scientific New Order is predicated on the cardinal principles of unified global health collaboration and partnership.[111] This precept was previously identified as categorically paramount by the GHSI, by the Indonesian protest (discussed below), and by the Oslo Declaration.

109 Hotez, P. J. (2010). Editorial.
110 Fidler (2005), 190, 192.
111 See also Fidler, D. P. (2004). Constitutional outlines, 252, 275–76; Fidler (2005), 191–92.

Timeline of Key Global Health Initiatives : From Partnership to Action

2001 — **Global Health Security Initiative (GHSI)**
An international effort established to strengthen global health security and tackle health threats collaboratively.

2007 — **Indonesian Protest**
Indonesia's objection to sharing H5N1 virus samples with the World Health Organization (WHO), citing sovereignty and equitable access to vaccines.

2007 — **Oslo Ministerial Declaration**
A declaration aimed at reducing maternal and child mortality, emphasizing the need for global partnership and targeted health interventions.

2007 — **International Health Partnership+ (IHP+)**
A collaboration to improve health services and outcomes in developing countries through better coordination and funding alignment.

Figure III-1. Timeline of Key Global Health Initiatives and Milestones.

Signatory nations to these treaties ceded authority out of concern for national health security, following a playbook that anticipated such coordinated alliances. Overall, an examination of these treaties shows that they encompass all nations working collaboratively to enhance global health security. This occurred between 2003–2007. Bill Gates's Organisation, *Global Alliance for Vaccines and Immunization* ("GAVI") is redefining international health governance and policy-making. Moreover, the discussion of GAVI involves an interconnected discussion of the Bill & Melinda Gates Foundation ("the Gates Foundation"). Hotez (2014) outlines the scope of vaccine diplomacy, particularly highlighting the intricate distribution networks and the pivotal role played by the Gates Foundation in the formative stages of GAVI. He asserts—

> Vaccine diplomacy refers to almost any aspect of global health diplomacy that relies on the use or delivery of vaccines and encompasses the important work of the GAVI Alliance, as well as elements of the WHO, the Gates Foundation, and other important international organisations.[112]

GAVI was established in 2000 and was later renamed to '*GAVI, the Vaccine Alliance*,' with the Gates Foundation as a founding partner and major donor.[113] In fact, Fidler (2004) noted that the Gates Foundation, "*. . . are pumping unprecedented sums of money into public health preparedness and response in the developing world,*" while acting as the chief donor to the WHO.[114] Small wonder, given the Gates Foundation's extensive philanthropic contributions, it arguably exerts considerable leverage over the strategic direction and policy decisions of both GAVI and the WHO. Vaccine diplomacy provides developing countries access to vital vaccines that would otherwise remain unattainable due to the to the high costs imposed by major pharmaceutical corporations, such as GlaxoSmithKline and Pfizer.[115] For example, in January 2007, Indonesia refused to share H5N1 viral samples, arguing patented vaccines were later, "*. . . [sold] back to developing countries at unaffordable prices,*" disproportionately benefiting wealthier nations.[116] Sedyaningsih et al. (2008) succinctly highlighted the imbalance of negotiating power between developed and developing countries concerning virus access and benefit sharing. Indonesia's appeal for change demanded a fairer and more equitable system—

> Poor countries have no bargaining position because their participation in the production of these products are not valued as they are "just" natural resources (clinical specimens, viruses, and other microbes) . . . If the world continues to operate in this way, the discrepancies will become wider and wider. . . . It is the responsibility of all nations to change this situation. Indonesia believes that the world must work in unity against the H5N1 virus infection and other diseases, and not taking advantage of the misery of others. The work must be conducted side by side with mutual trust, transparency and equity as global citizens and professionals . . . .[117]

112 Hotez (2014), 2. See also Benyera, 198.
113 See the Gavi website > Donors > Bill & Melinda Gates Foundation, accessed as at Dec. 3, 2024.
114 Fidler (2004), 281; Benyera, 207.
115 Fidler, D. P. (2010). Negotiating equitable access; Hotez, P. J. (2010). Editorial; Hotez (2014), 4.
116 Garrett & Fidler, 1712; Sedyaningsih et al., 486; Labonté & Gagnon, 8–9.
117 Sedyaningsih et al., 487. See also Kickbusch et al. (2007a), 231 and Benyera's, 202, general notation about Covid sales to Africa.

What began with the GHSI in 2001 reached a pivotal moment with the Indonesian issue in 2007. This event fostered greater integration into a global health-sharing approach and international cooperation,[118] promoting, in particular, a more equitable distribution of health resources. While historically significant, this is now an academic discussion. Generally, such diplomatic matters are now increasingly managed within GAVI, which has strategically expanded its remit to encompass the realm of Global Health Diplomacy. What was once the exclusive domain of diplomats has evolved enormously, with GAVI leveraging its position to negotiate more affordable prices for pharmaceutical drugs worldwide and to broaden access to essential medicines and vaccines on a global scale.

Accordingly, in response to Covid, the Gates Foundation contributed an initial $4.4 billion and pledged an additional $1.6 billion for R&D, supporting vaccine development and global distribution. These financial commitments enabled the development of new Covid vaccines, cost reductions through manufacturing efficiencies, and ensured widespread access in low-income countries. The GAVI board comprises 18 representative seats, including the Gates Foundation, the WHO, UNICEF, and the World Bank.[119] This level of influence raises a critical question about the Foundation's potential role, whether contributory or culpable, in the light of reported vaccine injuries worldwide. Ironically, the Global Vaccine Action Plan 2011–2020 ("GVAP"), endorsed by the 194 Member States of the World Health Assembly, targeted "*. . . efforts to introduce new vaccines, if undertaken together, [can] avert millions of future deaths, as well as hundreds of millions of cases of disease [by 2020] . . . .*"[120] Interestingly, the signatories of GVAP included—

i. Dr. Seth Berkley, GAVI executive
ii. Dr. Christopher Elias, President of the Gates Foundation
iii. Dr. Margaret Chan, Director General of the WHO
iv. Dr. Anthony Fauci, Director of NIAID

Ironically, the outcomes of Covid starkly contrasted with the stated objectives of these initiatives. Under the auspices of "*philanthropy*" provision, deaths and adverse health conditions worldwide were attributed to both the virus and the vaccines. Furthermore, Forbes (2021) reported that the pandemic enriched billionaires generally and, in particular, generated billions of dollars for pharmaceutical companies, "*. . . at the expense and interests of working and middle classes.*"[121] Beneficiaries allegedly included Fauci, Bill Gates, existing billionaires, and other entities across the supply chain.[122] For example, Gilead Sciences set the price of the experimental drug remdesivir at $3,120 per vial for 500,000 treatment courses despite reported production costs of $0.93/day, whereas the FDA-approved drug, hydroxychloroquine, costing $20.00 for 200mg oral tablet, $0.08/day.[123] The WHO's comprehensive vaccination plan, GVAP (promulgated in 2011), explicitly reflected this economic cost-benefit approach to vaccine deployment. Fast-forward to Covid vaccine production, the GVAP plan achieved its stated objectives—"*. . . generating hundreds of billions of dollars in economic impact over the*

[118] He & Chen, 74, 78.
[119] See the Gavi website.
[120] WHO. (2013). Global vaccine action plan, 28, 72, 77, accessed as at Dec. 3, 2024. See also Hotez (2014), 4.
[121] Anisin, 30, 37–38.
[122] Anisin, 4; Petersen-Withorn, C. (Apr 30, 2021). How Much Money America's Billionaires Have Made. See also Kennedy, "Enriching the Wealthy," unpaginated. See also Hirschler, B. (2016). GSK and Google.
[123] Denham, H. et al. (2020). Gilead sets price; Hill, A. et al. (2020). Minimum costs.

decade."[124] Benyera (2021) asserts that "*. . . Bill Gates was at the forefront of denying countries of the Global South permission to produce the vaccine.*"[125]

In hindsight, this paradox invites a critical examination of potential premeditated collusion as early as 2011, with the aim of designating 2020 as the year of a global pandemic. This coincided with the GVAP initiative, a nine-year preparedness strategy launched in 2011 that required an unprecedented global stress test of pandemic response systems. The principle of "*preparedness*" indicates readiness to manage a pandemic scenario broadly, not specifically for Covid. However, in this instance, Covid was the metric by which the preparedness plan was assessed.

Moreover, the Gates Foundation was on the GVAP board from 2011 and funded Covid vaccine distribution. According to official press releases, the Gates Foundation allocated $70 million toward global vaccine development and distribution of Covid vaccines and pledged an additional $50 million to support the marketing efforts of GAVI's COVAX Advance Market Commitment ("AMC").[126] Fauci also served on the GVAP board during this period as an advisor and head of the newly established interim Covid government in 2020. While an interesting coincidence, this correlation is hypothetical and unproven; correlation alone does not imply causation.[127]

ii. Future Global Preparedness Initiatives Refinement

THE INITIATIVES DELINEATED in Table III-2 critically address deficiencies in their predecessors. Key aspects include: (i) the pathogenic characteristics of the virus, (ii) the globally coordinated response structure, and (iii) the mitigation of dissent among country leaders while ensuring the equitable distribution of vaccines. The Ebola outbreak of 2014, notable for its severity and visibility, revealed significant weaknesses in preparedness planning, resulting in a lethargic response.[128] Each successive initiative, reflecting the dynamics of pathogen severity, has retrospectively unveiled gaps and loopholes in cooperation and the coordinated response. The WHO (2024) acknowledged that these issues post-Covid identified opportunities for improvement.[129] Subsequent initiatives, informed by lessons from previous programmes and new target areas, have been strategically designed to construct a comprehensive global framework capable of anticipating every conceivable scenario—

Table III-2. Key Global Health Initiatives and Their Impact (2010-Present)

Year	Initiative	Description	Addressing Gaps and Flaws
2011-2020	Global Vaccine Action Plan (GVAP)	A framework to reduce vaccine-preventable deaths by improving equitable access to existing vaccines globally.	Exposed uneven delivery capacity and limited scalability in global immunisation systems.[a]

[124] WHO (2013), 28, 72–79.
[125] Benyera, 207.
[126] Bill & Melinda Gates Foundation. (2020). Gates Foundation announces new funds to develop COVID-19 vaccines (Bill & Melinda Gates Foundation > Media Center > Gates Foundation announces new funds to develop COVID 19 vaccines), accessed as at Dec. 3, 2024.
[127] Apart from the official sources already presented, there is little or no other evidence or perspectives available to negate confirmation bias. What it does provide us with, without further information, is critical thinking and a healthy scepticism to question whether such high-level coincidences are purely coincidental.
[128] Holmes, E. C. et al. (2018). Pandemics, 180; See brief notation on Ebola in Kampmark, 61; WHO. (2007). The world health report, 61.
[129] WHO. (2024). Pandemic influenza, 8, 14–16.

Table III–2. Key Global Health Initiatives and Their Impact (2010-Present)

Year	Initiative	Description	Addressing Gaps and Flaws
2014	Global Health Security Agenda (GHSA)	Launched to accelerate implementation of international health regulations and strengthen global health security.	Exposed delays and weak compliance in implementing international health regulations across national health systems.[b]
2014	Ebola Response	International efforts to contain and manage the Ebola outbreak in West Africa.	Revealed weaknesses in global outbreak preparedness and the need for faster, coordinated international response mechanisms.[c]
2015	Sustainable Development Goals ("SDGs")	Included health targets to end epidemics, reduce maternal and child mortality, and achieve universal health coverage.	Exposed the lack of binding enforcement mechanisms and fragmented accountability in global health commitments.[d]
2019	Global Action Plan for Healthy Lives and Well-being for All	A roadmap to accelerate progress toward health-related SDGs.	Addressed fragmentation in SDG implementation by promoting cross-sector coordination across health, education, and infrastructure.[e]
2019	Gavi 5.0 Strategy	Focused on reaching zero-dose children and underserved communities, and strengthening immunisation programme sustainability.	Tackled the challenge of reaching unvaccinated populations and improving immunisation sustainability.[f]
2020	COVAX Initiative	A global effort to ensure equitable access to Covid vaccines.	Sought to address unequal vaccine distribution between high-income and low-income countries during the Covid response.[g]
2023	Health Emergency Preparedness, Response, and Resilience (HEPR)	Focuses on strengthening global health surveillance through improved technical and operational capacity and cooperation between governments, NGOs,[h] and electronic surveillance systems.	Addresses weaknesses in early warning systems and fragmented data-sharing by improving real-time public health intelligence and coordination.[i]
May 2025	Pandemic Agreement	A legally binding global framework to strengthen pandemic prevention, preparedness, and response while respecting national sovereignty, promoting early detection, coordinated action, and equitable access to vaccines, diagnostics, and treatments.	Responds to COVID-19 failures including fragmented coordination, unilateral state action, inequitable resource distribution, poor data transparency, and delayed response systems by strengthening preparedness, supply chains, and collective accountability.

Table III-2. Key Global Health Initiatives and Their Impact (2010-Present)

Year	Initiative	Description	Addressing Gaps and Flaws

Notes

[a] GVAP was developed by the WHO in partnership with UNICEF and the NIH.
[b] GHSA was launched in February 2014 by the CDC in collaboration with 44 countries and organisations, including the WHO.
[c] Ebola's impact was exacerbated by limited national capacity for early detection, prevention, and response. Although GHSA aimed to address such gaps, implementation was constrained by weak infrastructure, workforce shortages, and limited infection control capacity.
[d] The SDGs are a set of 17 global goals adopted by the United Nations in 2015.
[d] The Global Action Plan launched at the UN General Assembly in September 2019, bringing together 12 health, development, and humanitarian agencies.
[f] Gavi 5.0 (2021–2025) is a five-year strategy focused on strengthening health systems, introducing innovative financing, and ensuring no child is left behind in immunisation.
[g] COVAX is a global initiative co-led by Gavi, the Coalition for Epidemic Preparedness Innovations (CEPI), and WHO.
[h] Non-governmental organisations (NGOs)
[i] HEPR (WHO Framework for Strengthening Global Architecture for Health Emergency Preparedness, Response, and Resilience) enhances surveillance, early detection, information sharing, rapid response, and integration of new technologies with standardised data systems.

Five years on after Covid, what did the scientific community learn from Covid? First, from just a cursory observation, it was the first global pandemic of its kind since HIV/AIDS in the 1980s, to initiate a unified and collaborative surveillance response.[130] Second, it was unparalleled collaboration between world governments, the WHO, Centers for Disease Control ("CDC") industry, conglomerate media agencies and multinational companies, social platforms, pharmaceutical companies, and key figures such as Dr. Anthony Fauci, Director of the National Institute of Allergy and Infectious Diseases ("NIAID"). These entities ("the collaboration") worked in concert to implement sweeping emergency measures. From the outset, international agencies articulated a unified message across regions and countries, emphasising coordinated global action to manage and mitigate the impact of Covid. This type of coalition is not without precedent. Baumslag (2005) aptly notes that similar coalitions were formed in Nazi Germany to combat disease epidemics and produce vaccines—

> During World War II there was also an international exchange between German scientists and the Allies for the production and testing of typhus vaccines. The threat of typhus was felt by all nations involved and played a key factor in Allied military plans. [131] (Internal Citations omitted)

In practice, vaccine nationalism prevailed over vaccine diplomacy during Covid, and the aforementioned appeals for equitable distribution, failed.[132] Furthermore, vaccine diplomacy operated coercively, compelling recipients country to accept the non-negotiable terms attached to the Covid vaccines.[133]

A more methodical analysis revealed serious inadequacies in global cooperation. Certain world leaders disregarded scientific counsel and deemed it extraneous to

130 See generally Bayer, R. & Fairchild, A. (2010). When Worlds Collide.
131 Baumslag, 8. See also Weindling (2000) 143–171.
132 Bollyky, T. J. & Bown, C. P. (2020). The Tragedy; Benyera, E. (2021). Covid-19; Sparke, M. & Levy, O. (2022). Competing Responses.
133 Benyera, 199.

the decision-making process.[134] The maxim "*never let a good crisis go to waste*" is often attributed to Winston Churchill, although the authenticity of this attribution is unverifiable. However, Rahm Emanuel, Chief of Staff to President-elect Barack Obama in 2008, is documented as saying: "*You never want a serious crisis to go to waste. [. . .] It's an opportunity to do things you think you could not do before.*"[135] Given this existential pandemic dread, Covid presented an opportunity to expose systemic flaws and realise new mechanisms for change. Consider the following overarching principles as a strategy to foster change—

i. **See problems in action and rectify them:** The pandemic exposed systemic vulnerabilities, creating an opportunity to address deficiencies.
ii. **Optimise and update policy for societal governance:** The crisis highlights the imperative for modernised policies that can adapt swiftly to unforeseen challenges and ensure effective governance.
iii. **Facilitate the establishment of a New International Order:** A coordinated, unified global approach is essential for managing future crises effectively, strengthening cooperation, and improving preparedness.
iv. **Introduce new, cogent international treaties:** Covid highlighted the critical need for world nations to foster greater international cooperation and clearer frameworks to minimise dissent, including treaties that encourage solidarity and coordinated crisis response.

As previously discussed in the section on Managing Dissent, many WHO Member States adopted autocratic approaches to governing the social sphere during Covid. Conversely, some leaders demonstrated inaction or responded in a more permissive, lethargic, or ill-informed manner. Anisin (2024) similarly noted, while not all countries adopted authoritarian measures nor placed their populations under stringent surveillance, the majority did.[136] This is precisely the point—there were divergent approaches identified across countries. Anisin produced a table, collating data from the Oxford Coronavirus Government Response Tracker (OxCGRT) project, which ranked countries by stringency index. The table reproduced below is ordered from least to most stringent—

Table III-3. Stringency Index (n=54)

	Country (Alphabetical)	Stringency Index		Country (Least to Most Stringent)	Stringency Index
1.	Argentina	62.32	1.	Ghana	50.92
2.	Australia	70.72	2.	Brazil	52.11
3.	Austria	73.89	3.	Senegal	53.29
4.	Barbados	62.08	4.	Cape Verde	54.65
5.	Belgium	78.21	5.	Peru	57.32
6.	Brazil	52.11	6.	Mexico	57.84
7.	Bulgaria	65.28	7.	South Africa	58.39
8.	Canada	68.92	8.	Paraguay	58.76
9.	Cape Verde	54.65	9.	Jamaica	61.49
10.	Chile	67.22	10.	Trinidad and Tobago	61.97
11.	Costa Rica	63.84	11.	Barbados	62.08

[134] Fairchild, A. L. et al. (2007). Searching Eyes, 228–33; Forster, T. & Heinzel, M. (2021). Reacting.
[135] Friedman, S. D. (November 25, 2008). Recessions; Stoller, J. K. (2020). Reflections.
[136] Anisin, 5.

Table III-3. Stringency Index (n=54)

Country (Alphabetical)	Stringency Index	Country (Least to Most Stringent)	Stringency Index
12. Croatia	62.56	12. Panama	62.17
13. Cyprus	72.30	13. Argentina	62.32
14. Czech Republic	75.01	14. Romania	62.47
15. Denmark	77.84	15. Croatia	62.56
16. Estonia	70.15	16. Costa Rica	63.84
17. Finland	72.03	17. Poland	64.99
18. France	77.75	18. Bulgaria	65.28
19. Germany	75.89	19. United States	66.43
20. Ghana	50.92	20. Chile	67.22
21. Greece	70.95	21. Uruguay	67.67
22. Hungary	68.71	22. Sweden	68.15
23. Iceland	68.78	23. Slovak Republic	68.48
24. Ireland	75.41	24. Hungary	68.71
25. Israel	73.22	25. Taiwan	68.73
26. Italy	73.93	26. Iceland	68.78
27. Jamaica	61.49	27. Canada	68.92
28. Japan	69.07	28. Japan	69.07
29. Latvia	71.01	29. Estonia	70.15
30. Lithuania	71.67	30. South Korea	70.53
31. Malta	74.02	31. Australia	70.72
32. Mexico	57.84	32. Greece	70.95
33. Netherlands	76.22	33. Latvia	71.01
34. New Zealand	71.63	34. New Zealand	71.63
35. Norway	73.45	35. Lithuania	71.67
36. Panama	62.17	36. Slovenia	71.78
37. Paraguay	58.76	37. Finland	72.03
38. Peru	57.32	38. Portugal	72.12
39. Poland	64.99	39. Cyprus	72.30
40. Portugal	72.12	40. Israel	73.22
41. Romania	62.47	41. Norway	73.45
42. Senegal	53.29	42. Austria	73.89
43. Slovak Republic	68.48	43. Italy	73.93
44. Slovenia	71.78	44. Malta	74.02
45. South Africa	58.39	45. Spain	74.16
46. South Korea	70.53	46. Czech Republic	75.01
47. Spain	74.16	47. Switzerland	75.21
48. Sweden	68.15	48. Ireland	75.41
49. Switzerland	75.21	49. Germany	75.89
50. Taiwan	68.73	50. Netherlands	76.22
51. Trinidad and Tobago	61.97	51. United Kingdom	76.85
52. United Kingdom	76.85	52. France	77.75
53. United States	66.43	53. Denmark	77.84
54. Uruguay	67.67	54. Belgium	78.21
Margin of Difference			27.29

Source data: Anisin, A. (2024). Authoritarian Liberal Surveillance, 46–47, extracted from OxCGRT.

The margin between the most stringent response (78.21 for Belgium) and the least stringent (50.92 for Ghana) is a failure differential of 27.29. This substantial disparity, despite what was envisaged as a globally coordinated and uniform response to Covid revealed a patchwork of heterogeneous governmental approaches and a fragmented coalition for managing the pandemic, resulting in disparate outcomes.[137] Yet, who wants a non-conformist dystopian regime! As aforementioned, the disparity demonstrates underlying objective behind successive global health initiatives, as summarised in Table III-2: the development of an ironclad, cohesive and integrated "*one-global strategy.*" Löwy and Zylberman (2000) make this explicit, noting that the Rockefeller Foundation's overarching aim for world nations and Europe in the 1930s was—

> [R]ationalisation and homogenisation: of populations, of scientific approaches, and of administrative methods . . . [Rockefeller Foundation] directors wished . . . to improve the fate of local populations and to make the world a more hospitable place for North American elites.[138]

This approach seeks to eliminate divergent or competing strategies, in favour of a singular, unified effort. Variations in Covid responses may therefore be interpreted as 'departures from adherence to recommended stringent measures.' Several underlying factors contributed to this non-compliance. First, was the belief that Covid had been orchestrated and, second, was the view that it did not constitute the high-level threat it was portrayed to be. Indeed, critics argued that—

> Fauci and the public health establishment portrayed COVID-19 as a universal threat, when in reality it was a risk for those with pre-existing conditions and elderly segments of the population.[139]

This is not without precedent. According to Feldbaum et al. (2010) between 2000 and 2005, HIV/AIDS pandemic was touted as a "*threat*" but was misleading—

> [M]uch of the evidence . . . including the prevalence of HIV among African militaries and its potential to cause instability in 'next wave' states has been shown to be inaccurate.[140] (Footnote call numbers omitted)

Another example was H1N1. In 2009, the H1N1 influenza virus was identified. On May 4, 2009, one month prior to declaring it a pandemic, the WHO revised its definition of a "*pandemic*" by lowering the threshold for classification. Consequently, the WHO labelled the H1N1 outbreak a pandemic. Doshi (2009, 2011) maintained that the definition had been present on the WHO's Pandemic Preparedness homepage since 2003. He documented and archived the definition, and read as follows—

> An influenza pandemic occurs when a new subtype of influenza virus to which humans have no immunity enters the population, begins spreading widely, and causes severe illness.[141]

---

[137] He & Chen, 68.
[138] Löwy, I. & Zylberman, P. (2000). Medicine, 371. See also Biometric Technologies, 317, for further discussion about the Rockefeller Foundation.
[139] Anisin, 104.
[140] Feldbaum et al., 88, citing Singer, P. W. (2002). AIDS.
[141] Doshi, P. (2009). Calibrated response; Doshi, P. (2011). The elusive definition, 532.

Bloylock (2022) furnishes a similar definition—

> [T]he virus must have a high mortality rate for the vast majority of people . . . and it must have no known existing treatments . . . .[142]

The revised definition read as follows—

> An influenza pandemic may occur when a new influenza virus appears against which the human population has no immunity.[143]

Doshi (2011) reported that there was substantial debate. First, on whether the H1N1 influenza outbreak should have been labelled a pandemic at all and, second, whether the WHO's revision effectively ramped-up the definition, making it easier to declare future pandemics. While the WHO and other authorities critics disagreed, describing it as a "*. . . change in the 'description' but not in the 'definition' of pandemic influenza.*"[144] In 2007, Indonesia protested the WHO's inequitable vaccine distribution. One may plausibly infer that this dispute was a catalyst for a global reassessment of risk, which, in turn, reframed the definition of a "*pandemic*" two years later, in 2009. See discussion under Establishing the Narrative for Future Pandemics, 232. The other key factors included—

i. **Dissenting Actors**: Different countries and organisations pursued conflicting approaches and priorities, resulting in inconsistent and fragmented responses.[145]
ii. **Lack of Global Coordination and Poor Decisive Leadership**: The absence of a centralised, coordinated, and decisive governance structure produced inefficiencies and duplication of effort, undermining health-system resilience and weakening the capacity to adapt and respond effectively, resulting in further fragmentation.[146]
iii. **Inequitable Resource Distribution**: Vaccines, medical supplies, and funding were distributed inequitably, disproportionately affecting low- and middle-income countries.
iv. **Transparency in Reporting**: Delays and inconsistencies in data reporting, combined with limited information sharing, hindered the timely dissemination of critical information.
v. **Slow Response Mechanisms**: The absence of a robust control strategy resulted in ". . . substantial variation in the rate and speed at which stricter . . ." pandemic countermeasures were implemented. This produced divergent strategies, bureaucratic hurdles, and delayed decision-making, highlighting deficiencies in preparedness, readiness, and crisis management.[147]

---

[142] Blaylock, 4.
[143] Doshi (2009); Doshi (2011), 532;
[144] Doshi (2011), 532.
[145] Al Saidi, A. M. O. et al. (2020). Decisive leadership; Kaul, V. et al. (2020). Leadership; WHO. (2021). World health statistics; Lal, A. et al. (2021). Fragmented health; Ortega, F. & Orsini, M. (2020). Governing COVID.
[146] Al Saidi et al., 296; Assefa, Y. et al. (2022). Analysis of the COVID; Forman, R. et al. (2020). 12 Lessons; Gibney, E. (2020). Whose coronavirus strategy; Wu, P. & Sun, Z. (2024). Culture, 2, 9.
[147] Forman, R. et al. (2020); Peleg, K. et al. (2021). The COVID; Wu & Sun, 2.

As elaborated in Table III-2, the pervasive global failures witnessed during the Covid pandemic are addressed in the 2025 initiative the Pandemic Agreement, ratified by the WHO's 194 Member States. The preamble, available on the WHO website, states that the Agreement is intended as a legally binding framework, encompassing the following key aspects—

i. **Standardised Protocols:** Promoting coordinated approaches for data sharing, pathogen exchange, and emergency response, to enhance consistency and efficiency across countries while respecting national sovereignty.
ii. **Global Coordination**: Strengthening WHO-led coordination to facilitate guidance, technical support, and collaboration among Member States during pandemics.
iii. **Equitable Access**: Ensuring fair and timely access to vaccines, therapeutics, diagnostics, and other critical resources, particularly for low- and middle-income countries, to promote global inclusivity in pandemic response.
iv. **Transparent Communication**: Supporting timely, open, and reliable information sharing among Member States and with WHO to improve situational awareness and decision-making.
v. **Enhanced Preparedness**: Encouraging clearer processes, joint planning, and resource mobilisation to enable faster, more effective responses, without overriding national decision-making.

The "*Global Coordination*" posited in point ii., suggests that the WHO and GAVI support the emergence of a single centralised, quasi-authoritative biopolitical governance body, that steers global health policy and intervention. This framework fosters solidarity, across diverse governance systems and political ideologies.[148] In practical terms, vaccine diplomacy has become a significant instrument of global health policy. It enables interventions that intrude upon national sovereignty, yet falling short of being compelling or coercive,[149] it encourages compliance with global health directives "*. . . justified on the grounds of 'global health security,*"[150] be it through commission or omission.

Benyera (2021) broadly defines hard power as "*coercive (military and economic)*" and forceful, whereas soft power refers to the capacity to attract and co-opt.[151] Vaccine diplomacy can therefore be understood as a soft-power mechanism through which global actors exert influence over sovereign states, shaping health governance agendas and encouraging alignment with this emerging centralised governance architecture. In sum, it continues to grow in importance as a dynamic and evolving field,[152] adapting to new framings of pandemic preparedness and response through multilateral partnerships.

A parallel can be drawn from the section, National Awakening, 121, to the institutional restructuring that occurred at the inception of the Third Reich in 1933 under the policy of *Gleichschaltung*. These reforms centralised medical and scientific professions by abolishing institutional autonomy of health agencies, and consolidating oversight within a unified governing structure over public and private health practices and hospitals. From my analyses of the Weimar period, the pre-existing institutional and administrative structures were already in place, facilitating rapid post-1933 reorganisation. As previously discussed, the reforms were embedded within

[148] Forman, R. et al., 578–79.
[149] Feldbaum et al., 84, 86; Michaud & Kates, 26.
[150] Aldis, 372. See also Feldbaum et al., 88.
[151] Benyera, E. (2021). Covid-19, 197.
[152] Michaud & Kates, 26.

Germany's nationalist framework, characterised by singular, centralised coordination and standardised governance. As previously mentioned, Hobhouse (1911) argues that a collective that is in agreement is better than where there is variety of opinion.[153] Consequently, contemporary global health security initiatives in the era of globalisation privilege a unified voice, will marginalise dissenting actors and pluralistic alternative approaches. Hobhouse's construct compares to the transition from Nazi Germany's imperialist Europe approach to the global health framework evident during the Covid pandemic.[154] This model epitomises the paradigm shift toward transnational cooperation predicated on interdependence that "*. . . coordinate[s] actions at local, national, regional and global levels.*"[155]

153 Hobhouse (1911), 187.

154 See He & Chen (2021) for a comparative study on distinctions between nationalist and globalist actions during Covid. See also Greenfeld, L. (2011). The Globalization; Hearn, J. (2015). Nationalism and Globalization. These works are authority for the premise that national integration and interdependence have blurred national boundaries, fostering a sense of community.

155 Assefa, Y. et al. (2021). Reimagining global health, 2; He & Chen, 68; Kickbusch, I. & Szabo, M. M. (2014). A new governance, 2.

PART IV. PRESENT AND FUTURE TRAJECTORIES

> Ordinary sorts of people, using ordinary behaviour, have contributed to extraordinary evil, creating major horrors. As we understand more accurately how the ordinary can be harnessed for doing evil, we make progress toward a new freedom: the freedom to stop manufacturing extraordinary evil. There is hope.
>
> —Katz, Ordinary people and extraordinary evil

> All is god. . . . In the oldest religion, everything was alive, not supernaturally but naturally alive. . . . Rocks were alive, but a mountain had a deeper, vaster life . . . The whole life-effort of man was to get his life into direct contact with the elemental life of the cosmos . . . To come into the immediate felt contact, and so derive energy, power, and a dark sort of joy. This effort into sheer naked contact . . . is the root meaning of religion.
>
> —Lawrence, D. H. (1936). Phoenix : the posthumous papers of D. H. Lawrence

CHAPTER 9. INSTITUTIONAL ARCHITECTURE AND PRECEDING APPARATUS OF THE NEW GLOBAL ORDER

Introduction

In the section on SS education and training, readers saw how obedience to authority was systematically cultivated to produce perpetrators of mass violence. Within the same framework, Romanticist principles were applied to foster state-sanctioned societal violence. This discussion is revisited here to assess its relevance to future regimes. To understand the traits of tyrannical regimes, it is necessary to examining the nature of evil, with the Nazi regime serving as a historical case study. This analysis articulates the preamble for foundations of a Global World Order or New International Order, clarifying the origins and mechanisms of evil in totalitarian contexts and the socio-political and psychological malevolent forces that sustain them. Although grounded in the context of Nazi Germany, these lessons are relevant today. Before the discussion of the continuum of destruction and the operational mechanisms of mass violence, an interlude examines Romanticist doctrine as a catalyst for societal aggression.

A. Romanticism Dynamics in Restructuring Society to Forge a National Identity

i. Emotional Moral Disengagement in Ingroup Romanticist Conditioning

SOCIETAL GROUP VIOLENCE both coincides with and overlaps significantly with the doctrine of Romanticism. The Romantic movement, with its emphasis on emotion, individualism, and the sublime, can be viewed as both a catalyst for and a reflection of the turbulent and violent social upheavals that accompanied it. Romanticism's privileging of emotion and passion over reason corresponds with the fervour and zeal that animate collective violence. Webb (1974) specifies that Romanticism rests on two core principles: first, an intense focus on the self, and second, "*a pleasurable form of escapism [that] results from this concentration on the self.*"[1]

Romanticism was neither a link to nor a construct of Darwinism. It was a vector that, once racial hygiene policies became operational, enabled the brutal, savage aggression those policies demanded at a physiological level, disengaging moral control through displacement of responsibility, without affecting conscience or accountability for the harm caused.[2] Desmond's (1997) extraordinary statement in his biography of T. H. Huxley highlights this dynamic, emphasising the application of Darwin's theory, specifically the struggle for existence, as a justification for genocidal practices.

1 Webb, J. (1974). The Occult Underground, 10.
2 Bandura, A. (1999b). Moral Disengagement, 196.

Huxley asserted, "*. . . only from death on a genocidal scale could the few progress [by the destruction of the majority].*"[3]

The creation of a new "*normal*" within communities, under a regressive globalisation restructuring, could potentially trigger societal violence. A restatement of normal individual conduct, which Slight (1937) terms 'normality,' can be loosely defined as individual urges, impulses, and desires that are self-controlled, restrained, and filtered before being expressed, in ways acceptable within the legal, cultural, and social milieu.[4] However, as this section emphasises, the concept becomes fundamentally flawed if societal morals, values, norms, and institutions retrogress—permitting individual impulses and desires to become the new collective norm—a process Staub (1990, 1999) identifies as part of the continuum of destruction.[5]

In such a scenario, the majority's behaviour becomes the new standard, as witnessed in Nazi Germany. Schuman (1937) articulates a similar point that the psychological appeal to mass social groups relies on the premise that "*[u]nity and enthusiasm are functions of collective stimuli evoking intense and abiding collective reactions.*"[6] Mosse (1975) was on point expressing that "*. . . newly awakened national consciousness performed this function of citizenship.*"[7] Once the unconscionable behaviour becomes accepted as normal *in toto* the selection of new perpetrators becomes possible, and "*[i]nstitutions can change or are created to serve [the new regressive violent apparatus].*"[8] Galton (1998), alongside Cornford's (1955) commentary on Plato, traces such reasoning back to Aristotle laws, who proposed that infanticide laws should be applied to "*. . . children born with deformities.*"[9] Weikart wrote that Darwin saw human conduct as tending toward love and altruism, yet Nazi behaviour represented a regression to primitive instincts.[10]

The principles of beguilement outlined in Table IV-2 above, suggest that the doctrine of Romanticism seeks to rein in disobedience within the populace while maintaining a peripheral form of Romanticism. For example, in Nazi Germany, the integration of the masses and the discouragement of isolation and non-acceptance was achieved by fostering a strong sense of 'community' within the movement, which Hitler defined as "*esprit de corps.*"[11] Accordingly, Hitler employed a herd mentality propaganda strategy, specifically using the fear of isolation from the group, or the fear of not being part of it—

> An adherent of a new movement, [who] feels lonely and is easily seized with the fear of being alone, receives for the first time the picture of a greater community, something that has a strengthening and an encouraging effect on most people . . . If he steps for the first time out of his small work shop or out of the big enterprise, in which he feels very small, into the mass meeting and is now surrounded by thousands and thousands of people with the same conviction . . . he himself succumbs to the magic influence of what we call 'mass suggestion.'[12] (Omission in original)

[3] Desmond, A. J. (1997). Huxley, 271 in Weikart (2004), 74, and 80, quoting Haeckel, E. (1878). Freie Wissenschaft, 74.
[4] Slight, D. (1937). Disorganization, 841; Ross, E. A. (1920). Foundations,
[5] Staub (1990), 48; Staub (1999), 182–89. See also Schuman, 100.
[6] Schuman, 98.
[7] Mosse (1975), 2.
[8] Staub (1995), 105–06; Staub (1999), 183; Waller, 32.
[9] Galton (1998), *265-66; Plato (1955), 156–57n.1.
[10] Weikart (2004), 75.
[11] Hitler ([1925] (1971)), 478.
[12] Mosse (1999), 92; Neumann, 439, quoting Hitler, Mein Kampf (1971), 715–16. Although, See Hitler, A. ([1925] (1971), 478–79.

Wertham (1966) made an excellent point establishing this connection that like-minded individuals within a social collective are volatile—

> The impulse is individual, the rationalization is social. This is where the individual and the social meet. If the environment itself is ruthless, it is easy to take a cue from that and find a plausible rationalization for violent behavior.[13]

Indeed, Figure IV-1 illustrates that once the national, regional, and local groups reach optimal numbers, the collective adopts a shared superego, enabling violence on behalf of the group's cause—violence individuals would not commit alone.[14] Peck (1983) labelled this regression as "*group immaturity,*" where human groups behave in ways that individuals would not, "*. . . at a level that is more primitive and immature than one might expect.*"[15] Hence, immature humans are more prone to evil than mature humans.

Staub's (1989, 1995) principles, outlined in Table I-3, 57 and Table IV-1, 260 reflect the educational, psychological, and societal preparations necessary to create aggressors.[16] This process involves generalised conditioning and desensitisation, a concept echoed in Lt. Col. Grossman's (2017) observations examined in heading Manufacturing the Spartan, 65, which highlight the systemic training and conditioning that produce desensitised individuals.[17] Although Romanticism differs from the general indoctrination apparatus outlined in Education of Society Matrix, 55, it accomplishes an identical objective: compensating for deficiencies in general education by acting as a 'gap filler' that facilitates the desensitisation of conscience.

General educational conditioning and desensitisation are preparatory steps, gradually fostering the rise of a totalitarian regime. Romanticism intervenes at a critical juncture, (i) immediately prior to, and (ii) at the onset of any prospective totalitarian regime. It manifests as a state of ecstatic unity: hands joined, singing, and marching in synchrony, consumed by indescribable emotionalism.[18] If aesthetic, rational methods grounded in left-brain reasoning fail to instil the 'new man' mindset, Romanticism targets those dominated by right-brain thinking, exploiting hemispheric asymmetries in perception and cognition.[19] In the Nazi era, this pattern emerged—an impromptu response deeply intertwined with prior education.

A contemporary desensitising phenomenon is the portrayal of violence on television as entertainment, in which fantasies are derived from sexually stimulating and exciting contexts. Grossman (2017) posits that society forms vicarious pleasure rather than revulsion from such depictions, suggesting that people are conditioned to enjoy sadistic acts. This parallels Eisner's (1969) analysis of post-WWI Germany,

---

13 Wertham, 41.
14 Waller, 31, 33.
15 Peck, M. S. (1983). People of the lie, 216, 222, generally, 212–53; Waller, 32–33. For an opposing view, see Gregg, R. B. (1966). The power of nonviolence, 118. For further viewpoints, see Waller,'s analysis, 32-36.
16 Staub (1989), 13–18; Staub (1995), 101, 103.
17 Grossman, D. (2017). On Killing, 143–44, *149.
18 Reichmann, 187, 194, 202, 212.
19 Corballis, M. C. (2014). Left brain, right brain; Funnell, M. G. et al. (1999). A deficit in perceptual matching; Hammer, 231; Hartikainen, K. M. (2021). Emotion-Attention; Ornstein, R. E. (1977). The Psychology; Ross, E. D. (2021). Differential Hemispheric Lateralization of Emotions.

where a cultural proclivity for sadism emerged.[20] Grossman's observation illustrates a normalisation process, linking sadistic acts to broader societal consequences. The locus of this phenomenon reflects a regression in general standards of reason, dismantling established value systems and promoting reliance on raw, uncontrolled emotion. This manifests in both positive and negative ways. On the positive side, it may lead to emotion-driven behaviour, personal gratification, or what Grossman terms 'collective narcissistic gratification.' On the negative side, it can encourage—

i. Unrestricted primitive impulses;
ii. Self-indulgence of one's innermost cravings;
iii. Rebellion against higher laws of social existence;
iv. The desire to be free from obligations and responsibilities to humanity;
v. Lack of empathy or sympathy, and a tendency ". . . to give free rein to their instinctive affection for their own kinsfolk . . . ."[21]

The most famous book burning event in Nazi Germany was a clear outcome of adherence to Romanticist doctrine. Intellectual growth: life development and learning, including critical thinking, broad worldview points, ethical reflection, and advanced problem-solving—are antithetical to Romanticism's core attitudes. In fact, Schuman (1936) states that he discovered that Nazi leaders, including Hitler, boasted of their hysterical tendencies—

> [R]epeatedly proclaimed themselves to be not only non-rational but *anti*-rational. Anti-intellectualism and *Blut gegen Geist* (Spengler) are fundamental principles of National socialism. Hitler himself has always given feeling, emotion, fanaticism, and even hysteria precedence over calm ratiocination.[22] (Emphasis in original)

ii. Emotional Moral Exclusion and Antisemitic Affect Intensification

Romantist doctrine had a particular application to the Jews: the rehabilitation of Jewish hatred.[23] This ideological element was one of several motivations driving the brutal treatment of Jews, who were increasingly viewed as an existential threat to the nation. Reichmann (1951) emphasises that Hitler presented this hatred as a pseudo-religion to the masses—

> Where thinking was . . . no longer allowed, where criticism had to be avoided by all means, and a dim religious haze had to conceal the deficiency in real

---

[20] Eisner, 95–97; Fromm, 169–201; Grodin & Annas, 644; Grossman, 151 52; Staub (1989), 71.
Grossman states that, ". . . as of 1945, there have been more than 200 studies demonstrating the connection between television and violence. The American Psychological Association's commission on violence. The American Psychological Association's commission on violence and youth concluded in 1993 that '[t]here is absolutely no doubt that higher levels of viewing violence on television are correlated with increased acceptance of aggressive attitudes and increased aggressive behaviour' (153). See his reservation in n.4.

[21] Le Bon, G. (1903). The Crowd, 40–66; Slight, 841; Reichmann, *87, 89, 192; Rensmann, L. (2017). Guilt, 10–11, 23–24, references postwar German study Group Experiment carried out in the early 1950s, as to why they supported the genocidal crimes of Nazis in Adorno, T. W. et al. (2010). "The Meaning of Working Through the Past."

[22] Schuman, 97.

[23] Reichmann, 201.

> religion, a common enemy was the appropriate remedy . . . the Jew was particularly suitable as an anti-symbol for a creed based on authority and the renunciation of reason. The Jew was considered as intelligent, indeed as the typical intellectual.[24]

In line with Romanticism principles, Reichmann (1951) identified the salient points necessary to dismantle intellectual reasoning and rational systems—

> They will have to replace political guidance by mass propaganda. They will have to appeal to emotions and desires, and the more completely they apply their reason to the task of influencing these irrational motives, the more successful will they be.[25]

This was Hitler's oratorical approach. A gifted demagogue, he skilfully whipped crowds into a hypnotic state with patriotic fervour and emotional appeal, making "*rowdy*" antisemitism acceptable to thousands in the 1920s.[26] His rhetoric resonated with the masses, much like Joseph Goebbels, who employed mass, irrational propaganda to which "*. . . the people became hopelessly susceptible . . . .*"[27] This propaganda was a "*. . . combined, inconsistent conglomeration of ideas and suggestions . . . .*" Because the masses lacked the intelligence to control their elementary affective states, they experienced that their emotions had been put into words.[28] Accordingly, Arendt (2006), author of Banality of Evil, wrote that the "*. . . voice of conscience tells everybody not 'Thou Shalt not Kill' . . . the law of Hitler . . . demanded that the voice of conscience tell everybody: 'Thou shalt kill.'*"[29]

Ironically, such behaviour contradicted the early-1900s social 'hygiene' ideals, which emphasised self-control.[30] Negative conduct, on the other hand, involved renouncing all rationality— "*. . . lowering of the barriers of conscience, which was supposed to keep the primitive impulses under control.*"[31] In other words, intelligence, personal and collective inhibitions, and restraints that normally prevent indulgence in inherent desires were thus liberated, allowing pursuit of instinctive urges and provided "*peace for the conscience.*"[32] Whatever the extent of this peace, the medical term for the resulting affliction was neurosis, and in more extreme cases, psychosis.[33]

From both psychiatric and social-psychological perspectives vantage points, Romanticism is recognised as a cognitive disorder, characterised by the "*. . . disorganization of individual personality structure, to the collective behaviour of social groups,*" resulting in mental contagion.[34] Le Bon (1903) describes this state as one in which "*[h]e is no longer himself, but has become an automaton who has ceased to be guided*

[24] Reichamnn, 196, *198, citing Hitler, A. (1939). Mein Kampf, 110 See ; Michels, R. (1926). Psychologie, 352; Simmel, G. (1908). Soziologie, 473–74.
[25] Reichmann, *116, 117–119.
[26] Massing, 148; Schafft, 64. See also Moscovici, 81–95, for an analysis on mass hypnosis.
[27] Palvi, M. (1941). Economic Foundations, 471, 477, cited in Reichmann, 219; See also Reichmann, 176–77, 184; Weindling (1993), 320.
[28] Moscovici, 16–*17; Reichmann, 184, 187.
[29] Arendt (2006), 150; Kressel, 163. See also Milgram, S. (1974). Obedience to Authority, 7.
[30] Weindling (1993), 227.
[31] Reichmann, 194.
[32] Ibid., 202. See generally Staub (1990), 47–48.
[33] Schuman, 99.
[34] Ibid., 98–*99. See also Ross, 103, *105. See mainly, Le Bon (1903), 34, 35–36, and also, Moscovici, 93.

by his will."[35] Simply put, it was a mechanism that targeted "*. . . habits of thinking and feeling . . . ,*" infused with existing prejudices and latent antisemitism. This transformed anti-Jewish sentiment into a dynamic hatred, fuelled by propaganda, creating a climate conducive to unchallenged Nazi aggression toward Jews.[36] Indeed, Ginsberg (1947) wrote, Romanticism "*. . . openly glorifies violence and inculcates harshness and cruelty as ends desirable in themselves.*"[37] Hobhouse (1916) further expands on this, to paint a picture of a human being reduced to a carnivorous animal—

> The idea of violence was in the air . . . Men were being taught not to look too far ahead, not to wait till they could see where they were going, not to follow deliberately a reasoned policy, but rather to throw themselves on instinct . . . . They were above all things to be strong, and not let the native hue of resolution be sicklied o'er by the pale cast of thought. This was to be guided by the vital impulse, which was above rationality . . . . To feel this impulse in one was to be at the centre of things . . . .
>
> . . . .
>
> This philosophy carried us a step further on the line of retrogression from will to the instincts, emotions, and impulses which man shares with the brute creation. . . . The old order of values was reversed. Reason, that we were taught for ages to place at the summit of human faculty, was degraded, and instinct, which men spent painful generations in seeking to subdue, was set upon the throne.[38] (Emphasis added)

German Romanticism had a categorical role in uniting Germany, with Reichmann (1951) noting that "*. . . Romanticism played a special part.*" She argued that it was the progenitor of Anti-Semitism—the bedrock for National Socialism.[39] Goldhagen (2009, 1996) recognises that the social milieu, independent of coercion or obedience to immoral authority, zealously chose to perpetrate the extermination of Jews. However, he stopped short of explaining the framework through which the majority of the German populace aggregated in their propensity for genocide.[40] The esoteric significance of Romanticism lies in its power to escalate societies toward mass violence, as it did with nationalism during Nazi Germany and now amid global shifts following Covid. In sum, Romanticism is a state in which individuals are dominated by euphoria and unchecked emotion.

The practicalities, in this context, was somewhat of a hybrid. The irony is, on one hand, Aryan ideology framed non-Aryans as "*inferior*" savages—atavistic evolutionary throwbacks.[41] On the other, under Social Darwinist hierarchical logic, there is a reversal of Nazi values, where the 'supermen' became "*self-controlled*" savages. They were hybrids embodying both the primitive and the civilised. See cross-link The Superman Doctrine, 72. Simultaneously, they were conditioned to show no emotion toward non-Aryans,

35 Le Bon (1903), 36; Moscovici, 93.

36 Reichamnn, 186–87, 198; Kershaw, I. (1981). The Persecution, 264, 286–88; Waller, 43.
See also Reichmann, 202–43, for summation about what the people wanted and what Hitler gave. She is not concerned with exposing the evil of Nazism. Rather, unveiling the "secret" of Nazi success, namely, the ruthlessly applied Romanticism doctrine, 214.

37 Le Bon (1903), 133–34; Ginsberg, M. (1947). Reason, 2:173.

38 Hobhouse, L. T. (1916). The World, 50–51; See also Reichmann, 89.

39 Reichmann, *83–84, 214. See also Schuman, F. L. (1936). The Nazi Dictatorship, 109.

40 Goldhagen (2009), 145–150, 155–56; Goldhagen (1996), 9; Waller, 39.

41 Weikart (2004), 95.

while exhibiting affection and "*brotherly love*" toward fellow Aryans. This accords with Romanticist philosophy, which echoes Darwinian ideas on animal behaviour.

B. The Role of Social Psychology in Perpetrating Inhumanities

i. The Genesis of Evil in Totalitarian Regimes: A Psychological Perspective

THE TERM "*EVIL*" is incredibly difficult to conceptualise because it encompasses a broad spectrum of meanings and interpretations. Baumeister (1997) and Staub (1989) note that the term "*evil,*" does not have an agreed meaning, "*[t]here is wide variation in the use and definition of the word.*"[42] Darley (1992) aptly remarked, it may not be easy to define, but when we see it, we have no trouble recognising it.[43] The term transcends mere physical acts, and also extends into philosophical, mystical, supernatural, and esoteric dimensions.[44] In contrast to more precise terms such as "*genocide*" or "*murder,*" they denote the manifestations and outcomes of evil. Burleigh (2002) offers a more pragmatic and generalised approach, one that emphasises an intuitive understanding or instinctive ethical discernment of what constitutes evil—

> [H]uman beings are . . . moral actors, equipped with such senses as guilt and shame, good and evil, right and wrong, which seem to be of the essence of any account of human behavior.[45]

This framing may be flawed because it overlooks the simultaneity of behaviour, in which individuals can act virtuously and malevolently at the same time. Waller (2007) presents a related paradox that challenges the ease of categorising evil. He contends that "*[i]n reality, a purely evil person is just as much an artificial construct as a person who is purely good.*"[46] His analysis indicates the fluidity of the moral spectrum, revealing it to be far less clear-cut than theoretical discussions suggest.

The Bible, while not a conventional dictionary, provides a concise framework for distinguishing between good and evil behaviour.[47] By interlacing religious and psychological perspectives, it offers a more holistic understanding of human conduct and the behaviours exhibited in tyrannical regimes. Although some may find reference to the Bible undesirable, it contains foundational insights relevant to conceptualising evil. Staub (1999) argues that the concept of evil is not confined to religious discourse but also a secular term used to express profound "*. . . aversion to certain actions and the human beings or natural forces from which they originate.*"[48] In contemporary usage, however, 'evil' is regarded as archaic or conceptually burdensome, largely absent from everyday discourse and frequently supplanted by terms such as 'moral wrongness' or simply 'bad.'[49]

At its core, the Bible appears to define evil as cognitive "*hatred*" toward another. A milder form of hatred is dislike, which does not imply malicious intent or negative thoughts. See further, Figure IV-1 below. However, when performing a good deed or representing the disliked positively to a third party, the response is typically negative. It

42 Baumeister, R. F. (1997). Evil, 6; Staub (1989), 25.
43 Darley, J. M. (1992). Social Organization, 200.
44 Baumeister, 7.
45 Burleigh (2002), 121.
46 Waller, 18.
47 For an opposing view, see Huxley (1957), 220, "ethical absolutes like 'good' and 'evil' requires reformulation in the light of psychology and history."
48 Staub, (1999), 179.
49 Waller, 10–11.

is important to note that anger does not necessarily constitute dislike or hatred; a parent, for instance, may be justly angry at a child for wrongdoing without harbouring hatred. By contrast, love for another, expressed through empathy, compassion, or understanding, scarcely motivates harm or killing.

Hatred, as the next level of negative emotion beyond dislike, can create the *mens rea* (mental state) necessary to form the intent to commit evil acts, although this is not invariably the case. It may manifest as passive hatred, in which negative feelings are harboured without being acted upon and are therefore non-violent. One can hate someone without harbouring ill intent or engaging in harmful behaviour. By contrast, the next level, a subset of hatred, is active hatred, which involves aggression and anger that manifest in immediate harmful actions. In Sternberg's (2003) discourse on hate, he summarises Allport's (1954) definition of hatred as "*. . . an emotion of extreme dislike or aggressive impulses toward a person or group of persons.*"[50] Sternberg further notes that hatred is one of the most powerful drivers of killing and is linked to terrorism, massacres, and genocide.[51] The Biblical narrative of Haman illustrates this dynamic: his intense hatred for the Jews drove him to actively pursue genocide. According to Sternberg's taxonomy of hate, ranging from 'cool hate' to 'burning hate',[52] Haman would fall within the latter category.

50 Allport, G. W. (1979). The nature of prejudice, 363; Sternberg, R. J. (2003). A Duplex Theory, 305.
51 Sternberg, 304–05. See also Waller, 184.
52 Sternberg, 311. See Waller's discussion, 184.

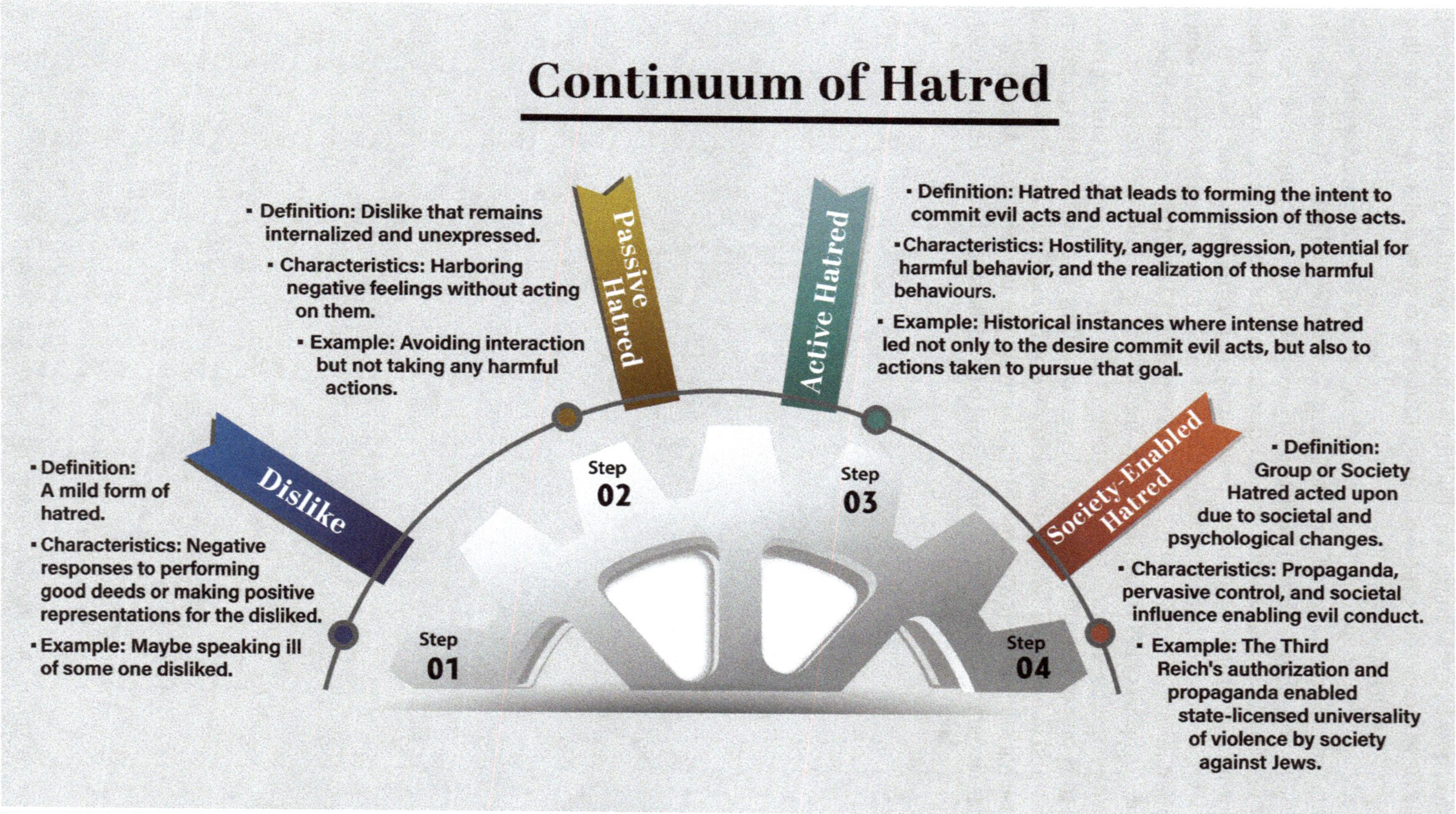

Figure IV–1. Continuum of Hatred: A Diagram of Ideological Escalation
This diagram shows the gradual escalation of hatred from social prejudices to violent extremism, mapping out the ideological continuum that enables systemic discrimination and violence.

The next level is societal group hatred. It materialises when pervasive psychological and societal transformations create an environment conducive to hatred enacted at scale, designated as collective group violence[53] and further incited by propaganda. The diagrammatic model in Figure IV-1 maps these strata of hatred. While it does not amount to a hard and fast rule or deterministic sequence, it categorically illustrates the progression from dislike to more intensified forms of hatred, and how societal and psychological factors enable individuals to act upon such sentiments. That said, the model does not account for every possible configuration or interpretive argument. See also Historical Anti-Semitism, 180.

Moscovici (1985) asserts the essential building-block principle that as soon as a person "*. . . joins a group and is caught up into a mass . . . becomes capable of excessive violence.*"[54] A prime example of this is English football matches. Moscovici cites the Roman adage, *Senatores omnes boni viri, senatus romanus mala bestia*—translated as "*[t]he senators are all good men, the Roman Senate is an evil beast.*"[55] This accentuates the contrast between individual virtue and the potential malevolence of collective behaviour. Waller (2007) adopts a broad, inclusive definition, asserting that "*. . . evil is anything detrimental to the well-being of living things.*"[56] Katz (1993) offers a marginally more specific formulation, adding that it involves "*. . . deliberately depriv[ing] innocent people of their humanity . . . .*"[57] This framing leaves open the possibility that those deemed "*innocent*" who retaliate against acts of evil may not themselves fall within the remit of evil.

Both Waller (2007) and Staub distinguish these forms of evil from natural forces, which are neutral and arise independently of human actions in their impact.[58] Evil need not always manifest through direct actions; it can also be rooted in passivity or omission, evidenced by the behaviour of bystanders. For example, in Nazi Germany, the regime's extensive control created conditions in which bystanders—despite witnessing the murder of non-Aryans or the deportation of Jewish people to death camps—were passive. Staub (1989, 1990, 1999) elaborates on this dynamic, suggesting that such passivity encompasses not only inaction but also the conscious choice to remain uninvolved, to tacitly support through inaction, or to exhibit indifference, all of which constitute a stance of neutrality. Strategic considerations also played a significant role in this behaviour.[59] Hitler, for instance, never personally killed anyone, but he was the mastermind and chief orchestrator behind the systemic genocide plan, prefaced by his in-house directives to subordinates to implement his agenda.[60]

ii. Techniques of Moral Disengagement in Group and Institutional Settings

THE BYSTANDER DISCUSSION naturally segues into group violence, as illustrated in Figure IV-1. Perpetrators need not be trained killers, ideological fanatics or party activists (Barnett, 1999); ordinary citizens and passive bystanders can become perpetrators. Staub (1989), cited by Barnett, notes that Nazi perpetrators were "*. . . normal by conventional standards of mental health.*"[61]

53 See Lifton, 480, 487; Staub (1999), 182–83, 187; Sternberg, 306–07, 320–21.
54 Moscovici, S. (1985). The Age of the Crowd, 13.
55 Ibid., 14.
56 Waller, 12.
57 Katz, 5.
58 Staub (1999), 180; Waller, 12.
59 Staub (1989), 20; Staub (1990), 47, 59–60; Staub (1999), 184–86. See also Barnett, 9–13.
60 Lifton, 480.
61 Barnett, 23; Staub (1989), 91;

Psychologically, Staub (1989, 1990, 1995, 1999) defines group violence as organised acts committed by a crowd, whether across society or within smaller communities such as schools, gangs, or families. Such violence is sometimes justified under the guise of higher ideals—Aryan culture, noble causes, or purported benefits to humanity—and driven by belief in a 'higher morality.' The underlying aspiration is to create a better nation, world, or other ostensibly worthy ends, akin to German Aryan ideology and the concept of the master race.[62] Bandura (1999b) demonstrates that at the societal level, moral disengagement accelerates and reduces guilt when state laws introduce new moral imperatives with institutional backing, "*. . . cognitively redefining the morality of killing so that it can be done free from self-censure.*"[63]

This analysis is consistent with Staub's (1989) distinction between evil groups and evil individuals. Staub's predisposing characteristics that heighten a society's propensity for evil are further developed by Mynatt & Sherman (1975) who remarked that group members relinquish rational judgement and moral restraint in collective settings, a phenomenon termed 'diffusion of responsibility.'[64] Ohana (2022) furnishes a broader account of this dynamic, tracing its conceptual origins to Darwin's hypothesis. As discussed in Part I., Darwinian theory arguably weakened the moral injunction embodied in "*thou shalt not kill,*" a foundational ethical constraint on human conduct.[65] This erosion is compounded by empirical findings that group interaction significantly increases risk-taking behaviour, with groups exhibiting greater risk shifts than individuals acting alone.[66] The following table highlights these contrasts, mapping progression along the continuum of destruction—

Table IV-1. Predisposing Characteristics in the Continuum of Destruction: Group vs. Individual Dynamics

Predisposing Characteristics of Society's Propensity for the Commission of Evil		
Characteristic	**Groups (Greater Propensity)**	**Individuals (Less Propensity)**
Nature of Propensity	Inherently selfish and uncaring, prioritising collective interests over ethical considerations	Individuals are capable of goodness and morality, guided by personal ethics and empathy
Concern for Power	Driven by power and self-interest, disregarding human consequences in their pursuit	Motivated by personal gain but generally operate within moral boundaries
Fear as a Deterrent	Fear is commonly the primary deterrent against unethical actions, with less emphasis on moral considerations	Fear and conscience both act as deterrents against unethical actions
Moral Constraints	Serve their own interests without moral constraints or obligations to others, resulting in ethical compromises	More likely to adhere to personal moral obligations and ethical guidelines
Diffusion of Responsibility	Members relinquish authority and responsibility to leaders, diffusing accountability	Retain personal responsibility for their actions, maintaining a sense of accountability

62 Staub (1989), 13, 17, 82, 94–95; Staub (1990), 48, 54, 58; Staub (1995), 101–08; *Staub (1999), 181–82;
63 Bandura (1999b), 195; Waller, 187–88.
64 Staub (1989), 28; Mynatt & Sherman, 1111.
65 Ohana, D. (2022). Jacob L. Talmon, 198.
66 Wallach, M. A. et al. (1962). GROUP INFLUENCE, 75–76, 85. See also Henningsen, M. L. M. et al. (2003). Social influence; Mynatt, C. & Sherman, S. J. (1975). Responsibility attribution.

Table IV-1. Predisposing Characteristics in the Continuum of Destruction: Group vs. Individual Dynamics

Predisposing Characteristics of Society's Propensity for the Commission of Evil		
Characteristic	**Groups (Greater Propensity)**	**Individuals (Less Propensity)**
Abandonment to Group	Members may exhibit enhanced commitment, leading to self-sacrifice for the group's goals and ideologies	Limited to personal commitments and sacrifices, with less inclination towards collective self-sacrifice
Repression of Dissent	Greater power to repress dissent, which can enhance the potential for evil by silencing opposition	Limited power to repress dissent, allowing for greater diversity of thought
Dynamic Nature	In groups, characteristics and processes are dynamic but tend to become institutionalised, creating persistent patterns of behaviour	Characteristics and processes are dynamic, more flexible, and open to change
Influences and Changes	Influenced by external factors that can amplify collective thoughts, feelings, motivations, and actions	Influenced by personal experiences and interactions, which can alter their thoughts, feelings, motivations, and actions
Propensity for Evil	Greater propensity for evil due to the organized characteristics and psychological processes, combined with collective dynamics and institutionalisation	Have a lesser propensity for evil due to personal responsibility and adherence to moral constraints

Source: Staub (1989). The roots of evil, 27–28. See also Le Bon, G. (1903). The Crowd, 25–37 and Mynatt, C. & Sherman, S. J. (1975). Responsibility attribution, 1111–1118.

In group settings, the psychological transformation ordinarily required to commit extraordinary evil is markedly reduced once such behaviour becomes socially sanctioned. Bandura (1999b) explains that individuals need not alter their ". . . *personality structures, aggressive drives, or moral standards because their actions are supported by law.*"[67] Diffusion of responsibility and moral disengagement thus enable atrocities to be committed with a clear conscience, as group dynamics and institutional backing provide moral justification. Cognitive dissonance studies indicate that enduring behavioural change rests on two key processes—

i. **Internalisation of commitment,** whereby individuals adopt a wilful mindset by finding intrinsic value in, or personal identification with, the behaviour;
ii. **Incremental compliance through external inducements,** in which extrinsic rewards or pressures encourage initial concessions, exemplified by the foot-in-the-door effect, where agreement to minor requests increases compliance with progressively larger demands.[68]

The logic of the foot-in-the-door effect, as a model of incremental compliance, is exemplary of how minor initial agreements can escalate into profound behavioural

[67] Bandura (1999b), 195.
[68] Freedman, J. L. & Fraser, S. C. (1966). Compliance, 195; Kressel, 152, 154; Wallach et al., 75. See Festinger, L. & Carlsmith, J. M. (1959). Cognitive consequences, 203, for opposing finding and operating conditions for rewards.

change. Kressel (1996) offers a compelling historical illustration in his account of Hitler's antisemitic strategy.[69] First, public acquiescence was secured for minor antisemitic acts and, second, he escalated measures to more severe policies, among many other factors, culminating in genocide. Empirical support for this pattern is provided by Freedman & Fraser (1966), who show that moderate external pressure induces compliance, with the critical caveat that excessive pressure undermines internalisation. Festinger & Carlsmith's (1959) findings further demonstrate the efficacy of minimal pressure in producing enduring attitudinal change.[70] In group contexts, the individual checks and balances that might prevent such behaviour are overridden by collective authority that normalises and endorses it. Collective dynamics intensify this process: the suppression of dissent and reinforcement of shared motivations erode individual restraint. This normalisation accelerates movement along the continuum of destruction, creating conditions in which extraordinary evil can be committed with minimal resistance. Indeed, Staub (1989, 1990) asserts that perpetrators are—

> [H]elped along by group processes, by shared values of a subgroup of society and shared views of its members . . . [which] makes it easier to engage in moral equilibration, devaluation . . . in the service of the shared goal of eliminating victims."[71]

This dynamic was evident in Nazi Germany, where hatred of the Jews, reinforced by the strategic use of propaganda, contributed to the escalation of Hitler's measures. The outcome was a systemic process of government-sanctioned, nationwide collective violence underpinned by a monolithic ideology of cultural and societal superiority.[72] Le Bon ([1895] 2002, 1903), the founder of crowd psychology, and Holocaust historian Rubenstein (1975) argue that once individuals are absorbed into a properly organised collective, their psychology becomes subordinated to the aggregate mentality. Under such conditions, it can exercise total domination over any group it designates. Accordingly, Le Bon thus foregrounds the subordination of the individual within the group dynamic—

> He is no longer himself, but has become an automaton who has ceased to be guided by his will.
>
> . . . .
>
> Isolated, he may be a cultivated individual; in a crowd, he is a barbarian—that is, a creature acting by instinct. He possesses the spontaneity, the violence, the ferocity, and also the enthusiasm and heroism of primitive beings . . . .[73]

A key characteristic of aggregation is unquestioning obedience to authority, a central predisposing factor. Kelman & Hamilton (1989) define the conditions under which obedience becomes criminal, arguing that culpability arises when actors knew, or should

69 Kressel, 155.
70 Freedman & Fraser, 195; Festinger & Carlsmith (1959), 210; Kressel, 152; See DeJong, W. (1979). An examination of self-perception mediation; Kelman & Hamilton, 103–16; Staub (1989), 80–81;
71 Staub (1990), 52, 59. See Staub (1989), 14–17, 19, 80–81.
72 Waller, 32.
73 Le Bon, G. ([1895] 2002). The Crowd, *8, 12–13; Le Bon (1903), 36–37; Waller, 30. See also Moscovici, 74–76; Rubenstein, R. L. (1975). The Cunning of History, 7, 75–76, 91.

have known, that their orders were illegal or inconsistent with general moral principles, within the "*. . . ordinary sense and understanding [of what actors are] reasonably expected to know.*"[74] Staub (1989, 1999) and Kren & Rappoport's (1980) similarly maintain that a strong authority orientation—characterised by subservience to position and power within hierarchical structures—constitutes a pivotal predisposing condition. Bialas (2013) underlines this dynamic, noting that it closely mirrors the obedience imperatives institutionalised in the Third Reich—

> Nazi morality was conceived as a procedural ethic of virtue. It combined absolute virtues that demanded unconditional obedience with the attitude of social engineering and replaced common sense intuitions with an ideological catalog of virtues and commands.[75] (Footnote call numbers omitted)

This element was deeply ingrained through the authoritarian upbringing of German children and buttressed by rigid school systems, conditioning ordinary individuals from an early age with the psychological discipline and mental fortitude necessary to facilitate participation in killing. This entrenchment created fertile ground for the emergence of genocidal behaviour.[76] Staub (1989, 1990) combines authority orientation and monolithic culture into a general precondition, observing that "*. . . there are no counteractions that might inhibit a progressive increase in its mistreatment.*"[77] Lang & Sibyll (1983), drawing on Israeli interview transcripts with Adolf Eichmann, found that during interrogation he described a childhood shaped by authoritarian virtues and blind obedience, which he referred to as the "*obedience of corpses*"—

> All my life I have been accustomed to obedience, from early childhood . . . an obedience which in my years of membership in the SS became blind and unconditional.
>
> . . . .
>
> We grew into them, all we knew was obedience to orders.[78]

Craig (1982) quoting the Württemberg publisher Karl Friedrich Moser (1758) wrote: "*[i]n Germany, it is obedience; in England, freedom; in Holland, trade; in France, the honor of the King.*"[79] This insight helps explain why the concept of unquestioning obedience, embedded in the Führer principle was assimilated with relative ease once the Nazi regime became operational. It demanded total loyalty to Hitler. This idea is integral not only to Nazi ideology but also to hierarchical leadership in occult doctrines and fraternal movements. At mass rallies, Hitler captivated the crowds, who succumbed to the spell of primal, instinctive desires. He likened the masses to women in a rapturous, orgiastic state, requiring unequivocal submission to the "*ruthless force and brutality*" of a strong

74 Kelman, H. C. & Hamilton, V. L. (1989). Crimes of Obedience, 46, 47, respectively.
75 Bialas (2013), 6.
76 Kren & Rappoport, 19 et seq; Sanford, N. (1973). Authoritarian Personality, 146–47, 157; Staub (1989), 19, 73–74, 109–11; Staub (1999), 184; See also Waller, 180–182, for analysis on Staub's discourse.
77 Staub (1989), 152; Staub (1990), 55–56, 233.
78 Lang, J. v. & Sibyll, C. (Eds.). (1983). Eichmann Interrogated, 291, 159. See also Arendt, H. (2006). Eichmann in Jerusalem, 120, 135 (obedience of corpses); Dicks, H. V. (1972). Licensed Mass Murder, 261. Waller, 181.
79 Craig, G. A. (1982). The Germans, 23; Staub (1989) 108.

man.[80] As a result, the German populace became impulsive, aggressive, stereotypical, and uncritical in their reasoning.[81]

Refusing or evading orders is exceptionally difficult due to the pervasive fear of reprisal, as "*[f]ew dare to disobey or leave.*"[82] Although obedience is theoretically a choice, Crelinsten (1995) argues that given the social, situational, and training contexts, meaningful choice is effectively absent. Indeed, Waller (2007) emphasises that this can lead ordinary people to commit extraordinary evil.[83] According to Crelinsten, this imposed moral order equates disobedience with punishment, disgrace, humiliation, expulsion, and even death.[84] Staub (1999) supporting Grodin's (2010) analysis, maintains that obedience to immoral orders is fundamentally rooted in deference to authority; without it, Staub asserts, "*group life is impossible.*"[85]

Milgram's (1974) seminal shock-experiment studies significantly advanced the understanding of obedience to authority.[86] He concluded that individuals will comply with orders regardless of content and without clear limits of conscience,[87] a fulcrum repeatedly observed in cases of genocide. Milgram's findings illuminate how ordinary individuals can be driven to commit atrocities, revealing the psychological mechanisms that facilitate such compliance.[88] The Nazi regime's project of creating "*supermen*" was preceded by a formal vow of loyalty to Hitler, reinforced through ritualised initiation practices. See heading Education of Society Matrix, 55; Manufacturing the Spartan, 65.

These mechanisms became institutionalised within military structures,[89] binding recruits, both explicitly and implicitly, to an authority of violence they were required to obey.[90] Staub (1989, 1999) contends, however, that while Milgram's research demonstrates the ease with which individuals can be induced to obey, obedience alone does not constitute the primary motive for mass killing or genocide. Staub states, "*. . . it is not the true motive for mass killing or genocide.*" He reformulates this distinction asserting that "*. . . those who willingly accept the authority of leaders are likely to have also accepted their views, ideology [and wishes].*"[91] Staub summarises this differentiation as—

---

80 McDougall, W. (1920). The Group Mind, 410; Le Bon, G. (1903). The Crowd, 41–43; Reichmann, 215n.429, citing Hitler (1939), 48, 161, 380, 381, 383, 395. Cf. Weiss (1987b), 139, re Schallmayer analogy of a woman's reproduction role with horse-breeding; Fromm, E. (1969). Escape from Freedom, 163, 165, re masochistic and sadistic striving as a basis for submission.

See also Kren, G. M. & Rappoport, L. (1980). The Holocaust, 20, for a critique of Fromm regarding the origins of modern authoritarianism.

81 Le Bon, 40; Reichmann, 226; Tenenbaum, 7.

82 Crelinsten, 58–*59. See also Darley, 212 and Kren & Rappoport, 100–02.

83 Crelinsten, 60; Waller, 108.

84 Crelinsten, 60.

85 Staub (1999), 184. See also Waller, 180.

86 Gourevitch, P. (2015). We wish to inform you; Kressel, 124, 142 164, 169–70 passim; Staub, 184.

87 Milgram, 188–*89. See also Haritos-Fatouros, 130; Kressel, 9–10 passim; Staub, 184; Waller, 102–11.

In Garbett's (1975) review, he criticizes Milgram's (1974) work for focusing on extreme authority structures, such as military and prison contexts, which overlook the complexities of negotiation and bargaining processes in everyday obedience. Wenglinsky (1975) argued that because the participants were informed that it was an experiment involving electric shocks, the experiment and the statistical results were rendered invalid as they were no longer naive (615-16). In other words, the subjects knew what to expect; they knew that no actual shocks would be administered. Wenglinsky's review argues that, much like in experiments, real life presents the challenge of uncertainty. We can never be sure when a situation or authority is different; whether one dictator differs from another or one war from another (616).

88 Grodin & Annas, 645; Haritos-Fatouros, 130; Kressel, 124, 142 164, 169–70 passim; Staub, 184. Cf. Crelinsten, 58–59. See also Barnett, 24–27.

89 Cf. Haritos-Fatouros (1995), 137; Waller, 33. See also Aronson, E. & Mills, J. (1959). The effect; Dixon, N. F. (1976). On the psychology Gerard, H. B. & Mathewson, G. C. (1966). The effects.

90 Haritos-Fatouros (1995), 137, 141.

91 Staub (1989), *29, 31, *43; Staub (1999), 184.

> A society's strong respect for authority is one source of genocidal violence. A tendency to like and obey authority is one characteristic of perpetrators.[92]

Indeed, Katz (1993) did not dissent from Milgram's obedience postulate; rather, he argued that Milgram did not go far enough in considering the possibility of "*new values*" or emerging norms, especially where disobedience has become commonplace in contemporary culture. Katz proposed four additional forms of obedience, rooted in the incremental corruption of moral values through subtlety, deception, and beguilement. The following table summarises Katz's points—

Table IV-2. Institutional Mechanisms of Systemic Repression

This table outlines the institutional mechanisms that normalise and sustain organised harm. Through information control, enforced complicity, suppression of dissent, and bureaucratic concealment, wrongdoing becomes embedded within governance structures rather than dependent on individual persuasion.

Forms of Beguilement	Description
Packaging of Evil	Making evil desirable to individuals who are not predisposed to doing evil. Repackaging harmful behaviours that appears to be morally acceptable or even appealing without understanding their malevolence.
Careerism	Pursuit of career advancement, unknowingly leading down a path to evil. A series of small incremental, innocent decisions accumulate, gradually eroding ethical standards until withdrawal becomes impossible.
Bureaucratization of Evil	Bureaucratic procedures are harnessed to produce evil outcomes. Systematic procedures that facilitate evildoing while masking the moral consequences.
Culture of Cruelty	Creation of a separate and distinct culture of cruelty that becomes enjoyable and rewarding. Individuals derive pleasure and a sense of fulfilment from engaging in acts of cruelty, further entrenching malevolent values within the group.

Source: Katz, F. E. (1993). Ordinary People and Extraordinary Evil, 6–7, 31–33, 36–37, 135–38.

92 Staub (1989), 30.

C. Reengagement of Past Connections: Occultism and the Foundations of Atrocity

i. Understanding the Parameters of Conspiracy: A Definition and Analysis

THE OBJECTIVE OF this section is not to indulge in wild conspiracy theories. Rather, it calls for discernment: critically examining occult material without dismissing valuable insights, while also confronting its problematic aspects. The reality is that conspiracies do exist. Ironically, as analysed in Part 1., WWII was shaped in part by a global conspiracy narrative, specifically, Nazi leadership belief in the alleged Jewish world domination plot outlined in the Protocols—a theory that was inherently false, and continues to be accepted as truth on a global scale today, 2026! Review heading Contemporary Anti-Semitism, 204.

Modern court dockets are replete with indictments conspiracy-related offences, including conspiracy to commit murder, corporate fraud, bribery, racketeering, airline collusion to regulate airfares, price-fixing, and drug trafficking. The federal definition under 18 U.S.C. § 371 (2023) states: "*[i]f two or more persons conspire . . . to commit any offense . . . and one or more of such persons do any act to effect the object of the conspiracy . . .*"[93] Pipes (1997) exhaustive work documents both sides of conspiracy theory. He examines its lore, pitfalls, historically verified conspiracies in both historical and contemporary life, while questioning whether broader conspiracy claims possess any validity.[94] He makes an excellent distinction how—

> Conspiracy theories have . . . become a way of seeing life itself. . . . In complete form, it takes over the lives of the faithful and becomes a prism through which they see all existence.[95]

Some may argue that Pipes's cautious and ambivalent approach goes too far and dismisses valuable truths, though he is justified in dismissing Robertson's book as a mixed-bag of conspiracy material.[96] Robertson fails to adequately handle the serious and problematic aspects of occult matters—

> Conspiracist texts often come packaged as solid-looking books with introductions, forewords, acknowledgements, quotations, footnotes, bibliographies, indexes, and the other conventional trappings of learnedness. Even forgeries come wrapped in a mock academic paraphernalia. By replicating these elements of academic authority, the pseudo expert finds it easier to convince the gullible or inexperienced reader to accept his pet theories. In addition, a profusion of references serves as a shield against criticism. When defending his book The New World Order from charges of antisemitism, Pat Robertson raised the irrelevancy that it 'was carefully researched and contains seven single-spaced pages of bibliography from original historical sources.'[97]

---

93 See also Pipes, 20.
94 Ibid., 17, 20–21.
95 Ibid., 22.
96 Ibid., 10.
97 Ibid., 33–*34. See also comments by Goodrick-Clarke (2002), 280.

Forming a reliable, hard-and-fast rule to decipher a correlation between occultism and antisemitism is difficult and risks quackery. A more practical approach is to analyse correlations in public policy and observable patterns in behaviour, worldviews, writings, speeches, and associations, using concrete evidence to assess alleged "*hidden agendas.*"[98] Through deductive reasoning, critical thinking, and empirical comparison of historical and contemporary data, we can assess whether a theory is consistent with its narrative. A clear grasp of current developments helps pinpoint claims with evidential grounding.[99] Jacob Katz (1904–1998), the acclaimed Israeli historian, offered a cautious statement on this—

> The antisemitic movement of the nineteenth century, which culminated in the Nazi period, stands revealed as the product of a peculiar constellation of historical circumstances. . . . The attempt to foresee the future on the basis of its analogies with the past is a futile undertaking. What the historian can do, beyond the critically controlled reconstruction of the past, is to assess the extent to which the elements involved in a past constellation of events are still operative in the present and how the reactions to its consequences seem to shape up—without pretending to know what new combination may emerge out of these and other unapprehended elements.[100]

At this point, some overarching principles are in order. The following framework strengthens the analysis, in line with the scientific method, for assessing the credibility of conspiracy theories generally—

i. **Correlation vs. Causation**: Conduct a comparative review of historical and contemporary developments. Distinguishing between correlation and causation is vital to avoid flawed conclusions. Correlation indicates a statistical association between two variables, meaning they tend to occur together. Causation, however, implies that one variable directly affects the other, creating a cause-and-effect relationship.
ii. **Multiple Perspectives**: Consider diverse perspectives and information sources to develop a well-rounded understanding. A realistic outline of how developments unfold helps avoid confirmation bias and strengthens analysis.
iii. **Critical Thinking**: Apply a rigorous, evidence-based approach to assess arguments. Scrutinise patterns and correlations in public policy while avoiding assumptions and bias, ensuring conclusions are grounded in sound reasoning and empirical data.
iv. **Healthy Scepticism**: Maintain disciplined questioning and verification through critical evaluation of sources. Identify red flags such as ambiguity, falsified documents, and contradictions. Integrate diverse perspectives and sources to build a more complete picture.

A prime example of a conspiracy is the most widely published and believed conspiracy of all: the Protocols. They were categorically proven to be fabrications thorough investigative analysis. As previously shown, their vagueness—omitting specific

98 Goodrick-Clarke (2004), 1.
99 Cf. Pipes, 38.
100 Katz (1970), 325–26.

names, dates, and events—was a hallmark of their fraudulent nature.[101] Those persuaded by them failed to clearly apply basic standards of scrutiny, lacking rigour to assess their credibility. Pipes (1997) noted—

> When the topic is conspiracy, it is often difficult to distinguish truth from falsehood. In part, this is because objective criteria—logic and evidence—do not clearly demarcate one from the other. Further muddying the waters, individuals in the know sometimes make untruthful claims, serious scholars endorse wild claims, and pseudo scholars pass off their work as genuine research.[102]

In rebuttal, and in line with the aforementioned logical approach, hypothetically there would be compelling reasons to maintain the secrecy of coordinated global agenda, which would necessarily remain concealed. The widely circulated notion that elite, clandestine groups pursuing a Global New Order, inherently highlights the need for caution. Logistically, disclosing such an ambitious agenda would be imprudent, since disclosure would risk public alarm, institutional resistance, dissent from authorities, and civil unrest. These practical considerations justify maintaining secrecy and explain why such an agenda would not be openly acknowledged. Claims that fraternal or subversive organisations are conspiring to create a global government are scarcely novel.

Conversely, let us not shy away from discussing unpalatable topics such as the occult simply because they are unpleasant. Similar to Edmund Burke's adage, adapted, reminds us: 'good men who do nothing until it is too late are like broken cisterns, allowing falsehoods and injustices to flourish unchecked.' Rather, as Fred Katz (1993) poignantly observes—

> [W]e must create the scientific tools for understanding how evil is produced—facts do not speak for themselves . . . To create that science requires a delicate balancing act. One must be immersed in the facts, in thoroughly knowing actual evil deeds. Yet one must stand back and, as a dispassionate outsider, devise ways of dissecting the facts in ways that make them understood in an effective way.[103]

ii. Structural Occult Beliefs and Anti-Semitism

THIS SECTION MOVES into largely uncharted territory within the present study. Building on the preceding section, The Genesis of Evil in Totalitarian Regimes, 256, this discussion examines the thematic nexus between the occult foundations of the Nazi regime,[104] antisemitism, and the formation of networks intended to reactivate these ideas within a putative Global New Order. The question is: why does the occult warrant attention? It is because the deliberate reengagement of historical narratives continues to inform contemporary belief systems, underpinning this continuity and linking prior analysis into a clear conceptual bridge forward.

This inquiry constitutes a serious scholarly examination, frequently undermined by sensationalism and misinformation. François (2023) critiques such tendencies as

101 Pipes, 85.
102 Ibid., 30.
103 Katz (1993), 137.
104 See generally Webb (1976), 219–237.

irrational, pseudo-academic speculations that flout the fundamental principles of academic research.[105] On the other hand, Staudenmaier (2013) wrestles with similar matters of occult inquiry, stating that, "*[s]uch questions enter controversial territory*"[106] Staudenmaier (2009) elaborates on the reason why, stating that "*[t]he occultist predilection for conspiracy theories has inevitably compromised the topic as a whole.*" Staudenmaier (2015) adds nuance to François's critique, cautioning that "*[t]his ambiguous situation warrants particular caution in dealing with 'the grey area of publications halfway between occultism and scholarly research.*"[107]

Yet this apparent disciplinary caution rests on the tension in which academia simultaneously continues to engage with, theorise, and critique esoteric and occult subject matter. Accordingly, Hanegraaff (2012) observes that academia customarily treats discussions of 'esoteric' or 'occult' topics with disdain.[108] Paradoxically, any fresh analytical engagement with these tensions must navigate an already substantial body of scholarship; to do so risks reinventing the wheel and duplicating prior research.

Indeed, historian Robert Galbreath has criticised scholars for their antagonism toward serious analysis of modern occult studies.[109] To borrow Goodrick-Clarke's (2002) sentiments, esoteric discourse is a provocative field that, "*. . . may well offend politically correct sensibilities by posing the very [content] which liberal elites prefer to ignore or suppress.*"[110] The subsequent discussion is framed neither as conjecture nor speculation, nor as derivative of conspiracy theories. Rather, it is grounded in meticulous study and careful observation. While academic sources are used wherever possible, it must be acknowledged that, given the nature of the subject, there will come a point at which no further formal sources are available.

We begin with an intriguing dictum drawn from Moorhouse's (2010) review of Adrian Weale's The SS: A New History (2010). Moorhouse identifies two unresolved issues regarding the SS. For present purposes, however, I will focus solely on the first question, which he argues remains unaddressed: how did a small elite protection squad evolve during wartime into a massive force of nearly a million men in the early 1930s?[111] This postulation corresponds closely with the thematic focus of the present discussion. The antecedents of the SS were examined separately in Education of Society Matrix, 55, while situated in the context of Nazi Germany, continues to retain contemporary relevance. Reichmann (1951) and Kurlander (2017a) advanced a profound foundational assertion regarding the nexus between Nazism and the catastrophic event of the Holocaust—

> Without Nazism this catastrophe [of the Holocaust] would never have occurred, but without the tendencies on which this analysis tries to throw light, Nazism would never have come into being.[112]
>
> Without the supernatural figuring of the monstrous Jew, the highly technical process of genocide could never have been applied as widely

[105] François, 4.
[106] Staudenmaier (2013), 26.
[107] Hanegraaff, W. J. (2012). Esotericism, 238n.326; Staudenmaier, P. (2009), 56; Staudenmaier (2015), 32.
[108] Hanegraaff, 3, 162.
[109] See rebuttal: Galbreath, R. (1986). Explaining Modern Occultism, 11–32; Staudenmaier, P. (2009). Occultism, 49.
[110] Goodrick-Clarke's (2002), 6.
[111] Moorhouse, R. (2010). [The SS]; Weale, A. (2010). The SS.
[112] Reichmann, 101.

> or vociferously as it was.[113]

If the insights of Reichmann and Kurlander are extended further, a restatement could read, without overstating causality: without the occult and esoteric dimensions of Nazi ideology, the leadership's promotion of mystical doctrines under the Führer principle, and the ensuing acts of barbarity, the Holocaust would not have materialised. When juxtaposed with the spiritual and historical trajectory of the Jewish people, a compelling raison d'être emerges for the Holocaust and the four-thousand-year history of Jewish persecution, understood not as isolated historical events, but as recurrent manifestations of hatred.

In Silverman's (1992) *fin-de-siècle* study, he identified substantial evidence involvement in occult movements often correlated with antisemitic attitudes. Édouard Drumont (1844–1917) fused anti-Jewish rhetoric with occult themes to captivate a prejudiced readership, while his collaborator, Gaston Méry (1866–1909), editor of *La Libre Parole* and founder of the occult-leaning periodical *L'Écho du merveilleux*, similarly conflated antisemitism with occult lore. Webb (1976) contended that "*[t]he connection between occultism and antisemitism obeys certain rules.*" Regrettably, he did not specify what those rules were, making it difficult to discern a clear pattern.[114]

Occultism, then, is not merely a spiritual or philosophical current; it is intricately intertwined with malevolent elements of outward practice that "*. . . continues to confront the world in new and varying forms.*"[115] As previously quoted in the Covid analysis, Frank-Walter Steinmeier famously articulated that evil historical malicious characters have disappeared; evil continues to be a perpetual force of human affairs —

> Of course, our age is a different age. The words are not the same. The perpetrators are not the same. But it is the same evil.[116]

When analysed through the prism of biblical and Tanakhic teachings on the dichotomy of good and evil—grounded in the relationship between God and adversarial non-human entities such as Satan or the Devil—this worldview holds that such forces are intrinsically linked to acts of evil. Without the Devil as the archetypal source, no central organising principle would exist for occult practitioners or antisemitic literature to channel malign aims. Consequently, the material application of occult doctrines would be untenable. In this vein, Webb (1976) argues—

> [T]he occult played a larger role in the genesis and propagation of [the] forgery of the [Protocols of the Elder of Zion against the Jews] than has been admitted, and the recognition of this fact can make plainer the devil-ridden universe in which such imaginings originate.[117]

[113] Kurlander (2017a), 252, citing Hale, C. (2003). Himmler's Crusade, 11, 19–27, Trevor-Roper, H. R., Ed. & Trans. (1953). Hitler's Secret Conversations, Rupnow, D. et al. (Eds.). (2008). Pseudowissenschaft, 292. Cf. Picker, H. (2003). Hitlers Tischgespräche, 78–79, Hitler, A. (1943). Mein Kampf, 305.
[114] Silverman, W. Z. (1992). Anti-Semitism, 155–63.
[115] Grodin (2010), 49. See also discourse by Berenbaum, 2, and Goodrick-Clarke's (2002), 3, comments on repacking ". . . the old ideology of Aryan racism in new cultic guises . . . ."
[116] Steinmeier, F.-W. (2020). Address by President of Germany to the Fifth World Holocaust Forum.
[117] Webb (1976), 214. See also Goodrick-Clarke (2002), 3.

Webb further contended that the Protocols were an accelerant for Nazi antisemitism.[118] Far from advocating the preservation of Jewish life, occult practitioners instead articulated a historical vision predicated on eradication. Nazi history intrinsically illustrates the close affiliation between antisemitism, elitist occult practices, and policies of annihilation.[119] Mosse (1964) consolidates this link by tracing Nazi aspirations to god-like status to occult ideologies. He cited the Thule Society (Atlantis) occultist Ellegaard Ellerbek, who declared in a lecture to students—

> "Do you know that you are gods?" he exhorted his audience, and followed with the accusation that the Jews were the malevolent exterminators of Germanism, that they prevented the Aryans from assuming their godlike stature, and that, consequently, they must be eradicated.[120]

As established, Hitler maintained close associations with occult figures such as Fritsch and Wagner, both of whom articulated virulent antisemitism. These ideologies were codified in policies and legislation that targeted the Jewish population. Occultism thus spread pernicious ideas, furnishing an ideological basis for broader racist and antisemitic doctrines, frequently intertwining with conspiratorial narratives. Webb (1976) broadly affirms this interpretation, stating—

> Theosophy, along with its doctrine, its foundress [(Blavatsky)], and its adherents, were implicated in racism, conspiracy theories, and at least one antisemitic tract.[121]

Similarly, individuals such as Aleister Crowley and Anton LaVey—prominent proponents of Luciferian or Satanist traditions, known for their deliberate inversion ("*inversionists*") of Judeo-Christian ethical norms—exhibit little or no substantive evidence of antisemitism in their writings or conduct. The distinction between occult practice and antisemitism is substantially blurred when one considers the Islamic world. While most Muslims do not engage in occult traditions, extremist interpretations of Islamic texts have articulated annihilationist attitudes towards Jews echoing Hitlerian logic. However, what binds these thematic aspects together is a comparative pattern between, on the one hand, Nazism and Satanist or occult currents associated with antisemitism, and, on the other, Islamist discourse and conduct marked by antisemitism. The commonality across them is hatred of Jews. Belief systems thus normalise the erosion of respect-for-life values, thereby enabling morally repugnant behaviour.[122] Notably, some Islamist actors have explicitly compared themselves to the Third Reich's persecution of Jews.

These inversionists illustrate the broader continuum of occult and quasi-religious literature, spanning nineteenth-century Germanic and Aryan ideological currents through to Nazi leaders. This body of thought reflects a complex interplay between occultism and social norms, in which occult symbolism exerted a subversive ideological

118 Webb (1976), 214.
119 See Goldstein, J. A. (2011). On Racism and Anti-Semitism in Occultism; Goodrick-Clarke (2002), 4–5.
120 Mosse (1964), *82–83. See also Luhrssen, D. (2012). Hammer of the Gods, 224n.59.
121 Webb (1976), 226.
122 Black (2012), 9; Goodrick-Clarke, 216.

influence. As Cohn (1981), Kurlander (2012, 2015, 2017), Mosse (1961–2000), Poliakov (1965), Webb (1974, 1976), and others demonstrate, the figure of the Devil and associated occult motifs were appropriated as symbolic frameworks for constructing ideological and social hierarchies. This intellectual trajectory helped pave the way for Social Darwinism, which dehumanised individuals as animals while elevating the Nazis, the alleged master, race, to divine demigod status.

The rationale for such dehumanisation—and for the emergence of perpetrators of atrocity—lies in the psychological mechanisms identified by Grodin & Annas (2007), which enable individuals to override the constraints of social norms. One such mechanism is dehumanisation, defined as a "*. . . key psychoanalytic defense mechanism that allows individuals to avoid fully processing troubling events.*"[123] This matrix was a psychological coping strategy. Grodin & Annas highlighted "*[u]ltimately, dehumanization allows the perpetrator to go beyond hatred and anger.*"[124] Within this construct, they identify two distinct processes: self-directed and object-directed dehumanisation. Self-directed dehumanisation acts as a protective mechanism, preventing mental collapse such as complex post-traumatic stress disorder (CPTSD). Object-directed dehumanisation reduces victims to objects devoid of human attributes. Staub (1990) notes, this process involves—

> [O]bjectify[ing] their orientation to their victims, [with] their organised brutality . . . [helped] to maintain the view of victims as less than human and to make the perpetrators' 'job' easier. [125]

Other diffusion of responsibility coping mechanisms[126] highlighted by Grodin & Annas, alongside dehumanisation, include the following—

i. **Splitting**: Believing one is doing good or acting heroically.
ii. **Numbing**: Gradual desensitisation to violence, producing emotional disconnection from the harm caused.
iii. **Omnipotence**: Belief in absolute power and infallibility, justifying harmful actions as necessary or righteous.
iv. **Medicalisation**: Framing harmful behaviour as therapeutic, rationalising pain or killing as a form of treatment or intervention.

These mechanisms retrofit dehumanisation by allowing individuals to bypass moral and ethical constraints, thereby facilitating the perpetration of atrocities. It is important to correct misconceptions about fraternal organisations associated with the occult—such as Luciferian groups, freemasonry, Skull and Bones, and the Bilderbergers. While regularly perceived as ghastly or malevolent, the opposite is typically true in public conduct. They present themselves with impeccable manners, behave with utmost respectability, and appear polished, articulate, and quite ordinary. Yet, while the 'same evil' attributed to Nazi Germany may apply, their harmful actions behind closed doors remain camouflaged and hidden from public view. The following table depicts Hilberg's (1985) process—

123 Grodin & Annas, 639.
124 Ibid., 640.
125 Grodin & Annas, 640; Staub (1990), 58.
126 Grodin & Annas, 640–42.

Table IV-3. Systemic Processes Facilitating the Routine Execution of Harmful Acts

Five-Stage Process of Repression	
Stage	**Description**
1. **Control Information Flow**	Restrict the dissemination of information to those with a legitimate need to know.
2. **Mandate Participation**	Ensure all insiders participate, creating collective complicity and preventing whistleblowing.
3. **Prohibition of Criticism**	Prohibit any criticism especially of the destructive process itself (often termed "groupthink"). This maintains secrecy and minimizes opposition.
4. **Eliminate Social Discussions**	Prevent discussion of the destructive process in social contexts to avoid public awareness and scrutiny. Bad form to talk about the killings. Once brought into the open, they become subject to scrutiny, criticism, and moral judgment, resulting in discontinuance.
5. **Employ Euphemistic Language**	Omit mention of "killings" or "killing installations" even in secret correspondences. Use euphemisms and coded language in all communications to obscure the true nature of the actions, even when conveyed privately.

This table outlines the psychological and social mechanisms that draw individuals into harmful behaviour. Through cognitive reframing, social influence, and cultural reinforcement, ordinary actors are gradually morally disengaged, making wrongdoing seem permissible, necessary, or desirable.
Source: Hilberg, R. (1985). The destruction of the European Jews, 279, 281–282. See also Janis, I. L. (1982). Groupthink; Staub (1989), 28–29; Waller, 208.

The dehumanisation process outlined above functions to protect cognitive faculties from physiological harm and to enable "*. . . moral disengagement through mental restructuring . . . .*"[127] Whereas Table IV-3 however, differs in that it represents an enhanced apparatus that reinforces training and dehumanisation through a process in which Waller (2007) classifies as—

> Professional socialization into a culture of cruelty, fostered by escalating commitments and reinforced by ritual conduct, carries with it a repression of conscience where 'outside' values are excluded and locally generated values dominate.[128]

This resembles peer-conformity strata, where group socialisation defines the acceptable and unacceptable behaviours within a cohort.[129] Waller explains how euphemistic language can render extraordinary acts of evil respectable—a phenomenon amplified when language is used to conceal, mystify, or obscure these acts.[130] This mirrors the obfuscation tactics employed by the Nazis to shield their euthanasia and gas chamber programmes, such as such as the T4 euthanasia programme '*Community Foundation for Institutional Care*' or the '*Reich Association for Hospitals.*' See also Covid Virus Outcomes, 227.

This parallels the Nazi killing regimens. The Third Reich deliberately concealed the truth of these programmes on Himmler's orders, following Hilberg's (1985) five-stage

127 Grodin (2007), 651.
128 Waller, 208.
129 Ibid., 205–12.
130 Ibid., 188.

psychological professional socialisation process for killing repression. In fact, this process is substantiated by Hilberg (1985) when he quotes Himmler on October 4, 1943—

> I want to mention here very candidly a particularly difficult chapter. Among us it should be mentioned once, quite openly, but in public we will never talk about it. Just as little as we hesitated on June 30, 1934, to do our duty and to put comrades who had transgressed [the brownshirts] to the wall, so little have we talked about it and will ever talk about it. It was with us, thank God, an inborn gift of tactfulness, that we have never conversed about this matter, never spoken about it. Every one of us was horrified, and yet every one of us knew that we would do it again if it were ordered and if it were necessary. I am referring to the evacuation of the Jews, to the extermination of the Jewish people.[131]

Reflecting on Broadening the Scope of Racial Hygiene, 113 section, demonstrates how German scientific advancements, fused with Aryan ideology at the *fin-de-siècle*, foreshadowed the Holocaust. The transmission of anthropological knowledge, combined with racial ideology, might have enabled perceptive analysts to predict a future Holocaust—had they recognised the writing on the wall. While such warnings would have been uncomfortable, they would surely have been invaluable. Consider an analogy: if an experienced stockbroker had foreseen the global markets crashing a year hence, any investor would have welcomed a reasoned warning. Similarly, during the Weimar years, twenty years after the *fin-de-siècle*, the German New Order racial hygiene ideology had crystallised into its final form. Recognising its imminent implementation within a ten-year window, how many would have dared to alert the world to the impending totalitarian regime?

iii. Occult Practice: Social Conditioning, and the Architecture of Systemic Violence

The following constitutes a rigorous analysis. Despite the veneer of polished etiquette, opulence, and philanthropic benevolence associated with the occult organisations, a candid assessment indicates that clandestine killings occur at the highest echelons of fraternal groups. Such claims are predictably denied, as few in the modern age would readily accept that these activities persist. Yet, when these activities are deliberately concealed, there is no way for outsiders to observe or verify them, leaving their existence effectively unknowable. The analysis that follows builds on the data presented in Table IV-3 above. Upon accepting an invitation, the process begins with initiation rites and oaths of loyalty. Participants are primarily drawn by the inducements listed in Table IV-2, 265: accelerated career advancement, fame, or financial rewards.[132] As with Nazism, occult organisations recruit the uninitiated, corrupt them through calculated incentives, isolate them from outgroups, and assimilate them into a closed brotherhood.

Once seduction begins, advancement through the ranks is gradual and largely unconscious. Any initial light or innocence held by the recruit at initiation shifts imperceptibly towards darkness. Like the transition from day to night, the change is incremental, progressive, and rarely noticed as it occurs. By full integration, it is too

[131] Hilberg, 281.
[132] Staub (1989), 82, cites Eichmann as an example of career progression.

late: no light remains, only darkness. This trajectory mirrors Staub's (1989) continuum of destruction, where perpetrators undergo a progressive psychological transformation.

The emergence of group violence emerges gradually, intensifying only after deeper integration into higher Degrees and Orders. First, in outline form, the process begins with passive bystanding, marked by the tacit acceptance of mistreatment. Second, it advances to participation in minor transgressions, gradually increasing both the capacity and willingness to harm victims.[133] Third, early socialisation through cultural activities and *hors d'œuvres* receptions provides emotional relief, masking the underlying moral erosion. In occult traditions, this symbolism is inverted: progression is framed as moving from darkness (ignorance), toward light (knowledge and enlightenment), illustrated as ascending a stairwell from a dark basement into daylight at ground level. In reality, the inverse holds true.

The killings are prescribed, it is desired, required or asked for, but as ritual requirements tracing back to Baal worship, in Middle and Far Eastern antiquity. Comparative analysis reveals striking parallels between the ritualised hierarchy of Nazism and contemporary occult practices. Among the higher ranks, the worship of Wotan or Odin was especially prominent. As discussed under heading, The Superman Doctrine, 72, scholars emphasise that chivalric and fraternal orders, oaths of loyalty, initiation rites, rituals, national symbols, imagery, and strict hierarchical obedience (hereafter "*esoteric elements*") were core elements of the Nazi cultural rubric, as well as in contemporary occultism. See heading The Role of Social Psychology. Fundamentally, these esoteric elements operate in a dual mode: they bind members both to the occult practices and to the group, whether in historical Nazism or contemporary occult structures, operates as a 'support group' or 'family network.'

Bialas (2013) affirms this position, asserting that "*Nazi morality . . . combined absolute virtues that demanded unconditional obedience with social engineering and replaced common sense intuitions . . . ,*" effectively ensuring compliance.[134] Moreover, Moscovici (1985) reiterated a basic law of social behaviour: "*. . . when [people] come together in a group, they instinctively submit to the authority of one of their number.*"[135] Authority-driven conformity structures thus cultivated communities bound by shared mystical values, fostering camaraderie and unwavering commitment to organisational ideology. Grodin & Annas (2007) and Grodin (2010) note, "*[t]his binding prevents members from resisting participation in torture.*"[136] The 'brotherhood' was therefore crucial to the psychological conditioning required for occult killings, providing mutual support, shared values, and collective perspectives that facilitated participation in murder. See Table IV-1, 260. References to antiquity, the old Teutonic Order, and chivalric ideals framed desired behaviours in mythical terms, buttressing aggressive, martial conduct.[137] Essentially, these are distinct expressions operating under the same overarching occult framework.

Fromm (1969) argues that obedience to authority is a psychological mechanism or managing fear and isolation, with escape from freedom achieved through total submission and self-humiliation.[138] Complementing the necessity of isolation, Staub (1989) notes that "*[t]he preparation of official . . . murderers often includes creation of a*

[133] Staub (1989), 17–18, 68.
[134] Bialas (2013), 6.
[135] Moscovici, 172.
[136] Grodin & Annas (2007), 645; Grodin (2010), 57;
[137] Grodin, M. & Annas, G. (2007). Physicians and torture, 645, 651; Grodin, M. A. (2010). Mad, Bad, or Evil; Mosse (1966), 95–96; Mosse (1975), 80–82, 136, chs.. 4 & 9. Peukert, 40. See also Staub (1989), 113.
[138] Fromm, 106–07; Moscovici, 180–81.

strong in-group bond and differentiation [or separation] from the rest of the world."[139] As discussed earlier, through deliberate cognitive restructuring, voluntary submission, and minimal coercion, the perpetrator acquiesces, coalescing around the behaviour. Within this paradigm of familial intimacy, subtle pressure is applied to secure conformity among members. Bialas (2013) contextualizing this dynamic in the Nazi regime, reaffirming the pivotal role of group composition, asserting that—

> [P]erpetrators did not act on their own initiative or responsibility, but as members . . . within hierarchies of command and decision-making.[140]

The sole viable recourse is unquestioning submission to the group 'family' leader through which the individual gains a renewed sense of security. Both Fromm (1969) and Moscovici (1985) identify an intrinsic drive for submission, providing protection, direction, a sense of belonging, and contentment—while replacing the existential anxiety of freedom with prescribed thoughts and behaviours. As noted above, the process of dehumanisation initially acts as a protective mechanism against psychological harm.

However, with constant repetition, Goodrick-Clarke (2002) observes that ". . . *the charisma of power, sexual and material pleasure are experienced through active practice.*"[141] This dynamic was exemplified by the Nazi SS, rigorously trained for obedience, violence, and unflinching brutality,[142] and it continues to be true in contemporary occult contexts. As cited in the Romanticism section, Hitler defined the brotherhood community as "*esprit de corps.*" Moreover, in *Mein Kampf*, where he similarly advocated a herd-mentality approach, instilling fear of isolation outside the group.[143]

Killings at higher levels are concealed from lower ranks and disclosed strictly on a need-to-know basis, even among participants and invitees. Secrecy is total, within and without. This pervasive concealment explains the uniquely clandestine, "*hush-hush*" character of these practices, as shown in Table IV-3 above and Table IV-4, and largely accounts for why they remain perennially unknown. Structural secrecy, combined with the euphemistic language typical of occult systems, further obscures these acts, ensuring they stay hidden from public scrutiny. Only discerning, critical analysis can uncover these realities and grasp their full implications.

Accordingly, exposure would result in condemnation and discountenance. Stage 4 in Table IV-3 thus accounts for the insistence on secrecy and the imposition of stringent decrees. This directly leads to Stage 5, wherein euphemisms and coded terminology have long been systematically employed within fraternal organisations to disguise the true nature of actions. This strategy is not novel. Eminent Holocaust historian Hilberg (1985) emphasises that the primary purpose of euphemistic and coded language is twofold: (i) to conceal the destructive process from outsiders, and (ii) to shield the perpetrators themselves, thereby suppressing moral awareness and censoring conscience.[144]

Arendt (2006) concurs with Hilberg's analysis, clarifying that Hitler's orders for mass extermination became "*bearers of secrets,*" entrusted solely to the elite SS, bound by a solemn oath of secrecy. These orders governed the subtle yet rigid "*language*

139 Staub (1989), 60, 68.
140 Bialas (2013), 11.
141 Goodrick-Clarke (2002), 219.
142 Fromm passim; Moscovici, 179. See also Staub (1989), 30, 128–50.
143 Neumann, 439, quoting Hitler, Mein Kampf, 715–16 (year provided). Although, See Hitler, A. ([1925] (1971), 478–79.
144 Hilberg (1985), 278. See also Table IV-1, 260. See also Staub (1989), 29.

rules" (Sprachregelung), were euphemisms carefully crafted to deceive. Dicks (1972) characterises this as "*psychological denial,*" designed not only to protect the external façade and the perpetrator's family, but, more importantly, to preserve the perpetrator's mental integrity by alleviating guilt and suppressing fear.[145]

In addition, this parallels the covert nomenclature employed for Nazi euthanasia medical procedures. Numerous euphemisms masked the reality of these medical crimes, for reasons already outlined in the Nazi Euthanasia, 173 heading: to enable reciprocal discussion among perpetrators while shielding the activities from scrutiny that might provoke outrage and lead to discontinuance. Hitler, for instance, replaced "*murder*" with "*mercy death,*" thereby easing perpetrators' consciences. Prescribed stock phrases such as "*final solution,*" "*evacuation*" (Aussiedlung), and "*special treatment*" (Sonderbehandlung) were likewise integral to the killing process.[146] Furthermore, as detailed in Nazi Euthanasia Schafft reported that euthanasia killings were often carried out under the guise of medical provision, official transport vehicles, mobile ambulances, or standard ward healthcare.

Arendt (2006) further explains that the term "*language rules*" was a code word in itself, understood internally to signify a deliberate lie.[147] This mirrors modern occult structures, which present exoteric "*truths*" to outgroups (deceptions), while reserving esoteric meanings (actual truths) exclusively for ingroups, the alleged enlightened. Although the following table focuses on leadership, Grodin & Annas (2007) emphasise that virtually anyone can be shaped into a perpetrator. See also Education of Society Matrix, 55. The procedural stages of indoctrination used to produce a torturer are therefore reproduced below—

Table IV-4. Psychosocial and Structural Factors Contributing to the Formation of Perpetrators of Violence

This table illustrates how individuals are drawn into violence through a combination of institutional recruitment strategies and psychological conditioning. It differentiates between external structural forces (e.g., careerism, bureaucratization) and internal traits or predispositions (e.g., obedience, dehumanization), providing a dual lens on perpetrator formation.

Category	Elements
1. **Recruitment Function**	Packaging of Evil
	Careerism
	Bureaucratization of Evil
	Culture of Cruelty
2. **Individual Psychology**	Predispositional Trait ▪ Personal History of Obedience to Authority ▪ Screen for intellect, physical strength, and positive identification with politics

[145] Arendt (2006), 85; Dicks, 89–90; Staub (1989), 29, 60.
[146] Arendt, 108, 145; Staub (1990), 58.
[147] Arendt, 85–86.

Table IV-4. Psychosocial and Structural Factors Contributing to the Formation of Perpetrators of Violence

	Developed Traits ▪ Dehumanize and Blame ▪ Use Elitist language ▪ Splitting ▪ Numbing ▪ Omnipotence
3. Group Psychology	Diffusion of Responsibility Theories of Aggression Training — Obedience to Authority ▪ Bind with initiation rites, new rules, new values ▪ Swearing allegiance, developing group adherence ▪ Isolation from outgroups ▪ Elitist attitude/in-group language ▪ Harass, intimidate, desensitize, promote instinctive responses ▪ Reward obedience ▪ Demonstrate social modelling of group violence ▪ Regularize and routinise violence ▪ Practice controlled violence Uniqueness of Group
4. Social Context	Control Information Flow (Five-Stage Process of Repression) Prohibition of Criticism (Five-Stage Process) Eliminate Social Discussions (Five-Stage Process) Employ Euphemistic Language (Five-Stage Process) Mandate Participation (Five-Stage Process)

Source: Bialas, W. (2013). Nazi Ethics : Perpetrators with a Clear Conscience; Grodin & Annas (2007), 651, generally, 639–50; Grodin (2010). Mad, Bad, or Evil 57; Katz, F. E. (1993). Ordinary People and Extraordinary Evil; Hilberg, R. (1985). The destruction of the European Jews.

So far, we have examined obedience to authority, esoteric elements, coded "*language rules*" and the process of beguilement. Grodin (2007) seamlessly demonstrates how these components interlock to shape individuals into compliant participants within occult and authoritarian systems—

> The decision to obey authority figures is enhanced by binding factors – taking oaths, swearing allegiance, developing group adherence and the creation of special language and rituals. When individuals are slowly pushed over the line of decency and where violence comes to be seen as a normal occurrence, anyone can become a torturer.[148]

More importantly, Bialas (2013) frames Grodin's observation as a step in the procedural moral reconditioning of recruits, demonstrating how easily ordinary individuals can be transformed into remorseless perpetrators capable of killing designated enemies without hesitation or conscience. Pollefeyt (1999) captures this process of Nazi moral disengagement succinctly "*[t]he Nazis knew what they were doing . . . and they acted consciously and creatively in accord with this new moral sense.*"[149]

[148] Grodin (2007), 651.
[149] Bialas (2013), 7; Pollefeyt, D. (1999). The Kafkaesque, 29.

This coincides with Figure I-1, 28, which depicts the general societal process of moral erosion, portraying how interconnected forces undermine decency and catalyse the perpetration of extraordinary evil. The process is analogous to the incremental rise of Nazi totalitarian regime, as discussed in the Education of Society Matrix, 55. Indeed, Staub (1989, 1999) argues that the path to genocide develops through a sequence of repressive and harmful acts that progressively become accepted, or even regarded as 'normal,' corresponding with the five-stage process in Table IV-3. For example, SS officer Suchomel forced Jewish prisoners in Treblinka to learn and sing the following song—

> Looking squarely ahead, brave and joyous,
> at the world,
> the squads march to work.
> All that matters to us now is Treblinka.
> It is our destiny.
> That's why we've become one with Treblinka
> in no time at all.
> We know only the word of our commander,
> we know only obedience and duty,
> we want to serve, to go on serving,
> until a little luck ends it all. Hurray!"[150]

This song illustrates the power of indoctrination in fostering duty and unwavering obedience among prisoners. Through repetitive reinforcement of loyalty to the commander and acceptance of their fate at Treblinka, individual resistance was gradually eroded and replaced by collective identity.

Where does this discourse fit in the bigger picture today? This author contends that global occult fraternities who represent these infrastructures, will debut in a gestalt. The argument gains traction in light of the post-October 7, 2023, prevalence of antisemitism among many national leaders during that period. Staub (1989) explains that institutions intending to commit genocide cannot fully develop without the machinery of destruction: "*[t]is arises from a pattern or gestalt, rather than from a single source.*"[151] In other words, where state institutions contribute to the enabling, legitimisation, or normalisation of such a machinery, they become implicated within its systemic formation. Within that structure, national leaders may be understood as operative participants insofar as their actions and rhetoric accord with and sustain its functioning. Arendt (1973) made a remarkable and somewhat ponderous statement, linking the demand for unwavering loyalty to the preparatory phase for establishing a totalitarian apparatus—

> Totalitarian movements are mass organisations of atomized, isolated individuals. Compared with all other parties and movements, their most conspicuous external characteristic is their demand for total, unrestricted, unconditional, and unalterable loyalty of the individual member. This demand is made by the leaders of totalitarian movements even before they seize power. It usually precedes the total organization of the country under their actual rule and it

[150] Lanzmann, C. (1985). Shoah : An Oral History, 105–06. See also Staub (1989), 68. Retained Lanzmann's formatting.
[151] Staub (1989), 20, 23.

> follows from the claim of their ideologies that their organization will encompass, in due course, the entire human race.[152]

A consistent pattern over the last two decades is that major crises followed by dramatic societal change. Recent examples include the 9/11 attacks, the 2008 global financial crash, and the Covid pandemic. This does not imply that contemporary society is directly involved in the occult, nor does it suggest universal participation. Rather, individuals are either active participants or passive bystanders, tacitly permitting actions to escalate along a continuum of destructive outcomes. The rise of the New Global Order maybe analogous to the ascent of Hitler's regime, facilitated by elite coteries engaged in occult practices. See The Superman Doctrine, 72. As previously mentioned, in the Third Reich's consolidation of power, authority to kill was systematically issued top down through the Führer principle, on a piecemeal basis, cemented by broader systemic compliance.[153] Kressel (1996), referring to the general populace, notes many historians and psychologists have pointed to the role of the "*tens of thousands of Nazis*" who carried out the mass murders—

> Though these Germans might have failed a test of democratic responsibility, neighborliness, or moral development, their motives cannot be described accurately as sociopathic, satanic, or evil to the core.[154]

This psychological dichotomy answers Moorhouse's (2010) question raised in his review of Adrian Weale's book. Hilberg's (1992) analysis concurs with Staub's observation, asserting that "*. . . every agency was contributor; every specialisation was utilised.*"[155] Hilberg's (1992) analysis, articulates that the SS's selection and training uniquely predisposed its members to carry out the majority of the killings—

> Nazi Genocide was also carried out by tens perhaps hundreds of thousands of bureaucrats, professionals, businessmen, police-men, workers, and soldiers-people from every walk of life.[156]

According to Kren & Rappoport (1980)SS membership increased from 280 members in 1929 to 30,000 by 1932. The Nazi Aryan elite, including the SS, were the primary practitioners of the occult from which Nazism emerged. See The Superman Doctrine, 72 analysis. After joining the Nazi Party, some members were funnelled into the SS, and subsequently integrated into medical, police, and military organisations.[157] As discussed, the SS shares a similar structural features and manifesto with the modern-day occult 'brotherhood.' Segev (1987) said as much, describing it as a religious order, a secret society, and a family all in one—encompassing a distinct way of life.[158] Analogously, in a

[152] Arendt (1973), 323.
[153] Kressel, 155–156; Dicks, 255.
[154] Kressel, 156.
[155] Hilberg, R. (1992). Perpetrators, *20–24 passim; Kressel,125.
[156] Kressel, 125, 152.
[157] See Lifton, R. J. (1986). The Nazi Doctors; Staub (1989, 141–44; Weindling, P. J. (1993). Health; Weindling, P. J. (2000). Epidemics; Weindling, P. (Ed.). (2017). From Clinic to Concentration Camp.
[158] Segev, 62, 216. See also Kren & Rappoport, 43, and notation on 168; Staub (1989), 130.
Segev's phrase 'secret society' does not appear in the (1987) version. It was from his dissertation titled the same, published by Boston University in 1977; University Microfilms, 77-21, 618.

future totalitarian system, a contemporary global occult fraternity could easily mobilise an SS-like entity within the broader social milieu. The historical link between occult ideologies and antisemitism has now evolved into a global phenomenon, intricately tied to the expansion of globalisation and reactivated through the concept of a New Global Order or Nazism 2.0.

CHAPTER 10. SCIENTIFIC IDEOLOGIES AND GOVERNANCE TECHNOLOGIES

A. Enlightenment Legacies and the Rise of Evolutionary Governance

Introduction

In Part I., we examined the historical foundations of genetics and anthropology. In Part III., Covid analysis, we explored how scientific metrics are applied in contemporary technologies, in particular, anthropological applications such as facial recognition. These technologies highlight the growing sophistication of surveillance systems, which are not only precursors to a potential New Global Order but also harbingers of future pandemic governance networks, commonly referred to as Disease X. This analysis also demonstrated the intersection of genetics and immunology, with genomic research and biotechnology playing a critical role in disease prevention, vaccine development, and global health security. The Covid analysis preemptively addressed the ideological framework sustaining this system, where scientific and medical domains integrate multiple methodologies converge to manage and restructure society. However, the conceptualisation of these connections has yet to be fully explored. Now, we pick up this theme.

Burleigh & Wippermann's written in 1991, analysis raised critical questions concerning the long-term intentionality underpinning Nazi policy, in particular its aspiration for a "*. . . new, global, racially-based social order . . . .*" They concluded that this question has been "*. . . neglected, or inadequately answered, by existing research.*"[1] The present section seeks to remedy this omission through a comparative historical analysis of Nazi policy alongside a philosophical examination of classical texts, drawing parallels with emerging concepts of a prospective International New Order that display striking ideological affinities with the Nazi vision. Such a trajectory was anticipated by Jodl (1891) who articulated this prescient dictum—

> [T]o use the past as an instructive object-lesson for the future, so that what Bacon and Hobbes set up as the aim of science in general shall apply also to history, namely, that the end of knowledge is power.[2]

Jodl's point places the past as a foundational substrate, with successive generations building upon and refining inherited knowledge. Proctor (1990) characterises this intellectual apparatus as a "*. . . universal ontogeny of science, according to which scientific*

[1] Burleigh & Wippermann, 2.
[2] Jodl, F. (1891). Morals, 204.

theories succeed one another through a well-defined and universal sequence of stages."[3] This cumulative transmission of knowledge has driven the exponential growth of scientific and technological capacity in the modern era, fuelling breakthroughs across the sciences, surveillance, transportation, monitoring, and related domains. Enlightenment rationalism redefined the boundaries between science, religion, and the occult. Rather than extinguishing esoteric thought, the period reframed it through empirical inquiry and universal ethics. The result was not rejection but reinvention, where mystery and reason were no longer mutually exclusive. The following premises outline the intellectual context in which this transformation occurred—

- **Reason as Foundation**: Rational inquiry replaces inherited authority as the starting point for knowledge.
- **Science over Superstition**: Naturalistic explanation supersedes supernatural belief as the standard for truth.
- **Universalism as Ethic**: Human unity grounded in shared principles rather than sectarian belief.
- **Knowledge as Progress**: The pursuit of truth is linked to the advancement of civilization and societal improvement.
- **Esotericism Rationalised**: Esoteric thought reinterpreted through the empirical language of science.
- **Brotherhood as Method**: Collaboration in inquiry is both a practical tool and a moral imperative.

i. Class-Based Stratification in a Technocratic Future Regime

CLASS-BASED STRATIFICATION PLAYS a crucial role in shaping the dynamics of a technocratic future regime, where power increasingly follows institutionalised rules and interests prioritising efficiency over democratic ideals. See Figure IV-3, 337. Mearsheimer (2019) defines an "*order*" as an organised set of international institutions and rules designed to govern state interactions, enforce compliance, regulate behaviour in ways that serve their objectives, and adjust or disregard those rules when they conflict with vital interests.[4] This definition is situated within broader debates on globalisation and its ideological roots, in which Outram (2013) draws a compelling connection between the emergence of a New International Order and the foundational structures of Enlightenment thought. The Enlightenment laid the groundwork for interconnectedness and the cultivation of global citizenship, exemplified by Kant and Locke, who championed universal human rights and the pursuit of knowledge across national and cultural boundaries.[5]

Stuurman (2017) similarly argues that intensified global communication, envisioned by Frederick Douglass (1818–1895), would foster mutual understanding and a universal sense of belonging to the brotherhood of humanity, grounded in a futurity of convergence and cooperation.[6] In a contrasting trajectory, Nagl (1974) cites Edward Kiss's fictional portrayal of the "*Ases*" establishing a Nordic-led global "*. . . organic community of a world-wide empire . . . with its centre and capital in Atlantis.*"[7] While fictional, this

3 Proctor, R. N. (1990). Nazi Biomedical Technologies, 17. Proctor attributes this idea to Thomas Kuhn, though without providing a specific reference or publication date, rendering the attribution somewhat abstract.
4 Mearsheimer, J. J. (2019). Bound to Fail, 9.
5 Outram, 8.
6 Stuurman (2017), 384–85, 415.
7 Nagl, M. (1974). SF, Occult Sciences, *189–92.

vision thematically overlaps with the Nazi regime's racial-utopian ambitions, it ultimately failed to extend beyond Europe. Stuurman highlights a critical tension within these Enlightenment ideals, describing them as double-barrelled. On one hand, they envisage a universal Enlightenment culture founded on the ideal of equality; on the other, they define equality as the right to pursue individual happiness, constrained only by respect for others' autonomy. This tension can give rise either to cultural imposition, where dominant groups enforce their values on others, or to unchecked individualism, in which personal freedoms take precedence over collective well-being.[8]

Consequently, the Enlightenment can entail the curtailment of certain individual rights, justified as serving the greater good, harmonising with the totalitarian regulatory controls of the International New Order, crafted to manage both checked and unchecked individual activities "*. . . to advance broad aims such as national security, public order, public health, and public morals.*"[9] Thus, the balancing act of individual rights and freedoms, when weighed against perceived societal security risks, if the latter is deemed greater, it wins. The greater will invariably triumph, especially when a case is presented favouring the majority. To this end, a 'proportionality test' is employed to determine whether a limitation of freedom is justified.

Although Gunatilleke (2021) maintains that the traditional proportionality test has both normative and practical weaknesses, it continues to face significant limitations.[10] He advocates a more robust framework than the traditional test; yet even such a refined regimen can be overridden by persuasive philosophical argument, particularly during existential crises such as Covid, when evidential claims can generate widespread public assent.[11] Gunatilleke's construct holds that a belief upheld by an overwhelming majority, such as viewing a minority group as "*culturally inferior,*" cannot constitute a publicly acceptable reason to deny that group equal citizenship.[12] Yet the global response to Covid provides a clear empirical example in which dissenting minority groups were rejected, overridden, and "*equalised*" in the name of the "*greater good*" or the "*protection*" of the global community, revealing that such ostensibly sacrosanct protections are easily bypassed in practice.[13] By inference, this dynamic accords with proposals for a new "*universal morality*" to mirror a new paradigm. Haas (1988) distilled a complex matrix explaining how a new 'moral code' could materialise and gain acceptance, if—

i. The ethic is coherent, consistent, and compatible with the dominant worldview—synchronising with individuals' intuitive sense of what is moral or immoral;
ii. The cultural thought reflects a shared community of values and draws on widely accepted linguistic conventions that are intelligible and meaningful.[14]

A clear application of this logic is seen in Christianity, Judaism, and Islam—each offering cohesive moral systems based on shared values. The Islamic ethic is held to be coherent, consistent, and compatible with the prevailing worldview. Consequently, Islam is incrementally positioned as a presaging moral benchmark, with other value systems seen as incompatible or immoral. In the West, Islamic beliefs are becoming more embedded, forming a cultural and communal framework that is, by and large, accepted

8 Stuurman (2017), 342; Weikart (2004), 104.
9 Gunatilleke, G. (2021). Justifying Limitations, 92.
10 Ibid., 92, 96.
11 Gaus, G. F. (1996). Justificatory Liberalism, 162.
12 Gunatilleke, 94.
13 Ibid., 95.
14 Haas, P. J. (1988). Morality, 38. See also Bialas (2013), 8.

as the new "*normal*" way of life across continents. Simultaneously, Judeo-Christian norms are being reinterpreted and misaligned with contemporary moral trends. This exemplifies how global moral codes are shifting rapidly.

For example, notable among these are the 2008 economic crash, the 9/11 terrorist attacks, and the hysteria surrounding infectious outbreaks such as Covid—each illustrating how security policy categorically changed overnight. This emerging moral framework could be termed '*the moral code balancing act.*' The equilibrium between individual rights and collective responsibilities continues to shift towards a progressive global equality, shaped by major events over the past decade. Stringent covert security measures imposed unprecedented constraints on freedom of movement across local, national, and international levels. The world has reached its zenith, where it is not hard to foresee a future calamity, though not necessarily a pandemic, in which rights to self-autonomy, lawful dissent, and free speech would be interpreted as misinformation or extremism. In the interests of the common good, public policy arguments prevail over individual autonomy. Bialas (2014) framed the totalitarian balance from the Nazi perspective with striking poetic clarity—

> Life does not concern itself with individuals and their trivial fates. . . . The slogan "the common good over the welfare of the individual" underscored the intention of bringing basic selfish instincts under the control of the national community. From the Nazi point of view, everything revolved around the duties to the people's community and not the rights of the individual.[15] (Footnote call numbers omitted)

From the perspective of the Nazi general public, Johnson (2011) indicates that "*. . . most ordinary German citizens freely gave their consent to a regime they thought was acting in their best interests.*"[16] See National Awakening, 121. This concept of 'surrender' inflects the tension between the perceived need for global cooperation and the preservation of individual liberties, highlighting the perpetual philosophical debate between security and freedom in an increasingly interconnected world. The utilitarian rationale underpinning this shift is embedded in a concept from Alice Bailey (1880–1949) known as "World Servers,' which surfaces in both National Socialist and liberal-progressive ideologies. In line with the principles of 19th- and early 20th-century, liberalism held considerable sway across Europe and beyond, to which Pulzer (1988) outlines the key tenets—

> [P]arliamentary government, the rule of law, the absence of legally established class privileges, a laissez-faire economy, and freedom of speech and association. . . . It was a set of moral qualities, those of rationalism, humanism . . . To be "liberal" means to be . . . broadminded and generous.[17]

There is arguably a reversal of values in contemporary socialist ideology. Readers will recall the discussion of Plato's egalitarian utopia at the outset of this paper; he foresees a classless society closely akin to socialist or communist ideals. National socialists

[15] Bialas (2014), 30.
[16] Johnson (2011), 601.
[17] Pulzer, 29–30.

are evolving into global socialists, embracing totalitarian goals oriented toward the architecture of an anticipated class-hierarchical global order. Bailey's 'World Servers' can be read as part of a longer transmission of esoteric social thought, arguably foreshadowing the emergence of a unified global system. This idea can be traced back to one of the early German occult architects of Nazism, Rudolf Steiner. Review Table I–5, 75.

Galbreath (1971) who observed, "*. . . spiritual occultism [is] represented by Theosophy, Rudolf Steiner, Gurdjieff and Ouspensky, and a ritualistic-magical wing . . . .*"[18] Within this intellectual milieu, Steiner articulated the notion of the "*Universal Human*" during his 1909–1916 lecture series, a concept that Bailey would later reframe as "*world servers.*" Bailey excluded race as a defining factor, a position synonymous with Steiner's progressivist conception of "*Universal Human*" and its framing of evolutionary spiritual development. Both Steiner and Bailey's writings fall within the genres of occultism and New Age spirituality, emphasising interrelatedness, the fulfilment of universal Enlightenment, and harmony. Indeed, Bailey imagines a unified global system underpinned by collaborative service, refracted through John Lennon's idealistic vision of a perfect world united in harmony and belief—

> Imagine there's no countries
> It isn't hard to do
> Nothing to kill or die for
> And no religion, too
> Imagine all the people
> Livin' life in peace
>
> You may say I'm a dreamer
> But I'm not the only one
> I hope someday you'll join us
> And the world will be as one
> Imagine no possessions
> I wonder if you can
> No need for greed or hunger
> A brotherhood of man

Although Bailey writes in cryptic and allegorical language, she portrays the 'World Servers' as essential cogs and gears in the Global New Order.[19] Put another way, they are the boots on the ground enforcers. In contrast, Steiner's process was based on his reconstruction of the *New Atlantis*—not merely as a mythical continent but as a pivotal phase in humanity's psychological and spiritual evolution.[20] The appropriate ontological framing for this field of inquiry is perhaps best understood as a form of "*perennial philosophy*" or "*spiritual science.*"[21]

Steiner is representative of his peers' shared convictions that the Aryans, both among the elite and the general populace, were "*. . . specially chosen population of Atlantis . . . ,*" while supposed inferiors were regarded as spiritually and evolutionarily regressive.[22] Yet Steiner rejected race as a criterion, warning that those who clung to racial distinctions would degenerate and be left behind, whereas those who embraced his 'Universal Human' matrix would advance—a process he and later followers described as

---

18 Galbreath, R. (1971). The History of Modern Occultism, 728 in original; 100 in bound volume. Galbreath's article was a bibliographical survey of the broader occult literature, but is now dated.
19 Bailey, A. A. (1971). Ponder; Bailey, A. A. (1976). Discipleship; Bailey, A. A. (1995). Initiation. For background on Bailey, see Hammer, O. (2003). Claiming Knowledge, 65 et seq.
20 Godwin (2010), 103–12.
21 *Hammer, 344; Leijenhorst, I:82; Staudenmaier (2008), 12, 29n.56, citing Steiner, R. (1990). The Universal Human, 42, 76, 93; Staudenmaier (2014), 37, 42. See also McDermott, R. A. (1987). Anthroposophy. I:320–21; McDermott, R. A. (1992). Rudolf Steiner, 288–310.
22 Staudenmaier (2008), 13; citing Steiner (1985b), 80.

"*cosmic eugenics.*"[23] The continuity of eugenic terminology is therefore not incidental but indicative of a deliberate repurposing and elevation to a metaphysical register, in which biology, spirituality, and destiny converge in a new form of esoteric 'natural selection.' Hobhouse (1911) affirmed this as a foremost doctrine, arguing that social progress arose from modifications in tradition and societal interaction rather than from biological or racial determinants.[24] The recalibration of societal values to the principles of the aforementioned 'moral code balancing act', are predicated on virtues essential for collective advancement—such as productivity, duty, and the cultivation of a "*natural elite*"—ties directly to the concepts of the 'World Server' and the 'Universal Human.' This brings us to the next point of discussion.

ii. The Quantification and Assessment of Human Economic Value

THIS HEADING RELATES to the reduction of human value and worth to the quantitative management of populations under the moral logic of a New Global Order. The older Darwinist framework is pithily expressed by Zmarzlik (1972, 2011)—

> Social Darwinism posits a causal or final relationship between the human individual and natural (or more precisely, scientifically interpreted) development. His personal worth is derived first from his biological origin and then from his biological efficiency.[25]

Under the Third Reich, personal value was conditional and racially defined, first, by race or biological origin and, second, by biological efficiency or genetic fitness. Failure on the latter criterion, even where racial origin was deemed "*satisfactory*", ultimately rendered the individual a total failure. Race was therefore a fundamental organising principle within the racial hygiene programme. By contrast, within the emerging global regime, this logic is reconstituted through a repurposed genetic framework in which race is no longer treated as a determining factor. Individual attributes and comparative worth are instead assessed in relation to broader social and economic structures, operating through what Allen (1983) characterises as a rank-order, value-judgement hierarchy of social stratification.[26] Huxley (1962a, 1962b) in his Galton Lecture to the Eugenics Society, prophetically articulated the future role of eugenics. He explicitly states, "*I would prophesy*", before advocating quantitative population control as a prerequisite for population reduction in "*. . . overcrowded countries like Britain, India and China, Japan, Java and Jamaica in particular.*"[27]

In this regard, while Ploetz held the Nordic race as the gold standard of fitness, Weiss's (1987a) annotation reveals the broader dimension of his position: "*. . . fitness, as defined in terms of social and cultural productivity, was the true measure of worth . . . .*"[28] On this reading, race-based framework is rendered redundant, with evaluative emphasis

[23] Steiner, R. (1981). Menschheitsentwicklung und Christus-Erkenntnis, 186, quoted in Staudenmaier (2008), 13–14, 21 29n.65; Staudenmaier (2014), 158, 167, 171, *42n.41–47, citing Steiner, R. (1966). Theosophy of the Rosicrucian, 130. See also Hammer's, 419–27, explicit commentary on Steiner's overarching spiritual science process.
[24] Hobhouse, L. T. (1911). Social Evolution, 38–39.
[25] Zmarzlik (1972), 466; Zmarzlik (2011), 34.
[26] Allen (1983), 106.
[27] Huxley, J. (1962a). Eugenics, 227–28; Huxley, J. (1962b). Eugenics, 133.
[28] Weiss (1987a), 538, 541.

transferring to productivity and socio-economic contribution as the decisive criteria of fitness and value. This principle underpins Plato's hierarchical system, which Popper (1963) identified as the "*. . . economic principle of the division of labour.*"[29] Weikart's (2004, 2013) analysis of the Viennese sociologist Rudolf Goldscheid (1870–1931) lends support to this interpretation, showing how evolutionary progress positions "*lower races*" as occupying a transient place within a rank-order social hierarchy, with the possibility of upward mobility. The originator of this reasoning lies with Haeckel, the architect of Social Darwinism, who maintained that this "*. . . struggle is a progressive force in human history, since it contributes to the development of new and higher forms.*"[30] In key respects, this logic echoes Plato's utopian programme.

Popper (1963) isolated an anomaly within the higher-echelon apparatus, to a limited extent, where: educated auxiliaries, drawn largely from the military and educated classes, could be promoted to the ruling class, while all others are unequivocally inferior. He is silent, however, on whether capable members of the working class might ascend to auxiliary status, acquire education, and ultimately attain promotion to the ruling class.[31] Accordingly, the acquisition of economic value and enhanced social utility is the agency enabling upward mobility.[32] The contrapositive is equally instructive. Within the framework of a New Global Order, those deemed to possess minimal economic value or cultural productivity, under a 'living space' ethos, are subject to exclusion or elimination, relative to assessments of genetic fitness.

This genetic register also represents the flipside of the 'moral code balancing act' discussed earlier. Bruns (2014), citing Schallmayer (1909), indicates that the ideological foundations of inclusion and exclusion were fluid,[33] subsuming concepts of natural selection and congenital traits, all oriented toward the common good for "*. . . future generations, the term used was 'generational ethics.*'"[34] See heading Weimar Era Refinements, 131. Bruns (2014) appropriately comments that the Nazi slogan *Gemeinnutz geht vor Eigennutz*—"*. . . the common good comes before self-interest,*"[35] captures the subordination of individual interest to the collective inherent in this moral code balancing act. Moreover, alongside the 1925 developments, the 'new physician' ceased to focus on individual care and was reoriented toward the "*general welfare of society.*"[36]

The founder of the "*Blut und Boden*" (Blood and Soil) ideology, Walther Darré, was authorised and instructed by Hitler to organise the farmers for support of the Nazi party. In order to accomplish weeding out the weak and impure, Darré called upon Germans and advised that it will "*. . . often be necessary to ignore feelings and to employ crude methods, including extermination . . . .*"[37] This, in turn, helped create a killing apparatus that would "*. . . exterminate those unworthy of life with a clear conscience propelled the new Nazi morality to supersede an allegedly defunct one.*"[38] The same modus operandi was employed through the propaganda apparatus in German society,

29 Popper, 78.
30 Haeckel, E. (1868). Natürliche Schöpfungsgeschichte, 227–29, quoted in Weikart, R. (1993), 476.
31 Popper, K. R. (1963). The Open Society, 45–46, 48, generally, 45–54.
32 Weikart (2004), 101–02, 105; Weikart (2013), 538, 541.
33 Bruns (2014), 233–35.
34 Bruns (2014), 214, citing Schallmayer (1909), 199-231.
35 Bruns (2014), 220.
36 Bruns & Chelouche, 592. See also, Bruns (2014), 225.
37 Lovin, 285; See also Weindling (1993), 475–79, for more background. Darré was famed for fusing Nordic racism with the practicalities agrarian economics and eugenics (475). See also Flowers, 288–95. See generally Bramwell, A. C. (1985). Blood and soil.
38 Bialas (2014), 23.

including the establishment of a specialised society for eugenic propaganda within the medical community.[39]

Just as race and eugenics (renamed to genetics), became the 'new master' in Nazi science, the emerging New Global Order appears poised to implement a restructured class system—one that retains genetics as the 'new master' within contemporary equality models, while race, by contrast, is discarded. This adaptation from race to genetics is not a rupture, but a continuity cloaked in new ideological language. Revisiting the old Nazi health hygiene model in Table I–9, only Division II. and III., the Scientific–Genetic–Medical and social classes columns, will be relevant, with secondary attention given to genetic makeup in the New Global Order. On the surface, the application of the Enlightenment model appears to promote equality on a fair playing field, anchored in the value-judgment principles of the 'World Servers' or the 'Universal Human' ideal.

Hutton (2005) conceptualises a form of 'World Servers' infused with elements of Nazi eugenics. Initially, the Nazi regime targeted Germans through an edifice identical to the 'World Servers' or 'Universal Human' ideal, focused on socio-economic productivity and class supremacy, while rendering ethnicity largely irrelevant and insignificant. To reiterate, the Enlightenment equality ideals previously outlined, is not a Nazi racial apparatus, but a class-based system. These principles are reignited in the envisioned New Global Order regime, which prioritises virtues such as, "*. . . a will to work and a dutiful nature . . . lead to the creation of a natural elite . . . to serve the whole*" for the collective good.[40] This ethos transcends individual gain, commensurate with the 'World Servers' notion of societal cohesion and service.

Yet, borrowing Allen's (1983) insight, it is more accurate to say that a "*. . . value-laden hierarchy [is] masquerad[ing] as a taxonomic hierarchy.*" This recalls Nazi Germany's biopolitical calculations and policies, which subjectively targeted 'unproductive' populations.[41] In short, this is essentially a paradigmatic transition, in which the focus of conflict migrates from the defence of sovereignty to the management of populations and control of resources. Additionally, this migration coincides with the acts of violence discussed earlier in Reengagement of Past Connections, 266, where killings are rationalised as modalities to reduce competition over finite resources and territorial space. Foucault (1978) said as much, stating—

> Wars are no longer waged in the name of a sovereign who must be defended; they are waged on behalf of the existence of everyone; entire populations are mobilized for the purpose of wholesale slaughter in the name of life necessity . . . one has to be capable of killing in order to go on living—has become the principle that defines the strategy of states. But the existence in question is no longer the juridical existence of sovereignty; at stake is the biological existence of a population.[42]

Staudenmaier (2008, 2014), citing Steiner and his followers, envisaged a paradigm shift toward a universal brotherhood as a central value, oriented toward a utopian future, while positioning Jews outside this ideal of Steiner's 'Universal Humanity', declaring

---

39 Weindling (1985), 307.

40 Hutton, 125, citing Gruber, M. v. (1927). Volk und Rasse, 247. See also Hitler, A. ([1925] (1971)), 297. Cf. Huxley (1962b), 127, 133, planning a society that increase a person's genetic capabilities, including cooperation a sense of duty.

41 Allen (1983), 127.

42 Foucault, M. (1978). The History of Sexuality, 137. See also Savage, 419, citing Foucault.

their " . . . *dissolution as the only possible solution.*" Needless to say, this discourse derives from a dystopian utilitarian rationale, framed within a collective eugenicist survivalist paradigm. It coincides with a reformulation of the variants of Darwinian principles of competition and survival—principles which, as Zmarzlik (1972, 2011) reminds us, are buttressed explicitly by Darwinist terminology: "*struggle for existence,*" "*extermination,*" and "*selection.*"[43]

Thus, while the Enlightenment championed ideals of equality, its underlying aim was the universal unification of a controlled society, with productivity as the engine of social cohesion. At its core lay a clinical, exacting mindset—rational, disciplined, and uncompromising. Within this Enlightenment architecture, though frequently overlooked, resides the mindset itself: the wellspring of intent and ethics from which all else flows. This rational outlook forges the work ethic, cultivates the desire and resolve to transcend self-interest, dismantle inherited prejudices, embrace universalist ideals, and promote altruism within tangible social contexts. Those who embody these qualities are envisaged as advancing along the trajectory toward the 'Universal Human' or brotherhood.

François (2023) cites Pois (1986), connecting pantheism as an integral component of Socialist ideals.[44] Moreover, Bailey's concept of 'World Servers' in relation to pantheism, and in equivalence with this Socialist or Marxist lens, is nominally characterised as altruistic agents of service; in practice, however, they act as enforcers of the operational structure of the Global Order, ensuring its synergistic function. They guide humanity's survival and evolutionary progress along pantheistic and monistic lines. Thus, 'World Servers' are individuals who, motivated by pantheistic principles, serve as agents of change sustaining the coherence and advancement of a unified global system, emphasising both human survival and development within hierarchical social structures.

These ideals are again identical in register and emphasis to the stated purposes of Blavatsky's Theosophical Society. Indeed, under the heading The Superman Doctrine, 72, the Theosophical Society and its members actively promoted inclusive membership and universalist values, welcoming individuals irrespective of religion, race, nationality, class, creed, or gender, in service of an overarching objective: the realisation of a universal brotherhood of all human beings. These principles are operative within the recruitment ethos of the Universal Co-Masonry organisation (1975–2026). Articulated in its Declaration of Principles, the organisation admits members who embody these values and frames its mission as the moral, intellectual, and spiritual perfection of humanity.[45] It is the same operational schema is reflected in Steiner's (1985 ed.) words—

> Our own souls once lived within the Atlantean race, and they then developed themselves upward to a higher race. [. . .] In this way we can comprehend how to justify the principle, the core principle of universal brotherhood without regard to race, color, status, and so forth. . . .[46] (Omission in original)

Steiner's thought is intricately connected to Haeckel's Monism; after meeting Haeckel in 1894, he embraced his ideas and adopted a worldview constructed on the naturalistic unity of all existence. See The Superman Doctrine. This outlook encompassed pantheistic

[43] Zmarzlik (1972), 435–36; (2011), 3–4, 9 passim.
[44] François, 36; Pois, R. A. (1986). National socialism.
[45] See the 'Universal Co-Masonry' website accessed February 10, 2026, https://www.universalfreemasonry.org/. See also Hammer, 389, for the scope of brotherhood of humanity.
[46] Steiner, R. (1985a). Die Welträtsel und die Anthroposophie, 153–54, quoted in Staudenmaier (2008), 9, 28n.41.

and panpsychic doctrines, emphasising the interconnectedness of all matter, organic and inorganic.[47] The ideology undergirding the Global New Order utopia draws directly from occult antiquity, the same source that informed Nazi Germany's esoteric traditions. Its future iteration refines these earlier forms, incorporating elements of Nazism while reconfiguring them to serve contemporary ambitions, yet retaining its roots in mystical antiquity.

Simply put, the Global New Order is not a reformulated or modernised form of Nazi ideology. Burleigh & Wippermann (1991) explicitly repudiate claims made by some 1960s authors regarding the purported modernity of the Third Reich. That viewpoint holds that Nazi Germany produced "*... modern, or even social 'revolutionary,' results.*"[48] Both Nazism and the Global New Order, however, are a derivative of the same corrupt fountain. The latter is a distinct entity, drawing from the same esoteric well that informed Nazism. Review Part I.

iii. The Philosophy of Progress: Rational Inquiry and the Emergence of the Superman Ideal

THE ARCHETYPE OF the superman across modern history is not fixed, but forged at the intersection of scientific method, metaphysical speculation, and ideological aspiration. Occult traditions have applied scientific methodology within the hard sciences to investigate, comprehend, and master the laws of nature. Whether through atomic mysticism or genetic engineering, each manifestation reflects a deeper ambition: to transcend humanity through knowledge, biotechnology, and control—verging on transhumanist ideologies. Scientific authority, from Newtonian precision to quantum uncertainty, has been appropriated to legitimise esoteric worldviews—culminating in the vision of a scientifically governed, biologically enhanced superhuman: the New Atlantis Superman.

Esoteric speculation in the pursuit of "*wholeness*" or "*totality*" is to medicine and the sciences what conspiracist history is to the historian's establishment of fact. Yet if logic, rationality, and hard evidence are applied—subjecting claims to rigorous testing through statistical analysis and empirically verifiable methods—they are, in principle, better positioned for reconciliation.[49] One may argue that this should not be understood as openness without limits. Although this view slightly departs from the idea of the occult's wholesale adoption of scientific method, it nonetheless depicts the core point: quantum mechanics, strange as it seems, imposes a hard boundary and cannot legitimise metaphysical claims, such as faster-than-light communication, solely because the occult employs the language of science.[50] This dictum involves a two-tiered appropriation. The first involves what Hammer (2003) labels "*... terminological loans from the physical sciences ...*," whereby occultists adopt scientific vocabulary to endorse occult phenomena.[51] The second entails appropriating the scientific method itself—undertaking genuine investigations, reproducing phenomena under controlled experimental conditions, and incorporating error-correcting procedures. It is to this second tier that we now turn.

---

47 Hutton, 174–75, 179; Myers, 42–64.
See Part I., for references regarding the Nazis worshipping nature.

48 Burleigh & Wippermann, 1–2. See also Pendas, D. O., Roseman, M., & Wetzell, R. F. (Eds.). (2017). Beyond the racial state, 3.

49 See Hammer's, 249–51, where he comments on astrology's adoption of a scientistic methodological approach in making its claims.

50 Eberhard, P. H. & Ross, R. R. (1989). Quantum field theory; Hammer, 271.

51 Hammer, 236; Mohan, S. (2023). Cosmic Dance.

It is this author's position that, despite their legacy of pseudoscience, occult practitioners have wholesale adopted the scientific method—not merely to imitate its vocabulary, but to engage with the hard sciences and master their applied constructions, complete with the evidential rigour of observation, experimentation, and validation. In the post-Enlightenment context, the revised bylaws of Blavatsky's Theosophical Society, adopted in 1878, emphasised universal brotherhood while making clear that its deeper purpose was to unite members in collaborative scientific inquiry—

1. The formation of a universal brotherhood without distinction of race, creed, caste or colour;
2. The encouragement of studies in comparative religion, philosophy and science;
3. The investigation of unexplained laws of nature and the powers latent in man.[52]

Bacon's future-oriented, mystical outlook, predicated on his utopian *New Atlantis* schema in which scientific knowledge transcends physical limits, is not merely probable but presented as concretely knowable.[53] Faivre (2010) grasps the essence succinctly, emphasising the emergence of this new esoteric stance—

> [M]ore marked taste for "pragmatic evidence" ("scientific proofs") . . . to prove the existence of several orders of reality, and often for various forms of practical 'magic' . . . .[54]

Faivre's point distills the imperative of Enlightenment thought, further elaborated by Martin (2002) and Hammer (2003): the "*magical heritage*" was supplanted by Enlightenment values, and scientific endeavour replaced religion as the "*. . . via regia to knowledge*"—advancing naturalistic explanations through rationalist inquiry for phenomena as diverse as ghosts, witchcraft, and the genesis of humankind.[55] Accordingly, Wokler (1955) avers by "*. . . scientific exactitude as against conjectural imprecision . . . ,*" reflects a rigorous, evidence-based process, first glimpsed in the astronomical models of Nicolaus Copernicus (1473–1543) and later formalised by Isaac Newton.[56] In short, it is an elegant, systematic method for utilising data to make projections grounded in robustly tested evidence. Thus, the Enlightenment reinforced a scientific, progressive, and reductively rational ethos.

Einstein's field equation unites time, space, matter, and energy into a single continuum, where relativity and quantum theory intersect with ancient arts of transformation. Einstein's iconic formula, $E = mc^2$, which maintains that mass and energy are interchangeable, is to occultists, a metaphor for hidden power variables latent in matter—forces that, as Pagels (1982) suggests, are "*. . . hiding behind a veil of quantum reality . . . .*"[57] However, it is the field equation itself, linking the geometry of space with the energy of matter, that holds the deeper esoteric allure. Within this occult inquiry, drawing from Bacon's *New Atlantis,* which gives

$$G_{\mu\nu} = \frac{8\pi G}{c^4} T_{\mu\nu}$$

---

[52] Campbell, 28–29. See also Hammer, *218–19.
[53] Martin, C. G. (2002). The Ahistoricism, 33; Weinberger, J. (1985). Science, faith, 304.
[54] Faivre, 80.
[55] Hammer, *202–05; Martin, 25.
[56] Barbour, 108; Culver, R. B. & Ianna, P. A. (1984). The Gemini, 28-29; Wokler, 32.
[57] Pagels, H. R. (1982). The Cosmic Code, 142; Pais, A. (1991). Niels Bohr's Times: In Physics, Philosophy, and Polity. Abraham Pais, the acclaimed biographer of Albert Einstein, offers a lucid account of the intellectual climate that shaped early atomic theory and quantum mechanics.

these principles form, an epistemological synthesis emerged: the *Seven Liberal Arts and Sciences*. Subdivided into the *Trivium* (grammar, rhetoric, and logic) and the *Quadrivium* (arithmetic, geometry, music, astronomy, and mathematics), this system, rooted in the teachings of Plato and Aristotle, went on to configure later esoteric thought.[58]

The synthesis of the Trivium and Quadrivium extends far beyond the classical arts, encompassing every modern discipline. The most prominent of these include law, sociology, psychology, and political science, integrating them into a unified body of knowledge. Rossi (1968) paraphrases Bacon's ideal, concluding that the diffusion of the Seven Liberal Arts and Sciences, alongside new discoveries, forms the foundation for knowledge reform, which must precede any true mastery of nature.[59]

Thorndike (1923), a pioneering scholar who explored the historical interplay between magic, mathematics, geometry, science, and metaphysics, is a foundational, though now dated authority in this field. His monumental seven-volume work offers extensive commentary on Roger Bacon (c.1214–1292), an early advocate of experimental methods. Thorndike highlighted Bacon's view of science as an instrument for achieving profound, almost occult-like results, linking knowledge of the cosmos to hidden or mysterious forces.[60] This concurs with ancient esoteric and metaphysical assumptions that matter is not fixed but rather transmutable. Einstein's equation, now incorporating the cosmological constant (Λ), exemplifies this dynamic—

$$R_{\mu\nu} - \frac{1}{2}g_{\mu\nu}R + \Lambda g_{\mu\nu} = \frac{8\pi G}{c^4}T_{\mu\nu}$$

This reflects occult ideas of a plenum or "*subtle aether*"—an unseen, all-pervading substance or field (a complete, unbroken space; an unseen essence) from which form and energy arise. Thus, Λ acts as a modern counterpart to ancient views of the void as a dynamic, creative field—consistent with the connection between relativity and metaphysical concepts of hidden forces shaping the cosmos. The relevance of Newton lends a measure of credibility to this discussion, grounded in his alchemical engagements, which underpinned his evolving concept of force. Dobbs (1975) explains—

> The new concept of force . . . as it developed in Newton's mind, undoubtedly had roots in his alchemical studies . . . Although Newton never published a work fully devoted to alchemy, after 1692 he made available his mature thoughts on the sub-structure of matter and the forces acting there.[61]

This conflated framework offers fertile ground for esoteric exploration, where general relativity, atomic theory, and quantum mechanics intersect with long-standing occult notions of transformation, interconnectedness, and the manipulation of subtle forces. Johannison (1988) observes that this synthesis is rooted in hermetic traditions of alchemy and extends across the spectrum of mathematical disciplines—

58 Abelson, P. (1906). The Seven Liberal Arts, 119-20; Joseph, S. M. (2002). The Trivium; Leff, G. (1991). The Trivium; North, J. (1991). The Quadrivium; Martineau, J. (2020). Quadrivium; Sandys, J. E. (2009). A History of Classical Scholarship; Yates (1999), 57–58, 61, 63, 96, 179, 208.
59 Rossi, 2–*3, 7, *9.
60 Thorndike, L. (1923). A History of Magic, 655–56. See also Rossi, P. (1968). Francis Bacon, 21.
61 Dobbs, B. J. T. (1975). The Foundations of Newton's Alchemy, 199.

> The concrete content of magical science issued from the emphasis in the Hermetic writings on alchemy, geometry, physics, astrology, and medicine set in a mathematical system derived from number mysticism and the cabala. Experimentation was central-experiments, it is true, biased toward hoped-for results.
>
> Cornelius Agrippa's De occulta philosophia (1533), based on among others Ficino's and Pico della Mirandola's works, can be seen as a practical handbook for this type of magical activity.
>
> . . . .
>
> To summarize, magic as a scientific activity builds on a defined conception of knowledge-derived from the Hermetic tradition-stressing experiments and rationality in a mathematical sense, together with a visionary utopianism aiming at practical results.[62]

Dalton's (1808) mechanistic model of the atom and Bohr's (1958) quantum model posed profound questions about physical reality and observation, laying the foundation for both materialist metaphysics and subsequent esoteric reinterpretations.[63] Hammer (2003) and Stenger (1995) argue that occultists have repurposed these scientific paradigms, frequently under the umbrella of "*quantum and atomic metaphysics,*" to bolster esoteric ideologies. Such reinterpretations merge atomic theory, subatomic physics, and quantum mechanics with broader aspects of modern physics, stripping them of their empirical rigour in favour of metaphysical narratives. These metrics are routinely incorporated into healing practices, where they are linked to the scientific principles of electromagnetism.[64] This is congruent with Eastern mysticism, echoing Steiner's synthesis of scientific knowledge and esoteric traditions. Pagels (1982) humorously observes that the occultist, well-versed in the material constraints of physics, suggests that "*. . . quantum physicists and occultists should now join forces . . . ,*" applying both cutting-edge physics and ancient metaphysical principles to reimagine what reality itself might be.[65]

One esoteric intersection is visibly prominent at the high-energy subatomic physics facility of CERN's Large Hadron Collider in Switzerland, where a bronze statue of Shiva, presented as a gift by India, stands on site. In Hindu cosmology, Shiva as Nataraja performs the Ānanda Tāṇḍava (cosmic dance), signifying creation (sṛṣṭi), preservation (sthiti), and destruction (saṃhāra). CERN adopted this imagery as a metaphor for the universe's perpetual transformation, blending scientific discourse with mystical symbolism through the motif of the Dance of Shiva, later reformulated by Western physicists as the "*cosmic dance*" of subatomic particles.[66] Since CERN's official founding on 15 February 1952, its remit has been the advancement of nuclear science, with

62 Johannison, K. (1988). Magic, Science, 252–53, generally, 251-61; Dobbs, B. J. T. (1975). The Foundations of Newton's Alchemy. See Merkel, I., & Debus, A. G. (Eds.). (1988). Hermeticism and the Renaissance.

63 Bohr, N. (1963). Essays; Dalton, J. (1808). A New System. See also Stenger, V. J. (1995). The Unconscious Quantum, 50–51, for passing comments on Dalton and Bohr's contributions in the context of metaphysical interpretation

64 Babour, 188–89; Hammer, 277–320; Stenger, 10–11, 20–21–26 See also Brooke, 326–34, for a contextual analysis; and Jones, R. H. (1986). Science and Mysticism, 172–74, for critical analysis; See generally Barbour, 165–94.

65 Pagels, 156–57, 178; Hammer, 271.

66 Hammer, 291; Stenger, 110–11.

particular emphasis on particle physics. Although it was not the first particle accelerator, as comparable high-energy laboratories had already emerged in the United States, the Swiss facility, renowned for its experimental rigour, was the first to incorporate Eastern esoteric symbolism into its institutional narrative. Within concepts such as quantum entanglement and the equivalence of energy and matter emerges a shared symbolic language, one that gestures towards decoding, and perhaps even modulating the hidden forces governing both visible and invisible realms.

Readers may recall from The Superman Doctrine, 72 section, Nazi ideology did not merely parallel scientific rationalism; it actively co-opted the scientific method to foster a branch of applied occult science (references previously provided). This modern integration, however, is not unprecedented. Nazism became a crucible in which anthropology, Social Darwinism, genetics, Aryanism, and other scientific disciplines, initially amalgamated in the early 1890s—later converged with occult doctrines and Eastern esoteric traditions promoted by Steiner and Blavatsky's Theosophical Society. While these structures were radical for their era, their technical sophistication is sharply differentiated from present-day advanced emerging sciences. Present-day science, though far more technologically capable, still echoes the ambitions of Bacon's *New Atlantis* and, refracted through modern scientific and cultural narratives, restates some of the same ideological drives that informed Nazism. As previously demonstrated, early Germany during the Occult Revival (1880–1910) within the Wilhelmine period (1890–1914) fused pioneering scientific inquiry with mythic and occult worldviews.

Coincidentally, Haeckel's pantheistic reverence for nature and animals was assimilated into Nazi Germany's broader ideological framework, which suffused with elements of diverse occult and mystical practices.[67] This also cross-references the heading, The Superman Doctrine, 72, where Nazi nature worship was intertwined with sun worship, itself associated with the superman archetype and Apollo-related deities. This nexus of beliefs further extended to nudism (Lebensreform), a foremost feature within the broader pantheistic paradigm. While Blavatsky approached the laws of nature as "*science*,"[68] Hitler and his contemporaries appropriated her ideas, elevated "*nature*" itself to the status of an object of veneration. Mosse (1975, 1999) reminds us—

> Manifestly, only those people who were closest to nature could grasp through their souls the inner, cosmic life force that constituted the eternal. . . . Guido von List set the tone for this kind of argument and fused it with the glories of an Aryan past. Nature was the great divine guide and from her flowed the life force. Whatever was closest to nature would therefore be closest to the truth.[69]

This pantheistic outlook fits seamlessly within the broader notion of evolutionary progress or advancement. Spielvogel & Redles (2011) note that Hitler was conversant with ideas of psychic powers and their recovery, and understood "*human progress*" to mean "*[t]o have 'magic insight' was apparently Hitler's idea of the goal of human progress.*"[70] Thinkers such as Ploetz envisaged human reformation refracted through biological and genetic enhancement. In the view of Banton (1978), "*[t]he French word évolution was applied to the development of individual creatures, and the proper equivalent to Darwin's*

67 François, 36.
68 Blavatsky, H. P. (1877). Isis Unveiled, 3; Mosse (1999), 123.
69 Mosse (1975), 197; Mosse (1999), 121.
70 Spielvogel & Redles, 86, 97n.42, citing Hermann Rauschning, Hitler Speaks (London, 1939), 240.

notion was transformisme."[71] Long before the Nazi period, and even prior to Darwin, the idea of evolutionary progress was a synonym for the "*advancement of science,*" anticipating the themes later articulated in Bacon's *New Atlantis*, itself inspired by Plato's account of Atlantis. Johnson (1937) and Adams (1949) aptly describe this "*progress*" as—

> [E]nthusiastic belief in future progress through the "advancement of science," were part of the "publicly avowed creed" of English scientists between 1550 and 1600. In short, there was considerable preparation of the intellectual climate for an idea of progress before Bacon's time.[72]

What Zilsel (1945) terms "*scientific progress*" captures this emerging pre-Baconian belief in cumulative scientific advancement. Consistent with the present author's interpretation and application of the concept, Zilsel offers the following insights—

> (1) the insight that scientific knowledge is brought about step by step through contributions of generations of explorers building upon and gradually amending the findings of their predecessors; (2) the belief that this process is never completed.[73]

The point being, "*evolutionary advancement*" or "*scientific progress*" is a path toward ever-improving perfection, marked by the continual reproduction of error, the formulation of refined models, and unending subsequent correction. Thus, there is no perfection in perpetuity. Even when an ostensibly perfect paradigm or absolute framework emerges, it continues to be refined and may be superseded by more advanced, more enlightened iterations.[74] Much like the latest Ferrari, which surpasses all previous models in both performance and design, future iterations inevitably incorporate corrections, new technological innovations and aesthetic enhancements.

This stance of "*scientific progress*" underpinned the thinking of figures such as Ploetz, whose vision of human reformation rested on scientific and genetic intervention. Inscribed on the walls of the Thomas Jefferson Memorial in Washington, D.C., is an excerpt from Jefferson's 1816 correspondence with Samuel Kercheval, in which he famously articulated that laws and institutions must evolve in harmony with the advancement of human intellect and societal enlightenment—

> [L]aws and institutions must go hand in hand with the progress of the human mind. As that becomes more developed, more enlightened, as new discoveries are made, new truths discovered and manners and opinions change, with the change of circumstances, institutions must advance also to keep pace with the times.[75]

---

71 Banton (1978), 90–*91.
72 Adams (1949), 376 et seq; Johnson, F. R. (1937). Astronomical Thought, 296; See also, Metz, R. (1938). Bacon's Part, 30; Zilsel, E. (1945). The Genesis, 336–49. In this work, Zilsel traces early chronological contributions to scientific progress from the pre-Baconian era, made by artisans, engineers, and philosophers.
73 Zilsel, E. (1945). The Genesis, 326.
74 See Barbour, 115–19, 124–27, 165 and Jones, 26–28, *33.
75 Jefferson, T. (1899). The Writings of Thomas Jefferson, 42–43.

This distinction highlights the striking contrast between the Nazi regime, the Covid pandemic, and the anticipated future utopia. The Nazi apparatus epitomised, in broad terms, a rigid and centralised form of authoritarian control. Conversely, the Covid era exemplified a more fluid authoritarianism—one that engaged society and innovated to preserve dominance, relied on overt repressive measures. Covid was thus an introduction to a new form of scientific totalitarianism, best described as techno-adaptive authoritarianism. See Figure IV-3, 337.

In this context, Bacon's *New Atlantis*, renowned for its incompleteness due to his death in 1626, portrays a scientifically governed society supported by cutting-edge technologies, inadvertently reflecting the adaptive and operationally fluid utopia. Part Five is only partially extant, and Part Six is entirely absent.[76] Given its inherently adaptive apparatus, this omission may have been purposeful; it is plausible that the intended governing regime is meant to fill these gaps. This 'adaptive' notion is harmonious with the concept of "*scientific progress,*" which is paramount to Bacon's anticipated utopia—a markedly clear departure from both the fixed totalitarian model of Nazi Germany and the dynamic authoritarianism of the Covid pandemic. The Jefferson Memorial inscription reminds us, society must evolve and continually adapt, with its institutions and social structures keeping pace with the advancing times.

Huxley (1957) asserts that as humanity's "*. . . mental organisation continues to reach ever higher levels,*"[77] the urgency for belief systems to evolve and complement intellectual progress grows. This vantage point connects to the earlier discussion on Bailey's World Servers, bridging Huxley's vision, Jefferson's notion of institutional evolution from the Memorial excerpt, and Haeckel's monistic philosophy. Huxley terms this worldview "*scientific humanism,*" a concept he believed was catalysed by his readings of Lord Morley's essays, in which Morley contended that "*[t]he next great task of science will be to create a religion for humanity.*"[78] In Walsh's (1958) review of Huxley's revised 1957 edition, he notes that it considerably refines the 1927 version, stating that it "*. . . is the most winsome and persuasive case for scientific humanism . . . ,*" he had read.[79] Huxley & Huxley (1947) (grandfather and grandson), with Huxley as the younger, offers insight into his worldview and his conception of a "*religion for humanity.*" Tracing Huxley's (the younger) attraction to this idea, it can be seen to date back to his grandfather, T. H. Huxley. With his grandfather as a role model, he reflects on how he was "*. . . bred by nineteenth-century humanitarianism out of traditional Christian ethics . . . .*"[80]

In its earliest articulation, prior to expansion, this orientation was not overtly religious. Rather, it was premised upon a secular moral inheritance: humanitarian in tone, post-Christian in substance, and moulded by post-Enlightenment egalitarian sensibilities. Over time, however, this ethical residue was recalibrated through the progressive incorporation of scientific and medical ideology, predominantly eugenic theory. What began as a broadly moral posture was reconstructed as "*scientific humanism,*" later reformulated as evolutionary humanism or evolutionary naturalism. The original substrate absorbed biological determinism, clinical authority and, ultimately, eugenic logic, now rebranded in certain contemporary discourse as "*NewGenics.*"[81] See also Biometric Technologies, 317.

76 Weinberger (1976), 868.
77 Huxley (1957), 214.
78 Huxley (1927), 127; Huxley (1957), 85. See also Dunér, I. (2025). A Razor, 66; Phillips, P. T. (2007). One World, 614.
79 Walsh, C. (1958). [Religion . . .], 210.
80 Huxley, T. H. & Huxley, J. (1947). Evolution and Ethics, 103, *105. See also Phillips, 616.
81 See for example, Huxley, J. (1948). UNESCO, 1–21; Phillips, P. T. (2007). One World, 613, 617–19.

The resulting matrix is best understood as a subdiscipline within pantheism, characterised by an ontological commitment to a totalising natural order and underpinned by collectivist principles. It seeks to synthesise the intellectual developments of modernity into a unified worldview that aspires "*. . . to restore our unity with nature,*"[82] an ambition evocative of Steiner's "*Universal Human.*" The objective is not ethical reform but metaphysical reintegration. A parallel articulation emerges in Huxley's (1948) reflections during his tenure as Director-General of UNESCO. Confronting accelerating global interdependence, he reconstituted his aspiration as "*. . . scientific world humanism,' global in extent and evolutionary in background*" as part of UNESCO's organisational manifesto.[83] The alteration was strategic. Huxley formed a symbiosis between Scientific rationality, evolutionary theory and global governance, unified within a single universalising project.

When Myers (2015) writes that a "*. . . class of scientists will lead society with their 'scientifically' constructed knowledge of the world,*"[84] he is referring not solely to conventional disciplines but also to the annexation of applied occult sciences. A cursory examination of Myers' statement reveals its empirical interrogation of how the laws of nature operate, under an overarching logic of pantheism, of which deism forms a subsidiary strand. However, when we speak of 'nature' in this context, we are in fact referring to alchemy: a metaphysical framework for comprehending and manipulating reality, encompassing both theoretical reflection and practical experimentation, and presupposing its mutability and capacity for transmutation. The ultimate aim of alchemical praxis was not the creation of gold, but the forging of the superman.

In this context, the natural world is composed of the four elements: earth, water, air, and fire. These were first proposed by Empedocles and later developed into the foundational principles of Aristotelian physics.[85] Mastery over these elements is intimately linked to transmutation, a core tenet of alchemical thought reflecting its practical application. Hopkins ([1934] 1967), Read (1995) and Goodrick-Clarke (2008) explain the scope of alchemy: "*[b]y substituting the qualities (hot, cold, wet, and dry) for one another, an element could be transmuted into another.*"[86]

This is the enigmatic, central objective: uncovering the deeper truths of the natural world through systematic, evidence-based inquiry, "*. . . science [cannot] proceed if the world is seen as a cluster of chance coincidences.*"[87] Viewed through a pantheistic lens, the ultimate aim, articulated within Paracelsian alchemy and Hermetic philosophy, is the rigorous and methodical investigation of natural laws, enabling mastery over nature and the physical transmutation of the Self.[88] This aspect of transmutation accords with Nelson's (2010) discussion, in which he exposes Enlightenment-era thinkers'

---

82 Huxley (1957), 227.
83 Huxley (1948), 7–*8. See also Phillips, 629.
84 Myers, 52. See also Owen, A. (2004), 13.
85 Gutmann, J. (1978). III. Western Philosophical, 236.
86 Faivre, 43, 49, 61; Read, J. (1995). From Alchemy, 2–4; Goodrick-Clarke (2008), 71; Hopkins, A. J. ([1934] 1967). Alchemy : Child of Greek Philosophy. See also Stanislas, K. de Rola (1992). Alchemy: The Secret Art; Churton, T. (2003). The Golden Builders.

These works are rigorous, well-documented contributions to scholarship, intended for readers with a serious interest in the intellectual and chemical foundations of early science. They offer critical insights into the development of alchemy and its role in the early history and foundation of modern chemistry.
87 Jones, 35.
88 Baldwin, J. M. (1911). Dictionary of Philosophy, 199 et seq., entry for 'Occult'; Brooke, 68–69; Flowers, 61–62; Goodrick-Clarke (2008), 40; Hastings, J. et al. (Eds.). (2007 [1917]). Encyclopaedia of Religion and Ethics, 444–47, entry for 'Occultism'; Laqueur, T. (2006). Why the Margins, 111, 118–19; Rossi, 4–5; Yates, F. A. (1964). Giordano Bruno and the Hermetic Tradition. Cf. Rossi, 13–16, Francis Bacon's own view and contribution to alchemical research.

unfulfilled pseudo-eugenic ambitions to engineer society through controlled "*breeding.*" For the purposes of identification, transmutation itself becomes the visible mark of transformation, replacing reliance on external insignia or branding—

> Unlike [Jean-Baptiste] Esteve and others who fantasized about creating an external mark to be placed upon people, Hilliard d'Auberteuil wanted to change the people themselves by transforming their appearance until their bodies were immutable and transparent signs of their social and civil status.[89]

Hammer (2003) extends this analysis by offering a thematic generalisation derived from patterns he identifies across esoteric movements. He contends that, while still concerned with mastery over nature, these traditions progressively shifted their register from the material transmutation of metals to the spiritual transmutation of the Self, signalling a deeper focus on bodily and personal transformation.[90] Interestingly, Owen (2004) and Hammer further explore occultism's complex entanglement with scientific inquiry, a dynamic rooted in the Renaissance and, *a fortiori*, codified during the Enlightenment. The codification prompted occultists to adopt the language of science and rationalism, worldviews formerly at odds with esoteric thought, in order to articulate occult practices in terms of correspondence and causation.[91]

From the standpoint of religion, as examined in the section Crystallization of Racial Hygiene, 110, Huxley (1957, 1960) expanded this philosophy by pressing for the restructuring of organised, traditional religion. Similarly, Haeckel's influential work, The Riddle of the Universe (1899), called for a synthesis of science, religion, and the world's major spiritual traditions.[92] Building on Haeckel's construct, Huxley proposed a model society anchored in secular religious beliefs—a form of a "*natural religion*" or what some might term evolutionary humanism.

Hutton (2005) noted that, for many thinkers, Haeckel's teleology "*. . . drained meaning from the natural world.*"[93] This new creed reinterprets spiritual heritage through modern intellectual developments, scientific progress, and 'technological radicalism,' resonating with the *New Atlantis* narrative.[94] Outram (2013) points out that this Enlightenment vision as a conflict model is founded on the principle that human affairs should be guided by reason and rationality, best termed as a conflict model, "*. . . rather than faith, superstition or revelation, a world view based on science, and not tradition.*" This perspective underpins the critical use of reason to "*. . . change man's relationship with himself and society.*"[95]

This rests on the perception that human intellectual development has outpaced traditional belief systems. In response, Huxley (1957) recommends that society establish a "*new social organ*" to manage its evolving destiny. He further argues that religions and belief systems, once coping mechanisms for ignorance and fear, must adapt to

89 Nelson, W. M. (2010). Making men, 1387.
90 Hammer, 52.
91 Ibid., 51; Owen, A. (2004), 11.
92 Haeckel (1900), 350–64.
93 Hutton, 174.
94 Huxley (1957), 212–13; Huxley, J. (1960). The Evolutionary Vision, 260; Huxley (1962b), 127–28, 139–40; See also Phillips', 631, commentary.
95 Myers, 51, 53, 55; Outram, D. (2013). Enlightenment, 2–*3, *4 passim; Phillips, 613. See also Lucas, 118–19.

scientific advancement.[96] Huxley would likely endorse the conflict model, subordinating religion to science.

By contrast, Metz (1938) and Adams (1949), in their commentary on Bacon's *New Atlantis*, reject such subordination. They present Bacon as a purist intent on preserving a clear boundary between the two domains. His objective was separation in the "*. . . interests of science, not of religion. He wished to keep science pure from religion.*"[97] Barbour (1997) outlines a broader typology, identifying four models of the science–religion relationship: (i) conflict, (ii) independence, (iii) dialogue and (iv) integration. These categories are analytical metrics for interpreting modes of interaction, rather than prescribing a fixed doctrinal stance.[98] In contrast to Huxley, Brooke (1991), Barbour (1997) and Hammer (2003) stress from the outset the historical complexity of the science–religion relationship. Although the category itself is a modern construct, religious apologists frequently maintain that science must ultimately defer to revelation.[99] Between these poles lies a spiritualised experiential science—one that seeks not subordination but harmonious unity with nature and the cosmos, integrating empirical inquiry with spiritual insight.[100]

The central tension is this: religion is traditionally the final authority on matters of faith, life and practice. With the ascendancy of science as the new arbiter of truth, there is no room for two totalising worldviews at the top, each claiming exclusive authority to explain reality in its entirety. The displacement of religion establishes a paradigm in which the scientific establishment dictates religious, moral and ethical norms and, by putting on its medical-science hat, extends its authority to defining health and what constitutes human well-being as well.

Within this synthesis, Judge (1893) adopts a harmonising outlook, threading occultism, religion, and science together by propounding that "*. . . Theosophy is a scientific religion and a religious science*"—a formulation that fuses all three domains into a coalesced worldview.[101] Judge was not alone; this vision of a perfect union was later echoed in the works of Steiner and Bailey.[102] Building on these developments, Owen (2004) frames the relationship between occultism and rationalism as twofold: on one hand, the quest to rationalise and control cosmic forces in order to attain "*union with the divine*"; on the other, the drive to comprehend and master the unexplained laws of nature—a stated aim of the Theosophical movement and a defining ambition of both Renaissance magic and the *fin-de-siècle* Occult Revival.[103] This dual pursuit underpins a conception that facilitates the impartation of superhuman abilities, forging the conceptual roles of the superman and supermen. Owen emphasises this continuity, highlighting their mutual entanglement—

> Magic in particular played to the idea that magical powers can be clearly demonstrated, and occultism in general allied itself with the idea of scientific validation.[104]

96 Huxley (1957), 211–14, 235.
97 Adams (1949), 386; Metz, R. (1938). Bacon's Part, 31.
98 See Barbour, 77–103.
99 Barbour, 77–79; Brooke, J. H. (1991). Science and Religion, 1–12; Hammer, 203.
100 Hammer, chap. 6: 331–453; See generally Brooke, J. H. (1991). Science and Religion.
101 Judge, W. Q. (1893). The Ocean of Theosophy, 1. See also Hammer, 222, 237.
102 Hammer, 228–29.
103 Owen, A. (2004), 10–11, 36–37. See also Owen, D. S. et al. (Eds.). (2004). The Vocation Lectures, *16–17, for contrasting applications of rationalization and control; and Lassman, P. et al. (Eds.). (1989). Max Weber's, 17, 116 et seq, 195, 200.
104 Owen, A. (2004), 8. See also Poley, 57.

> . . . .
>
> In particular, and in common with all magical traditions, *fin-de-siècle* occultism was characterised by the will to both know and control the natural world.[105]

This simultaneous quest for mystical and scientific knowledge animated thinkers such as Heinrich Cornelius Agrippa (1486–1535)[106] and Marsilio Ficino (1433–1499). Ficino, in particular, a Neoplatonic magician, integrated practical magic with emerging scientific paradigms, imbuing it with a "*semblance of respectability.*"[107] However, Bacon might have disapproved of scientific inquiry into the occult, as *New Atlantis* does not acknowledge it as a legitimate area of study ("baseline argument"). Adams (1949) concurs, adopting a more cautious stance by suggesting that both More and Bacon would have opposed granting humanity unrestricted power over nature, especially in the absence of ethical safeguards and limits.[108]

One could argue that scientific conquest represents the ultimate goal of knowledge, operating in two modes: dominance over nature and, second, to borrow from Thomas Hobbes, mastery over the problem of human nature.[109] Rossi (1968) challenges this construct, contending that Bacon did endorse practical or natural magic, viewing it as "*. . . the servant of nature,*" which "*. . . follow[s] faithfully in the footsteps of nature . . . ,*" citing a passage from Agrippa that Bacon is believed to have meditated upon. According to Rossi, this view was shared by many of Bacon's Renaissance contemporaries, who saw practical magic as harmonising with nature's laws, rather than an attempt to dominate or manipulate them.[110]

This, however, does not necessarily contradict Bacon's broader disapproval of engaging with the occult, as there is arguably no direct connection to occult doctrines. One could, however, argue that the practice of any form of magic inevitably leads down a similar path—towards the occult. In contrast, Johannes Valentinus Andreae (1586–1654), in his Latin utopia, was "*happy to accord*" metaphysics and theosophy as legitimate fields for scientific investigation.[111] This occult thread, originating in Renaissance esotericism, was later revived during the *fin-de-siècle* Occult Revival and subsequently infiltrated the Nazi regime. The ascendancy of occult power structures among the elite reveals a persistent pattern—an eliminationist, totalitarian utopia characterised by techno-adaptive authoritarian control.

---

[105] Owen, A. (2004), 14, 20. See also Poley, 65.

[106] See Nettesheim, H. C. A. (1993). Three Books of Occult Philosophy. See also Nettesheim, H. C. A. (2009). The Fourth Book of Occult. These texts are cornerstone works of Western esotericism and foundational to academic studies of Renaissance magic, alchemy, and occult thought.

[107] Attrell, D. (2020). Honoring the Outermost; Brooke, 64; Tambiah, S. J. (1990). Magic, Science and Religion, 24–29; Yates, F. A. (1964). Giordano Bruno, 144–68. Walker, D. P. (1975). Spiritual and Demonic Magic, 30–59. See also Ficino, M. (2001). Platonic Theology. Vol. 1, Books 1–4, The emanation of the universe and the occultist pursuit of the divine order; Vol. 2, Books 5–8, Practical applications of platonic ideas to Renaissance magic and theurgy; Vol. 3, Books 9–11, Practical aspects of occultism, focusing on magic, astrology, and the use of divine knowledge for personal transformation. Ficino is a key figure in Renaissance philosophy and occultism frequently cited in scholarly works. See also Ficino, M. (1980). The Book of Life, which discusses Ficino's views on astrology, celestial influence, and the relationship between the human soul and divine forces—key elements of Renaissance magical practices.

[108] Adams (1949), 389.

[109] Weinberger (1976) 865, 867.

[110] Rossi, 18–*19, 21, 31, generally, 1–35.

[111] Andreae, J. V. (1916). Christianopolis, preface, 217–18; Lucas, 117; Williams, C. A. (1916). [Christianopolis . . .].

iv. Portrait of the New Atlantis Superman and Subservient Supermen: From Darwinism to Genetic Engineering in Evolutionary Governance

WHAT IS THE modern-day superman narrative? He is portrayed as a cinematic science-fictional superhero endowed with superhuman powers and abilities. Gillespie (2005) defines him as a "*. . . creative genius, that is, as someone who is one with the genesis of the world and the superhuman power that underlies it.*"[112] He is the epitome of racial superiority, a descendant of an extinct, extraterrestrial, superhuman race outside of Earth's solar system. His role emerges from a world where no officiating governing body is effective enough to maintain order, and no one else measures up to the supernatural abilities required to accomplish the task. A being of unparalleled strength, power, intellectual genius and prowess, superman's self-assigned duty is to enforce law and order across the entire world. His enhanced abilities, stemming from his superior nature, set him apart from ordinary mortals, positioning him as the ultimate enforcer in this superhero narrative.

Myers (2015) conceptualises the 'superman/men' as more than just a 'method of enquiry;' rather, it represents the attainment of an elevated state akin to divinity. The goal is predominantly the acquisition of gnosis—a term that Goodrick-Clarke (2008) and leading authority, Rudolph (1987) describe as the recovery of the "*divine spark*" and the apprehension of "*divine secrets,*" reserved for the initiated elite. Hanegraaff (2009) frames this Platonic concept as "*salvational gnosis,*" while Fowden (1993) refers to it as "*Hermetic-gnosis,*" associating it with the direct apprehension of the divine mind. However, Grese (1988) postulates an intriguing twist: on one hand, this enables the attainment of an exceptional, higher form of knowledge, necessary for the practice of magic; and on the other, in a reverse relationship, the practice of magic imparts this exceptional, higher form of knowledge, dovetailing with the principles outlined in the Corpus Hermeticum.[113]

Grese further stylises this transformative dimension, suggesting that "*. . . 'gnosis is the goal of epistēmē— particularly neat formulation of the cognitive transition that the 'way of Hermes' was designed to bring about.*"[114] In Hermetic tradition, the "*way of Hermes*" denotes the path from intellectual, rational knowledge (epistēmē) to a direct, transcendent understanding of divine, exceptional knowledge. Fowden also maintains that this omniscient knowledge is the ultimate goal sought by the Hermetic initiate.[115] While he does not provide direct evidence for this claim, the numerous references throughout his paper collectively reinforce and converge on the idea that contemporary initiates, much like their historical predecessors, are profoundly oriented toward the attainment of divine knowledge.

This understanding is facilitated through an interpretive engagement with spiritually charged symbols in the world, which, in turn, enables one to attain the status of a god-man. Pagel (1992) situates this superman individual within an occult setting, "*. . . God-Man unity, the central doctrine of theosophy and monism.*" Revisiting the analysis of Nazi antiquity in The Superman Doctrine, Kurlander (2012) draws a similar conclusion, citing List and Lanz as sources for the belief in "*. . . the extraterrestrial origins of the original*

[112] Gillespie, 51. See also Hannaford, 311; Lindsay, 249.
[113] Fowden, G. (1993). The Egyptian Hermes, 104; Goodrick-Clarke (2008), 18–*20; Grese, W. C. (1988). Magic, 53–*54; Hanegraaff, W. (2009). The Pagan, 34. See generally Merkel, I., & Debus, A. G. (Eds.). (1988). Hermeticism and the Renaissance.
[114] Fowden, 101. See Hanegraaff, W. (2008). Altered States, 133–36, where he discusses modern translations of "gnosis is the goal of epistēmē." He mentions that interpretations of this idea differ widely.
[115] Fowden, 104. See also Hanegraaff (2008), 130.

Aryan 'God Men," a concept catalysed by the works of Haeckel and Blavatsky.[116] Myers (2015) echoes this view, distilling the significance of Haeckel's monism and suggesting that, as a result of scientific conquest arguably foreshadowed by the Covid pandemic, humanity will eventually evolve into a future where scientific practitioners form an advanced technocratic society. Essentially, this vision is the restoration of an Atlantean-style New Global Order—governed by an elite technocratic class, with a centralised capital as the nexus of global governance and technological innovation. Fischer (1990) defines the *New Atlantis* technocracy as—

> [A] system of governance in which technically trained experts rule by virtue of their specialized knowledge and position in dominant political and economic institutions.[117]

Searle (1971) bridges the connection between technocratic governance (Figure IV-3, 337) and the "*social salvation*" achieved through the scientific community's inheritance of total power—

> [T]echnocracy [is a political ideology that views] science as a . . . self-sustaining force which provides objectively valid solutions to social and political problems, quite independently of the wishes and beliefs of the majority—hence, the authoritarianism, even totalitarianism, [concentrating power in the hands of experts and sidelining democratic participation.][118]

Searle went on to comment that social salvation will arise once intellectual elites in science, "*. . . whose modes of thought and behaviour conform to scientific criteria,*" assume control.[119] The quotation carries negligible probative weight, insofar as it a normative endorsement of technocratic governance without demonstrating its necessity or legitimacy. Its force lies not in evidence but in implication: legitimacy is said to derive from epistemic conformity to scientific rationality.

In parallel, Martin (2002) submits that Bacon confers "*. . . a godlike status upon the 'sons of science . . . ,'*" casting them as heirs to an exalted guardianship and custodians of a privileged epistemic order. As inheritors of that mantle, they occupy a consolidated power-knowledge nexus, asserting absolute power over both natural world and humankind. Myers likewise designates these elites as "*science gurus*" to which she writes—

> Science gurus . . . to replace theologians, priests, and pastors as the filter through which knowledge about the world, human identity, society, and culture can be reconstituted and governed.[120]

116 Kurlander (2012), 529; Pagel, 100; Rudolph, K. (1987). Gnosis, *55–58, 80, 86, 91, 113 passim.
117 Fischer, F. (1990). Technocracy and the Politics of Expertise, 17, quoted in Friend, J. M. (2014). The Biopolitics, 35.
118 Searle, G. R. (1971). The Quest for National Efficiency, 260.
119 Ibid. See also Woiak, 115.
120 Myers, *49–50.

These "*science gurus*" operate as a modern analogue to Huxley's "*World Controllers,*"[121] themselves a reworking of Plato's philosopher-rulers in Timaeus and Critias, where the Atlantean elite are styled "*Guardians*" and "*Shepherds of Men*" (see start of this paper). In *New Atlantis*, Bacon likewise designates them as "*guardians of knowledge*" or epistemic gatekeepers, "*. . . in ways not wholly distinct from the Guardians of Plato's Republic.*"[122] While the quotation is rhetorically suggestive, shared terminology does not establish ideological continuity.

Popper (1963), however, interprets this architecture as a blueprint for a dominant ruling caste, analogous, but not identical, to a proto-"*master race*" construct.[123] In Plato's hierarchical vision, as argued earlier, the Nazi elite similarly presented themselves as demi-gods, relegating subordinate groups to animalised inferiors. The comparison revives the Aryan "*supermen*" caricature—individuals poised for global ascendancy, animated by calculated brutality and Spartan austerity.[124]

Within this schema, Huxley's (1957) ideas line up with pantheism, celebrating the unity and divinity present in all existence. His vision for evolving belief systems reflects the pantheistic principle that humanity's intellectual and spiritual growth must harmonise with the interconnectedness of the universe, recognising the divine in both nature and progress. The intent behind Huxley's concept of 'evolving destiny' is to cultivate more efficient and effective humans, with genetics steering progress toward the actualisation of this vision.[125] Writing in 1957, Huxley observed that the modern age stood apart from all prior eras due to the heightened significance of science and its widespread application.[126] The level of human optimisation to which Huxley's thesis refers is—

> [Humanity's destiny lies in fully harnessing reason, imagination, and conceptual thought to drive progress, achieving greater] efficiency and control of environment, self-regulation and independence of outer changes . . . wholeness or harmony of working, extent of awareness and knowledge . . . [and degree of] mental organisation.[127]

Huxley's call for fusion across multiple ideological levels was met with ambivalence by some,[128] given that the timeframe and the state of genetics and genomic technologies were insufficient to fully realise his, and, originally, Francis Bacon's, utopian vision. This is not a head-scratcher when one considers the paradigm that Prior (1954) apprehended in order to achieve such an ambitious goal, which helps clarify how a concept of this scale could materialise—

> Real confidence and hope lay only in the realization that the true goal was distant and that it required not one man but many, not one lifetime but generations of men working with a common purpose.[129]

[121] Baker, R. S. (1990). Brave New World, 90–129; Woiak, 115, generally, 105–29.
[122] Martin, 46.
[123] Popper, 45–54.
[124] Dunlap, 296.
[125] Huxley (1957), 212.
[126] Ibid.
[127] Ibid., 217.
[128] Sontag, F. (1958); Walsh (1958).
[129] Prior, M. E. (1954). Bacon's Man, 352.

This is not dissimilar to the dual nature of *New Atlantis*, which not only founded the sciences but also transmitted advancements, continuously refining them through new discoveries and adding finesse across subsequent generations. In Ecce Homo [1908] (1992), Nietzsche was not merely philosophising on the ideals of the superman; rather, he envisaged a future where—

> [T]he higher breeding of humanity, together with the remorseless destruction of all degenerate and parasitic elements, will again make possible on earth that superfluity of life, out of which the dionysian condition must again proceed.[130]

With respect to the "*Dionysian condition,*" Nietzsche commonly links it to the rebirth of humanity—a transformative process in which individuals transcend limitations, actualise their full potential, and pave the way for the "*Übermensch*" (superman).[131] Freed from the constraints of outdated moral systems, humanity can create new values and reconnect with existence in a deeper, instinctual mode. The concept resonates with themes of universal Enlightenment and pantheistic harmony, previously explored in Class-Based Stratification, 283, which examined the eugenics overlap with the socio-economic productivity of 'World Servers.' The parallel nuance that Hutton (2005) emphasises is the ongoing significance of eugenics, now reframed as genetics, and the potential emergence of supermen as the next stage of human development—

> The best quality had to be selected and bred, wherever it was present, whereas the lowest quality genetic lines to be eliminated as painlessly but as thoroughly as possible. The aim of eugenics was the creation of people who — in addition to having great strength, talents, firm judgement, a will to work and a dutiful nature . . . This would lead to the creation of a natural elite which could serve the whole.[132] (Internal citation omitted)

Hutton (2005) recognised the flaws in German eugenics, which sought to create a future sublime Aryan superman/men derived solely from the Nordic population. He maintained that the Nordics' own weaknesses were attributable to DNA mutations caused by line-breeding, and he distinguished between "*. . . beneficial and harmful forms of racial mixing.*"[133] The proposed solution was up-breeding with diverse races to produce a hypothetical superman/men—a "*higher man or men,*" synonymous with supernatural evolution, a concept to which Webb (1976) states circulated at the same time as Darwin's theory in 1859.[134] Moreover, up-breeding targets a spectrum of preferred individuals who are exceptionally well-gifted.[135] Hutton elaborates further on this notion that—

[130] Nietzsche, F. W. (1992). Ecce Homo, 51–52; Savage, 412; Weikart, R. (2004). From Darwin, 49n.28, citing Nietzsche, Ecce Homo, §4 The Birth of Tragedy.
[131] See comment by Goodrick-Clarke (110) on Hitler, links the Nietzschean vision of Aryan of new species and the Aryan superman.
[132] Hutton, *125 and 199, citing Verschuer, O. F. v. (1944). Erbanlage als Schicksal, 21–22.
[133] Hutton, 136, citing Kuhn, P. & Kranz, H. W. (1936). Von deutschen Ahnen für deutsche Enkel, 57-60.
[134] Webb (1976), 18.
[135] Huxley (1962a), 227; Huxley (1962b), 127–29, 133, 139.

> This ideal humanity would take on elements from other races. From the coloured races one might get adaptability to the tropics; from the Jews, strength of will and a strong sense of family.[136]

Salter (1915) appears to synthesise up-breeding and racial diversity into a vision of supermen: the ennoblement of a heterogeneous group reconstituted as a new species, which, in his view, rests upon Darwinian evolution—

> Nietzsche had in mind a possible evolution in the future such as Darwin is supposed to have proved in the past, namely, the development of one distinct order of beings out of another."[137]
>
> Let it be proved whether apes can be developed into men, and lower races into higher races, and whether from the best mankind has at present to show, something still higher can be reared.[138]

Goodrick-Clarke's (2002) research into Indian antiquity situates India as the locus for the quest for the Aryan superman or "*New Adam,*" envisaged to emerge through "*. . . selective breeding and occult initiation would govern the New Atlantis . . . .*"[139] This up-breeding discourse reflects historical ambitions for genetic refinement while also signalling the modern laboratory's capacity to engineer an augmented human—an enhanced superman. This positioning stresses the link to Freeden's (1979) argument and contemporary eugenic aspirations. He contextualises this up-breeding dimension within the broader discourse on genetics, stating—

> [T]he question of the breeding of a single type of superman. Not only popular, but some scientific opinion as well, often regarded eugenics as aiming at moulding 'the whole human race . . . into a homogeneous society which shall progress toward a standard, previously determined, of a noble humanity.[140]

Today, turning genetic "*defects*" into "*gold*" is increasingly feasible, elevating the best of humanity through biotech enhancement, somatic gene therapy, and germline engineering. Huxley's (1962a, 1962b) theorisation of the moral imperative of up-breeding stressed that it is not random tinkering. Unlike Nazi racial hygiene, his approach to racial selection rejected any single ideal type.

Lindsay (1916) loosely interpreted Nietzsche's concept of the Übermensch superman as a regular individual regenerated through experimentation and the application of positive eugenics ("*up-breeding of humanity*"), while recasting humanity to ennoble individuals with superhuman abilities.[141] Leon Trotsky (1879–1940), writing in Literature and Revolution ([1924], 1925), projected a similar trajectory, synthesised here as a composite quotation—

---

[136] Hutton, 30, citing Ploetz, A. (1927). Rasse und Menschheit, 252; Marten, 29. See also Günther (1966) 305.
[137] Salter, 423–24. See also generally, Staub (1989), 111–12.
[138] Salter, 427.
[139] Goodrick-Clarke (2002), 82, 89, 109–10.
[140] Freeden, 648.
[141] Krüger, G. (1966). The Renovation, 305; Mueller, 186, 189–90.

> [A]rtificial selection and psycho-physical training. . . . thereby to raise himself to a new plane, to create a higher social biological type or, if you please, a superman.[142]

Alongside Salter's (1915) experiments, Kurzweil's technological vision, and Lindsay's (1916) reinterpretation of Nietzsche's Übermensch, these represent diverse yet synonymous expressions of the same interrelated approach to advancing humanity. Ideas once confined to science fiction are now augmented by advances in biotechnology and AI.

Unlike the reforms advocated by Haeckel, which gained widespread acceptance, Huxley's ideas match more closely with his brother Aldous Huxley's Brave New World (1932) and, more directly, Bacon's *New Atlantis.* In 2026, the era Aldous imagined, aesthetic and genetic enhancements are materialising through groundbreaking advances in genetic technologies. In this light, Ray Kurzweil's Singularity (2004, 2005, 2024) is a key metric in debates on social progress, signalling transhumanist configurations—the merging of humans and AI to achieve 'superintelligence.' Julian Huxley might view this as "*quantitative addition,*" where the expansion of human capacities through technology is gradual yet transformative. By contrast, up-breeding targets genetic vulnerabilities to disease, which Huxley would frame as "*qualitative optimisation.*"

Woiak (2007) rightly observes that Aldous Huxley exemplifies how intellectuals worldwide "*. . . were drawn to eugenics as a progressive, technocratic means of improving the health and fitness of populations.*" The authority of science thus becomes the critical locus for advancing and preserving social progress.[143] This corresponds with Ploetz's conclusions, which Weiss (1987a, 1990) corroborated following his six-month attaché in the United States, where he encountered what he described as 'low quality humans'—

> For this reason I must direct my efforts not merely toward preserving the race but also toward improving it. . . . My views . . . immediately led me to the field of medicine-which appeared to be relevant to the biological transformation of human beings.[144]

While it may seem a stretch to connect Ploetz's views to contemporary cutting-edge science, it is important to note that Ploetz, as the founder of the eugenics and racial hygiene movement, championed early scientific, medical, and genetic interventions, believing they would inaugurate human reformation. These ideas correspond with Kurzweil's Singularity's ultimate aim: creating 'supermen' through cognitive and technological integration—a process grounded in the aforementioned concept of "*quantitative addition.*"[145]

Weikart's (2004) commentary critiques Alexander Tille's (1866–1912) misreading of Nietzsche's superman, widely rejected by philosophers for portraying it as biologically or physically superior. Four key objections arise: (i) Nietzsche opposed biological determinism, emphasising creativity and will; (ii) the "*Übermensch*" was a philosophical ideal, not a biological model; (iii) Tille's association with Social Darwinism misrepresents a philosophy Nietzsche himself criticised; and (iv) ethical concerns arise from Tille's

[142] Trotsky, L. (1925). Literature and Revolution, 254–56. See also Dunér, *84, 88.
[143] Woiak, 109–*10.
[144] Weiss (1987a), 200; Weiss (1990), 15.
[145] Huxley (1957), 230; Kurzweil, R. (2004). Human Body Version 2.0; Kurzweil, R. (2005). The Singularity Is Near; Kurzweil, R. (2024). The Singularity Is Nearer.

association with the discredited doctrines of eugenics. Weikart (2004) clarifies that Nietzsche's superman refers not to a "*. . . biologically superior specimen . . . but rather a creative genius who would rise to dominate others . . . .*"[146] Lindsay (1916), by contrast, linked the concept more closely with the romanticised Aryan ideal as a menacing figure, and less with cinematic hero portrayals—

> He will point the way to an ever-expanding vision of life in its fulness and beauty. He will be ruthless and unpitying because hardness is necessary to him who would lead. He will inflict suffering because suffering is the root of much that is precious, of strength of character, of artistic creation. He will be a member of an aristocracy of health, energy and talent, an aristocracy to which only those will be admitted who are the creators of values which humanity worships, and each member shall gain admittance to this charmed circle only by deeds of daring and prowess, physical or intellectual.[147]

In both biblical and esoteric traditions, they routinely cinematically cast superman as a supernatural magician and monarch and apocalyptic saviour of the world, frequently likened to the figure of the Antichrist.[148] A reappraisal of The Superman Doctrine in antiquity, esoteric traditions depicted this being as the Atlantean Priest-King deity, ethnic affiliations with Greek and Indian civilisational origins. Although known by various names across world religions, this archetype is structurally consistent—a sacral ruler who fuses charisma with sovereignty. As discussed under the Nazi Racial Theory, 141 heading, similarly this archetype was appropriated and racialised, with Hitler viewed in quasi-messianic terms: "*the great physician of the German people,*" the "*father of physicians,*" and "*a demi-deity.*"

Such language fused medicalised rhetoric with salvational expectation, recasting Hitler as healer of a racially defined German body politic. As explored in the section Elevation of Hitler as Superman, 108, German adulation construed him as Germany's living superman. The overlap between occultism and early National Socialism was ideologically symbiotic. See also heading Counterarguments, 95. First, early nationalist occultists, within the Thule Society and Germanorden, supported Hitler's rise. Second, Sebottendorff's 1919 account documents Hitler's early engagement with Thule members, turning to Thule members for support.

In Islam, he was received favourably, even likened to Muhammad. Carl Jung described him as a shaman, a "*truly mystic medicine man*" within the medical-political movement, "*. . . who managed to manipulate the unconscious of 78 million Germans.*"[149] Thus, Future wise, Hitler's rise is a portrait of the future superman. His political ascendancy combined pseudo-medical authority with ideological charisma. His promised Thousand-Year Reich collapsed after twelve years, paralleled esoteric millenarian vision, "*. . . the thousand-year kingdom of Dionysus,*"[150] restored Atlantis and Indo-Aryan sacral kingship. A continuous thread from antiquity to the present, in accordance with present trends, Hitler prefigures a future political saviour—a recast, reborn Atlantean Indian or black styled Aryan-Sanskrit superman king.

146 Weikart (2004), 34–35, 45–*46, 68, 82.
147 Lindsay, *249, 250, 255.
148 Cohn (1993) 31.
149 Grodin (2010), 51, 53; Knickerbocker, H. R. Is Tomorrow Hitler's?, 45–47; Kurlander (2015a), 500; Kurlander (2017b), 81; Sickinger, 108, *114–15; Weyers, W. (1998). Death of medicine, 152.
150 Gillespie, 64, 66.

The overarching concept of a 1000-year kingdom presupposes the notion of immortality, a theme pre-eminent in Greek antiquity. Gordon (2016) writes that the "*. . . ancient Greeks perpetuated this concept of undying immortals.*"[151] He begins his paper by referencing Heraclitus's famous paradox on mortality and immortality—

> αθάνατοι θνητοί, θνητοί αθάνατοι
> Immortals are mortal, mortals are immortal.[152]

Graham (2008) states that the precise context of this statement is unclear; however, this author's interpretation suggests it refers to human beings who once possessed immortality, lost it, and later regained it. Alternatively, it could equally imply a prophecy that immortality, once lost, will be reclaimed in the future by mortals. On this point, Webb's (1976) elaboration on these esoteric notions, citing American anthropologist W. E. Evans-Wentz (1878–1965), conforms seamlessly with this interpretation. Evans-Wentz synthesised occult doctrine and Darwinian science, constructing a "'*. . . Celtic esoteric theory of evolution,' which embodied the belief that 'the gods are beings which once were men, and the actual race of men will in time become gods.*'"[153] Graham's concluding remarks loosely affirm this, suggesting "*. . . elevated to the status of gods via their super-human powers.*"[154]

Following antiquity antecedents a future *New Atlantis*, with an Indian Antichrist superman monarch paralleling Hitler, emerges as a synthesis of world religions' ancient deities, reprising esoteric traditions and restoring lost knowledge believed to have originated in ancient Atlantis. Under its aegis, these principles are reintroduced into modern society, reviving the Atlantis utopia[155] (The Superman Doctrine, 72). Simultaneously, Godwin (2010) and Johnson (1998) emphasise that this new iteration is anchored in cutting-edge science and technology, renowned for Atlantis's legendary mastery while projecting it into a modern revival.[156] Thus, this scientific-technological *New Atlantis* engages Heraclitus's paradox of the mortality of 'gods,' seeking to elevate humanity toward a god-like ideal through biotech innovation. This era blends occultism and social reform, where mystical ideas of evolution coexist with scientific ones, creating a tapestry of thought in which innovation is both spiritually inspired and practically oriented.

[151] Gordon, 212.
[152] Gordon 211, quoting Kranz, W. (1951). Die Fragmente quoting Heraclitus DK 22 B62:1; Graham, D. W. (2008). Heraclitus, 172;
[153] Webb (1976), 18.
[154] Graham, 230.
[155] Gillespie, 63.
[156] Godwin (2010), 254, and 59–60, citing Bernard, R. (1985). Les maisons, 264; Johnson, K. P. (1998). Edgar Cayce in context, 63.

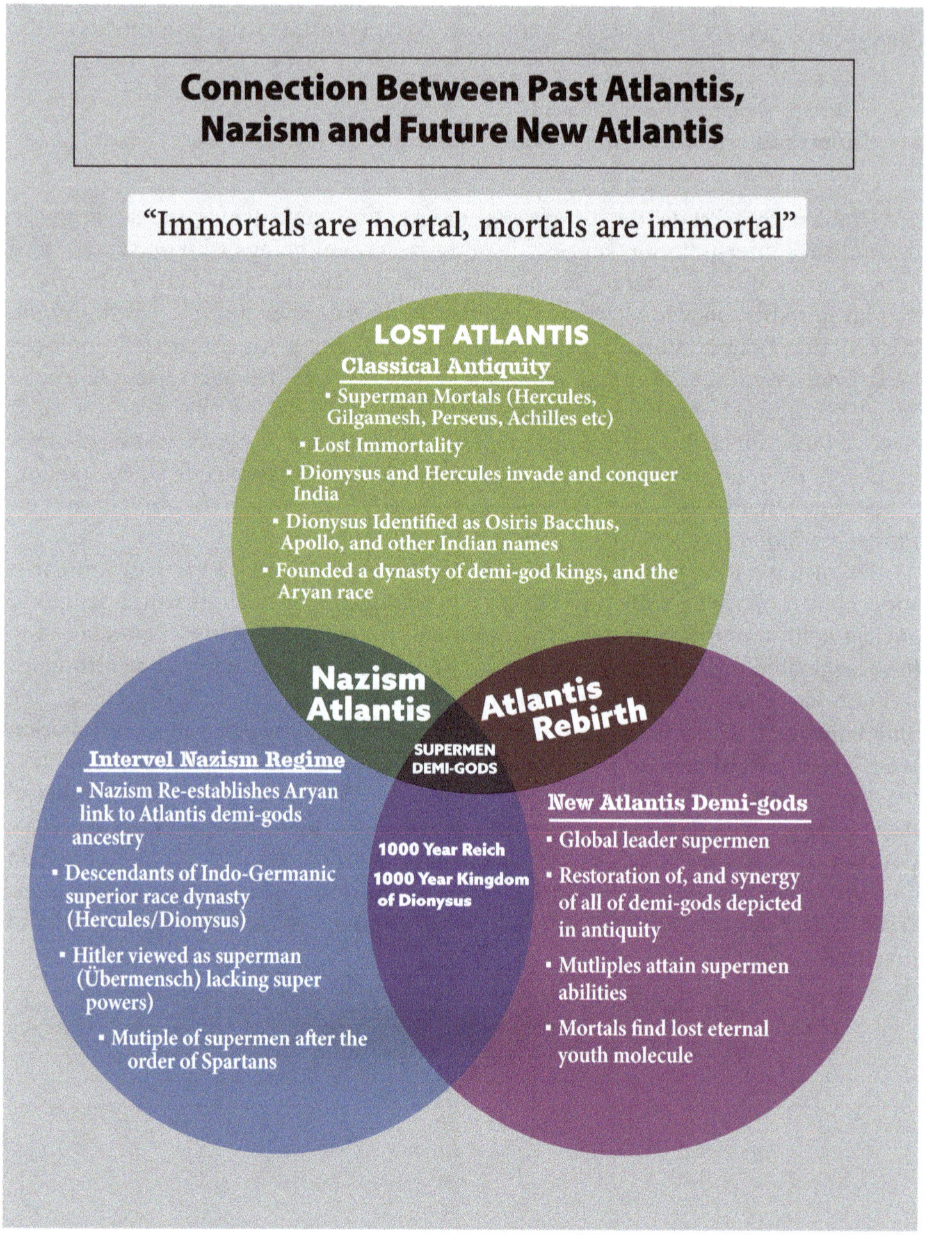

Figure IV-2. Conceptual Link Between Atlantis, Nazism, and the "New Atlantis" Vision

CHAPTER 11. DEVELOPMENT OF FUTURE PANDEMIC PREPAREDNESS STRATEGIES BASED ON THE COVID RESPONSE

Introduction

IMAGINE A BRIGHT, sunlit afternoon in a bustling metropolis—an ordinary day by all appearances. People hurrying to their jobs, street vendors peddling their wares, coffee shops and restaurants buzzing with activity, children laughing and playing in parks, and the familiar hum of city life pulsing through the streets. Then, without warning, a deadly contagion silently began to spread. Its initial signs were deceptively mild—coughing and sneezing marking the incubation period. But soon, the virus revealed its true, horrifying nature, combining the deadly characteristics of both Ebola and rabies. What began as a subtle threat rapidly escalated, gripping the city with terrifying speed and intensity. As the pathogen manifested, its effects were visible and immediate: grotesque bleeding from the eyes, animalistic behaviour, and a complete collapse of recognisable cognitive function. The scientific establishment had long warned that Disease X would one day come—a highly virulent pathogen capable of unleashing unimaginable devastation. We should have listened!

Unlike the Covid virus, whose pathogenicity was frequently clouded by political rhetoric and media manipulation, the above hypothetical would leave no room for doubt. Its visible terror would be evident to all. If a cure were developed but questions about the efficacy of a hastily produced vaccine came to light, fear and distrust would spread like wildfire. The situation would demand not just compliance with vaccine protocols but full, unyielding submission would be mandated. Fear would be weaponised, and control would be imposed with ruthless precision.

Under these conditions, it is no stretch to foresee that the spectre of mass compliance would become not just probable, but inevitable. The idea of a coercive, authoritarian regime enforcing conformity would resurface, echoing the horrors of the Nazi health regimen at its zenith. See Racial Hygiene at its Zenith, 160. Racial hygiene protocols, once a chilling reality, could once again find fertile ground under the guise of public health measures, as it did with Covid. In the face of Disease X, the synthesis of fear, compliance, and control would foster a chilling environment—one where the principles of a totalitarian, a health-driven agenda could be revived with terrifying efficiency, with a modern twist: a deployed technocratic, anthropologically-based adaptive regime.

In an era dominated by science and medicine, the appearance of novel viral pathogens and the rapid development of vaccines has become a critical global concern. Globalisation has intensified the threat of viruses capable of widespread devastation, necessitating a swift reassessment of public health preparedness strategies. Against this backdrop, the machinery of an international new global order comes into view. The interconnectedness of nations fosters a collective surveillance approach to manage health crises and wider social challenges, including crime rates. The proliferation of biometric technologies adds another layer to this complex landscape. Fingerprint scans, facial recognition, and electronic health monitoring systems play a central role in reinforcing public health measures and ensuring vaccination programme effectiveness. Yet, the collection and utilisation of such data provoke profound ethical and privacy dilemmas will have to be navigated carefully to preserve public trust.

Traditional theories of a New International Order typically intertwine with conspiracy narratives, that focuses on widespread disease, global governance, and biometric surveillance as harbingers of the approaching transformative era. However, these claims can be differentiated by discernible trends and emergent patterns, specifically, when examined through the lens of globalisation. Historically speaking, totalitarian regimes have arisen from an eclectic mix of authoritarianism, propaganda, economic instability, social homogenisation, scapegoating, militarisation, and ideological fanaticism. With nationalism receding, and progressive globalisation deepening—coupled with highly interconnected economies—the prospect of a shift in power towards a more authoritarian, centralised global governance model appears increasingly plausible.

A. The Role of Surveillance in Combating viruses: Contagion and Beyond

INFLUENZA OR VIRUS surveillance refers to the systematic collection, analysis, and interpretation of data related to viruses and Influenza activity. In efforts to contain the spread of Covid technologies were repurposed for surveillance. Kampmark (2020) points out, it "*. . . [gave] birth to what amounts to the pandemic surveillance state.*"[1] He duly noted that the rapid spread of Covid, produced a "*. . . fierce boom in surveillance technologies dedicated to monitoring whole populations.*"[2] We digress momentarily.

It was not until SARS (202–2003) that the grave and unmistakable "*. . . direct and continuing threat that transnational disease epidemics pose to health and economic interests*" became evident.[3] Fidler & Gostin (2006) found prior to this time, the old IHRs were ineffectual, as countries neglected to report "*. . . events that may constitute public health emergencies of international concern.*"[4] This limitation severely handicapped the WHO's ability to respond effectively to disease outbreaks. The SARS epidemic, however, generated the political impetus necessary to advance global health diplomacy. This momentum led to the adoption of the revised IHRs in 2005, a legally binding instrument of international law covering 196 countries, including all 194 WHO Member States.

These revised regulations, which expanded the scope of reportable diseases, came into force in 2007, concurrently with the launch of the IHP+, and were introduced under the heading "*A new code for international health security.*"[5] They reshaped global surveillance and accelerated the development of new international networks for disease surveillance and bioterrorism monitoring.[6] Consequently, the revised IHRs provided the WHO with a "*soft-power tool,*" obliging nations to comply with notification requirements. This enabled the WHO to draw on "*. . . surveillance reports from non-governmental organisations and electronic surveillance systems.*"[7]

1 Kampmark, 59.
2 Ibid, 62.
3 Feldbaum et al., 86.
4 Fidler, D. P. & Gostin, L. O. (2006). The New International Health Regulations, 88. See also Feldbaum et al., 84–86.
5 WHO. (2007). The world health report, 7–8. See also Kampmark, 61; Kelle, A. (2007). Securitization, 226–27; WHO (2024), 10.
6 Castillo-Salgado, C. (2010). Trends, 93.
7 Feldbaum et al., 86. See also Fidler & Gostin, 90, 93.

In the WHO's 2023 publication of the IHRs, informed by the pivotal lessons of Covid: "*[e]ffective national surveillance is a cornerstone of a strong global architecture for health emergency preparedness, response, and resilience (HEPR).*"[8] See abbreviated description in Table III-2, 239. Historically, health and wellness occupied a peripheral role in national foreign policy (as previously discussed), largely relegated to diplomatic negotiations conducted by government health envoys and now increasingly delegated to non-governmental actors ("NGOs") such as GAVI. Since Covid, however, the locus of health surveillance has taken centre stage in multinational and transnational efforts, becoming a cornerstone of contemporary and future global public health security agendas.[9] Accordingly, surveillance and global public health security are defined as—

> [Surveillance is] [t]he systematic, ongoing collection, collation, and analysis of data for public health purposes and the timely dissemination of public health information for assessment and public health response, as necessary.[10]
>
> [Global public health security are] the activities required . . . to minimize vulnerability to acute public health events that endanger the collective health of populations living across geographic regions and international boundaries.[11]

The purpose of virus surveillance is to monitor and track the spread of viruses, identify new strains, detect outbreaks early, assess impacts on public health, and inform public health decision-making. Outbreak prediction, a newer subset of health surveillance, emerged through pilot initiatives around 2009 to assess and pre-empt pandemic threats. The theory outlined by Holmes et al. (2018) posited that, with adequate data and AI, prediction could resemble weather forecasting.[12] However, they criticise this approach as flawed and instead advocate for proactive, real-time surveillance of human populations.[13] See Biotech Architectures, 317. Contact tracing is one such real-time monitoring measure. Integrating contact tracing into virus surveillance enhances outbreak control and mitigates viral spread, thereby protecting public health. This underpins the rationale for epidemiological surveillance and data collection.[14] Components of virus/flu surveillance are summarised as follows—

i. **Tracking virus cases and deaths**: Monitoring the number of cases, hospitalisations, and deaths to understand the severity and spread of the virus.
ii. **Monitoring virus-like illness**: Gathering data from healthcare providers about patients presenting with virus like symptoms.
iii. **Laboratory testing**: Identifying and analysing viruses to detect new strains and understand their characteristics.
iv. **Sentinel surveillance**: Using selected sites (sentinel sites) to collect detailed data on virus cases and strains.
v. **International collaboration**: Sharing data and collaborating with global health organisations, such as the WHO and GAVI, to track the spread of virus

8 WHO (2023). Defining collaborative surveillance, iv.
9 Ibid., viii.
10 Ibid., *vii, 4–5.
11 WHO (2007), ix. See also Aldis, W. (2008). Health security, 371.
12 Carroll, D. et al. (2018). The Global Virome Project; Holmes, E. C. et al. (2018). Pandemics, 181.
13 Holmes et al., 180.
14 WHO (2007), 6; WHO. (2023). Defining collaborative surveillance; Wolitski, R. J. et al. (2004). Ch. 40; Xie, R. et al. (2024). Integrating Contact Tracing.

internationally.

vi. **Contact Tracing Surveillance**: is an essential strategy within virus surveillance that involves identifying and monitoring individuals who have been in close contact with an infected person. This helps to:

   a. **Identify potential spreaders**: Quickly identify individuals who may have been exposed to the virus(es).
   b. **Interrupt transmission chains**: Isolate and treat contacts who develop symptoms to prevent further spread.
   c. **Provide targeted interventions**: Offer vaccines or antiviral treatments to contacts to reduce the risk of disease spread.
   d. **Gather data**: Collect information on the contacts' health status to understand transmission dynamics.

Baldwin (1997) poses critical questions pertaining to security: "*Security for whom? Security for which values? How much security? Security from what threats? Security by what means? Security at what cost? The value of security?*"[15] These questions apply to global health security and surveillance. Extrapolating from the remit of security services such as the CIA, MI5, and MI6, it is often assumed that there is organic harmony between 'health security', 'global public health security', national security, and foreign policy.[16] The 9/11 attacks and the contemporaneous anthrax incidents were pivotal catalysts for the causative integration of security measures as the natural response,[17] which had previously intersected only marginally with global health concerns. This development led to the assimilation of health into national security agendas and, according to Bengtsson & Rhinard (2019) "*[t]he 9/11 and anthrax attacks in the US prompted a further shift in discourse towards collective securitisation.*"[18]

Bunyan (2010) observed that 9/11 "*. . . remove[d] the democratic constraints on the use and development of widespread mass surveillance technologies*"[19]—marking a critical juncture in the war on terror. It provided the green light for the introduction of new surveillance technologies, thus, shaping a surveillance society, a key dimension of globalisation.[20] Despite perceived ambiguity surrounding health security, this amalgamation was mobilised to counter the ongoing threat of biowarfare and bioterrorism, including the acquisition of bioweapons by terrorist actors. This corresponds to a governance narrative linking medical surveillance and security services, enhancing the effectiveness of allied response measures.[21] Intelligence and reconnaissance play a crucial role enabling the early detection of infectious viruses. Advocates of totalitarian political regimes argue that such systems allow the swift implementation of coordinated, preemptive measures to protect global populations and contain epidemics. Tahir et al. (2022) acknowledge that the—

> [P]ossibility of an engineered pandemic pathogen also cannot be ignored. The release of such pathogens, either through laboratory accidents or as an act of

15 Baldwin, D. A. (1997). The Concept of Security, 12–26; See also Aldis, 371.
16 Aldis, 372–73.
17 Ingram, A. (2005). The New Geopolitics, 539.
18 Bengtsson, L. & Rhinard, M. (2019). Securitisation 355.
19 Bunyan, T. (2010). Just over the horizon, 3.
20 Ibid., 5–6.
21 Aldis, 372, 374; Baumslag, 211–12, generally, 202–16. See McInnes & Lee, 13–15.

> bioterrorism, might lead to a disastrous Disease X."[22]

Moreover, Baumslag's (2005) reference to Theodor Rosebury (1904-1976), a microbiologist at Camp Detrick during World War II, draws out this seminal point about public health collaboration generally concerning infectious diseases—

> [I]n 1949 [he] pleaded in his book *Peace and Pestilence* that expertise should be used for attacking infectious diseases since . . . The role of public health professionals and epidemiologists should be to collaborate in preparing for the aftermath of a biological attack. The early detection of chemical or biological agents, physical shelter from the agents, decontamination of exposed materials and clothing, and appropriate medical treatments may help to contain an epidemic and avoid further death and devastation.[23] (Internal citations omitted)

What other lessons were learned from Covid? Contemporary surveillance mechanisms, such as Bluetooth positioning systems, were adapted and proved somewhat effective in tracking potential carriers. However, they ultimately fell short of establishing a targeted framework capable of enforcing mass social-distancing measures tailored to pandemic conditions.[24] Emerging surveillance technologies, driven by advances in AI and automation, are poised to revolutionise and exceed current capabilities, potentially rendering direct human intervention in epidemic control increasingly obsolete. The principal motivation behind these developments is best termed cost reduction. Such innovations ease constraints on resource allocation, provide fiscal efficiencies, and conserve time and financial expenditures. Moreover, in the wake of Covid surveillance, Anisin (2024) citing Garrett (2021), savvily remarks that future forms of surveillance will become socially embedded as the 'new normal' akin to the sweeping security measures introduced after 9/11.[25]

Regarding data access, the Covid pandemic demonstrated that efforts to globally identify, as previously discussed, required explicit cooperation among multiple international and national agencies. These initiatives relied heavily on surveillance-capitalist structures, including technology and telecommunications companies that either monitored civilians directly or shared extensive behavioural data with the security services. The purpose was to enforce lockdowns, conduct contact tracing, and to monitor interactions—who individuals contacted, when, and where.[26] Accordingly, Anisin (2024) astutely observed that these policy measures and enforcement mechanisms constituted surveillance capitalism—

> [W]ould not have been possible without antecedent conditions that stem to surveillance capitalism[, and mass state-led surveillance of populations during the pandemic could not have been possible without collaboration, cooperation, and direct involvement of private telecommunication and technology companies.] Technological corporations shared their platforms with state security institutions

[22] Tahir, M. J. et al. (2022). Disease X.
[23] Baumslag, 213.
[24] Nguyen et al. (2020a), 153480–81.
[25] Anisin, 26 et seq; Bunyan, 5; Garrett, 1748 et seq.
[26] Anisin, 18, 26, *33–34, 50–51; Chumtong, 3; See also Garrett, 1758 and 1749, for data collection means.

> to monitor, quarantine, trace, and place entire populations under surveillance.[27] (Merged quote from same source)

Internationally, effective and cohesive disease control necessitates integration and harmonisation of systems, moving towards unification in practice. As previously explained, a relevant example is Australia's contact-tracing application, modelled on Singapore's TraceTogether.[28] The deployment of such technology is a relatively new development, characterised by its reliance on Bluetooth signals to record proximity interactions within a range of six feet or two metres through the exchange of temporary identifiers. By capturing and analysing vast datasets encompassing a "*. . . person's movements, associations and interactions . . .* ,"[29] the app assisted to identify potential exposures to infectious diseases. Its efficacy was contingent upon widespread public adoption. Health authorities utilised the collected data to trace contacts and mitigate transmission. The scope and intensity of such technologies varied substantially across countries. For instance, the European eHealth Network supported the development of applications operating on Google and Apple platforms, including SwissCovid (Switzerland), Immuni (Italy), and Corona-Warn-App (Germany).[30]

Comparable apps were implemented in Israel, India, South Korea, and China, where their use was compulsory.[31] In Taiwan and Russia, governments used cellphone location data and cellular triangulation to enforce compliance with lockdowns and social distancing mandates. By contrast, Europe, the UK, and South Korea were initially regarded as less intrusive, relying on anonymised shared location data. Subsequent reports, however, indicate that some European states later adopted more liberal surveillance practices, contravening established data protection protocols.

For example, in several European countries, contact-tracing applications collected data including "*. . . the day, time, and duration of a contact; whether the infected user is asymptomatic; the 1st day of illness; and the date of testing.*"[32] In Australia, the contact tracing application was ultimately decommissioned due to its limited effectiveness in identifying unique infections. Sathyamoorthy et al. (2021) argue that the app's predominant objectives were to reduce transmission below the basic reproduction number (an aim regarded as conceptually ambiguous), identify breaches of social-distancing measures, and enable compliance-enforcing countermeasures.[33] On this basis, Sathyamoorthy et al. advocate the development of more advanced technologies capable of addressing these limitations. Table IV-5, 322 outlines how prospective and longer-term applications of such developments could operate within a proposed Global New Order, highlighting differences in technological deployment, governance structures, societal implications, and enforcement metrics.

---

27 Anisin, *4, 5, *36, generally, 24–28.
28 Chumtong, J. (2020). Phone tracking, 4; Greenleaf, G. & Kemp, K. (2021). Australia's 'COVIDSafe,' 263.
29 Greenleaf & Kemp, 264. See also privacy protection concerns discussed in Kampmark, 65–66.
30 Anisin, 74; Blasimme, A. et al. (2021). Digital Contact Tracing, 2, 4; Kampmark, 65. Kaya, E. K. (2020). SAFETY AND PRIVACY, 3–4.
31 Greenleaf & Kemp, 259, 263; Nguyen, C. T. et al. (2020a). A Comprehensive Survey, 153493–94; Thomson & Ip (2020), 11.
32 Anisin, 74; Blasimme et al., 4; Boudreaux, B., et al. (2020). STRENGTHENING PRIVACY; Csernatoni, R. (2020). New states; Nguyen et al. (2020a), 153492. See generally Bunyan, T. (2010). Just Over the Horizon. The topic point is Europe's progressive transition to the creation of an stringent electronic surveillance society in everyday life.
33 Ferretti, L. et al. (2020). Quantifying SARS-CoV-2 transmission; Sathyamoorthy, A. J. et al. (2021). COVID surveillance robot, 2; Tamar, S. (2021). Blind-sided by privacy?, 546.

B. Biotech Architectures of the New Atlantis: Schematic for immortality

i. Biometric Technologies and Viral Surveillance: Implications and Applications

NOTED IN THE introduction, the Covid pandemic was a snapshot of the societal fallout attendant to a future Disease X event. Beyond its immediate health impacts, Covid catalysed the fortification and integration of global preparedness and response systems. However, it also briefly gave rise to a segregated society characterised by distrust, hostility, and unequal treatment, all justified under the guise of public health measures. Bialas (2013) compellingly argues that moral norms can be inverted during times of ideological or existential crisis. In such a scenario, it is easy to foresee how acts of homicide might be rationalised under the banner of public health imperatives, especially in a Disease X situation. Indeed, Bialas shows how easily such circumstance could arise—

> The unconditional and general prohibition on homicide that universal humanism fostered was to be replaced by the commandment of killing persons and groups identified as an incurable threat to the health of the people.[34]

This also revisits our earlier discussion of the 'World Server' or 'Universal Human', which redefined moral boundaries and justified violence as ethically permissible under the guise of economic value, cultural productivity, and collective well-being. The catalyst for such a regime, much like Covid, could be a 'Disease X' outbreak, a terrorist attack, or some other unforeseen event. Baumslag (2005) points to historical precedents we are all too familiar with—

> Biological warfare—the use of bacteria, viruses, or toxins during war to cause disease or death—long predates any scientific understanding . . . . In the earliest stage of biological warfare, diseased bodies were hurled at the enemy even when it was not known how disease spread. Later, as scientific knowledge increased, biological warfare became more sophisticated. . . . With the era of bacteriology in the nineteenth century came the knowledge that specific organisms caused certain diseases and that each disease was spread in its own way. This knowledge made it easy to cultivate organisms for epidemics.[35]

The severity of the pathogen would inevitably amplify these tensions, further exacerbating hatred, as did it with Covid. Benyera (2021) offers valuable insight into this trajectory, unintentionally highlighting the steps to the establishment of a regulatory surveillance society—

> With the intensification of vaccine nationalism, the Covid pandemic will [exacerbate inequalities] . . . Human movement is bound to be restricted without being vaccinated as the unvaccinated would become the 'dangerous other' who must not mix with the vaccinated. The human movement of the

[34] Bialas (2013), 17.
[35] Baumslag, 202. See also Marty, C. A. M. (2001). History, 429, quoted in Baumslag, 203.

> unvaccinated . . . curtailed, [or even] criminalised as spreaders of a virus hence threats to human security. . . . If we are to learn from our history, the vaccine will be tied to aid, democracy, transitional justice and other liberal constructs.[36]

This approach extends to both professionals and non-professionals within global agencies, medical establishments, and media platforms, all sharing a collective responsibility to enforce the government's and medical establishment's agenda. Zeidman (2020) and Alexander (1949) stipulate a stark truth from the Nazi era—

> [N]o law, even the stringent research laws in existence today, will prevent the repetition of the unethical atrocities that occurred in Nazi Europe, because medical science under dictatorship is subordinate to the prevailing agendas and policies of the of the totalitarian political regime.[37]

In hindsight, Covid inverted this dynamic. It demonstrated that, within a Global New Order system, the agendas and policies of a totalitarian political regime would fall subordinate to a surveillance-driven medical-science authority. In such a scenario, all political power would rest with this new body, whose primary mandate would be the protection of global citizens' health. The aforementioned agencies would shift into an authoritarian role, orchestrating a coordinated, unprecedented global effort to promote collective responsibility for immunisation.

Futuristically speaking, amid ongoing advances in robotics and AI, one notable development is Elon Musk's Tesla Optimus project, whose humanoid robots are expected to debut in 2026? In such a context, AI-driven automation systems—including GPS geolocation, Global Navigation Satellite Systems (GNSS), and Ultra-WideBand (UWB) technology for precise indoor positioning systems (IPSs)—could be further realised. If not through robotics, automated epidemic-control systems could instead be achieved through the integration of advanced AI algorithms and smart infrastructure technologies. Sathyamoorthy et al. (2021) said as much, retrospectively referring to Covid social distancing breaches and thermal-imaging health surveillance. Their paper argues that integrating such technology, along with thermal imaging cameras capable of virus detection—into its cognitive architecture, and interfacing them with diverse surveillance systems at public infrastructure hubs, are the boots-on-the-ground enforcers, implementing the aforementioned regimes.[38]

Contrary to perceptions of this technology as futuristic, the Chinese tech company *Megvii*, which specialises in facial biometrics, has already integrated such modules into its AI systems. This existing capability was a key factor in its deployment for pandemic monitoring, with these systems embedded into state surveillance infrastructures through national, city links and municipal partnerships, alongside the use of drones.[39] China was ahead of the curve when police patrol robots were deployed in major cities during Covid equipped with AI-enabled systems. Garrett (2021) states the purpose— "*. . . infra-red thermometers and high-resolution cameras to scan for anyone with a fever or not wearing*

---

[36] Benyera, E. (2021). Covid Vaccine, 196–97.
[37] Alexander (1949), 39; Zeidman, 2.
[38] Nguyen, C. T. et al. (2020b), 154215–219; Sathyamoorthy, A. J. et al. (2021). COVID surveillance robot.
[39] Kampmark, 62; Miyamoto, I. (2020). MASS SURVEILLANCE

a mask,"[40] alongside real-time data transmission capabilities to monitor compliance, identify potential infections, and support enforcement measures. In addition, Singapore's TraceTogether app, was designed for contact tracing, and had a unintended capability that transferred into broader categories of monitoring. This revelation underlines the dual-use nature of such technologies, whereby tools introduced primarily for public health monitoring purpose, can extend into broader surveillance functions. Garrett (2021) anecdotally remarks that where child welfare is in question—

> [You] do not have to have a dystopian sci-fi mindset to be able to recognise how surveillance technologies tend to become re-purposed and transfer into areas and practices for which they were not originally intended. For example, electronic surveillance and related testing for alcohol and drug use, formerly used on prisoners, are now applied to parents where professionals remain unconvinced about their trustfulness.[41]

Put simply, any proposed humanoid would likely go far beyond visual disease detection, incorporating wider surveillance and enforcement capabilities. Operationally, and given the agencies involved, the WHO's (2023) statement regarding "*. . . electronic data collection and real-time visualization tools . . . ,*"[42] is vague and subtly alludes to the utilisation of these technologies. The WHO stipulates that these technologies are driven by—

> [C]ollaborations with industry and technology sectors . . . public and private sector entities, security sector entities, [national health and emergency management bodies], academia, nongovernmental organisations, and civil society organisations.[43]

Two rebuttals are commonly advanced. First, the 'positive' case holds that such technologies represent effective pandemic management, enabling the early identification of infected individuals who might otherwise have continued undetected. Second, the checks-and-balances argument maintains that legislative safeguards and data protection standards can be instituted to prevent slippery-slope outcomes. However, as with any emergency crisis and as demonstrated during Covid, such safeguards and external institutional oversight were easily overridden.[44] (Review discussion at heading Managing Dissent, 210),

Evaluating the efficacy of social distancing measures, Nguyen et al. (2020) apply a mathematical model to assess infection probability and transmission rates during outbreaks. While such algorithms are effective for AI-based predictions of Covid infection and population migration, they may be less suitable for more severe outbreak scenarios, such as the hypothetical 'Disease X' outlined in the introduction, a term employed by the WHO since 2018.[45] In such cases, social distancing and isolation measures would be insufficient. Consequently, preparedness strategies would necessitate

40 Garrett (2021), 1759.
41 Ibid., 1748.
42 WHO (2023), 6.
43 Ibid., 10, 12.
44 Anisin, 73.
45 Honigsbaum, M. (2019). Disease X; Tahir, M. J. et al. (2022). Disease X; WHO (2023), 3.

immediate evacuation and the establishment of containment-zone perimeters to either exclude or confine infected populations.

In 2017, a report titled '*Emergency Preparedness, Resilience and Response*' was presented to the board of the UK National Health Service ("NHS"). The report marked the culmination of the NHS's participation in Exercise Cygnus, a three-day pandemic influenza simulation designed to test the preparedness and resilience of the healthcare system. Its findings and recommendations highlighted inadequate staffing levels and other systemic vulnerabilities in the management of a potential public health crisis. The report stated that "*[p]lans [were] revised to incorporate the learning from this exercise . . . .*"[46] The key takeaway is that the NHS was ill-prepared at a national level, pre-empting challenges that would later emerge on an unprecedented global scale during Covid. Exercise Cygnus was a crucial instrument for identifying and addressing systemic vulnerabilities, with a view to fortifying health systems against future pandemics, potentially under a model of authoritarian scientific governance. This is consistent with a future-oriented Baconian *New Atlantis* narrative, which presages a future government rooted in scientific authoritarianism. See Part IV., Biotech Architectures, 317.

A comparative analysis of surveillance measures within the totalitarian setting of Nazi Germany, as discussed in From Disease Legitimisation, 162, highlights how proposed contemporary surveillance technologies could facilitate an inescapable police state. For instance, how the typhus epidemic in Nazi Germany escalated to segregation ghettos. The commandant issued directives to control the disease and implemented a vaccination programme for SS personnel and staff. Echoing Rahm Emanuel's adage, "*never let a serious crisis go to waste,*" the Nazi regime engineered the epidemic and created the deplorable conditions that allowed it to flourish. This was then exploited as a pretext for isolation, with the Judenlager (trans. Jewish camp) isolated from the rest of the camp.[47] This historical example illustrates how stringent containment and monitoring measures were enforced against the Jewish population, both in contrived crises and real outbreaks, to control infectious diseases.

However, in comparison to the sophisticated surveillance technologies available today, the Nazi regime's capabilities for monitoring large populations was rudimentary at best. Baumslag (2005) commented that "*[p]ublic gatherings and cultural performances were forbidden, and schools were closed; contact with POWs was prohibited.*"[48] In the contemporary context of 2026, pervasive monitoring capabilities would prevent breaches and severely handicap any attempts by resistance groups to organise or operate covertly. As a result, participation in resistance movements would be virtually impossible—unlike during the Nazi era, when groups such as the French Resistance were able to operate more effectively despite the regime's rudimentary surveillance tools compared with today's technologies.

Bacon's *New Atlantis* envisions a society in which scientific knowledge and discovery guide governance, modernity, and social progress. The concept of "*speaking science to power*" is indicative of this vision, advocating for the use of scientific principles to inform and configure political decisions. Surveillance occupies a central role in *New Atlantis,* prioritised above political or ideological considerations. This approach is fundamental to understanding and monitoring health trends, detecting emerging diseases, and safeguarding population health. Fidler (2005) and Bayer & Fairchild (2010) argue that surveillance is not driven by ideology or politics but as a core scientific practice. They

---

46 Swindells, M. (March, 2017). Emergency Preparedness, para. 8; Thomson & Ip, 6–7.
47 Baumslag, 60, 67.
48 Ibid., 67.

outline the role of epidemiological research, stressing the necessity of identifying disease patterns across populations in order to inform assessments of morbidity and mortality.[49] Fairchild et al. (2007) conclude "*[t]hus the history of surveillance has been bounded by a promise of disease control and a spectre of intrusion.*"[50]

This technology now extends to biosensors, enabling precise diagnostics and real-time monitoring of influenza and other diseases. Specifically, silicon nanowire field-effect transistor (SiNWs-FET) biosensors[51] leverage nanostructures for the highly sensitive detection of biological molecules, offering a powerful platform for rapid disease monitoring. Tran et al. (2018) explain that the biosensor's sensing capabilities stem from biologically active field effects, where it is constructed on a nanowire substrate, and its conductivity is modulated by charged biomolecules (e.g., DNA) binding to surface probes.[52] Tran et al. report that—

> Semiconductor (III–V) [(i.e., composed of elements from groups III and V of the periodic table)] nanostructures are among the most promising nanotechnology building blocks and are actively being integrated into a range of nano-electronic applications.[53]

Nanoparticle applications have already been deployed in mRNA vaccines for Covid (as discussed and referenced), and in several *in vitro* studies. Case study on Patrick Doherty to follow. Indeed, more than 500 recent studies reveal a rapid shift toward nanotechnology-based sensors for real-time health monitoring *in vitro* applications.[54] In fact, this approach had already begun to transition toward practical use: in response to the Covid pandemic, DARPA (2020) funded research into biosensors for real-time monitoring of soldiers' health and biological threats, thereby paving the way for future developments in pandemic detection and response.[55]

Nguyen et al. (2020a, 2020b) and Sathyamoorthy et al. (2021) reviewed technological innovations addressing the shortcomings exposed by the Covid pandemic. Their review critically examined the roles of various technologies mobilised to enforce social distancing, finding significant overlap across Bluetooth, wireless, and satellite systems. These technologies monitored (i) adherence to social distancing, (ii) symptom

49 Bayer & Fairchild, 905; Fidler (2005), 186. See also Kampmark, 60.
50 Fairchild, A. L. et al. (2007). Searching Eyes, 1.
51 Tran, D. P. et al. (2018). CMOS-Compatible, 2.
52 Ibid.
53 Ibid., 1.
54 Barbosa, A. I. et al. (2020). Current nanotechnology; Goswami, A. et al. (2024). Review—Nanotechnology-Based Biosensors; Ji, X. et al. (2018). Silicon Nanomaterials for Biosensing; Malik, S. et al. (2023). Nanomaterials-based biosensor; Mishra, R. et al. (2025). Bioconvergence Ramesh, M. et al. (2022). Nanotechnology-Enabled Biosensors; Shen, L. et al. (2021). DNA Nanotechnology
55 Al-Deen, F. N. et al. (2014). Superparamagnetic Nanoparticle Delivery; Christiansen, M. G. et al. (2019). Magnetic Strategies; Diaz-Arévalo, D. & Zeng, M. (2020). Nanoparticle-based vaccines; Douglas, S. M. et al. (2012). A logic-gated nanorobot; Gualdani, R. et al. (2019). Superparamagnetic iron oxide nanoparticles; Iacovacci, V. et al. (2015). Untethered magnetic; Johnson, D. (22 December, 2015). Polymer Embedded; Liang, P. C. et al. (2016). Doxorubicin-modified magnetic nanoparticles; Liu, A. T. et al. (2021). Solvent-induced electrochemistry; Medhi, R. et al. (2020). Nanoparticle-Based Strategies; Mishra, S. R. et al. (2016). Selective and directional actuation; Nimpf, S. & Keays, D. A. (2017). Is magnetogenetics; Park, J. H. et al. (2021). Genetically engineered; Rao, S. et al. (2019). Remotely controlled; Shipman, M. (2015). Magnetic Nanoparticle; Tang, Y. S. et al. (2012). Bacterial magnetic; Tucker, P. (2020). A Military-Funded Biosensor; Wheeler, M. A. et al. (2016). Genetically targeted; Wu, K. et al. (2020). Magnetic-Nanosensor-Based Virus; Yao, J. et al. (2014). Nanowire nanocomputer.

prediction, (iii) quarantined individuals, (iv) contact tracing, and (v) AI-based algorithms identifying disease symptoms in public spaces.

These findings lay the groundwork for further advancements in surveillance, signalling a shift towards autonomous systems to manage future public health crises. The primary aim during Covid was to mitigate the transmission of infectious diseases by "*. . . reducing the frequency and closeness of human physical contacts . . . ,*" through the analysis of GPS data from mobile phones to "*. . . reconstruct[] the infection chains.*"[56] Chumtong (2020) reported that Chinese authorities traced over 3,000 contacts from a Covid-positive snack bar owner through GPS data, effectively mapping the infection chain.[57] The following table reproduces the original data and tabular presentation from Nguyen et al. (2020a, 2020b), summarised the functions, capabilities, and applications—

Table IV-5. Practical Social Distancing Scenarios and Matching Technology

Objective	Scenario	Description	Technologies
Keeping Distance	Keeping Distance Any Two People	Detect and monitor the distance between any two people	Bluetooth, Ultrasound Thermal, Inertial, Ultra-wideband
	Distance To/From Crowds	Alert when approaching a crowd	Alert when approaching a crowd
	Public Place Populations	Monitor and gauge the number of people inside at a public place	Wi-Fi, RFID, Zigbee, Cellular, GNSS, Ultrasound Computer Vision
	Physical Contact	Monitor physical contacts, e.g handshakes, hugs, between people	Computer Vision, Thermal Computer Vision, Thermal
	Sickness Symptoms	Detect and monitor sickness symptoms, e.g body temperature, coughs	AI
Real-Time Monitoring	Susceptible Groups	Monitor highly susceptible groups	Thermal
	Quarantines Violation	Detect and monitor quarantine people (e.g. for complying/violating the isolation/quarantine requirement)	Al, Cellular, GNSS, Ultrasound, Visible Lights, Computer Vision

56 Chumtong, 2; Nguyen, et al., 153480.
57 Chumtong, 2.

Table IV-5. Practical Social Distancing Scenarios and Matching Technology

Objective	Scenario	Description	Technologies
	Crowd Detection	Detect crowds/gatherings in public places	Wi-Fi, Bluetooth, RFID, Zigbee, AI, Ultra-wideband, Cellular, GNSS, Computer Vision, Visible Light
	Non-Essential Travel	Using location information to determine if the trip is essential (e.g. medical facilities and gasoline stations) or not (e.g. restaurants and cinemas)	Cellular, GNSS, Thermal
	Traffic Movement	Detect the vehicles on the street when isolation measures are in effect	GNSS, Computer Vision, Cellular
	Public Violation Detection	Detect violations of public place's closure or lockdown	GNSS, Cellular, Thermal
Information Systems	Infected Movement Data	Track the infected people's movement to notify susceptible people who were at the same places	Cellular, Blockchain, GNSS, Computer Vision
	Contact Tracing	Trace the contacts that an infected individual made	Bluetooth, Blockchain Ultra-wideband, Thermal, AI
Incentive	Location/Movement Sharing Encouragement	Encourage people to share their movement data	Bluetooth, Blockchain, incentive mechanism
	Stay-At-Home Encouragement	Incentivize people to stay home	Wi-Fi, Cellular, incentive mechanism, optimization

Table IV-5. Practical Social Distancing Scenarios and Matching Technology

Objective	Scenario	Description	Technologies
Scheduling	Workforce Scheduling	Limit the number of people at the workplaces	Scheduling
	Medical/Health Appointment Scheduling	Schedule medical appointments to reduce the number of patients	Scheduling
	Home Healthcare Scheduling	Optimize home healthcare services to reduce the number of patients at the hospitals	Scheduling
	Public Place/Building Access Scheduling	Control the number of people inside public buildings	Ultra-wideband, Wi-Fi, RFID, Zigbee
	Traffic Control	Regulate and reduce vehicles and pedestrians density	Visible Light, Scheduling
Automation	Robot-Assisted Social Distancing	Improve positioning and navigation of robots, especially medical robots inside hospitals	Ultra-wideband, GNSS, Visible Lights, Inertial, Ultrasound,
	Autonomous Delivery Systems (e.g. Uavs, ...)	Reduce the number of people going outside (food, merchandise, etc., delivery)	GNSS, Inertial
Modeling and Prediction	Infected Movement Prediction	Predict infected people's movement	AI
	Quarantined/At-Risk People Location Prediction	Predict quarantined and at-risk people's current location to enforce them stay at isolation/protection facility	Al
	People/Traffic Density Prediction	Predict people density and traffic density	Cellular, Al
	Sickness Trend Prediction	Predict sickness trends in specific areas	AI

Source: Nguyen, C. T. et al. (2020). A Comprehensive Survey of Enabling and Emerging Technologies for Social Distancing—Part I, p. 153487.

From the comparative analyses of Nazism, Schallmayer conceptualised the broad domain of genetic technologies as a form of "*generative ethics*" and "*biomedical technocracy.*"[58] It comes as no surprise, then, that 9/11 marked a defining moment in the rise of the biometrics era. In the abstract by Noah et al. (2002), indirectly referenced the catastrophic events of 9/11 and the anthrax attacks, expressing the following—

> [The] availability of biological warfare agents and supporting technologic infrastructure, coupled with the fact that there are many people motivated to do harm to the United States, means that America must be prepared to defend

58 Weiss (1987b), 58. See also Adams, M. B. (1990). The Wellborn Science, 4.

her homeland against biological agents.[59]

Noah et al. examined the global rise of biotechnology and the evolving dynamics of terrorist groups. They argued that such trends categorically led to the adoption of a strategy focused on enhanced surveillance systems, accelerated vaccine development programmes, and comprehensive training initiatives. While the authors' abstract acknowledged the importance of education and public participation, it curiously limits these to specific medical professionals, as the *sine qua non* of future defence preparedness.

Resurfacing themes from the works of Darwin and Galton, Provine's (1971) study traces the origins of biometrics to Galton, who applied statistical methods to address gaps in Darwin's natural selection hypothesis.[60] Allen (1976, 1983) explains that these methods later converged with Mendelian genetics, cementing modern population genetics as the successor to biometry.

Furthermore, Allen (1983) connects pre-Nazi anthropological thinking to early attempts to categorise social rank and value determinations through genetic traits and biometric cranial measurements.[61] Galton (1883) pioneered the use of composite portraits to classify criminal traits and types using facial construction and physiognomical profiles. This approach reflected a belief that physical features were indicators of intellectual deficiencies, linked to moral flaws and criminal tendencies.[62] Galton's classification system, grounded in natural selection, labelled criminals and marginalised groups as "*gregarious brute animals,*" a mindset that foreshadowed the eugenic policies exploited by the Nazis.[63]

One does not have to look far to see how this legacy has evolved. Distinguished by its anthropological origins, modern biometric techniques such as facial and iris scanning are uniquely designed to analyse facial patterns and geometric configurations, thereby facilitating increasingly precise identification.[64] Barkan (1992) seemingly first coined the term "*NewGenics*" to describe the postwar reconfiguration of eugenic thought within human genetics, but it was undefined. Black (2003, 2012) later expanded and popularised the term, applying it to the modern resurgence of eugenic ideology adapted to an era of advanced genetic technologies and pervasive biometric surveillance infrastructures. In today's technocratic age, a new protocol is emerging—one accompanied by anthropological, digital-biometric, and data-driven algorithmic profiling.

Accordingly, Barrett (2020) reported that biometrics can "*. . . quickly and efficiently identify many people in a crowd at once,*" demonstrating their exceptional utility in high-density scenarios.[65] Consequently, these advancements have significantly shaped and revolutionized criminal identification processes. While not foolproof, they have helped mitigate wrongful arrests and convictions. Genetic material is routinely collected through medical examinations, arrests, and forensic procedures. However, it is unclear whether such data is systematically aggregated from local institutions—surgeries, hospitals, criminal records, government databases, or schools. Indeed, the widespread use of DNA databanks—now routinely shared across international enforcement agencies—has become a cornerstone in global forensic investigations and genetic

59 Noah, D. L. et al. (2002). The history.
60 Galton, F. (1883). Inquiries, 14–19.
61 Allen (1976), 106–07, 109; Allen (1983), 107, 109–10; Provine, 25–35.
62 Galton (1883), 49–56, 61–68.
63 Ibid., 69.
64 Choudhary, V. & Kapoor, A. (2013). Role of Biometrics; Maguire, M. (2009). The Birth of Biometric Security; Wevers, R. (2018). Unmasking Biometrics.
65 Barrett, L. (2020). Ban Facial Recognition.

surveillance. Black (2003, 2012) documents the unprecedented scale of this initiative, highlighting the participation of nations including China, Canada, Austria, Holland, Germany, England, Australia, and the United States in expanding DNA profiling infrastructures. He observes—

> The network of DNA databases will soon become global. Interpol conducts a regular International DNA Users Conference to proliferate and link police DNA databank systems worldwide.[66]

Black first reported on emergent forms of global collaborative information-sharing in 2003. It is now 2026. The infrastructure has matured. In the United States, the Federal Bureau of Investigation recorded 1,101,629 violent crimes between April 2023 and April 2025 (aggravated assault, homicide, rape and robbery, excluding property offences).[67] Numbers of this scale provide political rationale for expanding integrated surveillance—not only for epidemiological identification and containment, but for predictive policing and behavioural management. Covid exposed, across the world, the depth and agility of covert state surveillance capabilities: a structurally precarious equilibrium between profiling criminals, monitoring health, and managing populations.

Merged with biometric systems and biotechnology, it forms a new governance model of adaptive algorithmic profiling (Figure IV-3, 337), a technocratic totalitarianism. This is the underside of the equality-based social rank-order hierarchy outlined in Class-Based Stratification, 283. Individuals categorised as possessing little or no economic or cultural value are managed as "*social problems.*" Review discussion under National Awakening, 121. Governance shifts from inclusion to management; from citizenship to optimisation. The central issue, therefore, is structural convergence: crime statistics, public health data, and digital identity systems operating within a single informational architecture. Surveillance becomes the organising principle of governance.

ii. Genetic Technologies and the Pursuit of Immortality

THE DISCUSSION OF biometrics and biosensors for monitoring health conditions is a segue into modern eugenics and the quest to eradicate health disorders. Another technology within the ambit of NewGenics is what Sparrow (2014) identifies *in vitro* positive eugenics—a colloquial term for germline editing or "*designer humans.*" Sparrow defines this process as selective breeding *in vitro* to produce genetically enhanced genome—

> [B]y fusing sperm and egg derived from different stem-cell lines to create an embryo and then deriving new gametes from stem cells derived from that embryo, which in turn might be used in the creation of another embryo. Repeated iterations of this process would allow scientists to proceed through multiple human generations in the laboratory.[68]

---

66 Chen, R. et al. (2003). DNA fingerprinting abnormalities. See also Black (2003, 2012), 430.
67 FBI's Crime Data Explorer, accessed as at April, 4, 2025.
This figure was obtained by conducting a search for violent crime statistics within the timeframe of April 2023 to April 2025. The platform provides comprehensive, official data customizable by category and period.
68 Sparrow, R. (2014). In vitro eugenics, 725. See also Wilkinson, S. (2010). On the Distinction, 120–122, for an analysis of the blurred distinctions between positive and negative eugenics.

However, Siegel (2014) challenges Sparrow's position on welfarist grounds, arguing that moral priority lies with the welfare of existing children, not hypothetical designer humans.[69] Siegal's concluding remarks, he expresses ambivalence toward *in vitro* eugenics, asserting that it lacks a robust *raison d'être*, especially considering the significant ethical and regulatory barriers he identifies earlier in his analysis. Macpherson et al. (2019) view reproductive germline enhancements as an imminent reality and stress the need for "*. . . effective expert panels and committees . . . universal ethical principles . . . ,*" before such interventions become widespread.[70]

In the 1970s, when IVF became a reality, first-generation bioethicists Leon Kass and Paul Ramsey, documented by Henig (2004) and Woiak (2007), warned it could lead to eugenic cloning, invoking Frankenstein as a cautionary tale.[71] Sixty years later, these concerns have been supplanted by debates on more invasive genetic technologies. A decade after Siegel's (2014) analysis, one might ask: has the ethical and welfare discourse on genetic enhancement shifted since then, and since Macpherson et al. (2019)?

Bruin et al. (2025) argue that with advancements in stem cell–derived gametes, the question is not if, but when the technology will emerge.[72] The recent development of *in vitro* gametogenesis ("IVG") marks a groundbreaking advancement in the field. Unlike *in vitro* eugenics, as posited by Sparrow (2014), which seeks to enhance genetic traits, IVG targets infertile gametes—reprogramming non-reproductive somatic cells into functional gametes (sperm or eggs) for assisted reproduction. Cohen (2025) calls IVG an exciting and potentially revolutionary technology that "*. . . intersects stem cell biology, genetics, and reproductive technology . . . .*"[73]

However, IVG differs from Preimplantation Genetic Diagnosis ("PGD") in key ways.[74] PGD is used post-fertilisation to select embryos based on their genetic makeup, while IVG is a pre-fertilisation technology that generates gametes from somatic cells. IVG allows for genetically screened gametes, expanding genetic selection before conception. Unlike PGD, which only selects embryos, IVG enables genetic screening and selection at the gametic level, broadening reproductive options. In sum, IVG holds the promise of creating genetically screened gametes from somatic cells, whereas PGD is a post-fertilisation technology focused on embryo selection based on genetic criteria.

The field is edging closer still with a next-gen refinement of CRISPR-Cas9. While CRISPR-Cas9 uses a 'cut-and-repair' mechanism, base editing chemically rewrites a DNA base pair with far greater precision. *In vivo* delivery methods include viral vectors and nanoparticles. Technologies like prime editing, epigenetic editing, CRISPR-Cas12 (Cpf1), CRISPR-Cas13, epitranscriptomic CRISPR therapeutics, and synthetic biology could complement or surpass base editing and gene circuits. Each innovation offers advantages in precision, versatility, and therapeutic potential, broadening the possibilities for treating genetic diseases. Together, they represent the next steps in gene editing. A summary of each, along with selected references, ranked from least to most ethically controversial, is as follows

69 Siegel, A. W. (2014). Some doubts.
70 Macpherson, I. et al. (2019). Ethical Challenges, 8.
71 Henig, R. M. (2004). Pandora's Baby, 70–71; Turney, *85–90 passim; Woiak, 122.
72 Bruin, I. J. d. et al. (2025). Stem cell-derived gametes.
73 Cohen, I. G. & Adashi, E. Y. (2025). Ethical and legal implications, 30.
74 Wilkinson (2010), 116.

Table IV-6. Comparison of Gene Editing Technologies

Most to Least Ethical	Technology	Summary	Key Difference	Refs
1.	**Epigenetic Editing**	Alters epigenetic marks (e.g., DNA methylation, histone modification) to regulate gene expression without changing the underlying DNA sequence.	Focuses on temporary or reversible gene regulation, rather than permanent genetic edits.	[75]
2.	**CRISPR-based Therapeutics for Epitranscriptomics**	Utilises Cas13 to target and modify RNA modifications (epitranscriptome) that control gene expression.	Focuses on reversible RNA modifications to regulate gene expression, providing therapeutic potential for diseases related to RNA processing, without altering the underlying DNA.	[76]
3.	**Synthetic Biology and Gene Circuits**	Designs gene circuits to control gene expression in response to specific signals.	Emphasises dynamic, real-time control over gene networks, rather than permanent edits to the genome.	[77]
4.	**CRISPR-Cas12 (Cpf1) and CRISPR-Cas13**	Cas12 cuts DNA more precisely, providing better precision in DNA editing; Cas13 targets RNA to modulate gene expression, enabling temporary RNA-based therapies for gene silencing.	Cas12: Permanent DNA changes (in the form of DNA cleavage and repair). Cas13: No permanent DNA changes; temporary RNA modifications for gene regulation.	[78]
5.	**Base Editing**	Directly converts one DNA base into another with high precision.	More accurate than CRISPR-Cas9, but limited to single-base changes. **Permanent DNA changes**	[79]
6.	**Traditional CRISPR-Cas9**	Uses Cas9 to create double-strand breaks in DNA, allowing gene edits through the cell's repair machinery.	Powerful but prone to off-target effects and unintended mutations, making it less precise than newer methods. **Permanent DNA changes**	

[75] Akatsuka, K. et al. (2020). Ethical Framework; Sgro, A. & Blancafort, P. (2020). Epigenome engineering; Smith, J. et al. (2025). Clinical epigenetics; Yuan, L. et al. (2025). Epigenome editing.
[76] Alum, E. U. et al. (2025). Epitranscriptomic alterations; Cerneckis, J. et al. (2024). The rise of epitranscriptomics; Athanasopoulou, K. et al. (2025). Exploring the m(5)C epitranscriptome.
[77] Cloney, R. (2019). Cooperating on synthetic gene circuits; Covert, M. & Press, C. R. C. (2015). Fundamentals; Nandagopal, N. & Elowitz, M. B. (2011). Synthetic Biology.
[78] Badon, I. W. et al. (2024). Recent application of CRISPR-Cas12; Bigini, F. et al. (2023). Unleashing the potential; Hillary, V. E. & Ceasar, S. A. (2023). A Review; Shi, P. & Wu, X. (2024). Programmable RNA.
[79] Porto, E. M. et al. (2020). Base editing; Xu, W. et al. (2024). From bench to bedside.

Table IV-6. Comparison of Gene Editing Technologies

Most to Least Ethical	Technology	Summary	Key Difference	Refs
7.	**Prime Editing**	"Search-and-replace" method with high precision, correcting a broad range of mutations.	More versatile than base editing. Corrects insertions, deletions, and point mutations. **Permanent DNA changes**	[80]
8.	*In Vitro* **Gametogenesis (IVG) /** *In Vitro* **Eugenics**	Uses lab-grown gametes (sperm or eggs) derived from somatic cells for reproductive purposes. IVG was primarily developed to address infertility, enabling couples unable to produce gametes naturally to have genetically related children.	Ethically controversial due to concerns about designer babies and permanent DNA changes at the embryonic stage. **Permanent DNA changes**	[81]
9.	*In Vitro* **Positive Eugenics (Sparrow's (2014) Concept))**	Fuses stem-cell-derived sperm and eggs to create embryos and derive new gametes for use in subsequent generations.	Involves iterative genetic selection and enhancement over multiple generations, creating "designer humans" through stem-cell-based gamete production. Effect: Controversial due to the potential for multi-generational genetic manipulation and permanent DNA changes. **Permanent DNA changes**	

This table ranks gene editing technologies on an ethical scale from most ethical (1) to least ethical (8). It summarizes their core functions, advantages, and limitations, highlighting the permanence of DNA changes and the ethical considerations associated with each technology. Technologies that enable temporary gene regulation (e.g. RNA editing, epigenetic editing) are ranked lower on the ethical concern scale, while those involving permanent genetic modifications (e.g. CRISPR-Cas9, *In Vitro* Eugenics) are ranked higher due to their ethical implications.

[80] Chen, P. J. & Liu, D. R. (2023). Prime editing; Krausz, S. L. et al. (2025). ProPE expands the prime editing; Newby, G. A. & Liu, D. R. (2021). *In vivo* somatic cell base editing and prime editing.
[81] Bigini, F. et al. (2023). Unleashing the potential; Cohen, I. G. & Adashi, E. Y. (2025). Ethical and legal implications; Taylor, H. S. et al. (2025). In-vitro gametogenesis

Table IV-6. Comparison of Gene Editing Technologies

Most to Least Ethical	Technology	Summary	Key Difference	Refs

Ethical Ranking Explanation:
Ranks 1: Technologies that do not involve permanent DNA changes, such as epigenetic editing, RNA editing, and synthetic biology, focus on reversible modifications or gene regulation, with minimal ethical concerns.
Ranks 2–6: CRISPR-based technologies (including CRISPR-Cas12, Prime Editing, and Base Editing) are more precise and targeted than older methods like CRISPR-Cas9. While they involve permanent DNA changes, they present lower risks of unintended consequences, reducing ethical concerns but still raising significant questions about their application.
Rank 7: *In Vitro* Gametogenesis (IVG) / *In Vitro* Eugenics: This technology uses lab-grown gametes for reproductive purposes. While it represents an advance beyond IVF, it raises significant ethical concerns due to the potential for designer babies and embryonic modifications, leading to permanent genetic changes.
Ranks 8–9: *In Vitro* Positive Eugenics (Sparrow's Concept): This technology involves stem-cell-derived gametes to create "designer humans" by iterating genetic selection over multiple generations. It poses significant ethical dilemmas, including concerns about eugenics, genetic inequality, and the fundamental nature of human identity. Given its potential for multi-generational genetic manipulation, it sits at the highest ethical concern.

Let us now turn to positive and negative biotech eugenics. Early research by David J. Galton (1998), a modern medical geneticist and bioethicist, foresaw the ability to predict predispositions to common illnesses such as cancer, heart disease, and diabetes through the multitude of genetic markers.[82] Wilkinson (2010) reasons that the distinction between "*negative*" and "*positive*" eugenics is a matter of perspective, guided by the specific goal—whether enhancing traits like athleticism, attractiveness, or intelligence. Wilkinson illustrates this: "*. . . for example, selecting out low(er) intelligence ('negative eugenics') might be the same practice as selecting in higher intelligence ('positive eugenics').*"[83]

Allen (2002) poses an evocative question: "*Are we likely to experience a true revival of eugenic thinking or true eugenic practice in the near future?*"[84] Given current trends, the answer is yes. He effectively argues, and Scott (2006) concurs, that the Human Genome Project is viewed as the "*Holy Grail,*" where fate is no longer guided by the stars, but is encoded within our genes.[85] Todorov, writing in 1993 foresaw the possibilities of such developments when he wrote that "*. . . male semen for artificial insemination can be chosen . . . breeding or cloning of superman does not seem far-fetched.*"[86] At the time, this idea seemed speculative, even fantastical; today, this is modern reality. Genetic-level interventions, including gene therapy and CRISPR/Cas9 technologies (see Table IV–6, 328), are already rewriting the human genome to treat or prevent disease. The medical community is already actively repairing damaged or diseased DNA. Todorov (1993) further elaborates—

82 Galton (1998), 263.
83 Wilkinson, 117.
84 Allen (2002), 37.
85 Ibid., 13–14; Scott, A. (2006). 'Like editing bits of ourselves.'
* The concept—"our fate is in our genes"—is commonly referred to as genetic determinism, a form of biological determinism that has been broadly rejected by the scientific community. See Allen (2002), 15–16; Marks, J. (2024). Our Fate Is Not in Our Genes. I use the term here critically to indicate that ongoing developments in genomics are predisposed toward such a deterministic framework. For example, see Saunders, G. R. B. et al. (2022), "Genetic Diversity"; and Stein, R. (June 2021), "He Inherited a Devastating Disease."
86 Harris, E. (2024). Sickle Cell Disease Approvals; Krüger (1999), 42.

> [In a] more insidious manner we are beginning to accept the legitimacy of genetic manipulations and other techniques of intervention in the process of human reproduction. We are not far from . . . eliminat[ing] imperfect embryos, or to choose the sex, [eye colour], and even the type of intelligence, of our children.[87]

In November 2018, two genetically modified babies were born in China after embryo editing using CRISPR technology. By June 2021, gene editing had expanded further, with four patients in London and New Zealand volunteering for CRISPR procedures. One notable case was Patrick Doherty in London, who received gene therapy for a hereditary liver disease that had killed his father. The procedure involved delivering billions of lipid nanoparticles carrying CRISPR–Cas nucleases into Doherty's liver, where they precisely targeted and disabled faulty genes. This intervention halted harmful protein production and repaired the underlying genetic damage.[88] Wong (2023) calls this a turning point, with the UK becoming the first nation to approve such treatments, offering the potential to "*. . . potential to cure many [future] genetic diseases.*"[89] Both Black (2003, 2012) and Parhiz et al. (2024) describe this as a historic shift, framing "*mRNA-based medicine*" as a transformative moment in science—one that redefines what is possible and marks a pivotal moment in the path toward precision healing—

> No one should fear the benefits of human reengineering that can obliterate terrible diseases, such as cystic fibrosis and Tay-Sachs. . . . Every such medical advance is a long overdue miracle. Society should welcome corrective genetic therapies and improvements that will enhance life and better mankind.[90]

Beyond gene editing, injectable nanoparticles are now being employed *in vivo* for tumour targeting, infection treatment, drug delivery, and organ repair, as seen in the case of Patrick Doherty. Moreover, Prakash et al. (2016) and Parhiz et al. (2024) report that "*. . . 5000–8000 monogenic diseases [defined as inherited conditions arising from mutations on a single gene] have been identified,*" and may benefit from mRNA-, viral-, and DNA-based treatments, including mRNA-encoded genomic editing.[91]

These breakthroughs are to geneticists what 19th- and 20th-century scientific discoveries were to social scientists: a transformative force, revolutionising medicine and redefining the long-sought after ideal of eternal youth. Indeed, René Descartes (1596–1650)—the French philosopher, mathematician, and scientist hailed as the father of modern philosophy—shared this Baconian ambition to conquer death. A prominent aspect of this ambition was the pursuit of eternal youth, an aspiration that Gutmann (1978) highlighted in his analysis—

87 Todorov, 167–68.
88 Stein, R. (2021). He Inherited A Devastating Disease; Beers, B. C. v. (2020). Rewriting the human genome; Straiton, J. (2019). Genetically Modified Humans. See also Abdelnour, S. A. et al., (2021). The Potential of CRISPR; Bahal, R. et al. (2016). Precise Genome Modification; Baylis, F. et al. (2020). Human Germline; Parhiz, H. et al. (2024). mRNA-based therapeutics, 1198–99; Qiu, M. et al. (2021). Lipid nanoparticle-mediated codelivery; Rothgangl, T. et al. (2021). *In vivo* adenine, 949; Tran, D. et al., (2018). CMOS-Compatible Silicon Nanowire; Zheng, Y. et al. (2024). Precise genome-editing.
89 Wong, C. (2023). UK first to approve, 676.
90 *Black (2003, 2012), 428; Parhiz et al., 1197.
91 Prakash, V. et al. (2016). Current Progress, 465; Parhiz et al., 1197.

> As for Bacon, so for Descartes the sciences have a definitely practical aim, the harnessing of nature to the purposes of man. The will-o'-the-wisp of his life was the conquest of death not only for the soul but also for the body.[92]

Parhiz et al. (2024) elaborates that technologies have the "*. . . potential to treat the root cause of genetic diseases, leading to lifelong cure.*"[93] Gutmann (1978) resounds Bacon's emphasis on "*. . . the practical applications of knowledge as power, [and] the pragmatic mood in philosophy.*"[94] In short, knowledge is framed as the exercise of practical power, fostering a philosophical mindset focused on tangible results.

While Francis Galton is credited with founding eugenics, Burleigh (2002) and Kohler (1976, 1978, 1991) consider that it was wealthy philanthropists, such as the Carnegies, Harrimans, Kelloggs, Loebs, and Rockefellers, who truly popularised and institutionalised the eugenic movement.[95] Fast forward to the present: Bill Gates's coordination with the WHO and his foundation's philanthropic investments in the Covid response, vaccine development, and global distribution mirror the philanthropic strategies historically pursued by the Rockefeller Foundation. See original discussion under heading Weimar Era Refinements, 131.

However, as a sidenote, the roots of modern philanthropy can be traced back even further to Francis Bacon. Prior (1954) states that philanthropy is the "*. . . seed from which the new sciences must grow.*"[96] By the 1920s, the Rockefeller Foundation had transformed this idea to an unprecedented level, with Kohler (1978) stating that the Foundation elevated scientific research through large-scale research patronage "*. . . pioneered in establishing the general institutional traditions and the specific administrative techniques for the patronage of individual research on a large scale.*"[97] Warren Weaver, head of the Natural Sciences Division at the Rockefeller Foundation from 1932 to 1955, had already established the Foundation's research grant model by 1928, marking the moment when the Foundation formally committed itself to the "*. . . advancement of science its principal goal.*"[98]

According to Kohler (1978), Löwy & Zylberman (2000), and Weindling (1997), the Rockefeller Foundation was the first global philanthropic body, a mosaic of public health measures (including eugenics) under the auspices of improving mental and physical health. It pioneered international agencies based on the American model, operating without central control, laying the groundwork for future global health initiatives and serving as a precursor to both the WHO and the Gates Foundation.[99] Simply put, viral threats, vaccine development, population management, and even elimination strategies are all intertwined.[100] Weaver's biomedical reforms were catalysed by his ambition that

---

92 Gutmann, 240.

93 Parhiz et al. *1198–1201; Prakash, V. et al. (2016). Current Progress.

94 Gutmann, J. (1978). III. Western Philosophical Thought, 240. See generally Martin, C. G. (2002). The Ahistoricism of the New Historicism, 22–49.

95 Black (2003, 2012), 40, 43, 235 passim; Burleigh (2002), 117; Kohler, R. E. (1976). The Management of Science, 279; Kohler, R. E. (1978). A Policy for the Advancement; Kohler, R. E. (1991). Partners in Science; Richardson, M. (1990). Philanthropy, 22, 38.

96 Prior, 354.

97 Kohler, (1978), 480, 487.

98 Ibid., *481–82.

99 Ibid., 491, 495–96, 514 passim; Löwy & Zylberman, 366; Weindling, P. (1997). Philanthropy and World Health.

100 To clarify, this is not to suggest that Rockefeller engineered Covid or any other virus. Rather, the point to highlight is this, that there is an observable pattern across these interconnected domains.

genetics would one day become so expansive as to enable the breeding of a superior race, a belief he publicly expressed in 1933—

> Can we develop so sound and extensive a genetics that we can hope to breed a superior race of men? Can we solve the mysteries of various vitamins so that we can nurture a race sufficiently healthy and resistant? Can psychology be shaped into a tool effective for man's everyday use? In short, can we rationalise human behaviour and create a new science of man?[101]

The science Weaver alluded to is now capable of driving genetic-level changes to ennoble and refine humanity. Allen's (1976, 1983, 2002) research, alongside the Rockefeller Foundation's continuous support, reveals that eugenicists viewed the growing number of those deemed genetically "*defective*" or low IQ as the root cause of 'social problems.'[102] See Weimar Era Refinements, 131.

> [Rockefeller Foundation] officials sincerely believed that advance in hygiene and public health, improvement in the understanding of natural phenomena, and furtherance of the control of human behaviour would all lead to the development of a happier human society . . . .[103]

Moreover, as noted, it is unsurprising that, in the proto-Third Reich, the Rockefeller Foundation's 1923 backing of natural sciences in Germany synchronised with its 1923 'population policy', later becoming part of the Foundation's philanthropic portfolio.[104] While Allen (2002) does not directly address the Rockefeller Foundation, his critique is extant and relevant to later philanthropic models, where scientific initiatives masquerading as humanitarian efforts rested on flawed genetic assumptions, particularly regarding population control.[105]

Löwy and Zylberman (2000) somewhat uncritically, directly linked the vectors of population genetic control and the dissemination of scientific knowledge through human welfare efficiency strategies "*. . . to improve economic or political structures.*" This was the underlying intent for the creation of the Rockefeller Foundation with the "*. . . goal of developing a true 'science of public health.'*" This effort culminated in the founding of the first School of Public Health at Johns Hopkins University, Baltimore.[106] This framing is arguably synonymous with what Kohler (1976) recorded as being—

> [G]oals were consonant with the idea that human welfare was best promoted by the systematic and rational application of objective knowledge [to improve

[101] Rockefeller Foundation. (1933). The Rockefeller Foundation : Annual Report, 199; Kohler, R. E. (1976). The Management of Science, 291, quoting Weaver; Löwy, I. & Zylberman, P. (2000). Medicine, 374. The Annual Report (1933) is accessible from the Rockefeller Website.

[102] Allen (1976), 105–21; Allen (1983), 105–27; Allen (2002), 13–17, 29–38.

[103] Löwy & Zylberman, 367.

[104] For the United States parallel approach between 1910 and 1940, see Allen (2002), 31; Fosdick, R. B. (1952). The story of the Rockefeller; Gaudillière, J.-P. (2000). Rockefeller strategies; Löwy, I. & Zylberman, P. (2000). Medicine; Macrakis, K. (1989). The Rockefeller Foundation, 35.

[105] Allen (2002), 108; Black (2003, 2012), 95–102.

[106] Löwy & Zylberman, 368, 371; Kohler, R. E. (1991). Partners in Science, 44, 79; Weindling, P. (1997). Philanthropy and World Health, 274.

> human welfare or the welfare of mankind throughout the world].[107]

The Rockefeller Foundation's 1920s funding objectives, framed by terms like "*human welfare*" and "*the welfare of mankind,*" though not explicitly defined by Kohler, implicitly point to a grand-scale, top-down social engineering effort grounded in eugenic policies. These long-term goals extended far beyond Nazi Germany, conceived as a global initiative with an indeterminate timeline.[108] Arguably, this legacy is still current. The Foundation's ongoing role in addressing global health threats—through philanthropic responses to viral outbreaks and vaccine development—illustrates its historical vested interest in managing populations, in particular, those deemed to have low IQs.[109] These interpretations are not incidental. As noted earlier, Weaver, a Rockefeller director, not only affirmed its remit in 1933 but extended it beyond negative eugenic applications, actively promoting positive eugenics by the creation of a "*superior race.*"[110] This intersection raises profound, existential questions about the role of science in shaping global population policy—questions amplified and brought to light by the Covid pandemic.

Historically and in contemporary contexts, state-imposed restrictions on movement and social interaction—whether under Nazi Germany's totalitarian regime or during the Covid pandemic—have been justified through public health frameworks that redefine individual freedoms. Savage (2007) links this shift in governance, utilising germ theory to understand the changing dynamics between personal liberty and state authority—

> Germ theory redefined the concept of individual liberty, making it acceptable for governments to investigate citizens and restrict their movements, since no individual had the right to contaminate others. Furthermore, the elimination of illness through state surveillance and state control of the individual could be seen not only as a necessity, but also as humane.[111]

In pre-Nazi Germany, reprising Schallmayer's genetic reforms, he proposed: "*[h]ealth passports, medical genealogies, [and] marriage restrictions as the 'direct' hygienic tools by which human selection might be monitored.*" Under the Nazi regime, these ideas were operationalised to enforce genetic fitness and control reproduction.[112] This brings us full circle to modern anthropology applications. Today, the modern equivalent of these concepts is digital biometry, influenced by population genetics, which targets perceived 'social problems.' The Covid pandemic once again exposed the depths of technocratic surveillance, profiling mainstream society through digital tools and mandating health passports for travel. Savage (2007) further contextualises this dynamic, demonstrating how hygienic medicine evolved into a instrument for health governance, steering political regulation and population health management through surveillance practices—

[107] Kohler (1976), 280–81, 284, 300; Kohler (1978), 515.
[108] Kohler (1976), 294.
[109] Fosdick, R. B. (1952). The story of the Rockefeller Foundation, 54–56, *128–29; Gaudillière, J.-P. (2000). Rockefeller strategies.
[110] Rockefeller Foundation. (1933). The Rockefeller Foundation : Annual Report, 199.
[111] Savage, 418n.97, quoting Otis (1999), p. 34–35.
[112] Weiss (1987b), 58.

> [N]ew technology of population (that is, political science), encompassing the tools for internal measurement and regulation, was the only body equipped to deal with the necessary processes of discipline: identification, categorisation, containment, and (if necessary) elimination. The health of the population would be ensured by the "police" of the social body, and specifically by the new formation, "medical police."[113] (Footnote call numbers omitted)

Beyond forensic utility, DNA fingerprinting detects abnormalities through genetic profiling and inadvertently exposes predispositions to health conditions.[114] Genetic data, once aggregated into hereditary repositories, devolve into prediction; what begins as identification slides into prognostication. Black (2003, 2012) foresees their expansion into litigation, healthcare, employment and insurance, enabling hereditary screening on a globally interconnected scale.[115] This development is not unprecedented. Weindling (1993) shows that Nazi Germany operated hereditary registries and travel health passports to assess eligibility, evaluations of genetic propensity for criminality, and fitness for work—

> Hereditary data-banks would be used as the basis for research on the inheritance of diseases, mental disorders and crime, and of positive qualities such as leadership, fitness and genius. Doctors argued that it was not enough to locate only the sick and their parents, because the healthy could be carriers of latent pathogenic genes. Medical experts demanded the screening of total populations. For this the support of local, provincial, state and Reich authorities would be necessary.
>
> Demands for health passports and hereditary data banks date from before 1914, and were linked to a long-term process of expansive medical professionalization. Psychiatrists and eugenicists took a lead in instigating data banks as a means of controlling crime and delinquency.[116] (Footnote call number omitted)

A contemporary parallel emerged during Covid. As previously analysed in The Role of Surveillance, 312, Singapore's TraceTogether app, while initially designed for contact tracing, possessed the unintended capability of enabling police access and utilise aggregated movement data for criminal investigations unrelated to Covid.[117] What was framed as public health infrastructure was thus repurposed for broader state surveillance. The growing reliance on aggregated genetic databanks and digital profiling architectures indicates that such function creep is not an exception, but just the beginning.

Conclusion

The struggle for epistemic dominance between science and religion, exemplified by figures like Bacon and Andreae, has never been purely theoretical. As science ascends as the ultimate arbiter of truth, it determines not only what is true but also what is

[113] Savage, 417, citing Foucault, M. (1980). Power/knowledge, 169–170, 176; Foucault, M. (1991). Governmentality, 100–02; See generally Fox, C. (1995). Introduction, 3–4.
[114] Black (2003, 2012), 430.
[115] Ibid., 431–32, 440.
[116] Weindling (1993), 384.
[117] Anison, 51; Greenleaf and Kemp, 274.

morally permissible. Bacon not only anticipated such advancements but sketched the blueprint for a society governed by scientific authority. The *New Atlantis* narrative, which began with Bacon's vision, continues to shape the future of global governance, blending scientific humanism with pantheistic ideologies. Bacon's axiom—"*knowledge is power.*" Metz (1938) likened it to *regnum hominis*—a revolution of knowledge where scientific conquest becomes the ultimate goal.[118] This vision is now manifesting in a technocratic regime, underpinned by the potential for global communicative pandemics, where scientific and medical advancements legitimise authoritarian control.

Yet the enduring revival of occult forces, from Renaissance esotericism to the Nazi regime, reveals an ongoing hunger for supernatural knowledge and authority. The Nazi regime, believing itself descended from the original Atlanteans, only partially realised this vision. Its pursuit of Nietzsche's Übermensch, grounded in myths of immortal ancestors and aspirations to restore an ancestral demigod status through scientific experimentation, the Nazi regime ultimately failed to unlock the "*eternal youth*" gene. What eluded the Nazis, however, now appears technologically conceivable within the emerging biopolitics of health security.

Hooton (1935) cautioned that education and religion cannot transmute "*base metal into gold,*" a metaphor for the biologically "*inferior,*" exposing the perceived limits of social reform.[119] Today, however, advances in genetic engineering, longevity research, military enhancement programmes, and pandemic-driven health security frameworks suggest that these limits are being recast through a new scientific authority organised around the biopolitics of health security. Ambitions once dismissed as ideological fantasy are increasingly approaching technical feasibility. The pursuit of immortality and "*eternal youth*" is no longer merely speculative rhetoric, but an expanding domain of applied science.[120]

The progressive consolidation of scientific power raises critical questions about human identity, autonomy, and governance. The issue is not whether technological ambitions such as eradicating disease, extending life, and enhancing human capacity are achievable, but whether their realisation entails the erosion of individual rights and freedoms, construed as societal security risks and aligned with a "*moral code balancing act.*" In the aftermath of COVID-19, contagion management and pandemic health security increasingly operate as the organising logic of a new global order founded on technocratic adaptive authoritarianism and justified through the language of health security. Under the permanent possibility of Disease X, scientific authority acquires exceptional powers through biosecurity, surveillance, and the management of collective risk. The old divisions between science and religion, nature and human agency, are rapidly dissolving. Where this leads remains uncertain, but one thing is clear: the future will be shaped by the fusion of science, power, and the Atlantean utopian ideal. As a concluding note, readers are encouraged to move regularly between Parts I, II, III, and IV to facilitate analogous comparisons.

[118] Bury, 50–51; Metz, 30.
[119] Hooton, E. A. (1935). Homo Sapiens, 31. See also Krüger, A. (1999). Breeding, rearing and preparing the Aryan body. See Staub's (1989), 113, general comments.
[120] Bahal, R. et al., (2016). Precise Genome Modification; Corbyn, Z. (2015). Crispr: is it a good idea to 'upgrade' our DNA?; Kurzweil, R. (2004). Human Body Version 2.0; Regalado, A. (2015). Engineering the Perfect Baby; Sullenger, B. A. (2003). Targeted genetic repair; Wang, B. (2014). Ido Bachelet announces 2015 human trial.

COMPARISON OF TOTALITARIAN GOVERNANCE AND THE FUTURE

Nazi Regime

Collaborations and Regions
- Europe-wide.
- Extensions into parts of Middle East & Africa.
- **Fixed Authoritarianism:**
  - Centralized control, no societal freedom.
  - Imperial and Racial Totalitarian Regime.
  - Strict control over public and private life.

Health Governance & Disease Management
- Medical institutions co-opted to serve Nazi ideology.
- Forced sterilization of "inferior" groups.
- Racial hygiene laws to ensure Aryan purity.
- Engineered Typhus epidemic and cultivated conditions for contagion.
- Segregation designed to enable genocide.

Compliance Measures
- Suppression of opposition and free speech.
- Targeting of Jews and non-Aryans.
- Removal of genetically and socially inferior groups.
- Creation of a new racial, genetic class hierarchy.

Surveillance & Enforcement
- Rudimentary.
- Physical Policing by Secret Police and SS.
- Reliance on citizen informants to report dissenters.

Propaganda
- **Indoctrination & Media Control:**
  - State-run press and radio pushed Nazi messaging.
  - Censorship crushed dissent and truth.
  - Goebbels' ministry scripted the national narrative.
  - Fuelled by fear, repetition, and emotion.
- **Youth Indoctrination & Social Engineering**
  - Nazism hardwired into school lessons.
  - Hitler Youth molded loyalty from childhood.
  - Enemies vilified in classrooms.

Covid

Collaborations and Regions
- Global Collaboration
- **Dynamic Authoritarianism**
  - Introduction to future scientific totalitarianism.
  - Varied from strict to limited freedom.
  - National Interdependence.

Health Governance & Disease Management
- Global Medical Alliance.
- Household Segregation leading to hospitalisation
- Alleged Engineered Virus.
- Severe Flu & Contagious.
- Separation encouraged from unvaccinated.

Compliance Measures
- Varied suppression of opposition and free speech.
- Enforced lockdowns and surveillance compliance.
- Hierarchy granted access to social venues and participation in society.

Surveillance & Enforcement
- Bluetooth, Biometric facial recognition and drones.
- Physical police patrols and roadblocks.
- Stay-at-home orders.
- Contact Tracing.

Propaganda
- **Indoctrination & Media Control:**
  - World governments collaboration and social media platforms with claims of misinformation.
  - Utilised Actor(esses) and Sports to promote vaccines.
  - Transparency Limited.

Future

Collaborations and Regions
- Global, interconnected nations' collaboration.
- Unified, global governance model.
- **Techno–Adaptive Authoritarianism**
  - Centralized system utilizing data and technology globally.
  - Global alliances under scientific totalitarian principles.

Health Governance & Disease Management
- Surveillance-led medical-science body overseeing healthcare.
- Designed to enforce global compliance.
- Disease X (Ebola-Rabies combination?)
- Population control and societal restructuring.
- High Contagion.

Compliance Measures
- Restrictions, lockdowns.
- Enforced surveillance compliance.
- Societal divisions by enforcing obedience through fear of contagion.

Surveillance & Enforcement
- Behavioural Biometrics.
- Health and Temperature Monitoring.
- Drone Surveillance (crowd control, infection tracking, and density prediction).
- Facial Recognition Technology (public spaces and security checkpoints).
- Indoor Positioning Systems (IPS) for precise tracking in confined areas.
- Humanoid robots equipped with advanced surveillance technologies for enforcement and oversight.

Propaganda
- **Indoctrination & Media Control:**
  - Indoctrination by governments and social media.
  - Algorithmic censorship suppressing dissent under the pretense of combating misinformation.

Legend

- Dynamic is flexible and evolves with changing circumstances, utilizing technology for surveillance, propaganda, and control.
- Fixed rigid power structures rely on censorship, repression, and centralized authority. Their failure to adapt to societal or political shifts exposes them to instability and collapse.
- Technocratic Adaptive Authoritarianism combines dynamic adaptability, centralized control, and data-driven, technology-enabled governance.
- Drones, Humanoids, IPS, employ data aggregation and tracking: combines GPS, Wi-Fi, and Bluetooth data to monitor movements and activities across digital and physical spaces.
- Health Monitoring: Thermal Imaging and Infra-red Thermometer Monitoring are used to detect temperature variations, monitor crowds, and identify fever.

Figure IV–3. Comparison of Past, Present, and Future Totalitarian Governance Models

Conflict of Interest and Funding Disclosure

The author declares no conflicts of interest with respect to the research, authorship, and/or publication of this article and received no funding associated with the content featured in this paper.

Index of Figures

Index of Tables

References

Fedorenko v. United States, 449 U.S. 490 (1981).

A'lam, J., Thanthow, S. A., & Namer, M. (1987). Dendam Barat & Yahudi terhadap Islam (M. Thalib & M. Mahdamy, Trans.). Pustaka Mantiq.

Abdelnour, S. A., Xie, L., Hassanin, A. A., Zuo, E., & Lu, Y. (2021). The Potential of CRISPR/Cas9 Gene Editing as a Treatment Strategy for Inherited Diseases [Review]. Front Cell Dev Biol, 9, 699597. https://doi.org/10.3389/fcell.2021.699597

Abelson, P. (1966). The Seven Liberal Arts : A Study in Mediæval Culture. Michigan, Teachers' College, Columbia University.

Abolmaali, M., Rezania, F., Behnagh, A. K., Hamidabad, N. M., Gorji, A., & Mirzaasgari, Z. (2022). Guillain-Barré syndrome in association with COVID-19 vaccination : a systematic review. Immunologic Research, 70(6), 752–764. https://doi.org/10.1007/s12026-022-09316-6

Ach, M. (1977). Hitlers "Religion" pseudoreligiöse Elemente im nationalsozialist. Sprachgebrauch. München, Arbeitsgemeinschaft für Religions. Weltanschauungsfragen.

Ackerknecht, E. H. (1953). Rudolf Virchow : Doctor, Statesman, Anthropologist. Madison, University of Wisconsin Press.

Adams, M. B. (1990). The Wellborn Science : Eugenics in Germany, France, Brazil, and Russia. Oxford, United States, Oxford University Press, Incorporated.

Adams, R. P. (1949). The Social Responsibilities of Science in Utopia, New Atlantis and after. Journal of the History of Ideas, 10(3), 374–398. https://doi.org/10.2307/2707043

Adorno, T. W., Pickford, H. W., Olick, J. K., & Perrin, A. J. (2010). "The Meaning of Working Through the Past". In Guilt and Defense (pp. 213–228). Harvard University Press. https://doi.org/10.2307/j.ctv1nzfgd5.12

Agassi, J. (2012). The Very Idea of Modern Science : Francis Bacon and Robert Boyle. Dordrecht, NETHERLANDS, THE, Springer Netherlands.

Aginam, O. (2005). Bio-terrorism, human security and public health : Can international law bring them together in an age of globalization? Medicine and law, 24(3), 455–462.

Ahlwardt, H. (1890). Der Verzweiflungskampf der arischen Völker mit dem Judentum. Grobhäuser.

Ahmed, S., Karim, M. M., Ross, A. G., Hossain, M. S., Clemens, J. D., Sumiya, M. K., Phru, C. S., Rahman, M., Zaman, K., Somani, J., Yasmin, R., Hasnat, M. A., Kabir, A., Aziz, A. B., & Khan, W. A. (2021). A five-day course of ivermectin for the treatment of COVID-19 may reduce the duration of illness. International Journal of Infectious Diseases, 103, 214–216. https://doi.org/10.1016/j.ijid.2020.11.191

Aiken, L., & Michaels, I. (2007). Genesis : The Untold Story : A Refreshing Look at Our Owner's Manual. Beverley Hills, CA, USA, Rossi Publications.

Akatsuka, K., Sasaki-Honda, M., & Sawai, T. (2020). Ethical Framework for Next-Generation Genome and Epigenome Editing. American journal of bioethics, 20(8), 32–36. https://doi.org/10.1080/15265161.2020.1782524

Al Saidi, A. M. O., Nur, F. A., Al-Mandhari, A. S., El Rabbat, M., Hafeez, A., & Abubakar, A. (2020). Decisive leadership is a necessity in the COVID-19 response. The Lancet, 396(10247), 295–298. https://doi.org/https://doi.org/10.1016/S0140-6736(20)31493-8

Al-Deen, F. N., Selomulya, C., Ma, C., & Coppel, R. L. (2014). Superparamagnetic Nanoparticle Delivery of DNA Vaccine. In M. Rinaldi, et al. (Eds.), DNA Vaccines: Methods and Protocols (pp. 181–194). Springer New York. https://doi.org/10.1007/978-1-4939-0410-5_12

Albanese, D. (1990). The New Atlantis and the Uses of Utopia. ELH, 57(3), 503–528. https://doi.org/10.2307/2873232

Aldis, W. (2008). Health security as a public health concept : a critical analysis. Health Policy and Planning, 23(6), 369–375. https://doi.org/10.1093/heapol/czn030

Alexander, L. (1945). Combined Intelligence Objectives Sub-committee (CIOS) report : Public Mental Health Practices in Germany : Sterilization and Execution of Patients Suffering from Nervous and Mental Diseases. Item No. 24, File No. XXVIII- 50. National Archives, Washington, DC, 1– 173.

_____. (1949). Medical Science under Dictatorship. New England Journal of Medicine, 241(2), 39–47. https://doi.org/doi:10.1056/NEJM194907142410201

Alexander, N. (2019). Race in a Godless World : Atheism, Race, and Civilization, 1850-1914. Manchester, UK, Manchester University Press.

Alexander, P. E., Armstrong, R., Fareed, G., Lotus, J., Oskoui, R., Prodromos, C., Risch, H. A., Tenenbaum, H. C., Wax, C. M., Dara, P., McCullough, P. A., & Gill, K. K. (2021). Early multidrug treatment of SARS-CoV-2 infection (COVID-19) and reduced mortality among nursing home (or outpatient/ambulatory) residents. Medical hypotheses, 153, 110622–110622. https://doi.org/10.1016/j.mehy.2021.110622

Allen, G. E. (1976). Genetics, Eugenics and Society : Internalists and Externalists in Contemporary History of Science. Soc Stud Sci, 6(1), 105–122. https://doi.org/10.1177/030631277600600107

_____. (1983). The Misuse of Biological Hierarchies : The American Eugenics Movement, 1900-1940. Hist Philos Life Sci, 5(2), 105–128.

_____. (2002). The Ideology of Eliminination : American and German Eugenics, 1900–1945. In Medicine and Medical Ethics in Nazi Germany : Origins, Practices, Legacies (pp. 13 39). Berghahn Books, Incorporated.

Allen, P. M., & Pietzner, C. (1974). A Christian Rosenkreutz Anthology. Rudolf Steiner Publications.

Allport, G. W. (1979). The Nature of Prejudice (First Edition, second printing ed.). Reading, Mass, Addison-Wesley Pub. Co.

Alum, E. U., Ejemot-Nwadiaro, R. I., Basajja, M., Uti, D. E., Ugwu, O. P.-C., & Aja, P. M. (2025). Epitranscriptomic alterations induced by environmental toxins : implications for RNA modifications and disease. Genes and Environment, 47(1), 14. https://doi.org/10.1186/s41021-025-00337-9

Ammon, O. (1895). Die Gesellschaftsordnung und ihre natürlichen Grundlagen. Jena, G. Fischer.

_____. (1900). Die Gesellschaftsordnung und ihre natürlichen Grundlagen; Entwurf einer Sozial-Anthropologie zum Gebrauch für alle Gebildeten, die sich mit sozialen Fragen befassen: Mit 6 Figuren im Text. G. Fischer.

Amorim, C., Douste-Blazy, P., Wirayuda, H., Jonas Gahr, S., & et al. (2007). Oslo Ministerial Declaration-global health : a pressing foreign policy issue of our time. The Lancet, 369(9570), 1373–1378. https://doi.org/https://doi.org/10.1016/S0140-6736(07)60498-X

Andreae, J. V. (1916). Christianopolis : An Ideal State of the Seventeenth Century (F. E.

Held, Trans.). Oxford University Press.
Anisin, A. (2024). Authoritarian Liberal Surveillance and the COVID-19 Pandemic. Berlin, Boston, De Gruyter.
Annas, G. J., & Grodin, M. A. (1992). The Nazi Doctors and the Nuremberg Code : Human Rights in Human Experimentation. New York, Oxford University Press.
Anon. (1905). Geheime Figuren der Rosenkreuzer aus dem 16ten und 17 ten Jahrhundert'. Neue Metaphysische Rundschau, 8(12), 41–48, 92–48.
______. (2005). A New Syrian Edition of the Protocols of the Elders of Zion. Intelligence and Terrorism Information Center at the Center for Special Studies.
Appleby, J., Hunt, L., & Jacob, M. C. (1994). Telling The Truth About History. New York, Norton.
Arendt, H. (1944). Race-Thinking Before Racism. The Review of Politics, 6(1), 36–73.
______. (1973). The Origins of Totalitarianism (New edition with added prefaces. ed.). New York, New York : Harcourt Brace Jovanovich.
______. (2006). Eichmann in Jerusalem : A Report on the Banality of Evil. Penguin Publishing Group.
Aristotle. (1885). The Politics of Aristotle (B. Jowett, Trans.). Clarendon Press.
Aronson, E., & Mills, J. (1959). The effect of severity of initiation on liking for a group. The Journal of Abnormal and Social Psychology, 59(2), 177.
Arvidsson, S. (1999). Aryan Mythology As Science and Ideology. Journal of the American Academy of Religion, 67(2), 327–354.
Asch, S. E. (1951). Effects of Group Pressure Upon the Modification and Distortion of Judgments. In H. S. Guetzkow (Ed.), Groups, leadership and men ; research in human relations (pp. 177–190). Carnegie Press. (Original work published Asch, S. (1961). EFFECTS OF GROUP PRESSURE UPON THE MODIFICATION AND DISTORTION OF JUDGMENTS. In M. Henle (Ed.), Documents of Gestalt Psychology (pp. 222-236). Berkeley: University of California Press. https://doi.org/10.1525/9780520313514-017)
Assefa, Y., Gilks, C., F., van de Pas, R., Reid, S., Gete, D. G., & Damme, W. V. (2021). Reimagining global health systems for the 21st century : lessons from the COVID-19 pandemic. BMJ Global Health, 6(4), e004882. https://doi.org/10.1136/bmjgh-2020-004882
Assefa, Y., Gilks, C. F., Reid, S., van de Pas, R., Gete, D. G., & Van Damme, W. (2022). Analysis of the COVID-19 pandemic : lessons towards a more effective response to public health emergencies. Globalization and health, 18(1), 10. https://doi.org/10.1186/s12992-022-00805-9
Athanasopoulou, K., Adamopoulos, P. G., Tsiakanikas, P., & Scorilas, A. (2025). Exploring the m(5)C epitranscriptome of mRNAs in breast cancer cells through genome engineering and long-read sequencing approaches. Funct Integr Genomics, 25(1), 136. https://doi.org/10.1007/s10142-025-01648-4
Attrell, D. (2020). Honoring the Outermost : Saturn in Picatrix, Marsilio Ficino, and Renaissance Cosmology. Preternature: Critical and Historical Studies on the Preternatural, 9(2), 169–208. https://doi.org/10.5325/preternature.9.2.0169
Augstein, H. F. (1996). Race : The Origins of an Idea, 1760-1850 (Vol. 14). Bristol, England, Bristol, England : Thoemmes Press.
Aven, T., Cox, L. A., Flage, R., Guikema, S. D., & Haas, C. N. (2024). Special issue : Risk science foundations in light of COVID-19. Risk analysis, 44(12), 2759–2761. https://doi.org/10.1111/risa.17686

Ayçoberry, P. (1981). The Nazi Question : An Essay on The Interpretations of National Socialism (1922-1975) (R. Hurley, Trans.; First American edition. ed.). New York, New York : Pantheon Books.

Bacon, F. (1986). The Great Instauration ; And, New Atlantis (J. Weinberger, Ed.). Harlan Davidson.

Badon, I. W., Oh, Y., Kim, H.-J., & Lee, S. H. (2024). Recent application of CRISPR-Cas12 and OMEGA system for genome editing. Molecular Therapy, 32(1), 32–43. https://doi.org/https://doi.org/10.1016/j.ymthe.2023.11.013

Bahal, R., Gupta, A., & Glazer, P. M. (2016). Precise Genome Modification Using Triplex Forming Oligonucleotides and Peptide Nucleic Acids. In T. Cathomen, et al. (Eds.), Genome Editing (pp. 93–110). Springer New York. https://doi.org/10.1007/978-1-4939-3509-3_6

Bailey, A. A. (1971). Ponder on this : From the Writings of Alice A. Bailey and the Tibetan Master, Djwhal Khul. New York, NY, USA, Lucis Publishing Companies, 1971.

______. (1976). Discipleship in the New Age (Vol. I). New York, NY, USA, Lucis Publishing Company.

Bailey, A. A. (1995). Initiation and the Rays; To Serve. In The Seven Rays of Life (pp. 274–308; 309–347). Lucis Publishing Company.

Baird, J. W. (1990). To Die for Germany : Heroes in the Nazi Pantheon. Indiana University Press.

Baker, R. S. (1990). Brave New World : History, Science, and Dystopia. Twayne Publishers.

Baldwin, D. A. (1997). The Concept of Security. Review of International Studies, 23(1), 5–26.

Baldwin, J. M. (1911). Occult. In Dictionary of Philosophy and Psychology. https://books.google.com/books?id=c09AAQAAMAAJ

Balinska, M. A. (2000). The Rockefeller Foundation and the National Institute of Hygiene, Poland, 1918–45. Studies in History and Philosophy of Science Part C: Studies in History and Philosophy of Biological and Biomedical Sciences, 31(3), 419–432. https://doi.org/https://doi.org/10.1016/S1369-8486(00)00008-X

Bancel, N., David, T., & Thomas, D. (Eds.). (2014). The Invention of Race : Scientific and Popular Representations. Taylor & Francis Group.

Bandura, A. (1999a). Moral Disengagement. In I. W. Charny (Ed.), Encyclopaedia of Genocide (Vol. II: I–Y, pp. 415–418). Santa Barbara, Calif: ABC-CLIO.

______. (1999b). Moral Disengagement in the Perpetration of Inhumanities. Personality and Social Psychology Review, 3(3), 193–209. https://doi.org/10.1207/s15327957pspr0303_3

Banton, M. (1978). The Idea of Race. Boulder, Colorado, Westview Press.

______. (2015). What We Now Know about Race and Ethnicity. New York, NY, New York, NY: Berghahn Books.

Barbosa, A. I., Rebelo, R., Reis, R. L., Bhattacharya, M., & Correlo, V. M. (2020). Current nanotechnology advances in diagnostic biosensors. Medical Devices & Sensors, 4(1). https://doi.org/10.1002/mds3.10156

Barbour, I. G. (1997). Religion and Science : Historical and Contemporary Issues. HarperCollins.

Barkan, E. (1992). Retreat of Scientific Racism : Changing Concepts of Race in Britain and the United States Between the World Wars. Cambridge; New York,

Cambridge University Press.
Barlow, N. (1958). The Autobiography of Charles Darwin 1809-1882. Collins London.
Barnett, V. (1999). Bystanders : Conscience and Complicity During the Holocaust. Westport, Conn., Westport, Conn. : Greenwood Press.
Barrett, L. (2020). Ban Facial Recognition Technologies for Children—And for Everyone Else.
Bärsch, C.-E. (1998). Die politische Religion des Nationalsozialismus: Die religiöse Dimension der NS-Ideologie in den Schriften von Dietrich Eckart, Joseph Goebbels, Alfred Rosenberg und Adolf Hitler.
Bartov, O. (2000). Mirrors of Destruction : War, Genocide, and Modern Identity. Oxford; New York, Oxford University Press.
Barzun, J. (1938). Race : A Study in Superstition. London, Methuen.
_____. (1965). Race : A Study in Modern Superstition. Harper & Row.
Bauman, Z. (2000). Modernity, Racism, Extermination. In L. Back & J. Solomos (Eds.), Theories of Race and Racism : A Reader (Vol. Part 3: Racism and antisemitism, pp. 212–228). Routledge.
Baumeister, R. F. (1997). Evil : inside human cruelty and violence. New York, W.H. Freeman.
Baumeister, R. F., Bushman, B. J., & Campbell, W. K. (2000). Self-Esteem, Narcissism, and Aggression : Does Violence Result from Low Self-Esteem or from Threatened Egotism? Current Directions in Psychological Science, 9(1), 26–29.
Baumgarten, M., Kenez, P., & Thompson, B. A. (Eds.). (2009). Varieties of Antisemitism : History, Ideology, Discourse. University of Delaware Press.
Baumslag, N. (2005). Murderous Medicine : Nazi Doctors, Human Experimentation, and Typhus. Wesport, CT, Praeger Publishers.
Baumslag, N., & Shmookler, B. M. (2014). Typhus Epidemic Containment as Resistance to Nazi Genocide. In M. A. Grodin (Ed.), Jewish Medical Resistance in the Holocaust (pp. 39–48). Berghahn Books, Incorporated.
Baur, E., Fischer, E., & Lenz, F. (1921). Grundriss der menschlichen Erblichkeitslehre und *Rassenhygiene* (Vol. 1-2). Munich, Lehmann, 1921.
_____. (1931). Human Heredity (E. Paul & C. Paul, Trans.; Third ed.). The Macmillan Company.
Bayer, R., & Fairchild, A. (2010). When Worlds Collide : Health Surveillance, Privacy, and Public Policy. Social Research, 77(3), 905–928.
Baylis, F., Darnovsky, M., Hasson, K., & Krahn, T. M. (2020). Human Germline and Heritable Genome Editing : The Global Policy Landscape. The CRISPR Journal, 3(5), 365–377. https://doi.org/10.1089/crispr.2020.0082
Baynes, N. H. (Ed.). (1942). The Speeches of Adolf Hitler, April 1922-August 1939 (Vol. 1). Oxford University Press.
Beaumont, M. (2010). Socialism and Occultism at the "*Fin-de-siècle*" : Elective Affinities. Victorian Review, 36(1), 217–232.
Bebel, A. (1910). Woman and Socialism (M. L. Stern, Trans.; Jubilee 50th ed.). Socialist Literature Co.
Beers, B. C. v. (2020). Rewriting the human genome, rewriting human rights law? Human rights, human dignity, and human germline modification in the CRISPR era. Journal of Law and the Biosciences, 7(1). https://doi.org/10.1093/jlb/lsaa006
Bein, A. (1964). The Jewish Parasite : Notes on the Semantics of the Jewish Problem, with special Reference to Germany. The Leo Baeck Institute Yearbook, 9(1), 3–40.

https://doi.org/10.1093/leobaeck/9.1.3
Beinfeld, S. (2014). Health Care in the Vilna Ghetto. In Jewish Medical Resistance in the Holocaust. Berghahn Books, Incorporated.
Bejarano, L. G. (2012). Anti-Semitic Discourse or the Voice of a Disguised Converso in a Seventeenth-Century Spanish Treatise. In A. I. Aronson-Friedman & G. B. Kaplan (Eds.), Marginal Voices : Studies in Converso Literature of Medieval and Golden Age Spain (pp. 233–248). BRILL.
Beltran Gonzalez, J. L., González Gámez, M., Mendoza Enciso, E. A., Esparza Maldonado, R. J., Hernández Palacios, D., Dueñas Campos, S., Robles, I. O., Macías Guzmán, M. J., García Díaz, A. L., Gutiérrez Peña, C. M., Martinez Medina, L., Monroy Colin, V. A., & Arreola Guerra, J. M. (2022). Efficacy and Safety of Ivermectin and Hydroxychloroquine in Patients with Severe COVID-19 : A Randomized Controlled Trial. Infect Dis Rep, 14(2), 160–168. https://doi.org/10.3390/idr14020020
Benedict, R. (2000). Race : What it is not. In L. Back & J. Solomos (Eds.), Theories of Race and Racism : A Reader (Vol. Part 2: Sociology, Race and Social Theory, pp. 113–118). Routledge.
Bengtsson, L., & Rhinard, M. (2019). Securitisation across borders : the case of 'health security' cooperation in the European Union. West European Politics, 42(2), 346–368. https://doi.org/10.1080/01402382.2018.1510198
Benyera, E. (2021). Covid-19 Vaccine Nationalism and Vaccine Diplomacy : A New Currency in Soft Power? [Review]. Strategic Review for Southern Africa, 43, 193+.
Berenbaum, M. (2022). The Question of Relevance. In S. Gallin & I. Bedzow (Eds.), Bioethics and the Holocaust : A Comprehensive Study in How the Holocaust Continues to Shape the Ethics of Health, Medicine and Human Rights (Vol. 96, The International Library of Bioethics, pp. 1–11). Springer Nature; Cham : Springer International Publishing AG.
Bergen, B. J. (1998). The Banality of Evil : Hannah Arendt and "the final solution" (First ed.). Lanham, MD, The Rowman & Littlefield Publishing Group.
Berger, D. (1997). Anti-Semitism : An Overview. In D. Berger (Ed.), History and Hate : The Dimensions of Anti-Semitism (pp. 1–14). The Jewish Publication Society.
Bering, D. (2002). Gutachten über den antisemitischen Charakter einer namenpolemischen Passage aus der Rede Jörg Haiders, 28.2.2001. In A. Pelinka & R. Wodak (Eds.), "Dreck am Stecken" : Politik der Ausgrenzung (pp. 173–186). Czernin.
Bernard, L. (2024). Selling Natural Laws : The Structure and Function of Naturism in early 1950s France. Implicit Religion, 25.
Bernard, R. (1985). Les maisons secrètes de la Rose-Croix. Editions Rosicruciennes.
Beta, O. (1876). Darwin, Deutschland und die Juden oder Der Juda-Jesuitismus : dreiunddreißig Thesen nebst einer Nachschrift über einen vergessenen Factor der Volkswirthschaft. Selbstverl. des Verf.
Bethencourt, F. (2013). Racisms : From the Crusades to the Twentieth Century. Princeton, UNITED STATES, Princeton University Press.
Bettelheim, B. (1952). Remarks on the Psychological Appeal of Totalitarianism. The American Journal of Economics and Sociology, 12(1), 89–96.
Beusterien, J. L. (1999). Jewish Male Menstruation in Seventeenth-Century Spain. Bull Hist Med, 73(3), 447–456. https://doi.org/10.1353/bhm.1999.0097

Bevir, M. (1994). The West Turns Eastward : Madame Blavatsky and the Transformation of the Occult Tradition. Journal of the American Academy of Religion, 62(3), 747–767.

Beyerchen, A. (1977). Scientists under Hitler : Politics and the Physics Community in the Third Reich. New Haven, New Haven : Yale University Press.

Bialas, W. (2013). Nazi Ethics : Perpetrators with a Clear Conscience. Dapim : Studies on the Holocaust, 27(1), 3–25. https://doi.org/10.1080/23256249.2013.812821

______. (2014). Nazi Ethics And Morality : Ideas, Problems And Unanswered Questions. In W. Bialas & L. Fritze (Eds.), Nazi Ideology and Ethics (pp. 15–56). Cambridge Scholars Publishing.

Bialas, W., & Fritze, L. (Eds.). (2014). Nazi Ideology and Ethics. Cambridge Scholars Publishing.

Biddiss, M. (1997). History as Destiny : Gobineau, H. S. Chamberlain and Spengler. Transactions of the Royal Historical Society, 7, 73–100. https://doi.org/10.2307/3679271

Bierman, J. (2020). Science and Society in the New Atlantis and Other Renaissance Utopias. PMLA/Publications of the Modern Language Association of America, 78(5), 492–500. https://doi.org/10.2307/460726

Bigini, F., Lee, S. H., Sun, Y. J., Sun, Y., & Mahajan, V. B. (2023). Unleashing the potential of CRISPR multiplexing : Harnessing Cas12 and Cas13 for precise gene modulation in eye diseases. Vision Research, 213, 108317. https://doi.org/https://doi.org/10.1016/j.visres.2023.108317

Bildhauer, B. (2003). Blood, Jews, and Monsters in Medieval Culture. In B. Bildhauer & R. Mills (Eds.), The Monstrous Middle Ages (pp. 75–96). University of Toronto Press.

Bildhauer, B. (2020). Blood, Jews, and Monsters in Medieval Culture. In J. A. Weinstock (Ed.), The Monster Theory Reader (pp. 192–210). University of Minnesota Press.

Bill & Melinda Gates Foundation. (2020). Gates Foundation announces new funds to develop COVID-19 vaccines and increase access to affordable vaccines in low-income countries.

Billig, M. (1987). Anti-Semitic Themes and the British Far Left : Some Social-Psychological Observations on Indirect Aspects of the Conspiracy Tradition. In C. F. Graumann & S. Moscovici (Eds.), Changing Conceptions of Conspiracy (pp. 115–136). Springer New York. https://doi.org/10.1007/978-1-4612-4618-3_7

Binding, K., & Hoche, A. (1920). Die Freigabe der Vernichtung lebensunwerten Lebens : ihr Mass und ihre Form. F. Meiner.

Binding, K., & Hoche, A. (1975). The Release of the Destruction of Life Devoid of Value : It's Measure and It's Form.

Birn, A.-E. (2000). Wa(i)ves of influence : Rockefeller Public Health in Mexico, 1920–50. Studies in History and Philosophy of Science Part C: Studies in History and Philosophy of Biological and Biomedical Sciences, 31(3), 381–395. https://doi.org/https://doi.org/10.1016/S1369-8486(00)00009-1

Black, E. (2003). War Against the Weak : Eugenics and America's Campaign to Create a Master Race. New York, NY, USA, Four Walls Eight Windows.

______. (2012). War Against the Weak : Eugenics and America's Campaign to Create a Master Race. Washington, DC, USA, Dialog Press.

Black, M., & Kurlander, E. (Eds.). (2015). Revisiting the 'Nazi Occult' : Histories, Realities, Legacies [Book]. Camden House.

Black, S., & Evans, S. (2023). Serious adverse events following mRNA vaccination in randomized trials in adults. Vaccine, 41(23), 3473–3474. https://doi.org/10.1016/j.vaccine.2023.04.040

Blasimme, A., Ferretti, A., & Vayena, E. (2021). Digital Contact Tracing Against COVID-19 in Europe : Current Features and Ongoing Developments. Front Digit Health, 3, 660823. https://doi.org/10.3389/fdgth.2021.660823

Blavatsky, H. P. (1877). Isis Unveiled : A Master Key To The Mysteries Of Ancient And Modern Science And Theology (Vol. II: Theology). J. W. Bouton.

_____. (1888). The Secret Doctrine : The Synthesis of Science, Religion, and Philosophy (Vol. I: Cosmogenesis). Theosophical Publishing Company.

_____. (1893). The Secret Doctrine : The Synthesis of Science, Religion, and Philosophy (Vol. 2: Anthropogenesis). Theosophical Publishing Society.

_____. (1925). The Secret Doctrine (Facsimile of the original work published 1888) (Vol. II: Anthropogenesis). Theosophical Publishing Company.

Blaylock, R. L. (2022). COVID UPDATE : What is the truth? Surg Neurol Int, 13, 167. https://doi.org/10.25259/SNI_150_2022

Bledin, K. (2021). Further thoughts on "'If you prick me do I not bleed?" : Antisemitism, racism and group analysis - some thoughts.' A response to Bacha, Einhorn and Lieberman. Group Analysis, 54(4), 534–541. https://doi.org/10.1177/05333164211030460

Blumenbach, J. F. ([1865] (1978)). The Anthropological Treatises of Johann Friedrich Blumenbach. Massachusetts, Massachusetts: Longwood Press, 1978.

Blumenthal, K. G., Phadke, N. A., & Bates, D. W. (2021a). Safety Surveillance of COVID-19 mRNA Vaccines Through the Vaccine Safety Datalink. JAMA, 326(14). https://doi.org/10.1001/jama.2021.14808

Blumenthal, K. G., Robinson, L. B., Camargo, C. A., Shenoy, E. S., Banerji, A., Landman, A. B., & Wickner, P. (2021b). Acute Allergic Reactions to mRNA COVID-19 Vaccines. JAMA, 325(15). https://doi.org/10.1001/jama.2021.3976

Bock, G. (1983). Racism and Sexism in Nazi Germany : Motherhood, Compulsory Sterilization, and the State. Signs, 8(3), 400–421.

Bohr, N. (1963). Essays, 1958-1962, on Atomic Physics and Human Knowledge (Vol. 3). Interscience Publishers.

Bollyky, T. J., & Bown, C. P. (2020). The Tragedy of Vaccine Nationalism : Only Cooperation Can End the Pandemic. Foreign Affairs, 99(5), 96+.

Bölsche, W., & McCabe, J. (1906). Haeckel, His Life and Work. T. F. Unwin.

Bor, A., Jørgensen, F., & Petersen, M. B. (2023). Discriminatory attitudes against unvaccinated people during the pandemic. Nature, 613(7945), 704–711. https://doi.org/10.1038/s41586-022-05607-y

Borowski, T. (1976). [This Way for the Gas, Ladies and Gentlemen] (B. Vedder & M. Kandel Trans.). USA, Penguin Books, Inc.

Boudreaux, B., DeNardo, M., Denton, S. W., Sanchez, R., Feistel, K., & Dayalani, H. (2020). STRENGTHENING PRIVACY PROTECTIONS : in COVID-19 Mobile Phone-Enhanced Surveillance Programs. http://www.jstor.org/stable/resrep25384

Bracher, K. D. (1970). The German Dictatorship : The Origins, Structure, and Effects of National Socialism. Wadsworth.

_____. (1972). Stages of Totalitarian "Integration" (Gleich-schaltung) : The Consolidation of National So-cialist Rule in 1933 and 1934. In H. Holborn & R. Manheim (Eds.), Republic to Reich : The Making of the Nazi Revolution; Ten

Essays (pp. 109–128). Pantheon Books.

Bramwell, A. C. (1985). Blood and soil : Richard Walther Darré and Hitler's 'Green Party'.

Brandis, E. (1936). Die Ehegesetze von 1935. Berlin, Verlug Fur Standesamtswesen.

Breitman, R. (1991). The Architect of Genocide : Himmler and The Final Solution (First edition. ed.). New York, Knopf.

British Medical Journal. (1940). Medical News. The British Medical Journal, 2(4148), 36–37.

Broca, P. (1861). Sur le volume et la forme du cerveau suivant les individus et suivant les races. Bulletins de la Societe d'Anthropologie, 2, 139–207, 301–321, 441–446.

Broek, R. v. d., Hanegraaff, W. J., & NetLibrary, I. (1998). Gnosis and Hermeticism from Antiquity to Modern Times. Albany, State University of New York Press.

Brooke, J. H. (1991). Science and Religion : Some Historical Perspectives (G. Basalla, Ed.). Cambridge [England] ;, Cambridge University Press.

Browder, G. C. (1996). Hitler's Enforcers : The Gestapo and the SS security service in the Nazi Revolution. New York, Oxford University Press.

Brown, T. S. (1955). The Reliability of Megasthenes. The American Journal of Philology, 76(1), 18–33. https://doi.org/10.2307/291703

Browning, C. (2004). The Origins of The Final Solution : The Evolution of Nazi Jewish Policy, September 1939-March 1942. USA / Israel, University of Nebraska Press, Lincoln; Yad Vashem, Jerusalem.

Bruin, I. J. d., Spaander, M. M., Harmsen, S., Edelenbosch, R., Ploem, M. C., Dartée, N., Cardoso Vaz Santos, M., Lakshmipathi, M., Mulder, C. L., van Pelt, A. M. M., Baarends, W. M., Chuva de Sousa Lopes, S. M., de Wert, G., Segers, S., Hamer, G., & Pereira Daoud, A. M. (2025). Stem cell-derived gametes : what to expect when expecting their clinical introduction. Hum Reprod, 40(9), 1605–1615. https://doi.org/10.1093/humrep/deaf123

Bruns, C. (2011). Toward a Transnational History of Racism : Wilhelm Marr and the Interrelationships between Colonial Racism and German Anti-Semitism. In M. Berg & S. Wendt (Eds.), Racism in the Modern World : Historical Perspectives on Cultural Transfer and Adaptation (pp. 122–139). Berghahn Books, Incorporated.

Bruns, F. (2014). Turning away from the Individual : Medicine and Morality under the Nazis. In W. Bialas & L. Fritze (Eds.), Nazi Ideology and Ethics (pp. 211–236). Cambridge Scholars Publishing.

Bruns, F., & Chelouche, T. (2017). Lectures on Inhumanity : Teaching Medical Ethics in German Medical Schools Under Nazism. Ann Intern Med, 166(8), 591–595. https://doi.org/10.7326/M16-2758

Buchanan, J. A., & Higley, E. T. (1921). The relationship of blood-groups to disease. British journal of experimental pathology, 2(6), 247.

Bullock, A. (1991). Hitler : A Study in Tyranny (Abridged edition, first HarperPerennial edition. ed.). New York, NY, HarperPerennial.

Bunyan, T. (2010). Just over the horizon — the surveillance society and the state in the EU. Race & Class, 51(3), 1–12. https://doi.org/10.1177/0306396809354162

Burkhardt, R. W., Jr. (2013). Lamarck, evolution, and the inheritance of acquired characters. Genetics, 194(4), 793–805. https://doi.org/10.1534/genetics.113.151852

Burleigh, M. (1994). Death and Deliverance : 'Euthanasia' in Germany, C.1900 to 1945. Cambridge University Press.

______. (2002). The Legacy of Nazi Medicine in Context. In Medicine and Medical Ethics in Nazi Germany : Origins, Practices, Legacies (pp. 112–127). Berghahn Books, Incorporated.

Burleigh, M., & Wippermann, W. (1991). The Racial State : Germany, 1933—1945. Cambridge; New York, Cambridge University Press.

Bury, J. B. (1920). The Idea of Progress : An Inquiry Into Its Origin and Growth. Macmillan and Company, limited.

Butler, R. d. O. (1941). The Roots of National Socialism, 1783-1933. Faber & Faber.

Cadegiani, F. A., Goren, A., Wambier, C. G., & McCoy, J. (2021). Early COVID-19 therapy with azithromycin plus nitazoxanide, ivermectin or hydroxychloroquine in outpatient settings significantly improved COVID-19 outcomes compared to known outcomes in untreated patients. New Microbes and New Infections, 43, 100915. https://doi.org/https://doi.org/10.1016/j.nmni.2021.100915

Calisher, C., Carroll, D., Colwell, R., Corley, R. B., Daszak, P., Drosten, C., Enjuanes, L., Farrar, J., Field, H., Golding, J., Gorbalenya, A., Haagmans, B., Hughes, J. M., Karesh, W. B., Keusch, G. T., Lam, S. K., Lubroth, J., Mackenzie, J. S., Madoff, L.,...Turner, M. (2020). Statement in support of the scientists, public health professionals, and medical professionals of China combatting COVID-19. The Lancet, 395(10226), e42–e43. https://doi.org/10.1016/s0140-6736(20)30418-9

Call, L. (2016). Anti-Darwin, Anti-Spencer : Friedrich Nietzsche's Critique of Darwin and "Darwinism". History of Science, 36(1), 1–22. https://doi.org/10.1177/007327539803600101

Campbell, B. F. (1980). Ancient Wisdom Revived : A History of the Theosophical Movement. University of California Press.

Caplan, A. (1994). The Relevance of the Holocaust to Bioethics Today. In E. Medicine, and the Third Reich: Historical and Contemporary Issues (Ed.), (pp. 3–12). Sheed & Ward.

Caplan, J. (1988). Government Without Administration : State and Civil Service in Weimar and Nazi Germany. Oxford : New York, Oxford : Clarendon Press; New York : Oxford University Press.

Capra, F. (1982). The Turning Point : Science, Society, and the Rising Culture. New York, Simon and Schuster.

Carabelli, A. M., Peacock, T. P., Thorne, L. G., Harvey, W. T., Hughes, J., de Silva, T. I., Peacock, S. J., Barclay, W. S., de Silva, T. I., Towers, G. J., Robertson, D. L., & Consortium, C.-G. U. (2023). SARS-CoV-2 variant biology : immune escape, transmission and fitness. Nature Reviews Microbiology, 21(3), 162–177. https://doi.org/10.1038/s41579-022-00841-7

Carli, G., Nichele, I., Ruggeri, M., Barra, S., & Tosetto, A. (2021). Deep vein thrombosis (DVT) occurring shortly after the second dose of mRNA SARS-CoV-2 vaccine. Internal and Emergency Medicine, 16(3), 803–804. https://doi.org/10.1007/s11739-021-02685-0

Carroll, D., Daszak, P., Wolfe, N. D., Gao, G. F., Morel, C. M., Morzaria, S., Pablos-Méndez, A., Tomori, O., & Mazet, J. A. K. (2018). The Global Virome Project. Science, 359(6378), 872–874.

Cassirer, E. (1955). The Myth of the State. Doubleday.

Castillo-Salgado, C. (2010). Trends and Directions of Global Public Health Surveillance. Epidemiologic Reviews, 32(1), 93–109. https://doi.org/10.1093/epirev/mxq008

Caulfield, T. (2020). Pseudoscience and COVID-19 — we've had enough already. Nature.

https://doi.org/10.1038/d41586-020-01266-z

Cavendish, R. (1983). Incantation. In Man, Myth & Magic : The Illustrated Encyclopaedia of Mythology, Religion and the Unknown. https://archive.org/details/man-myth-magic-illustrated-encyclopaedia-vol-1-11-and-index-editor-richard-cavendish-1983/page/1417/mode/2up

Cecil, R. (1972). The Myth of the Master Race : Alfred Rosenberg and Nazi Ideology. New York, New York : Dodd, Mead.

Cerneckis, J., Ming, G. L., Song, H., He, C., & Shi, Y. (2024). The rise of epitranscriptomics : recent developments and future directions. Trends Pharmacol Sci, 45(1), 24–38. https://doi.org/10.1016/j.tips.2023.11.002

Chamberlain, H. S. (1899). Die Grundlagen des neunzehnten Jahrhunderts (Vol. 2). Bruckmann.

______. (1905). Arische Weltanschauung. Bruckmann.

______. (1912). Foundations of the Nineteenth Century (J. Lees, Trans.; Vol. 2). J. Lane Company.

______. (1968). Foundations of the Nineteenth Century (J. Lees, Trans.; Vol. 1). New York, Howard Fertig.

Chase, A. (1980). The Legacy of Malthus : The Social Costs of the New Scientific Racism. University of Illinois Press.

Chazan, R. (1980). Church, State, and Jew in the Middle Ages. Behrman House.

______. (1997a). Medieval Antisemtism. In D. Berger (Ed.), History and Hate : The Dimensions of Anti-Semitism (pp. 49–65). The Jewish Publication Society.

______. (1997b). Medieval Stereotypes and Modern Antisemitism. Berkeley, University of California Press.

Chelouche, T. (2022). Teaching Medical Ethics in Nazi Germany : Debunking the Myth that the Nazi Physicians Abandoned Their Ethics. In S. Gallin & I. Bedzow (Eds.), Bioethics and the Holocaust : A Comprehensive Study in How the Holocaust Continues to Shape the Ethics of Health, Medicine and Human Rights (Vol. 96, The International Library of Bioethics, pp. 13–34). Springer Nature; Cham : Springer International Publishing AG.

Chen, P. J., & Liu, D. R. (2023). Prime editing for precise and highly versatile genome manipulation. Nature Reviews Genetics, 24(3), 161–177. https://doi.org/10.1038/s41576-022-00541-1

Chen, R., Rabinovitch, P. S., Crispin, D. A., Emond, M. J., Koprowicz, K. M., Bronner, M. P., & Brentnall, T. A. (2003). DNA fingerprinting abnormalities can distinguish ulcerative colitis patients with dysplasia and cancer from those who are dysplasia/cancer-free. The American journal of pathology, 162 2, 665–672.

Childers, T. (1983). The Nazi Voter : The Social Foundations of Fascism in Germany, 1919-1933. University of North Carolina Press.

Choudhary, R., & Sharma, A. K. (2020). Potential use of hydroxychloroquine, ivermectin and azithromycin drugs in fighting COVID-19 : trends, scope and relevance. New Microbes New Infect, 35, 100684. https://doi.org/10.1016/j.nmni.2020.100684

Choudhary, V., & Kapoor, A. (2013). Role of Biometrics in Anthropological Settings. International Research Journal of Social Sciences, 2(8), 55–57.

Choudhury, A., Das, N. C., Patra, R., Bhattacharya, M., Ghosh, P., Patra, B. C., & Mukherjee, S. (2021). Exploring the Binding Efficacy of Ivermectin Against the Key Proteins of SARS-CoV-2 Pathogenesis : an in silico Approach. Future

Virology, 16(4), 277–291. https://doi.org/10.2217/fvl-2020-0342
Christiansen, M. G., Senko, A. W., & Anikeeva, P. (2019). Magnetic Strategies for Nervous System Control. Annu Rev Neurosci, 42, 271–293. https://doi.org/10.1146/annurev-neuro-070918-050241
Chrystal, W. G. (1975). Nazi Party Election Films, 1927-1938. Cinema Journal, 15(1), 29–47. https://doi.org/10.2307/1225103
Chumtong, J. (2020). Phone tracking against COVID-19 Using digital technology against an analogue virus. http://www.jstor.org/stable/resrep25296
Churton, T. (2003). The Golden Builders : Alchemists, Rosicrucians and the First Free Masons (Second ed.). Signal Publishing.
Cirlot, J. E. (1971). Swastika. In A Dictionary of Symbols. https://archive.org/details/j.-e.-cirlot-dictionary-of-symbols/page/n3/mode/2up
Claeys, G. (2000). The "Survival of the Fittest" and the Origins of Social Darwinism. Journal of the History of Ideas, 61(2), 223–240. https://doi.org/10.2307/3654026
Clarke, I. F. (1972). 4. Bacon's New Atlantis : Blueprint for progress. Futures, 4(3), 273–279. https://doi.org/10.1016/0016-3287(72)90072-9
Claudy, M. C., Vijayakumar, S., & Campbell, N. (2022). Reckless spreader or blameless victim? How vaccination status affects responses to COVID-19 patients. Soc Sci Med, 305, 115089. https://doi.org/10.1016/j.socscimed.2022.115089
Cloney, R. (2019). Cooperating on synthetic gene circuits. Nature Biotechnology, 37(7), 729–729. https://doi.org/10.1038/s41587-019-0182-3
Cocks, G. C. (2012). The State of Health : Illness in Nazi Germany (First ed.). Oxford, Oxford University Press.
Cohen, I. G., & Adashi, E. Y. (2025). Ethical and legal implications of *in vitro* gametogenesis and germline editing—current status. Fertility and Sterility, 124(1), 30–36. https://doi.org/https://doi.org/10.1016/j.fertnstert.2025.02.024
Cohen, J. (1982). The Friars and the Jews : The Evolution of Medieval Anti-Judaism. Ithaca, Ithaca : Cornell University Press.
Cohen, N. W. (1984). The Challenges of Darwinism and Biblical Criticism to American Judaism. Modern Judaism, 4(No. 2), 121–157.
Cohn, N. (1967). Warrant for Genocide : The Myth of the Jewish World-conspiracy and the Protocols of the Elders of Zion (Reprint ed.). Harper & Row.
______. (1981). Warrant for Genocide : The Myth of the Jewish World-conspiracy and the Protocols of the Elders of Zion (J. Neusner, et al., Eds. Third ed.). Chico, CA, Scholars Press.
______. (1993). Europe's Inner Demons : The Demonization of Christians in Medieval Christendom (Revised ed.). Pimlico.
Cohn, S. K. (2007). The Black Death and the Burning of Jews. Past & Present(196), 3–36.
Coleridge, S. T. (2002). Opus Maximum : The Collected Works of Samuel Taylor Coleridge (Vol. 15 of Bollingen Series (General) Series). Princeton, NJ, USA, Princeton University Press.
Congressional Record. (2024). Concepts that are inconvenient to the country. Retrieved from https://www.govinfo.gov/content/pkg/CREC-2024-11-21/html/CREC-2024-11-21-pt1-PgH6177.htm
Cook, J. D., Harwood, P., Pollock, W. H., Harris, F., & Hodge, H. (Eds.). (1890). The Origin of the Aryans (Vol. 69). J. W. Parker and Son.
Corballis, M. C. (2014). Left brain, right brain : facts and fantasies. PLoS Biol, 12(1), e1001767. https://doi.org/10.1371/journal.pbio.1001767

Corbyn, Z. (2015). Crispr : Is it a good idea to 'upgrade' our DNA? http://www.theguardian.com/science/2015/may/10/crispr-genome-editing-dna-upgrade-technology-genetic-disease

Cornwell, J. (2003). Hitler's scientists : science, war, and the devil's pact. New York, N.Y, Viking.

Coustasse, A., Kimble, C., & Maxik, K. (2021). COVID-19 and Vaccine Hesitancy : A Challenge the United States Must Overcome. The Journal of Ambulatory Care Management, 44(1), 71–75. https://doi.org/10.1097/jac.0000000000000360

Covert, M., & Press, C. R. C. (2015). Fundamentals of systems biology : from synthetic circuits to whole-cell models (First ed.). Boca Raton, FL, CRC Press.

Craig, G. A. (1978). Germany, 1866-1945. New York, New York : Oxford University Press.

_____. (1982). The Germans. New York, Putnam.

Craig, T. L. (2010). On the Significance of the Literary Character of Francis Bacon's "New Atlantis" for an Understanding of His Political Thought. The Review of Politics, 72(2), 213–239.

Cranium. (1909). The Brain and Skull. The Phrenological Journal and Science of Health, 122(7), 206–216.

Crelinsten, R. D. (1995). In Their Own Words : The World of the Torturer. In R. D. Crelinsten & A. P. Schmid (Eds.), The Politics of Pain : Torturers and Their Masters (pp. 35–64). Westview Press.

Crelinsten, R. D., & Schmid, A. P. (1995). Introduction : The Politics of Pain. In R. D. Crelinsten & A. P. Schmid (Eds.), The Politics of Pain : Torturers and Their Masters (pp. 1–11). Westview Press.

Crossman, R. H. S. (1958). The Creed of a Modern Propagandist. In D. W.E. & M. Janowitz (Eds.), A Psychological Warfare Casebook (pp. 35–47). Johns Hopkins Press.

Csernatoni, R. (2020). New states of emergency : normalizing techno-surveillance in the time of COVID-19. Global Affairs, 6(3), 301–310. https://doi.org/10.1080/23340460.2020.1825108

Culver, R. B., & Ianna, P. A. (1984). The Gemini Syndrome : A Scientific Evaluation of Astrology. Prometheus Books.

d'Alviella, E. G. (1894). The Migration of Symbols. Archibald Constable and Company.

Daim, W. (1958). Der Mann : der Hitler die Ideen gab; von den religiösen Verirrungen eines Sektierers zum Rassenwahn des Diktators. [Aus dem Institut für Politische Psychologie Wien. Munich, Germany,

Dalton, J. (1808). A New System of Chemical Philosophy. By John Dalton. [Part I-II]. S. Russel, 125 Deansgate.

Dalton, T., Ed. & Trans. (2019). Goebbels on the Jews : The Complete Diary Entries - 1923 To 1945. Castle Hill Publishers.

Dams, C., & Stolle, M. (2014). The Gestapo : Power and Terror in the Third Reich (C. Ryland, Trans.). OUP Oxford.

Danchin, M., & Buttery, J. (2021). COVID-19 vaccine hesitancy : a unique set of challenges. Internal Medicine Journal, 51(12), 1987–1989. https://doi.org/10.1111/imj.15599

Darley, J. M. (1992). Social Organization for the Production of Evil [Crimes of Obedience: Towards a Social Psychology of Authority and Responsibility, Herbert C. Kelman, V. Lee Hamilton; The Nazi Doctors: Medical Killing and the Psychology of Genocide, Robert Jay Lifton; The Roots of Evil: The Origins of Genocide and Other Group Violence, Ervin Staub]. Psychological Inquiry, 3(2), 199–218.

Darouza, M. I. (1992). Mengungkap tentang Yahudi : watak, jejak, pijak dari kasus-kasus lama Bani Israel. Pustaka Progresif.

Darwin, C. (1876). The Descent of Man, and Selection in Relation to Sex (Second: Revised and Augmented ed.). D. Appleton and Company.

Daugherty, W. E., & Janowits, M. (1958). A Psychological Warfare Casebook. Md., Johns Hopkins Press.

David, E. (1909). Darwinismus und soziale Entwicklung. In W. Boelsche, et al. (Eds.), Darwin, seine Bedeutung im Ringen um Weltanschauung und Lebenswert: 6 Aufsätze (Vol. 4, pp. 45–65). Hilfe.

Davies, A. (1980). Racism and German Protestant Theology : A Prelude to the Holocaust. The Annals of the American Academy of Political and Social Science, 450, 20–34.

Davies, J. D. (1971). Phrenology : Fad and Science. A 19th-Century American Crusade. Hamden, Conn.], [Hamden, Conn.] : Archon Books.

De Camp, L. S. (1970). Lost continents : The Atlantis Theme in History, Science, and Literature. New York, Dover Publications.

De Jonge, A. (1979). The Weimar Chronicle : Prelude to Hitler. New York, New American Library.

DeJong, W. (1979). An examination of self-perception mediation of the foot-in-the-door effect. Journal of personality and social psychology, 37(12), 2221–2239. https://doi.org/10.1037/0022-3514.37.12.2221

Demirtaş-Madran, H. A. (2020). Exploring the Motivation Behind Discrimination and Stigmatization Related to COVID-19 : A Social Psychological Discussion Based on the Main Theoretical Explanations [Review]. Frontiers in Psychology, 11. https://doi.org/10.3389/fpsyg.2020.569528

Denham, H., Abutaleb, Y., & Rowland, C. (2020). Gilead sets price of coronavirus drug remdesivir at $3,120 as Trump administration secures supply for 500,000 patients. The Washington Post. Retrieved Dec 18. 2024 from

Derwand, R., Scholz, M., & Zelenko, V. (2020). COVID-19 outpatients : early risk-stratified treatment with zinc plus low-dose hydroxychloroquine and azithromycin : a retrospective case series study. International journal of antimicrobial agents, 56(6), 106214–106214. https://doi.org/10.1016/j.ijantimicag.2020.106214

Desmond, A. J. (1997). Huxley : Evolution's High Priest (Vol. 2). Michael Joseph.

Dethloff, C. (2011). Coming Up to Code : Ancient Divinities Revisited. In G. Kovacs & C. W. Marshall (Eds.), Classics and Comics (pp. 103–114). Oxford University Press, Incorporated.

Deuerlein, E. (1959). Hitlers Eintritt in die Politik und die Reichswehr. Vierteljahrshefte für Zeitgeschichte, 7(2), 177–227.

Dias, L., Soares-dos-Reis, R., Meira, J., Ferrão, D., Soares, P. R., Pastor, A., Gama, G., Fonseca, L., Fagundes, V., & Carvalho, M. (2021). Cerebral Venous Thrombosis after BNT162b2 mRNA SARS-CoV-2 vaccine. Journal of Stroke and Cerebrovascular Diseases, 30(8). https://doi.org/10.1016/j.jstrokecerebrovasdis.2021.105906

Diaz, G. A., Parsons, G. T., Gering, S. K., Meier, A. R., Hutchinson, I. V., & Robicsek, A.

(2021). Myocarditis and Pericarditis After Vaccination for COVID-19. JAMA, 326(12). https://doi.org/10.1001/jama.2021.13443

Diaz-Arévalo, D., & Zeng, M. (2020). Nanoparticle-based vaccines. In Nanopharmaceuticals (pp. 135–150). https://doi.org/10.1016/b978-0-12-817778-5.00007-5

Dicks, H. V. (1972). Licensed Mass Murder : A Socio-Psychological Study of Some SS killers. New York.

Dinter, A. (1920). Die Sünde wider das Blut : ein Zeitroman (Vol. 1). Matthes und Thost.

Dixon, N. F. (1976). On the psychology of military incompetence. New York, Basic Books.

Dobbs, B. J. T. (1975). The Foundations of Newton's Alchemy or "The Hunting of the Greene Lyon". Cambridge University Press.

Dodd, R. H., Pickles, K., Nickel, B., Cvejic, E., Ayre, J., Batcup, C., Bonner, C., Copp, T., Cornell, S., Dakin, T., Isautier, J., & McCaffery, K. J. (2021). Concerns and motivations about COVID-19 vaccination. The Lancet Infectious Diseases, 21(2), 161–163. https://doi.org/10.1016/s1473-3099(20)30926-9

Doerfler, W. (2021). Adenoviral Vector DNA- and SARS-CoV-2 mRNA-Based Covid-19 Vaccines : Possible Integration into the Human Genome - Are Adenoviral Genes Expressed in Vector-based Vaccines? Virus Research, 302, 198466. https://doi.org/https://doi.org/10.1016/j.virusres.2021.198466

Döllinger, F. (1920). Baldur und Bibel : Weltbewegende neue Enthüllungen über die Bibel : Germanische Kultur im biblischen Kanaan und Germanisches Christentum vor Christus. Lorenz Spindler Verlag.

Domarus, M. (1968). Der Reichstag und die Macht. Würzburg,

Donnelly, I. (1882). Atlantis : The Antediluvian World (Eleventh ed.). New York, Harper & Bros.

Doob, L. W. (1978). Panorama of Evil : Insights From the Behavioral Sciences. Westport, Conn., Greenwood Press.

Doshi, P. (2009). Calibrated response to emerging infections. BMJ : British Medical Journal (Online), 339. https://doi.org/https://doi.org/10.1136/bmj.b3471

______. (2011). The elusive definition of pandemic influenza. Bull World Health Organ, 89(7), 532–538. https://doi.org/10.2471/blt.11.086173

Douglas, B. (2005). Notes on 'Race' and the Biologisation of Human Difference. The Journal of Pacific History, 40(3), 331–338.

Douglas, S. M., Bachelet, I., & Church, G. M. (2012). A logic-gated nanorobot for targeted transport of molecular payloads. Science, 335(6070), 831–834. https://doi.org/10.1126/science.1214081

Drążkowski, D., & Trepanowski, R. (2024). "It's Your Fault!"—Said the Vaccinated to the Unvaccinated. The Effect of COVID-19 Vaccination on Responsibility Attribution. Sage Open, 14(2), 21582440241251471. https://doi.org/10.1177/21582440241251471

Dubé, E., Laberge, C., Guay, M., Bramadat, P., Roy, R., & Bettinger, J. A. (2014). Vaccine hesitancy. Human vaccines & immunotherapeutics, 9(8), 1763–1773. https://doi.org/10.4161/hv.24657

Dunbavin, P. (2003). Atlantis of the West : A Drowned Neolithic Civilisation. Carol & Craf Publishers.

Dundes, A., & Hauschild, T. (1983). Auschwitz Jokes. Western Folklore, 42(4), 249–260. https://doi.org/10.2307/1499500

Dunér, I. (2025). A Razor for the Divine : Constructing a New Religion. In Julian Huxley, Evolutionism and the History of Transhumanism (pp. 61–101). Springer Nature Switzerland. https://doi.org/10.1007/978-3-031-81720-5_3

Dunlap, K. (1944). The Great Aryan Myth. The Scientific Monthly, 59(4), 296–300.

Eberhard, P. H., & Ross, R. R. (1989). Quantum field theory cannot provide faster-than-light communication. Foundations of Physics Letters, 2(2), 127–149. https://doi.org/10.1007/BF00696109

Edighoffer, R. (2005). Rosicrucianism II : 18th Century. In Dictionary of Gnosis & Western Esotericism.

Ehrenreich, E. (2007). Nazi Ancestral Proof : Genealogy, Racial Science, and the Final Solution. Bloomington, UNITED STATES, Indiana University Press.

Eisner, L. H. (2008). The Haunted Screen ; Expressionism in the German Cinema and the Influence of Max Reinhardt. Berkeley, University of California Press.

Ellis, R. (1998). Imagining Atlantis (First edition. ed.). New York, Alfred A. Knopf.

Eltringham, N. (2017). “Nordics” and “Hamites” : Joseph Deniker and the Rise (and Fall) of Scientific Racism. In A. Morris-Reich & D. Rupnow (Eds.), Ideas of ‘Race’ in the History of the Humanities (pp. 251–272). Springer International Publishing. https://doi.org/10.1007/978-3-319-49953-6_10

European Medicines Agency (EMA). (2021). EMA finds possible links to blood clots with COVID-19 vaccines. Medical Laboratory Observer, 53(6), 7.

Endelman, T. M. (1997). Comparative Perspectives on Modern Anti-Semitism in the West. In D. Berger (Ed.), History and Hate : The Dimensions of Anti-Semitism (pp. 95–114). The Jewish Publication Society.

Engelsing, R. (1973). Analphabetentum und Lektüre. Germany: Stuttgart, Metzler.

Enserink, M. (2007). Indonesia Earns Flu Accord at World Health Assembly. Science, 316(5828), 1108–1108.

Ericksen, R. P. (1985). Theologians Under Hitler. USA, Yale University Press.

Essner, C. (2002). Die “Nürnberger Gesetze,” oder, Die Verwaltung des Rassenwahns 1933-1945. Paderborn: Ferdinand Schoningh.

Ettinger, S. (2011). The Origins of Modern Anti-Semitism. In M. R. Marrus (Ed.), The Nazi Holocaust (Vol. 2 The Origins of the Holocaust), Part Two : Anti-Semitic Background, pp. 208–244). De Gruyter. (Original work published In The Catastrophe of European Jewry : Antecedents--History-Reflections; Selected Papers, Yad Vashem, Jerusalem, 1976)

Evans, R. J. (2005). The Third Reich in power, 1933-1939. New York, Penguin Press.

Evola, J. (1995). Revolt against the modern world (G. Stucco, Trans.). Rochester, Vt, Inner Traditions International.

Fabre d’Olivet, A. (1915). Hermeneutic Interpretation of the Origin of the Social State of Man (N. L. Redfield, Trans.). New York; London, G. P. Putnam’s Sons.

Fairchild, A. L., Bayer, R., & Colgrove, J. (2007). Searching Eyes : Privacy, the State, and Disease Surveillance in America. Berkeley, UNITED STATES, University of California Press.

Faivre, A. (2010). Western Esotericism : A Concise History (C. Rhone, Trans.). Albany, UNITED STATES, State University of New York Press.

Faksova, K., Walsh, D., Jiang, Y., Griffin, J., Phillips, A., Gentile, A., Kwong, J. C., Macartney, K., Naus, M., Grange, Z., Escolano, S., Sepulveda, G., Shetty, A., Pillsbury, A., Sullivan, C., Naveed, Z., Janjua, N. Z., Giglio, N., Perälä, J.,… Hviid, A. (2024). COVID-19 vaccines and adverse events of special interest : A multinational Global Vaccine Data Network (GVDN) cohort study of 99 million vaccinated individuals. Vaccine, 42(9), 2200–2211. https://doi.org/https://doi.org/10.1016/j.vaccine.2024.01.100

Farmer, P. (1999). Infections and inequalities : the modern plagues ([Updated edition with a new preface]. ed.). Berkeley, University of California Press.

Fazio, L. K., Rand, D. G., & Pennycook, G. (2019). Repetition increases perceived truth equally for plausible and implausible statements. Psychon Bull Rev, 26(5), 1705–1710. https://doi.org/10.3758/s13423-019-01651-4

Feder, K. L. (2010). Encyclopaedia of Dubious Archaeology : From Atlantis to the Walam Olum. Greenwood.

Fein, H. (1979). Accounting for genocide : National Responses and Jewish Victimization during the Holocaust. New York, Free Press.

______. (1993). Accounting for genocide after 1945 : Theories and some findings. International Journal on Group Rights, 1(2), 79–106.

Feldbaum, H., Lee, K., & Michaud, J. (2010). Global Health and Foreign Policy. Epidemiologic Reviews, 32(1), 82–92. https://doi.org/10.1093/epirev/mxq006

Feldman, L. H. (1997). Anti-Semitism in the Ancient World. In D. Berger (Ed.), History and Hate : The Dimensions of Anti-Semitism (pp. 15–42). The Jewish Publication Society.

Feldscher, W. (1943). Rassen-und Erbpflege im deutschen Recht. Deutscher Rechtsverlag.

Ferranna, M. (2023). Causes and costs of global COVID-19 vaccine inequity. Seminars in Immunopathology, 45(4-6), 469–480. https://doi.org/10.1007/s00281-023-00998-0

Ferretti, L., Wymant, C., Kendall, M., Zhao, L., Nurtay, A., Abeler-Dörner, L., Parker, M., Bonsall, D., & Fraser, C. (2020). Quantifying SARS-CoV-2 transmission suggests epidemic control with digital contact tracing. Science, 368(6491), eabb6936. https://doi.org/doi:10.1126/science.abb6936

Fest, J. C. (1970). The Face of the Third Reich (M. Bullock, Trans.). London, Weidenfeld & Nicolson.

______. (1974). Hitler (R. Winston & C. Winston, Trans.; [First edition]. ed.). New York, Harcourt Brace Jovanovich.

Festinger, L., & Carlsmith, J. M. (1959). Cognitive consequences of forced compliance. The Journal of Abnormal and Social Psychology, 58(2), 203–210. https://doi.org/10.1037/h0041593

Ficino, M. (1980). The Book of Life (C. Boer, Trans.). Spring Publications.

______. (2001). Platonic Theology (J. Hankins, et al., Eds. Vol. 3: bks. 9-11). Harvard University Press.

Fidler, D. P. (2004). Constitutional outlines of public health's "new world order". Temple law review, 77(2), 247–289.

Fidler, D. P. (2005). Health as foreign policy : between principle and power. Whitehead J. Dipl. & Int'l Rel., 6, 179.

Fidler, D. P. (2010). Negotiating equitable access to influenza vaccines : global health diplomacy and the controversies surrounding avian influenza H5N1 and pandemic influenza H1N1 [Report]. PLoS Medicine, 7, NA.

______. (2020). Vaccine nationalism's politics. Science, 369(6505), 749. https://doi.org/10.1126/science.abe2275

Fidler, D. P., & Gostin, L. O. (2006). The New International Health Regulations : An

Historic Development for International Law and Public Health. The Journal of law, medicine & ethics, 34(1), 85–94. https://doi.org/10.1111/j.1748-720X.2006.00011.x

Field, G. G. (1981). Evangelist of Race : The Germanic Vision of Houston Stewart Chamberlain. New York, New York : Columbia University Press.

Filsinger, M., & Freitag, M. (2024). Asymmetric affective polarization regarding COVID-19 vaccination in six European countries. Sci Rep, 14(1), 15919. https://doi.org/10.1038/s41598-024-66756-w

Fischer, E. (1913). Die Rehobother Bastards und das Bastardierungsproblem beim Menschen. Akademische Verlagsanstalt, 1961.

Fischer, E., & Pearson, K. (1912). Zur Frage der Kreuzungen beim Menschen. Druck und Verlag von B.G. Teubner.

Fischer, F. (1990). Technocracy and the Politics of Expertise. SAGE Publications.

Fishberg, M. (1911). The Jews : A Study of Race and Environment. Scribner's.

Fisher, P. S. (1991). Fantasy and Politics : Visions of the Future in the Weimar Republic. Madison, Wis, University of Wisconsin Press.

Flowers, S. E. (2022). The Occult in National Socialism : The Symbolic, Scientific, and Magical Influences on the Third Reich. Simon and Schuster.

Forman, R., Atun, R., McKee, M., & Mossialos, E. (2020). 12 Lessons learned from the management of the coronavirus pandemic. Health Policy, 124(6), 577–580. https://doi.org/https://doi.org/10.1016/j.healthpol.2020.05.008

Forster, T., & Heinzel, M. (2021). Reacting, fast and slow : how world leaders shaped government responses to the COVID-19 pandemic. Journal of European Public Policy, 28(8), 1299–1320. https://doi.org/10.1080/13501763.2021.1942157

Forti, S. (2006). The Biopolitics of Souls : Racism, Nazism, and Plato. Political Theory, 34(1), 9–32.

Fortier, P. A. (1967). Gobineau and German Racism. Comparative Literature, 19(4), 341–350. https://doi.org/10.2307/1769493

Fosdick, R. B. (1952). The story of the Rockefeller Foundation ([First edition]. ed.). New York, Harper.

Foucault, M. (1978). The History of Sexuality : An introduction (Vol. I). Pantheon Books.

______. (1980). Power/knowledge : selected interviews and other writings, 1972-1977 (C. Gordon, et al., Trans.). In C. Gordon (Ed.), Selected interviews and other writings, 1972-1977. ([First American edition]. ed.). Pantheon Books.

______. (1991). Governmentality. In G. Burchell, et al. (Eds.), The Foucault Effect : Studies in Governmentality : with two lectures by and an interview with Michel Foucault (pp. 87–104). University of Chicago Press.

Fowden, G. (1993). The Egyptian Hermes : A Historical Approach to the Late Pagan Mind. Princeton University Press.

Fox, C. (1995). Introduction : How to Prepare a Noble Savage : The Spectacle of Human Science. In C. Fox, et al. (Eds.), Inventing Human Science : Eighteenth-Century Domains (pp. 1–30). Berkeley : University of California Press.

Fraiman, J., Erviti, J., Jones, M., Greenland, S., Whelan, P., Kaplan, R. M., & Doshi, P. (2022). Serious adverse events of special interest following mRNA COVID-19 vaccination in randomized trials in adults. Vaccine, 40(40), 5798–5805. https://doi.org/10.1016/j.vaccine.2022.08.036

Franck, M. J. (2013). Brave "New World", Plato's "Republic", and Our Scientific Regime.

The New Atlantis(40), 73–88.

François, S. (2023). Nazi Occultism : Between the SS and Esotericism (E. Uskalis, Trans.). Taylor & Francis.

Fredrickson, G. M., & Camarillo, A. (2002). Racism : A Short History. Princeton, USA, Princeton University Press.

_____. (2015). Racism : A Short History. Princeton, USA, Princeton University Press.

Freeden, M. (1979). Eugenics and progressive thought : a study in ideological affinity. Hist J, 22(3), 645–672. https://doi.org/10.1017/s0018246x00017027

Freedman, J. L., & Fraser, S. C. (1966). Compliance Without Pressure : The foot-in-the-door technique. Journal of personality and social psychology, 4(2), 195–202. https://doi.org/10.1037/h0023552

Fried, L. P., Bentley, M. E., Buekens, P., Burke, D. S., Frenk, J. J., Klag, M. J., & Spencer, H. C. (2010). Global health is public health. The Lancet (British edition), 375(9714), 535–537. https://doi.org/10.1016/S0140-6736(10)60203-6

Friedlander, H. (1995). Origins of Nazi Genocide : From Euthanasia to the Final Solution. Chapel Hill, UNITED STATES, The University of North Carolina Press.

_____. (2002). Physicians as Killers in Nazi Germany : Hadamar, Treblinka, and Auschwitz. In Medicine and Medical Ethics in Nazi Germany : Origins, Practices, Legacies (pp. 59–76). Berghahn Books, Incorporated.

Friedman, J. (1987). Jewish Conversion, the Spanish Pure Blood Laws and Reformation : A Revisionist View of Racial and Religious Antisemitism. Sixteenth Century Journal, 18(1), 3–30. https://doi.org/10.2307/2540627

Friedman, L. J., & Chase, A. (1977). The Legacy of Malthus : The Social Costs of the New Scientific Racism [The Legacy of Malthus: The Social Costs of the New Scientific Racism, Allan Chase]. The American Historical Review, 82(4), 923–923. https://doi.org/10.2307/1865122

Friedman, S. D. (November 25, 2008). Recessions : Do Not Waste This Crisis. Harvard Business Review.

Friend, J. M. (2014). The Biopolitics of Policy-Making : Medical Activism in the Century of the Brain. New Political Science, 36(1), 32–51. https://doi.org/10.1080/07393148.2013.859902

Fritsch, T. (1893). Antisemiten-Katechismus : eine Zusammenstellung des wichtigsten Materials zum Verständniss der Judenfrage. Fritsch.

Fromm, E. (1969). Escape from Freedom (Eighteenth reprinting ed.). Avon Books.

Fuchs, B. (2008). German Anthropology and Nazi GenocideFrom Racism to Genocide : Anthropology in the Third Reich. By Gretchen E. Schafft. Chicago: University of Illinois Press, 2004 [From Racism to Genocide: Anthropology in the Third Reich]. Current Anthropology, 49(2), 345–346. https://doi.org/10.1086/525859

Funnell, M. G., Corballis, P. M., & Gazzaniga, M. S. (1999). A deficit in perceptual matching in the left hemisphere of a callosotomy patient. Neuropsychologia, 37(10), 1143–1154. https://doi.org/https://doi.org/10.1016/S0028-3932(99)00033-0

Galbreath, R. (1971). The History of Modern Occultism : A Bibliographical Survey. The Journal of Popular Culture, 5(3), 726–754. https://doi.org/https://doi.org/10.1111/j.0022-3840.1971.0503_726.x

_____. (1986). Explaining Modern Occultism. In H. Kerr & C. L. Crow (Eds.), The Occult in America : New Historical Perspectives (Illini Books edition. ed., pp. 11–37). University of Illinois Press.

Gallagher, H. G. (1990). By Trust Betrayed : Patients, Physicians, and the License to Kill in the Third Reich (First edition ed.). New York, H. Holt.
Galton, D. J. (1998). Greek Theories on Eugenics. J Med Ethics, 24(4), 263–267. https://doi.org/10.1136/jme.24.4.263
Galton, F. (1876). The History of Twins, as a Criterion of the Relative Powers of Nature and Nurture. The Journal of the Anthropological Institute of Great Britain and Ireland, 5, 391–406. https://doi.org/10.2307/2840900
_____. (1883). Inquiries Into Human Faculty and Its Development. Macmillan.
Garbett, G. K. (1975). [Obedience to Authority: An Experimental View., Stanley Milgram]. Man, 10(1), 167–168. https://doi.org/10.2307/2801239
Garrett, L., & Fidler, D. P. (2007). Sharing H5N1 viruses to stop a global influenza pandemic. PLoS Med, 4(11), e330. https://doi.org/10.1371/journal.pmed.0040330
Garrett, P. M. (2021). 'Surveillance Capitalism, COVID-19 and Social Work' : A Note on Uncertain Future(s). The British Journal of Social Work, 52(3), 1747–1764. https://doi.org/10.1093/bjsw/bcab099
Gasman, D. (1971). The Scientific Origins of National Socialism : Social Darwinism in Ernst Haeckel and the German Monist League. London; New York, London: American Elsevier.
Gaudillière, J.-P. (2000). Rockefeller strategies for scientific medicine : molecular machines, viruses and vaccines. Studies in History and Philosophy of Science Part C: Studies in History and Philosophy of Biological and Biomedical Sciences, 31(3), 491–509. https://doi.org/https://doi.org/10.1016/S1369-8486(00)00017-0
Gaus, G. F. (1996). Justificatory Liberalism : An Essay on Epistemology and Political Theory. New York, UNITED STATES, Oxford University Press, Incorporated.
Geisenhainer, K. (2002). "Rasse ist Schicksal" : Otto Reche, 1879-1966 : ein Leben als Anthropologe und Völkerkundler. Evangelische Verlagsanstalt.
Gellately, R. (1988). The Gestapo and German Society : Political Denunciation in the Gestapo Case Files. The Journal of Modern History, 60(4), 654–694.
_____. (1996). Denunciations in Twentieth-Century Germany : Aspects of Self-Policing in the Third Reich and the German Democratic Republic. The Journal of Modern History, 68(4), 931–967.
_____. (2001). Backing Hitler : Consent and Coercion in Nazi Germany. Oxford University Press.
Gerard, H. B., & Mathewson, G. C. (1966). The effects of severity of initiation on liking for a group: A replication. Journal of Experimental Social Psychology, 2(3), 278–287. https://doi.org/https://doi.org/10.1016/0022-1031(66)90084-9
Gérard, R. (1963). L'Orient et la Pensée Romantique Allemande. Didier.
Gerdmar, A. (2009). Roots of Theological Anti-Semitism : German Biblical Interpretation and the Jews, from Herder and Semler to Kittel and Buttmann (Vol. 20). The Netherlands, The Netherlands, Brill.
Gerhard, U. (1998). The Discoursive Construction of National Stereotypes : Collective Imagination and Racist Concepts in Germany Before World War I. In N. Finzsch & D. Schirmer (Eds.), Identity and Intolerance : Nationalism, Racism, and Xenophobia in Germany and the United States (pp. 71–96). Cambridge University Press.
Germana, N. A. (2010). Self-Othering in German Orientalism : The Case of Friedrich Schlegel. The Comparatist, 34, 80–94.
Gerstenfeld, M. (2006). Anti-Semitic Conspiracy Theories in Contemporary Germany

[Antisemitische Verschwörungstheorien nach dem 11. September (Anti-Semitic Conspiracy Theories after September 11), Tobias Jaecker]. Jewish Political Studies Review, 18(3/4), 172–174.

Gibney, E. (2020). Whose coronavirus strategy worked best? Scientists hunt most effective policies. Nature, 581(7806), 15–16. https://doi.org/https://doi.org/10.1038/d41586-020-01248-1

Gilbhard, H. (1994). Die Thule-Gesellschaft : vom okkulten Mummenschanz zum Hakenkreuz. Munich, Kiessling.

Gillespie, M. A. (2005). "Slouching Toward Bethlehem to Be Born" : On the Nature and Meaning of Nietzsche's Superman. Journal of Nietzsche Studies(30), 49–69.

Gilman, S. L. (1987). The Struggle of Psychiatry with Psychoanalysis : Who Won? Crit Inq, 13(2), 293–313. https://doi.org/10.1086/448391

_____. (1990). Jewish Self-Hatred : Anti-Semitism and the Hidden Language of the Jews. Baltimore, Baltimore : Johns Hopkins University Press.

_____. (1991). The Jew's Body. London, UK, Taylor & Francis Group.

_____. (1992). The Jewish Body : A foot-note. In H. Eilberg-Schwartz (Ed.), People of the body : Jews and Judaism from an embodied perspective (pp. 223–242). Albany, N.Y. : State University of New York Press.

_____. (1998). Creating Beauty to Cure the Soul : Race and Psychology in the Shaping of Aesthetic Surgery. Duke University Press.

_____. (2000). Are Jews White? Or, The History of the Nose Job. In L. Back & J. Solomos (Eds.), Theories of Race and Racism : A Reader (Vol. Part 3: Racism and antisemitism, pp. 229–237). Routledge.

Ginsberg, M. (1947). Reason and Unreason in Society : Essay in Sociology and Social Philosophy (Vol. 2). London School of Economics and Political Science, University of London.

Gittelman, M. (2006). The Holocaust, Racism, Mental Illness : Editor's Introduction. International Journal of Mental Health, 35(3), 5–16.

Glass, B. (1981). A Hidden Chapter of German Eugenics between the Two World Wars. Proceedings of the American Philosophical Society, 125(5), 357–367.

Glass, J. M. (1997). Life Unworthy of Life : Racial Phobia And Mass Murder In Hitler's Germany (First edition. ed.). New York, Basic Books.

Glazer, S. (2010). Issues in Peace and Conflict Studies : Selections From CQ Researcher. In Anti-Semitism in Europe: Are Israel's Policies Spurring a New Wave of Hate Crimes? (pp. 109—142). Sage.

Glette-Iversen, I., Aven, T., & Flage, R. (2023). A risk science perspective on vaccines. Risk analysis, 44(12), 2780–2796. https://doi.org/10.1111/risa.14228

Glowka, H. J. (1981). Deutsche Okkultgruppen 1875-1937. Arbeitsgemeinschaft für Religions- und Weltanschauungsfragen.

Godwin, J. (1994). The Theosophical Enlightenment. Albany, State University of New York Press.

_____. (2010). Atlantis and The Cycles of Time : Prophecies, Traditions, and Occult Revelations. Simon and Schuster.

Goldhagen, D. J. (1996). Hitler's Willing Executioners : Ordinary Germans and the Holocaust (First ed.). New York, New York : Alfred A. Knopf.

_____. (2009). Worse Than War : Genocide, Eliminationism, and the Ongoing Assault on Humanity. New York, UNITED STATES, Public Affairs.

Goldstein, J. A. (2011). On Racism and Anti-Semitism in Occultism and Nazism. In M.

R. Marrus (Ed.), The Nazi Holocaust (Vol. 2 The Origins of the Holocaust), Part One: Racism, thre Occult, and Eugenics, pp. 59–78). De Gruyter.

Goodman, D. G. (2011). The Protocols in Japan. In S. T. Katz & R. Landes (Eds.), The Paranoid Apocalypse : A Hundred-Year Retrospective on The Protocols of the Elders of Zion (pp. 135–151). NYU Press.

Goodman, M. (2008). Rome and Jerusalem : The Clash of Ancient Civilizations. Knopf Doubleday Publishing Group.

Goodrick-Clarke, N. (1998). Hitler's Priestess : Savitri Devi, the Hindu-Aryan Myth, and Neo-Nazism. New York University Press.

_____. (2002). Black Sun : Aryan Cults, Esoteric Nazism, and the Politics of Identity (First ed.). New York, NYU Press.

_____. (2004). The Occult Roots of Nazism : Secret Aryan Cults and their Influence on Nazi Ideology. London, UK, Tauris Parke.

_____. (2005). List, Guido Karl Anton (von). In Dictionary of Gnosis & Western Esotericism. https://archive.org/details/dictionaryofgnos0000unse

_____. (2008). The Western Esoteric Traditions. Oxford University Press.

Gordon, J. (2016). When Superman smote Zeus : analysing violent deicide in popular culture. Classical Receptions Journal, 9(2), 211–236. https://doi.org/10.1093/crj/clw008

Gordon, S. A. (1984). Hitler, Germans, and the "Jewish question". Princeton, N.J., Princeton, N.J. : Princeton University Press.

Görlitz, W., & Quint, H. A. (1952). Adolf Hitler : eine Biographie.

Goswami, A., Garg, S., Bhatt, E., Chaudhary, V., & Dang, S. (2024). Review—Nanotechnology-Based Biosensors for Biomedical Applications. Journal of The Electrochemical Society, 171(9). https://doi.org/10.1149/1945-7111/ad7908

Gotzmann, A., & Wiese, C. (2007). Modern Judaism and Historical Consciousness : Identities, Encounters, Perspectives. Brill.

Gould, S. J. (1996). The Mismeasure of Man (Revised and expanded ed.). New York, New York : Norton.

Gourevitch, P. (2015). We wish to inform you that tomorrow we will be killed with our families (Vol. 24). Pan Macmillan.

Gow, A. C. (2003). 'Sanguis Naturalis' and 'Sanc de Miracle' : Ancient Medicine, 'Superstition' and the Metaphysics of Mediaeval Healing Mircles. Sudhoffs Archiv, 87(2), 129–158.

Graham, D. W. (2008). Heraclitus : Flux, Order, and Knowledge. In P. Curd & D. W. Graham (Eds.), Oxford Handbook of Presocratic Philosophy (pp. 169–188). Oxford University Press.

Graham, L. R. (1977). Science and Values : The Eugenics Movement in Germany and Russia in the 1920s. The American Historical Review, 82(5), 1133–1164. https://doi.org/10.2307/1856342

_____. (1981). Between Science and Values. Columbia University Press.

Greene, J. C. (1959). Darwin and Religion. Prooceedings of The American Philosophical Society, 103(No. 5), 716—725.

Greenfeld, L. (2011). The Globalization of Nationalism and the Future of the Nation-State. International Journal of Politics, Culture, and Society, 24(1/2), 5–9.

Greenleaf, G., & Kemp, K. (2021). Australia's 'COVIDSafe' law for contact tracing : an experiment in surveillance and trust. International Data Privacy Law, 11(3), 257–275. https://doi.org/10.1093/idpl/ipab009

Gregg, R. B. (1966). The power of nonviolence ([Second revised edition]. ed.). New York, Schocken Books.

Grese, W. C. (1988). Magic in Hellenistic Hermeticism. In I. Merkel & A. G. Debus (Eds.), Hermeticism and the Renaissance : Intellectual History and the Occult in Early Modern Europe (pp. 45–58). Folger Shakespeare Library.

Gripp, R. F. (1981). [The Release of the Destruction of Life Devoid of Value, Karl Binding, Alfred Hoche, Robert L. Sassone, Santa Ana]. The Journal of Mind and Behavior, 2(4), 453–456.

Grobman, A. (2011). License to Murder : The Enduring Threat of the Protocols of the Elders of Zion. Balfour Books; America-Israel Friendship League.

Grodin, M., & Annas, G. (2007). Physicians and torture : lessons from the Nazi doctors. International Review of the Red Cross, 89(867), 635–654.

Grodin, M. A. (2010). Mad, Bad, or Evil : How Physician Healers Turn to Torture and Murder. In S. Rubenfeld (Ed.), Medicine after the Holocaust (pp. 49–65). Palgrave Macmillan US. https://doi.org/10.1057/9780230102293_6

_____ (Ed.). (2014). Jewish Medical Resistance in the Holocaust. Berghahn Books, Incorporated.

Gross, A. G. (1990). Rhetoric of Science. Cambridge, Mass, Harvard University Press.

Gross, L. (1996). How Charles Nicolle of the Pasteur Institute Discovered that Epidemic Typhus is Transmitted by Lice : Reminiscences from My Years at the Pasteur Institute in Paris. Proceedings of the National Academy of Sciences - PNAS, 93(20), 10539–10540. https://doi.org/10.1073/pnas.93.20.10539

Gross Solomon, S. (2000). 'Through a Glass Darkly' : The Rockefeller Foundation's International Health Board and Soviet Public Health. Studies in History and Philosophy of Science Part C: Studies in History and Philosophy of Biological and Biomedical Sciences, 31(3), 409–418. https://doi.org/https://doi.org/10.1016/S1369-8486(00)00010-8

Grossman, D. (2017). On Killing : the psychological cost of learning to kill in war and society. In R. Tweedy (Ed.), The Political Self : Understanding the Social Context for Mental Illness (First ed., pp. 141–155). Routledge. https://doi.org/10.4324/9780429482762-7

Gruber, M. v. (1927). Volk und Rasse : Die Rassenfrage. Süddeutsche Monatshefte, 24(10), 244–248.

Grunberger, R. (1971). The 12-year Reich : A Social History of Nazi Germany, 1933-1945 (First ed.). New York, New York : Holt, Rinehart and Winston.

Gualdani, R., Guerrini, A., Fantechi, E., Tadini-Buoninsegni, F., Moncelli, M. R., & Sangregorio, C. (2019). Superparamagnetic iron oxide nanoparticles (SPIONs) modulate hERG ion channel activity. Nanotoxicology, 13(9), 1197–1209. https://doi.org/10.1080/17435390.2019.1650969

Gugenberger, E., & Schweidlenka, R. (1993). Die Fäden der Nornen : zur Macht der Mythen in politischen Bewegungen. Wien, Verlag für Gesellschaftskritik.

Gunatilleke, G. (2021). Justifying Limitations on the Freedom of Expression. Human Rights Review, 22(1), 91–108. https://doi.org/10.1007/s12142-020-00608-8

Günther, H. F. K. ([1922] (1934). Kleine Rassenkunde des deutschen Volkes. Lehmann.

Günther, H. F. K. (1927). The Racial Elements of European History (G. C. Wheeler., Trans.; Second German ed.). England, England: Methuen & co. ltd.

Günther, H. F. K. (1933). Kleine Rassenkunde des deutschen Volkes. Lehmann.

_____. (1934). Die nordische Rasse bei den Indogermanen Asiens : zugleich ein Beitrag zur Frage nach der Urheimat und Rassenherkunft der Indogermanen. J.F. Lehmann.

_____. (1966). The Nordic Race as "Ideal Type" (S. Attanasio, Trans.). In G. L. a. Mosse (Ed.), Nazi Culture : Intellectual, Cultural and Social Life in the Third Reich (pp. 61–65). Grosset & Dunlap.

Günther, M. (1982). Die Institutionalisierung der *Rassenhygiene* an den deutschen Hochschulen vor 1933

Gutmann, J. (1978). III. Western Philosophical Thought. In W. T. Reich (Ed.), Encyclopaedia of Bioethics (Vol. 1, pp. 235–242). Free Press.

Gutteridge, R. (1976). The German Evangelical Church and the Jews, 1879-1950. Barnes & Noble Books.

Haas, P. J. (1988). Morality After Auschwitz : The Radical Challenge of the Nazi Ethic. Philadelphia, Fortress Press.

Habib, M. A. R. (2005). A History of Literary Criticism and Theory : From Plato to the Present. USA / UK / Australia, Blackwell Publishing.

Hadomi, L. (1991). From Technological Dystopia to Intopia Brave New World And Homo Faber. Utopian Studies(3), 110–117.

Haeckel, E. (1868). Natürliche Schöpfungsgeschichte : Gemeinverständliche wissenschaftliche Vorträge über die Entwickelungslehre.

_____. (1876). The History of Creation : Or the Development of the Earth and Its Inhabitants by the Action of Natural Causes (E. R. Lankester, Trans.; Vol. I) [Monograph]. London, Henry S. King & Co.

_____. (1878). Freie Wissenschaft und freie Lehre : eine Entgegnung auf Rudolf Virchow's Münchener Rede über "Die Freiheit der Wissenschaft im modernen Staat". Schweizerbart.

_____. (1900). The Riddle of the Universe : At the Close of the Nineteenth Century. New York; London Harper & Brothers.

_____. (1905). The Wonders of Life : A Popular Study of Niological Philosophy (J. McCabe, Trans.). New York; London, New York: Harper & brothers.

Hagemeister, M. (1998). Die Protokolle der Weisen von Zion und der Basler Zionistenkongress von 1897 In H. Haumann (Ed.), Der Traum von Israel: die Ursprünge des modernen Zionismus (pp. 250–273). Beltz Athenäum.

_____. (2008). The Protocols of the Elders of Zion : Between History and Fiction. New German Critique(103), 83–95.

Hagen, W. (1973). Krieg, Hunger und Pestilenz in Warschau 1939–1943. In Gesundheitswesen und Desinfektion (Vol. 8:115–28; 9:9:129–43).

_____. (1978). Auftrag und Wirklichkeit : Sozialarzt im 20. Jahrhundert. Munich-Grafelfin.

Hale, C. (2003). Himmler's Crusade : The Nazi Expedition to Find the Origins of the Aryan Race. Wiley.

Halik, E. (1954). Keine Invasion aus dem Weltraum! Mensch und Schicksal, 8(9), 3–5.

Hall, C. M. (2009). An Army of Spies? The Gestapo Spy Network 1933-45. Journal of Contemporary History, 44(2), 247–265.

Hall, D. I. (2017). Wagner, Hitler, and Germany's Rebirth after the First World War. War in History, 24(2), 154–175. https://doi.org/10.1177/0968344515608664

Halliday, R. J. (1971). Social Darwinism : A Definition. Victorian Studies, 14(4), 389–405.

Hamann, B. (1996). Hitlers Wien : Lehrjahre eines Diktators. Piper.

______. (1999). Hitler's Vienna : A Dictator's Apprenticeship. New York, Oxford University Press.

Hammer, O. (2003). Claiming Knowledge : Strategies of Epistemology from Theosophy to the New Age. Boston, UNITED STATES, BRILL.

Hampton, L. M., Aggarwal, R., Evans, S. J. W., & Law, B. (2021). General determination of causation between Covid-19 vaccines and possible adverse events. Vaccine, 39(10), 1478–1480. https://doi.org/10.1016/j.vaccine.2021.01.057

Hanegraaff, W. (2008). Altered States of Knowledge : The Attainment of Gnōsis in the Hermetica. International journal of the Platonic tradition, 2(2), 128–163. https://doi.org/10.1163/187254708X335728

______. (2009). The Pagan Who Came from the East : George Gemistos Plethon and Platonic Orientalism. In W. Hanegraaff & J. Pijnenburg (Eds.), Hermes in the Academy : Ten Years' Study of Western Esotericism at the University of Amsterdam (pp. 33–49). Amsterdam University Press.

Hanegraaff, W., & Pijnenburg, J. (2009). Hermes in the Academy : Ten Years' Study of Western Esotericism at the University of Amsterdam. Amsterdam, NETHERLANDS, THE, Amsterdam University Press.

Hanegraaff, W. J. (2005). Occult/Occultism. In Dictionary of Gnosis & Western Esotericism.

______. (2012). Esotericism and the Academy : Rejected Knowledge in Western Culture (First ed.). Cambridge, Cambridge University Press.

Hannaford, I. (1996). Race : The History of an Idea in the West. Washington, D.C. : Baltimore, Md., Woodrow Wilson Center Press, Baltimore, Md.

Hansen, E. (Ed.). (1981). Seit über einem Jahrhundert—verschüttete Alternativen in der Sozialpolitik: sozialer Fortschritt, organisierte Dienstleistermacht und nationalsozialistische Machtergreifung, der Fall der Ambulatorien in den Unterweserstädten und Berlin: 100 Jahre kaiserliche Botschaft zur Sozialversicherung, eine Festschrift.

Hargie, O. (2010). Skilled Interpersonal Communication : Research, Theory and Practice (Fifth ed.). Florence, United States, Taylor & Francis Group.

Haritos-Fatouros, M. (1995). The Official Torturer : A Learning Model for Obedience to the Authority of Violence. In R. D. Crelinsten & A. P. Schmid (Eds.), The Politics of Pain : Torturers and Their Masters (pp. 131–146). Westview Press. (Original work published Journal of applied social psychology (1988), 18(13), 1107-1120. Oxford.UK. Blackwell Publishing Ltd. https://doi.org/10.1111/j.1559-1816.1988.tb01196.x)

Harrington, A. (1996). Reenchanted Science : Holism in German Culture from Wilhelm II to Hitler. Princeton University Press.

Harris, E. (2024). Sickle Cell Disease Approvals Include First CRISPR Gene Editing Therapy. JAMA, 331(4), 280–280. https://doi.org/10.1001/jama.2023.26113

Harris, S. H. (1994). Factories of death : Japanese biological warfare, 1932-45, and the American cover-up. London ;, Routledge.

Harten, H.-C. (2014). Himmlers Lehrer : die weltanschauliche Schulung in der SS, 1933-1945. Paderborn, Ferdinand Schöningh.

Hartikainen, K. M. (2021). Emotion-Attention Interaction in the Right Hemisphere. Brain Sci, 11(8). https://doi.org/10.3390/brainsci11081006

Harwood, J. (1987). National Styles in Science : Genetics in Germany and the United States between the World Wars. Isis, 78(3), 390–414.

Haslem, W., Ndalianis, A., & Mackie, C. J. (2007). Super/heroes : From Hercules to Superman. New Academia Pub.

Hassan, A., & Barber, S. J. (2021). The effects of repetition frequency on the illusory truth effect : Principles and Implications. Cognitive Research, 6(1). https://doi.org/https://doi.org/10.1186/s41235-021-00301-5

Hastings, D. (2009). Catholicism and the Roots of Nazism : Religious Identity and National Socialism. Oxford, UNITED STATES, Oxford University Press, Incorporated.

Hastings, J., Selbie, J. A., & Gray, L. H. (Eds.). (1917). Encyclopaedia of Religion and Ethics (Vol. 9, Mundas—Phrygians). New York : Charles Scribner's Sons.

_____ (Eds.). (1918). Encyclopaedia of Religion and Ethics (Vol. X: PICTS-SACRAMENTS). Creative Media Partners, LLC.

Hayton, J. (2015). Beyond Good and Evil : Nazis and the Supernatural in Video Games. In M. Black & E. Kurlander (Eds.), Revisiting the 'Nazi Occult' : Histories, Realities, Legacies (pp. 248–269). Camden House.

He, Z., & Chen, Z. (2021). The Social Group Distinction of Nationalists and Globalists amid COVID-19 Pandemic. Fudan journal of the humanities and social sciences, 14(1), 67–85. https://doi.org/10.1007/s40647-020-00310-6

Hearn, J. (2015). Nationalism and Globalization Challenging Assumptions. The SAIS Review of International Affairs, 35(2), 5–11.

Heberer, G. (1943). Rassengeschichtliche Forschungen im indogermanischen Urheimatgebiet. G. Fischer.

Hecht, J. M. (2000). Vacher de Lapouge and the Rise of Nazi Science. Journal of the History of Ideas, 61(2), 285–304. https://doi.org/10.1353/jhi.2000.0018

Heidary, F., & Gharebaghi, R. (2020). Ivermectin : a systematic review from antiviral effects to COVID-19 complementary regimen. The Journal of Antibiotics, 73(9), 593–602. https://doi.org/10.1038/s41429-020-0336-z

Heiden, K. (1935). A History of National Socialism. A.A. Knopf.

Heilbronner, O. (2015). The Wewelsburg Effect : Nazi Myth and Paganism in Postwar European Popular Music. In M. Black & E. Kurlander (Eds.), Revisiting the 'Nazi Occult' : Histories, Realities, Legacies (pp. 270–286). Camden House.

Helmreich, E. C. (1980). The German Churches Under Hitler : Background, Struggle, and Epilogue. Detroit, Detroit : Wayne State University Press.

Heng, G. (2011). The Invention of Race in the European Middle Ages II : Locations of Medieval Race. Literature Compass, 8(5), 332–350. https://doi.org/https://doi.org/10.1111/j.1741-4113.2011.00795.x

Henig, R. M. (2004). Pandora's Baby : How the first Test Tube Babies Sparked the Reproductive Revolution. Boston, Houghton Mifflin.

Henningsen, M. L. M., Henningsen, D. D., Cruz, M. G., & Morrill, J. (2003). Social influence in groups : A comparative application of relational framing theory and the elaboration likelihood model of persuasion. Communication Monographs, 70(3), 1–1. https://doi.org/10.1080/0363775032000167398

Herf, J. (2009). Nazi Propaganda for the Arab World. New Haven, UNITED STATES, Yale University Press.

Hertz, D. (1997). The Genealogy Bureaucracy in the Third Reich [03/09/24]. Jewish History, 11(2), 53–78.

Hilberg, R. (1985). The Destruction of the European Jews (Revised and definitive edition. ed.). New York, Holmes & Meier.

______. (1992). Perpetrators, Victims, Bystanders : The Jewish Catastrophe, 1933-1945 (First edition. ed.). New York, Aaron Asher Books.

Hill, A., Wang, J., Levi, J., Heath, K., & Fortunak, J. (2020). Minimum costs to manufacture new treatments for COVID-19. Journal of Virus Eradication, 6(2), 61–69. https://doi.org/https://doi.org/10.1016/S2055-6640(20)30018-2

Hill, A. B. (1965). The Environment and Disease : Association or Causation? Proceedings of the Royal Society of Medicine, 58(5), 295–300. https://doi.org/10.1177/003591576505800503

Hillary, V. E., & Ceasar, S. A. (2023). A Review on the Mechanism and Applications of CRISPR/Cas9/Cas12/Cas13/Cas14 Proteins Utilized for Genome Engineering. Mol Biotechnol, 65(3), 311–325. https://doi.org/10.1007/s12033-022-00567-0

Hillgruber, A. (1984). War in the East and the Extermination of the Jews. Yad Vashem, 103–132. (Original work published The Nazi Holocaust : Historical Articles on the Destruction of European Jews, Westport, CT; London: Meckler, 1: 106–111)

______. (1989). War in the East and the Extermination of the Jews. In M. Michael Robert (Ed.), Volume 1 (pp. 85–114). K. G. Saur. https://doi.org/doi:10.1515/9783110970487-008

Himmelfarb, G. (2014). Evolution and Ethics, Revisited. The New Atlantis(42), 81–87.

Hirschler, B. (2016). GSK and Google parent forge $715 million bioelectronic medicines firm. Business News, Aug, 1.

Hislop, A. (1903). The Two Babylon's : Romanishm and its Origins. London, S. W. Partridge & Co.

Hitler, A. ([1925] (1971)). Mein Kampf (R. Manheim, Trans.). Boston, Boston : Houghton Mifflin.

______. (1971). Mein Kampf (R. Manheim, Trans.; R. Manheim, Ed.). Houghton Mifflin.

______. (2000). Hitler's Table Talk, 1941-1944 : His Private Conversations (N. Cameron & R. H. Stevens, Trans.). Enigma Books.

Hobhouse, L. T. (1911). Social Evolution and Political Theory. New York, New York : The Columbia university press.

______. (1916). The World in Conflict. T. F. Unwin.

Hochhuth, R. (1980). A German Love Story. Little, Brown.

Hoff, R. v. (1934). Rasen Materialismus? Rasse : Monatsschrift der Nordischen Bewegung, 1, 145–151.

Hofstadter, R. (1955). Social Darwinism in American Thought (Revised edition. ed.). Boston, Beacon Press.

Holmes, E. C., Rambaut, A., & Andersen, K. G. (2018). Pandemics : spend on surveillance, not prediction [Article]. Nature, 558, 180+.

Honigsbaum, M. (2019). Disease X and other unknowns. Lancet, 393(10180), 1496–1497. https://doi.org/10.1016/s0140-6736(19)30803-7

Hooton, E. A. (1935). Homo Sapiens—Whence and Whither. Science, 82(2115), 19–31. https://doi.org/10.1126/science.82.2115.19

Hopkins, A. J. ([1934] 1967). Alchemy : Child of Greek Philosophy. AMS Press.

Hoquet, T. (2014). Biologization of Race and Racialization of the Human Bernier, Buffon, Linnaeus. In N. Bancel, et al. (Eds.), The Invention of Race : Scientific and Popular Representations (pp. 17–32). Taylor & Francis Group.

Horkheimer, M., & Adorno, T. W. (2002). Dialectic of Enlightenment. New York, Seabury Press.

Hotez, P. J. (2010). Editorial : Peace through Vaccine Diplomacy. Science (American

Association for the Advancement of Science), 327(5971), 1301–1301. https://doi.org/10.1126/science.1189028
_____. (2014). "Vaccine diplomacy" : Historical Perspectives and Future Directions [Viewpoint essay]. PLoS Neglected Tropical Diseases, 8.
How, W. W. (1912). A commentary on Herodotus, with introduction and appendixes (Vol. I). England, Clarendon Press.
Howe, E. (1967). Urania's Children : The Strange World of the Astrologers. Kimber.
_____. (1983). *Germanenorden.* In Man, Myth & Magic : The Illustrated Encyclopaedia of Mythology, Religion and the Unknown. https://archive.org/details/man-myth-magic-illustrated-encyclopaedia-vol-1-11-and-index-editor-richard-cavendish-1983/page/1081/mode/2up?q=Hitler
Hudson, N. (1996). From "Nation to "Race" : The Origin of Racial Classification in Eighteenth-Century Thought. Eighteenth-Century Studies, 29(3), 247–264.
Hughes, E. C. (1962). Good People and Dirty Work. Social Problems, 10(1), 3–11. https://doi.org/10.2307/799402
Hutchinson, J. (2003). Nationalism, Globalism, and the Conflict of Civilizations. In U. Özkırımlı (Ed.), Nationalism and its Futures (pp. 71–92). Palgrave Macmillan UK. https://doi.org/10.1057/9780230524187_5
Hutton, C. (2005). Race and the Third Reich : Linguistics, Racial Anthropology and Genetics in the Dialectic of Volk. Cambridge, UK & Malden, MA, Cambridge, UK & Malden, MA : Polity.
Huxley, J. (1927). Religion Without Revelation. New York, New York : Harper.
_____. (1948). UNESCO : Its Purpose and Its Philosophy. Washington, Public Affairs Press.
_____. (1957). Religion Without Revelation (New and revised ed.). New York, New York : New American Library.
_____. (1960). The Evolutionary Vision. In S. Tax & C. Callender (Eds.), Evolution After Darwin. The University of Chicago Centennial. (Vol. III: Issues in Evolution, pp. 249–261). University of Chicago Press.
_____. (1962a). Eugenics in Evolutionary Perspective. Nature, 195(4838), 227–228. https://doi.org/10.1038/195227a0
_____. (1962b). Eugenics in Evolutionary Perspective. Eugen Rev, 54(3), 123–141.
Huxley, T. H., & Huxley, J. (1947). Evolution and Ethics : 1893-1943. Pilot Press.
Iacovacci, V., Lucarini, G., Ricotti, L., Dario, P., Dupont, P. E., & Menciassi, A. (2015). Untethered magnetic millirobot for targeted drug delivery. Biomed Microdevices, 17(3), 9962. https://doi.org/10.1007/s10544-015-9962-9
Ingram, A. (2005). The New Geopolitics of Disease : Between Global Health and Global Security. Geopolitics, 10(3), 522–545. https://doi.org/10.1080/14650040591003516
Institute for Propaganda Analysis. (1938). How to Detect Propaganda. In Institute for Propaganda Analysis : Propaganda Analysis (Vol. 1, pp. 5–8). USA: Institute for Propaganda Analysis.
Introvign, M. (2005). Rosicrucianism III : 19th-2oth Century. In Dictionary of Gnosis & Western Esotericism. https://archive.org/details/dictionaryofgnos0000unse
Isaac, B. (2006). The Invention of Racism in Classical Antiquity. Princeton, UNITED STATES, Princeton University Press.
Islam, M. S., Kamal, A.-H. M., Kabir, A., Southern, D. L., Khan, S. H., Hasan, S. M. M., Sarkar, T., Sharmin, S., Das, S., Roy, T., Harun, M. G. D., Chughtai, A. A.,

Homaira, N., & Seale, H. (2021). COVID-19 vaccine rumors and conspiracy theories : The need for cognitive inoculation against misinformation to improve vaccine adherence. PloS one, 16(5), e0251605–e0251605. https://doi.org/10.1371/journal.pone.0251605

Israel, J. I. (2006). Enlightenment Contested : Philosophy, Modernity, and the Emancipation of Man 1670-1752. Oxford, UK, Oxford University Press, Incorporated.

Italia, M., Della Rossa, F., & Dercole, F. (2023). Model-informed health and socio-economic benefits of enhancing global equity and access to Covid-19 vaccines. Scientific Reports, 13(1). https://doi.org/10.1038/s41598-023-48465-y

Jäckel, E. (1972). Hitler's Weltanschauung : A Blueprint for Power (H. Arnold, Trans.). Wesleyan University Press.

Jäckel, E., & Kuhn, A. (Eds.). (1980). Hitler : sämtliche Aufzeichnungen 1905-1924. Deutsche Verlags-Anstalt.

Jacobs, E. (1879). Der Brocken in Geschichte und Sage. Pfeffer.

Jaiswal, S. (1967). THE IDENTITY OF HERAKLES AND DIONYSOS IN MEGASTHENES. Proceedings of the Indian History Congress, 29, 107–113.

James, P. (1996). The Sunken Kingdom : The Atlantis Mystery Solved. Pimlico.

Janis, I. L. (1982). Groupthink : Psychological Studies of Policy Decisions and Fiascoes (Second ed.). Boston, Houghton Mifflin.

Jefferson, T. (1899). The Writings of Thomas Jefferson : 1816-1826 (P. L. Ford, Ed. Vol. X). G.P. Putnam's Sons.

Jessopp, A., & James, M. R. (1896). The Life and Miracles of St. William of Norwich. University Press.

Ji, X., Wang, H., Song, B., Chu, B., & He, Y. (2018). Silicon Nanomaterials for Biosensing and Bioimaging Analysis. Front Chem, 6, 38. https://doi.org/10.3389/fchem.2018.00038

Jikeli, G. (2009). "Jew" as a Slur in German and French Today. Journal for the Study of Antisemitism, 1(2), 209—232.

Jodl, F. (1891). Morals in History. International Journal of Ethics, 1(2), 204–223.

Johannison, K. (1988). Magic, Science, and Institutionalization in the Seventeenth and Eighteenth Centuries. In I. Merkel & A. G. Debus (Eds.), Hermeticism and the Renaissance : Intellectual History and the Occult in Early Modern Europe (pp. 251–261). Folger Shakespeare Library.

Johnson, D. (22 December, 2015). Polymer Embedded With Metallic Nanoparticles Enables Soft Robotics : New production technique for nanoparticle-enabled polymers could lead to better control of soft robots. Retrieved Nov. 25 from https://spectrum.ieee.org/nanoclast/semiconductors/materials/polymer-embedded-with-metallic-nanoparticles-enables-soft-robotics;

Johnson, E. A. (2000). Nazi Terror : The Gestapo, Jews, and Ordinary Germans. Basic Books.

______. (2011). CRIMINAL JUSTICE, COERCION AND CONSENT IN 'TOTALITARIAN' SOCIETY : The Case of National Socialist Germany. The British Journal of Criminology, 51(3), 599–614.

Johnson, F. R. (1937). Astronomical Thought in Renaissance England : A Study of the English Scientific Writing from 1500 to 1645. Baltimore, The Johns Hopkins press.

Johnson, K. P. (1998). Edgar Cayce in context : the Readings, truth and fiction. Albany,

N.Y, State University of New York Press.
Johnson, W. (2012). The myth of Jewish male menses. Journal of Medieval History, 24(3), 273–295. https://doi.org/10.1016/s0304-4181(98)00009-8
Jones, R. H. (1986). Science and Mysticism : A Comparative Study of Western Natural Science, Theravāda Buddhism, and Advaita Vedānta. Lewisburg, Bucknell University Press.
Joseph, S. M. (2002). The Trivium : The Liberal Arts of Logic, Grammar, and Rhetoric : Understanding the Nature and Function of Language (M. McGlinn, Ed. First Paul Dry Books edition. ed.). Philadelphia, PA, Paul Dry Books.
Jowett, G., & O'Donnell, V. (2006). Propaganda and Persuasion (Fourth ed.). Sage.
Judge, W. Q. (1893). The Ocean of Theosophy. The Path.
Jung, C. G. (1970). "Wotan," Neue Schweizer Rundschau (R. F. C. Hull, Trans.). In The Collected Works of C. G. Jung : Civilization in Transition (Second ed., Vol. 10, pp. 179–193). Princeton University Press. (Original work published "Wotan," Neue Schweizer Rundschau (Zurich), n.s., III (March, 1936), pp. 657–69.)
Junginger, H. (2007). From Buddha To Adolf Hitler : Walther Wüst And The Aryan Tradition. In H. Junginger (Ed.), The Study of Religion under the Impact of Fascism (pp. 105–177). Brill. https://doi.org/https://doi.org/10.1163/ej.9789004163263.i-663
Junginger, H., & Åkerlund, A. (2013). Nordic Ideology Between Religion and Scholarship. Peter Lang.
Kailany, M., & Baraja, A. (1988). Bahaya zionisme terhadap dunia islam. Mantiq.
Kaminsky, U. (2014). "Mercy Killing" and Economism : On Ethical Patterns of Justification for Nazi Euthanasia. In W. Bialas & L. Fritze (Eds.), Nazi Ideology and Ethics (pp. 237–274). Cambridge Scholars Publishing.
Kampmark, B. (2020). The pandemic surveillance state : an enduring legacy of COVID-19. Journal of Global Faultlines, 7(1), 59–70. https://doi.org/10.13169/jglobfaul.7.1.0059
Kant, I. (2007). On the use of teleological principles in philosophy (1788). In R. B. Louden & G. Zöller (Eds.), Anthropology, History, and Education (pp. 192–218). Cambridge University Press. https://doi.org/DOI: 10.1017/CBO9780511791925.016
Kaplan, G. B. (2012). The Inception of Limpieza de Sangre (Purity of Blood) and its Impact in Medieval and Golden Age Spain. In A. I. Aronson-Friedman & G. B. Kaplan (Eds.), Marginal Voices : Studies in Converso Literature of Medieval and Golden Age Spain (pp. 19–42). BRILL.
Karmarkar, R. D. (1950). THE FIRST GREEK CONQUEROR OF INDIA. Annals of the Bhandarkar Oriental Research Institute, 31(1/4), 239–249.
Kater, M. H. (1983). The Nazi Party : A Social Profile of Members and Leaders, 1919-1945. Cambridge, Mass., Cambridge, Mass. : Harvard University Press.
______. (1987). Hitler's Early Doctors : Nazi Physicians in Predepression Germany. The Journal of Modern History, 59(1), 25–52.
______. (1989). Doctors Under Hitler. University of North Carolina Press.
Katz, F. E. (1993). Ordinary People and Extraordinary Evil : A Report on the Beguilings of Evil. Albany, State University of New York Press.
Katz, J. (1970). Jews and freemasons in Europe 1723-1939 (L. Oschry, Trans.). Harvard University Press.
______. (1980). From Prejudice to Destruction : Anti-Semitism, 1700-1933. Harvard

University Press.

Katz, R., Kornblet, S., Arnold, G., Lief, E., & Fischer, J. E. (2011). Defining Health Diplomacy : Changing Demands in the Era of Globalization. The Milbank quarterly, 89(3), 503–523. https://doi.org/10.1111/j.1468-0009.2011.00637.x

Kaufmann, J. R., & Feldbaum, H. (2009). Diplomacy and The Polio Immunization Boycott In Northern Nigeria. Health Affairs, 28(4), 1091–1101. https://doi.org/https://doi.org/10.1377/hlthaff.28.4.1091

Kaufmann, W. (2009). Das Dritte Reich und Tibet : die Heimat des "östlichen Hakenkreuzes" im Blickfeld der Nationalsozialisten. Ludwigsfelder Verlagshaus.

Kaul, V., Shah, V. H., & El-Serag, H. (2020). Leadership During Crisis : Lessons and Applications from the COVID-19 Pandemic. Gastroenterology, 159(3), 809–812. https://doi.org/https://doi.org/10.1053/j.gastro.2020.04.076

Kaya, E. K. (2020). SAFETY AND PRIVACY IN THE TIME OF COVID-19 : CONTACT TRACING APPLICATIONS. http://www.jstor.org/stable/resrep26089

Kayser, V., & Ramzan, I. (2021). Vaccines and vaccination : history and emerging issues. Human vaccines & immunotherapeutics, 17(12), 5255–5268. https://doi.org/10.1080/21645515.2021.1977057

Keith, A. (1947). Evolution and Ethics. New York, New York : G. P. Putnam's sons.

Keith, A. B. (1933). The Origins of the Aryan Gods. Journal of the Royal Asiatic Society of Great Britain and Ireland(4), 813–820.

Kelle, A. (2007). Securitization of International Public Health : Implications for Global Health Governance and the Biological Weapons Prohibition Regime. Global Governance, 13(2), 217–235.

Kelly, A. (1981). The Descent of Darwin : The Popularization of Darwinism in Germany, 1860-1914. Chapel Hill, UNITED STATES, University of North Carolina Press.

Kelman, H. C., & Hamilton, V. L. (1989). Crimes of Obedience : Toward a Social Psychology of Authority and Responsibility. New Haven, Yale University Press.

Kennedy, G. (1998). Children of the sun : a pictorial anthology, from Germany to California 1883-1949. Ojai, Calif, Nivaria Press.

Kennedy, R. F. (2021). The Real Anthony Fauci : Bill Gates, Big Pharma, and the Global War on Democracy and Public Health. Skyhorse Publishing, Inc.

Kershaw, I. (1981). The Persecution of the Jews and German Popular Opinion in the Third Reich. The Leo Baeck Institute Year Book, 26(1), 261–289. https://doi.org/10.1093/leobaeck/26.1.261

______. (1987). The 'Hitler myth' : Image and Reality in the Third Reich. Oxford; New York, Oxford : Clarendon Press.

______. (1998). Hitler, 1889-1936 : Hubris. Allen Lane.

______. (2000). Hitler, 1936-45 : Nemesis (First American ed.). New York, New York : W. W. Norton.

Kevles, D. J. (1985). In the Name of Eugenics Genetics and the Uses of Human Heredity. New York, Knopf.

Kickbusch, I., & Kökény, M. (2013). Global health diplomacy : five years on. World Health Organization. Bulletin of the World Health Organization, 91(3), 159–159A. https://doi.org/https://doi.org/10.2471/BLT.13.118596

Kickbusch, I., Novotny, T. E., Drager, N., Silberschmidt, G., & Alcazar, S. (2007b). Global health diplomacy : training across disciplines. World Health Organization. Bulletin of the World Health Organization, 85(12), 971–973. https://doi.org/

https://doi.org/10.2471/BLT.07.045856
Kickbusch, I., Silberschmidt, G., & Buss, P. (2007a). Global health diplomacy : the need for new perspectives, strategic approaches and skills in global health. Bulletin of the World Health Organization, 85(3), 230–232. https://doi.org/10.1590/S0042-96862007000300018
Kickbusch, I., & Szabo, M. M. (2014). A new governance space for health. Glob Health Action, 7, 23507. https://doi.org/10.3402/gha.v7.23507
Kidd, B. (1924). The Science of Power. New York (State), G. P. Putnam's sons.
Kim, K. H., Kim, S.-H., Park, N. Y., Hyun, J.-W., & Kim, H. J. (2022). Onset of various CNS inflammatory demyelination diseases following COVID-19 vaccinations. Multiple Sclerosis and Related Disorders, 68. https://doi.org/10.1016/j.msard.2022.104141
King, D. (2005). Finding Atlantis : A True Story of Genius, Madness, and an Extraordinary Quest for a Lost World. Crown.
King, R. C., Mulligan, P. K., & Stansfield, W. D. (2014). Eugenics. In A Dictionary of Genetics. https://www.oxfordreference.com/view/10.1093/acref/9780199766444.001.0001/acref-9780199766444-e-2199
Kiple, K. F. (1993). The Cambridge World History of Human Disease. Cambridge, Cambridge University Press.
Kleingeld, P. (2007). Kant's Second Thoughts on Race. The Philosophical Quarterly (1950-), 57(229), 573–592.
_____. (2024). Anti-Racism and Kant Scholarship : A Critical Notice of Kant, Race, and Racism: Views from Somewhere, by Huaping Lu-Adler. Mind. https://doi.org/10.1093/mind/fzae022
Knickerbocker, H. R. (1941). Is Tomorrow Hitler's? 200 Questions on the Battle of Mankind. Project Gutenberg.
Knox, R. (1850). The Races of Men : A Fragment. H. Renshaw.
Koehl, R. L. (1983). The Black Corps : The Structure and Power Struggles of the Nazi SS. University of Wisconsin Press.
Koehne, S. (2014). The Racial Yardstick : "Ethnotheism" and Official Nazi Views on Religion. German Studies Review, 37(3), 575–596.
_____. (2015). Were the National Socialists a Völkisch Party? Paganism, Christianity, and the Nazi Christmas. Central European History, 47(4), 760–790. https://doi.org/10.1017/s0008938914001897
Koellreutter, O. (1934). Der deutsche Führerstaat. Mohr.
Kogon, E. (1980). The Theory and Practice of Hell : The German Concentration Camps and the System Behind Them. Berkley Books.
_____. (1998). The Theory and Practice of Hell : The German Concentration Camps and the System Behind Them. Berkley Books.
Koh, K. P., Sasse, K., Mahabir, R., Gkountouna, O., Crooks, A., & Croitoru, A. (2024). Understanding the determinants of vaccine hesitancy in the United States : A comparison of social surveys and social media. PloS one, 19(6). https://doi.org/10.1371/journal.pone.0301488
Kohler, R. E. (1976). The Management of Science : The Experience of Warren Weaver and the Rockefeller Foundation Programme in Molecular Biology. Minerva, 14(3), 279–306.
_____. (1978). A Policy for the Advancement of Science : The Rockefeller Foundation, 1924-29. Minerva, 16(4), 480–515.

______. (1991). Partners in Science : Foundations and Natural Scientists, 1900-1945. University of Chicago Press.

Kohn, H. (1960). The Idea of Nationalism : A Study in Its Origins and Background. New York, The Macmillan Company.

Kohn-Behrens, C. (1934). Was ist Rasse? Gespräche mit den grössten deutschen Forschern der Gegenwart. Munchen: Eher.

Koonz, C. (2003). The Nazi Conscience. Cambridge, Mass., Cambridge, Mass. : Belknap Press.

Koplan, J. P. P., Bond, T. C. P., Merson, M. H. P., Reddy, K. S. P., Rodriguez, M. H. P., Sewankambo, N. K. P., & Wasserheit, J. N. P. (2009). Towards a common definition of global health. The Lancet (British edition), 373(9679), 1993–1995. https://doi.org/10.1016/S0140-6736(09)60332-9

Koyama, T., Weeraratne, D., Snowdon, J. L., & Parida, L. (2020). Emergence of Drift Variants That May Affect COVID-19 Vaccine Development and Antibody Treatment. Pathogens, 9(5). https://doi.org/10.3390/pathogens9050324

Kranz, W. (1951). Die Fragmente der Vorsokratiker (Vol. I). Weidmann.

Kratz, R. G. (2009). Eyes and Spectacles : Wellhausen's Method of Higher Criticism. The Journal of Theological Studie, 60(2), 381–402.

Krausnick, H. (1968). The Persecution of the Jews. In Anatomy of the SS state. Walker.

______. (1970). The Persecution of the Jews (D. L. a. M. Jackson, Trans.). In Anatomy of the SS State. Paladin.

Krausz, S. L., Simon, D. A., Bartos, Z., Biczók, Z., Varga, É., Huszár, K., Kulcsár, P. I., Tálas, A., Ligeti, Z., & Welker, E. (2025). ProPE expands the prime editing window and enhances gene editing efficiency where prime editing is inefficient. Nature Catalysis, 8(10), 1100–1116. https://doi.org/10.1038/s41929-025-01406-6

Kren, G. M., & Rappoport, L. (1980). The Holocaust and the Crisis of Human Behavior. New York, Holmes & Meier.

Kreniske, J. (1989). [Racial Hygiene : Medicine under the Nazis, Robert N. Proctor]. Journal of Public Health Policy, 10(3), 410–414. https://doi.org/10.2307/3342803

Kressel, N. J. (1996). Mass Hate : The Global Rise of Genocide and Terror. New York, Plenum Press.

Kröner, W. (1938). Die Wiedergeburt des Magischen : das Magische als natürliche Funktion des Unbewussten. Leipzig, Richard Hummel Verlag.

Krüger, A. (1972). Die Olympischen Spiele 1936 und die Weltmeinung : Ihre aussenpolit. Bedeutg. unter bes. Berücks. d. USA. Bartels & Wernitz.

______. (1999). Breeding, rearing and preparing the Aryan body : creating supermen the Nazi way. The International Journal of the History of Sport, 16(2), 42–68. https://doi.org/10.1080/09523369908714070

Krüger, G. (1966). The Renovation of the Academic Community (S. Attanasio, Trans.). In G. L. Mosse (Ed.), Nazi Culture : Intellectual, Cultural and Social Life in the Third Reich (pp. 304–308). Grosset & Dunlap.

Kruger, S. F. (1992). The bodies of Jews in the late Middle Ages. In D. R. Howard, et al. (Eds.), The Idea of Medieval Literature: New Essays on Chaucer and Medieval Culture in Honor of Donald R. Howard (pp. 301–323). University of Delaware Press.

Kubizek, A. (2011). The Young Hitler I Knew : The Definitive Inside Look at the Artist Who Became a Monster (G. Brooks, Trans.). Arcade.

Kubo, Y., & Okada, I. (2022). COVID-19 health certification reduces outgroup bias :

evidence from a conjoint experiment in Japan. Humanities and Social Sciences Communications, 9(1), 306. https://doi.org/10.1057/s41599-022-01324-z

Kuhl, S. (2002). The Nazi Connection : Eugenics, American Racism, and German National Socialism. Cary, UNITED STATES, Oxford University Press, Incorporated.

Kuhn, A. B. (1930). Theosophy, a Modern Revival of Ancient Wisdom (Vol. New York). H. Hold and Company.

Kuhn, P., & Kranz, H. W. (1936). Von deutschen Ahnen für deutsche Enkel : Allgemeinverständliche Darstellung der Erblichkeitslehre, der Rassenkunde und der *Rassenhygiene*. J.F. Lehmann.

Küntzel, M. (2007). Jihad and Jew-hatred : Islamism, Nazism and the Roots of 9/11. Telos Press Pub.

Kurlander, E. (2012). Hitler's Monsters : The Occult Roots of Nazism and the Emergence of the Nazi 'Supernatural Imaginary'. German History, 30(4), 528–549. https://doi.org/10.1093/gerhis/ghs073

______. (2013). The Orientalist Roots of National Socialism? In J. M. Cho, et al. (Eds.), Transcultural Encounters Between Germany and India : Kindred Spirits in the 19th and 20th Centuries (pp. 155–169). Taylor & Francis Group.

______. (2015a). The Nazi Magicians' Controversy : Enlightenment, "Border Science," and Occultism in the Third Reich. Central European History, 48(4), 498–522.

______. (2015b). Hitler's Supernatural Sciences : Astrology, Anthroposophy, and World Ice Theory in the Third Reich. In M. Black & E. Kurlander (Eds.), Revisiting the 'Nazi Occult' : Histories, Realities, Legacies (pp. 132–156). Camden House.

______. (2017a). Hitler's Monsters : A Supernatural History of the Third Reich. Yale University Press.

______. (2017b). 'One Foot in Atlantis, One in Tibet' : The Roots and Legacies of Nazi Theories on Atlantis, 1890-1945. Leidschrift, Jaargang 32(1), 81–101.

______. (2019). Nazism and Religion. https://doi.org/10.1093/acrefore/9780199340378.013.680

Kurzweil, R. (2004). Human Body Version 2.0. In B. Klein & S. Sethe (Eds.), The Scientific Conquest of Death: Essays on Infinite Lifespans (pp. 93—106). LibrosEnRed.

______. (2005). The Singularity Is Near : When Humans Transcend Biology. Penguin Publishing Group.

______. (2024). The Singularity Is Nearer : When We Merge with AI. Penguin Publishing Group.

Labonté, R., & Gagnon, M. L. (2010). Framing health and foreign policy : lessons for global health diplomacy. Globalization and health, 6(1), 14–14. https://doi.org/10.1186/1744-8603-6-14

Ladapo, J. A., McKinnon, J. E., McCullough, P. A., & Risch, H. A. (2020). Randomized Controlled Trials of Early Ambulatory Hydroxychloroquine in the Prevention of COVID-19 Infection, Hospitalisation, and Death: Meta-Analysis. medRxiv, 2020.2009.2030.20204693. https://doi.org/10.1101/2020.09.30.20204693

Lal, A., Erondu, N. A., Heymann, D. L., Gitahi, G., & Yates, R. (2021). Fragmented health systems in COVID-19 : rectifying the misalignment between global health security and universal health coverage. The Lancet, 397(10268), 61–67. https://doi.org/10.1016/S0140-6736(20)32228-5

Landheer, B., & Massing, P. (1950). Rehearsal for Destruction : A Study of Political Anti-

Semitism in Imperial Germany [Rehearsal for Destruction: A Study of Political Anti-Semitism in Imperial Germany., Paul Massing]. American Sociological Review, 15(2), 319–319. https://doi.org/10.2307/2086810
Landig, W. (1971). Götzen gegen Thule : ein Roman voller Wirklichkeiten. Pfeiffer.
Lang, J. v., & Sibyll, C. (Eds.). (1983). Eichmann Interrogated : Transcripts from the Archives of the Israeli Police (R. Manheim, Trans.). Farrar, Straus, Giroux.
Langbehn, J. (1892). Der Rembrandtdeutsche. Von einem Wahrheitsfreund. Dresden,
Langbein, H. (1972). Menschen in Auschwitz. Europaverl.
Lange, H.-J. (Ed.). (1999). Otto Rahn und die Suche nach dem Gral: Biografie und Quellen. Arun.
Lange, H.-J. r. (Ed.). (1995). Otto Rahn : Leben & Werk. Arun.
Langmuir, G. I. (1990). History, Religion, and Antisemitism. Berkeley, University of California Press.
Lanzmann, C. (1985). Shoah : An Oral History of the Holocaust : The Complete Text of the Film (First American edition. ed.). New York, Pantheon Books.
Lapouge, G. V. d. (1899). L'Aryen; son rôle social. A. Fontemoing.
Laqueur, T. (2006). Why the Margins Matter : Occultism and the Making of Modernity. Modern Intellectual History, 3(1), 111–135. https://doi.org/10.1017/S1479244305000648
Laqueur, W. (1984). Young Germany : A History of the German Youth Movement. Transaction Books.
Lassman, P., Velody, I., & Martins, H. (Eds.). (1989). Max Weber's 'science as a Vocation'. Unwin Hyman.
Lasswell, H. D. (1927). Propaganda Technique in the World War. K. Paul, Trench, Trubner & Company, Limited.
Latour, B. (1988). The Pasteurization of France (A. Sheridan & J. Law, Trans.). Cambridge, Mass, Cambridge, Mass: Harvard University Press.
Lavoie, J. D. (2018). The philosophy of Guido von List (1848–1919) : connecting Gnosticism to Nazism. In G. W. Trompf, et al. (Eds.), The Gnostic World (pp. 499–509). Taylor & Francis.
_____. (2021). Theosophical Chronology in the Writings of Guido von List (1848–1919) : A Link Between H.P. Blavatsky's Philosophy and the Nazi Movement. In G. D. Hedesan & T. Rudbøg (Eds.), Innovation in Esotericism from the Renaissance to the Present (pp. 255–279). Springer International Publishing. https://doi.org/10.1007/978-3-030-67906-4_10
Lawrence, A. (2015). This Way for the Gas, Ladies and Gentlemen : And Other Stories. In T. Riggs (Ed.), Histories of Everyday Life in Totalitarian Regimes (Vol. 2, pp. 107–110). St. James Press.
Lawrence, D. H. (1936). Phoenix : the posthumous papers of D. H. Lawrence (E. D. McDonald, Ed.). New York, The Viking press.
Le Bon, G. ([1895] 2002). The Crowd : A Study of the Popular Mind (Reprint ed.). Dover Publications.
_____. (1903). The Crowd : A Study of the Popular Mind. T.F. Unwin.
Le Goff, A., Jeffries Hein, R., Hart, A. N., Roberson, I., & Landecker, H. L. (2024). Anticipating in vitro gametogenesis : Hopes and concerns for IVG among diverse stakeholders. Stem Cell Reports, 19(7), 933–945. https://doi.org/https://doi.org/10.1016/j.stemcr.2024.05.002
Le Goff, J. (1988). Medieval Civilization 400-1500 (J. Barrow, Trans.). Basil Blackwell.

Lee, A. M. (1952). How to Understand Propaganda. Rinehart.
Lee, J., Woodruff, M. C., Kim, E. H., & Nam, J.-H. (2023). Knife's edge : Balancing immunogenicity and reactogenicity in mRNA vaccines. Experimental & Molecular Medicine, 55(7), 1305–1313. https://doi.org/10.1038/s12276-023-00999-x
Leff, G. (1991). The Trivium and The Three Philosophies. In H. d. Ridder-Symoens (Ed.), A History of the University in Europe: Volume 1: Universities in the Middle Ages (Vol. 1, pp. 307–336). Cambridge University Press. https://doi.org/DOI: 10.1017/CBO9780511599507.012
Leijenhorst, C. (2005). Anthroposophy. In Dictionary of Gnosis & Western Esotericism. https://archive.org/details/dictionaryofgnos0000unse_d1x1/mode/2up
Leiss, W., & Chociolko, C. (1994). Risk and Responsibility. Montreal, CANADA, McGill-Queen's University Press.
Leunissen, M. (2017). Eugenics and the Production of Good Natural Character. In From Natural Character to Moral Virtue in Aristotle (pp. 81–104). Oxford University Press. https://doi.org/10.1093/acprof:oso/9780190602215.003.0004
Levenda, P. (2002). Unholy Alliance : A History of Nazi Involvement with the Occult (Second ed.). Continuum.
Levy, R. S. (1975). The Downfall of the Anti-Semitic Political Parties in Imperial Germany. New Haven, New Haven : Yale University Press.
Lewis, A. (1934). German Eugenic Legislation : An examination of Fact and Theory. Eugen Rev, 26(3), 183–191.
Lewkowicz, J., Woźniak, M., & Wrzesiński, M. (2022). COVID-19 and erosion of democracy. Economic Modelling, 106, 105682. https://doi.org/https://doi.org/10.1016/j.econmod.2021.105682
Liang, P. C., Chen, Y. C., Chiang, C. F., Mo, L. R., Wei, S. Y., Hsieh, W. Y., & Lin, W. L. (2016). Doxorubicin-modified magnetic nanoparticles as a drug delivery system for magnetic resonance imaging-monitoring magnet-enhancing tumor chemotherapy. Int J Nanomedicine, 11, 2021–2037. https://doi.org/10.2147/IJN.S94139
Lifton, R. J. (1986). The Nazi Doctors : Medical Killing and the Psychology of Genocide. New York, Basic Books.
Limb, M. (2011). "Vaccine hesitancy" means scientists need to be more honest about risks. BMJ, 342(apr15 2), d2479–d2479. https://doi.org/10.1136/bmj.d2479
Lindenthal, J. J. (2014). The Epidemiological Status and Health-Care Administration of the Jews before and during the Holocaust. In M. A. Grodin (Ed.), Jewish Medical Resistance in the Holocaust (pp. 13–38). Berghahn Books, Incorporated.
Lindsay, J. A. (1916). Eugenics and the Doctrine of the Super-man. Eugen Rev, 7(4), 247–262.
Lindsay, J. O. (Ed.). (1957). The New Cambridge Modern History (Vol. 7 of Old Regime, 1713-1763). Cambridge University Press.
Linebarger, P. M. A. (1954). Psychological Warfare (Second ed.). New York, USA, Duell, Sloan and Pearce.
Link, F. (2010). Der Mythos Burg im Nationalsozialismus. In G. U. Grossmann & H. Ottomeyer (Eds.), Die Burg (pp. 302–311). Sandstein Verlag.
_____. (2014). Burgen und Burgenforschung im Nationalsozialismus : Wissenschaft und Weltanschauung 1933-1945. Böhlau.
Link, F., & Hare, L. J. (2015). Pseudoscience Reconsidered : SS Research and the

Archaeology of Haithabu. In Revisiting the 'Nazi Occult' : Histories, Realities, Legacies (pp. 105–131). Camden House.

Linke, U., & Dundes, A. (1988). More on Auschwitz Jokes. Folklore, 99(1), 3–10.

List, G. v. (1908). Das Geheimnis der Runen (1. - 2. Tsd ed.). Leipzig, Steinacker.

_____. (1910). Die Bilderschrift der Ario-Germanen : Ario-Germanische Hieroglyphik. Guido-von-List-Gesellschaft.

_____. (1913). Deutsch-Mythologische Landschaftsbilder (Second ed., Vol. 2). Österreichische Verlagsanstalt, Vienna; Leipzig.

Liu, A. T., Kunai, Y., Cottrill, A. L., Kaplan, A., Zhang, G., Kim, H., Mollah, R. S., Eatmon, Y. L., & Strano, M. S. (2021). Solvent-induced electrochemistry at an electrically asymmetric carbon Janus particle. Nature Communications, 12(1), 3415. https://doi.org/10.1038/s41467-021-23038-7

Llewellyn, S. (2020). Covid-19 : how to be careful with trust and expertise on social media. BMJ. https://doi.org/10.1136/bmj.m1160

Lockwood, E. (2020). The Antisemitic Backlash to Financial Power : Conspiracy Theory as a Response to Financial Complexity and Crisis. New Political Economy, 26(2), 261–270. https://doi.org/10.1080/13563467.2020.1841141

Löffler, P. (2021). Review : Vaccine Myth-Buster – Cleaning Up With Prejudices and Dangerous Misinformation. Frontiers in Immunology, 12. https://doi.org/10.3389/fimmu.2021.663280

Longerich, P. (2012). Heinrich Himmler (J. Noakes & L. Sharpe, Trans.). Oxford, UK, Oxford University Press.

Lovejoy, A. O. (1959). Kant and Evolution. In B. Glass, et al. (Eds.), Forerunners of Darwin: 1745-1859 (pp. 173–206). Baltimore : Johns Hopkins Press.

Lovin, C. R. (1967). Blut Und Boden : The Ideological Basis of the Nazi Agricultural Program. Journal of the History of Ideas, 28(2), 279–288. https://doi.org/10.2307/2708423

Löwy, I., & Zylberman, P. (2000). Medicine as a social instrument : Rockefeller Foundation, 1913–45. Studies in History and Philosophy of Science Part C: Studies in History and Philosophy of Biological and Biomedical Sciences, 31(3), 365–379. https://doi.org/https://doi.org/10.1016/S1369-8486(00)00011-X

Lu-Adler, H. (2023). Kant, Race, and Racism : Views from Somewhere. Oxford University Press.

Lucas, P. (2018). Bacon's New Atlantis and the Fictional Origins of Organised Science. Open Cultural Studies, 2(1), 114–121. https://doi.org/10.1515/culture-2018-0011

Ludmerer, K. M., & Lappé, M. (1978). Eugenics. In W. T. Reich (Ed.), Encyclopaedia of Bioethics (Vol. 1, pp. 457–468). Free Press.

Luhrssen, D. (2012). Hammer of the Gods : The Thule Society and the Birth of Nazism. Potomac Books Incorporated.

Ma'oz, M. (2011). Muslim Attitudes to Jews and Israel: The Ambivalences of Rejection, Antagonism, Tolerance & Co-Operation. Liverpool University Press.

Machado, B. A. S., Hodel, K. V. S., Fonseca, L. M. d. S., Pires, V. C., Mascarenhas, L. A. B., da Silva Andrade, L. P. C., Moret, M. A., & Badaró, R. (2022). The Importance of Vaccination in the Context of the COVID-19 Pandemic : A Brief Update Regarding the Use of Vaccines. Vaccines (Basel), 10(4), 591. https://doi.org/10.3390/vaccines10040591

MacMaster, N. (2001). Racism in Europe, 1870-2000. Hampshire; New York : Palgrave.

Macpherson, I., Roqué, M. V., & Segarra, I. (2019). Ethical Challenges of Germline

Genetic Enhancement. Front Genet, 10, 767. https://doi.org/10.3389/fgene.2019.00767

Macrakis, K. (1989). The Rockefeller Foundation and German Physics under National Socialism. Minerva, 27(1), 33–57. https://doi.org/10.1007/BF01096714

_____. (1993). Surviving the Swastika : Scientific Research in Nazi Germany. Oxford University Press.

Maguire, M. (2009). The Birth of Biometric Security. Anthropology Today, 25(2), 9–14.

Maitland, C. (1846). The Church in the Catacombs. London, Longman, Brown, Green, And Longmans.

Male, E. (1958). The Gothic Image : Religious Art in France of The Thirteenth Century (D. Nussey, Trans.). Avalon Publishing.

Malik, S., Singh, J., Goyat, R., Saharan, Y., Chaudhry, V., Umar, A., Ibrahim, A. A., Akbar, S., Ameen, S., & Baskoutas, S. (2023). Nanomaterials-based biosensor and their applications : A review. Heliyon, 9(9), e19929. https://doi.org/https://doi.org/10.1016/j.heliyon.2023.e19929

Mallory, J. P. (1989). In Search of the Indo-Europeans : Language, Archaeology, and Myth. Thames and Hudson.

Manjapra, K. (2014). Age of Entanglement : German and Indian intellectuals Across Empire. Cambridge, Massachusetts, Harvard University Press.

Marchand, S. L. (2009). German Orientalism in the Age of Empire : Religion, Race, and Scholarship. Washington, D.C, German Historical Institute.

Marcus, I., & Crook, B. (2011). The Protocols of the Elders of Zion : An Authentic Document in Palestinian Authority Ideology. In L. Richard (Ed.), The Paranoid Apocalypse : A Hundred-Year Retrospective on The Protocols of the Elders of Zion (pp. 152–160). New York University Press. https://doi.org/doi:10.18574/nyu/9780814748923.003.0011

Marcus, J. R., & Saperstein, M. (1999). The Jew in the Medieval World : A Source Book : 315-1791 (Revised ed.). Hebrew Union College Press.

Marks, J. (2024). Our Fate Is Not in Our Genes. Cambridge, Cambridge University Press.

Marr, W. (1879). Der Sieg des Judenthums über das Germanenthum: vom nicht confessionellen Standpunkt aus betrachtet. Verlag nicht ermittelbar.

Marten, H. G. (1999). Racism, social darwinism, anti-semitism and Aryan supremacy. The International Journal of the History of Sport, 16(2), 23–41. https://doi.org/10.1080/09523369908714069

Martin, C. G. (2002). The Ahistoricism of the New Historicism : Knowledge as Power versus Power as Knowledge in Bacon's New Atlantis. In C. J. Summers & T.-L. Pebworth (Eds.), Fault Lines and Controversies in the Study of Seventeenth-Century English Literature (pp. 22–49). University of Missouri Press.

Martineau, J. (2020). Quadrivium. Wooden Books.

Marty, C. A. M. (2001). History of the Development and Use of Biological Weapons. Clinics in Laboratory Medicine, 21(3), 421–434. https://doi.org/https://doi.org/10.1016/S0272-2712(18)30017-9

Maser, W. (1970). Hitler's 'Mein Kampf' : an Analysis (R. H. Barry, Trans.). Faber.

Mass, S. (2022). The 'Volkskörper' in Fear : Gender, Race and Sexuality in the Weimar Republic. In (Vol. 13, pp. 233–250). Berghahn Books. https://doi.org/10.1515/9781845459765-015

Massin, B. (1996). From Virchow to Fischer : Physical Anthropology and "Modern Race Theories" in Wilhelmine Germany. In G. W. Stocking, Jr. & G. W. Stocking

(Eds.), Volksgeist As Method and Ethic : Essays on Boasian Ethnography and the German Anthropological Tradition (Vol. 8, pp. 79–154). United States: University of Wisconsin Press.

Massing, P. W. (1949). Rehearsal For Destruction : A Study of Political Anti-Semitism in Imperial Germany (M. Horkheimer & S. Flowerman, H., Eds.). New York, New York : Harper.

Matsumura, J. (1996). Internal Security in Wartime Japan (1937-45) and the Creation of Internal Insecurity. Canadian journal of history, 31(3), 395–412. https://doi.org/10.3138/cjh.31.3.395

Matteoni, F. (2008). The Jew, the Blood and the Body in Late Medieval and Early Modern Europe. Folklore, 119(2), 182–200.

Matthews, S. (2008). Theology and Science in the Thought of Francis Bacon. Abingdon, UK, Taylor & Francis Group.

Maymind, I. (2016). On the Concept of Self-Hatred : A Misnomer. Journal of Jewish identities, 9(1), 19–37. https://doi.org/10.1353/jji.2016.0003

Mazumdar, P. M. (1990). Blood and soil : the serology of the Aryan racial state. Bull Hist Med, 64(2), 187–219.

McCullough, P. A., Alexander, P. E., Armstrong, R., Arvinte, C., Bain, A. F., Bartlett, R. P., Berkowitz, R. L., Berry, A. C., Borody, T. J., Brewer, J. H., Brufsky, A. M., Clarke, T., Derwand, R., Eck, A., Eck, J., Eisner, R. A., Fareed, G. C., Farella, A., Fonseca, S. N. S.,…Zelenko, V. (2020). Multifaceted highly targeted sequential multidrug treatment of early ambulatory high-risk SARS-CoV-2 infection (COVID-19). RCM, 21(4), 517–530. https://doi.org/10.31083/j.rcm.2020.04.264

McCullough, P. A., Kelly, R. J., Ruocco, G., Lerma, E., Tumlin, J., Wheelan, K. R., Katz, N., Lepor, N. E., Vijay, K., Carter, H., Singh, B., McCullough, S. P., Bhambi, B. K., Palazzuoli, A., De Ferrari, G. M., Milligan, G. P., Safder, T., Tecson, K. M., Wang, D. D.,…Risch, H. A. (2021). Pathophysiological Basis and Rationale for Early Outpatient Treatment of SARS-CoV-2 (COVID-19) Infection. The American journal of medicine, 134(1), 16–22. https://doi.org/10.1016/j.amjmed.2020.07.003

McDermott, R. A. (1987). Anthroposophy. In The Encyclopaedia of Religion. https://archive.org/details/encyclopaediaofre01elia/page/320/mode/2up

______. (1992). Rudolf Steiner and Anthroposopay. In A. Faivre & J. Needleman (Eds.), Modern Esoteric Spirituality (Vol. 21 of World Spirituality : An Encyclopedic History of the Religious Quest). Crossroad.

McDougall, W. (1920). The Group Mind : A Sketch of the Principles of Collective Psychology, with Some Attempt to Apply Them to the Interpretation of National Life and Character. G.P. Putnam's Sons.

McGetchin, D. T. (2009). Indology, Indomania, and orientalism : ancient India's rebirth in modern Germany. Madison [N.J, Fairleigh Dickinson University Press.

McGovern, J. B. (1904). Concerning the "Chi Rho" Monogram. In E. Walford, et al. (Eds.), The Antiquary : A Magazine Devoted to the Study of the Past (pp. 5–9). Elliot Stock.

McGuire, W., & Hull, R. F. C. (1993). C.G. Jung speaking : interviews and encounters. Princeton, N.J., Princeton University Press.

McInnes, C., & Lee, K. (2006). Health, Security and Foreign Policy. Review of International Studies, 32(1), 5–23.

McIntosh, C. (1980). The Rosy Cross Unveiled : The History, Mythology and Rituals of

an Occult Order. Wellingborough, Northamptonshire. UK, Aquarian Press.
_____. (2011). The Rose Cross and the Age of Reason : Eighteenth-Century Rosicrucianism in Central Europe and Its Relationship to the Enlightenment. Albany, UNITED STATES, State University of New York Press.
McLaughlin, P. (2007). Kant on Heredity and Adaptation. In Heredity Produced : At the Crossroads of Biology, Politics, and Culture, 1500-1870 (pp. 277–291). MIT Press.
Meade, M. (1980). Madame Blavatsky, the Woman Behind the Myth. Putnam.
Mearsheimer, J. J. (2019). Bound to Fail : The Rise and Fall of the Liberal International Order. International Security, 43(4), 7–50. https://doi.org/10.1162/isec_a_00342
Medhi, R., Srinoi, P., Ngo, N., Tran, H. V., & Lee, T. R. (2020). Nanoparticle-Based Strategies to Combat COVID-19. ACS Appl Nano Mater, 3(9), 8557–8580. https://doi.org/10.1021/acsanm.0c01978
Mees, B. (2004). Hitler and Germanentum. Journal of Contemporary History, 39(2), 255–270.
_____. (2008). The Science of the Swastika. Amsterdam, NETHERLANDS, THE, Central European University Press.
Mehrtens, H. (1987). The Social System of Mathematics and National Socialism : A Survey. Sociological Inquiry, 57(2), 159–182. https://doi.org/10.1111/j.1475-682X.1987.tb01040.x
Meijer, M. C. (2014). Cranial Varieties in the Human and Orangutan Species. In N. Bancel, et al. (Eds.), The Invention of Race : Scientific and Popular Representations (pp. 33–47). Taylor & Francis Group.
Merkel, I., & Debus, A. G. (Eds.). (1988). Hermeticism and the Renaissance : Intellectual History and the Occult in Early Modern Europe. Folger Shakespeare Library.
Merle, W. (1936). Guttmanns Medizinische Terminologie : Ableitung und Erklärung der gebräuchlichsten Fachausdrücke aller Zweige der Medizin und ihrer Hilfswissenschaften. Urban & Schwarzenberg.
Merton, R. K. (1938). Science and the Social Order. Philosophy of Science, 5(3), 321–337.
Metcalfe, N. (2002). A Short History of Biological Warfare. Medicine, Conflict and Survival, 18(3), 271–282.
Metz, R. (1938). Bacon's Part in the Intellectual Movements of His Time. In J. D. Wilson (Ed.), Seventeenth Century Studies Presented to Sir Herbert Grierson (pp. 21–32). Octagon Book, Inc. (Reprinted from 1967)
Michael, J. S. (2017). Nuance Lost in Translation : Interpretations of J. F. Blumenbach's Anthropology in the English Speaking World. NTM, 25(3), 281–309. https://doi.org/10.1007/s00048-017-0173-8 (Original work published Nuancen, die in der Ubersetzung verloren gegangen sind : Interpretationen von J. F. Blumenbachs Anthropologie in der englischsprachigen Welt.)
Michael, R. (2006). Holy Hatred : Christianity, Antisemitism, and the Holocaust. New York, UNITED STATES, Palgrave Macmillan US.
Michaud, J., & Kates, J. (2013). Global health diplomacy : advancing foreign policy and global health interests. Global health science and practice, 1(1), 24–28. https://doi.org/10.9745/GHSP-D-12-00048
Michell, J. F. (2001). The Dimensions of Paradise : The Proportions and Symbolic Numbers of Ancient Cosmology. Adventures Unlimited Press.
Michels, R. (1926). Psychologie der antikapitalistischen Massen-bewegungen. In

Grundriss der Sozialökonomik (Vol. IX. Abteilung, Part I, pp. 241–359). J.C.B. Mohr (P. Siebeck).
Mikkeli, H. (1999). Hygiene in the Early Modern Medical Tradition. Finnish Academy of Science and Letters.
Milgram, S. (1974). Obedience to Authority : An experimental view ([First edition]. ed.). New York, Harper & Row.
Miller, J., Engelberg, S., & Broad, W. J. (2001). Germs : Biological weapons and America's secret war. New York, Simon & Schuster.
Milton, S. (1991). Gypsies and the Holocaust. The History Teacher, 24(4), 375–387. https://doi.org/10.2307/494697
Mishra, R., Minocha, S., Goel, R., Gaur, P. K., Lata, K., & Singh, R. (2025). Bioconvergence : advancing biosensors with nanotechnology for real-time health monitoring. Bulletin of the National Research Centre, 49(1). https://doi.org/10.1186/s42269-025-01308-4
Mishra, S. R., Dickey, M. D., Velev, O. D., & Tracy, J. B. (2016). Selective and directional actuation of elastomer films using chained magnetic nanoparticles [10.1039/C5NR07410J]. Nanoscale, 8(3), 1309–1313. https://doi.org/10.1039/C5NR07410J
Miyamoto, I. (2020). MASS SURVEILLANCE AND INDIVIDUAL PRIVACY. http://www.jstor.org/stable/resrep24871
Mnookin, S. (2011). The Panic Virus : A True Story of Medicine, Science, and Fear. Simon & Schuster.
Mohammadi, S., Sisay, M. M., Saraswati, P. W., Osman, A. K., Zuithoff, N. P. A., Weibel, D., Sturkenboom, M., & Ahmadizar, F. (2025). COVID-19 vaccine safety studies among special populations : A systematic review and meta-analysis of 120 observational studies and randomized clinical trials. Vaccine, 61. https://doi.org/10.1016/j.vaccine.2025.127342
Mohan, S. (2023). Cosmic Dance and The Universe. Shanlax International Journal of Arts Science and Humanities, 10, 11–17. https://doi.org/10.34293/sijash.v10i4.6005
Mommsen, H. (1996). The Rise and Fall of Weimar Democracy (L. E. Jones & E. Forster, Trans.). USA, University of North Carolina Press.
Mommsen, W. J. (1974). Max Weber und die deutsche Politik, 1890-1920. Mohr.
______. (1990). Max Weber and German Politics, 1890-1920 (M. S. Steinberg, Trans.). University of Chicago Press.
Montagu, A. (1974). Man's Most Dangerous Myth : The Fallacy of Race (Fifth edition revised and enlarged ed.). London; New York, London; New York : Oxford University Press.
Montgomery, J. A. (1914). 'Present Tendencies in Old Testament Criticism'. The Biblical World, 43(No. 5), 310–320.
Montriou, J. A. L. (1788). Elements of Universal History.
Moore, P. (2024). "It isn't a book to be read through" : The Germans and Mein Kampf, 1925-1945. The Historian, 86(2-3), 209–237. https://doi.org/10.1080/00182370.2024.2464406
Moorhouse, R. (2010). The SS : A New History, by Adrian Weale, Little, Brown. Financial Times. https://www.ft.com/content/651bf2d8-c76a-11df-aeb1-00144feab49a
Moran, B. T. (1994). Alchemy, Prophecy, and the Rosicrucians: Raphael Eglinus and Mystical Currents of the Early Seventeenth Century. In P. Rattansi & A.

Clericuzio (Eds.), Alchemy and Chemistry in the 16th and 17th Centuries (Vol. 140, pp. 103—119). Springer-Science & Business Media, B.V.

Morens, D. M., Folkers, G. K., & Fauci, A. S. (2022). The Concept of Classical Herd Immunity May Not Apply to COVID-19. J Infect Dis, 226(2), 195–198. https://doi.org/10.1093/infdis/jiac109

Morrisson, M. S. (2007). The Periodical Culture of the Occult Revival : Esoteric Wisdom, Modernity and Counter-Public Spheres. Journal of Modern Literature, 31(2), 1–22.

Moscovici, S. (1985). The Age of the Crowd : A Historical Treatise on Mass Psychology (J. C. Whitehouse, Trans.). Cambridge [Cambridgeshire] ;, Cambridge University Press.

Mosse, G. L. (1961). The Mystical Origins of National Socialism. Journal of the History of Ideas, 22(1), 81–96. https://doi.org/10.2307/2707875

_____. (1964). The Crisis of German Ideology : Intellectual Origins of the Third Reich (First ed.). New York, New York : Grosset & Dunlap.

_____. (1966). Nazi Culture : Intellectual, Cultural and Social Life in the Third Reich (S. Attanasio, Trans.). Grosset & Dunlap.

_____. (1970). Germans and Jews : the Right, the Left, and the search for a "Third Force" in pre-Nazi Germany ([First edition]. ed.). New York, H. Fertig.

_____. (1975). The Nationalization of the Masses : Political Symbolism and Mass Movements in Germany from the Napoleonic Wars Through the Third Reich ([First American edition]. ed.). New York, New York : H. Fertig.

_____. (1978). Toward the Final Solution : A History of European Racism (First ed.). Howard Fertig.

_____. (1998). The Crisis of German Ideology : Intellectual Origins of the Third Reich. H. Fertig.

_____. (1999). The Fascist Revolution : Toward a General Theory of Fascism (First edition. ed.). New York, H. Fertig.

_____. (2000). The Jews : Myth and Counter-Myth. In L. Back & J. Solomos (Eds.), Theories of Race and Racism : A Reader (Vol. Part 3: Racism and antisemitism, pp. 195–205). Routledge.

_____. (2020). Toward the Final Solution : A History of European Racism. Columbia, UNITED STATES, University of Wisconsin Press.

Motadel, D. (2014). Islam and Nazi Germany's war. Cambridge, Massachusetts, The Belknap Press of Harvard University Press.

Moustaki, N., & Borman, G. (2011). CliffsNotes on Orwell's 1984. NY, USA, Houghton Mifflin Harcourt.

Mueller, R. G. (2019). The "Breeding of Humanity" : Nietzsche and Shaw's Man and Superman [Jul 16, 24]. Shaw, 39(2), 183–203. https://doi.org/10.5325/shaw.39.2.0183

Mueller, T., & Beddies, T. (2006). "The Destruction of Life Unworthy of Living" in National Socialist Germany. International Journal of Mental Health, 35(3), 94–104.

Mühlberger, D. (2004). Hitler's Voice : The Völkischer Beobachter, 1920–1933 (Vol. 1 : Hitler's Voice : Organisation & development of the Nazi Party). P. Lang.

Muller, H. J. (1933). Book Reviews [Review]. The Birth Control Review, XVII, 18+.

_____. (1962). Studies in genetics. Bloomington, Indiana University Press.

Müller-Hill, B. (1998). Murderous Science : Elimination by Scientific Selection of Jews,

Gypsies, and Others in Germany, 1933-1945 (G. R. Fraser, Trans.; English ed.). Plainview, N.Y, Plainview, N.Y: Cold Spring Harbor Laboratory Press.

Müller-Wille, S., & Rheinberger, H.-J. (2007). Heredity—The Formation of an Epistemic Space. In S. Müller-Wille & H.-J. Rheinberger (Eds.), Heredity Produced : At the Crossroads of Biology, Politics, and Culture, 1500-1870 (pp. 3–34). MIT Press.

Mund, R. J. (1976). Jörg Lanz v. Liebenfels und der Neue Templer Orden : d. Esoterik d. Christentums (First ed.). Stuttgart, Rudolf Arnold Spieth Verlag.

Mundy, R. (2006). Musical Evolution and the Making of Hierarchy. The World of Music, 48(3), 13–27.

Murard, L., & Zylberman, P. (2000). Seeds for French health care : did the Rockefeller Foundation plant the seeds between the two World Wars? Studies in History and Philosophy of Science Part C: Studies in History and Philosophy of Biological and Biomedical Sciences, 31(3), 463–475. https://doi.org/https://doi.org/10.1016/S1369-8486(00)00016-9

Myers, P. (2015). An Elite Class of Thinkers : Monism between Science and the Spiritual in Wilhelmine Germany. In M. Black & E. Kurlander (Eds.), Revisiting the 'Nazi Occult' : Histories, Realities, Legacies (pp. 42–64). Camden House.

Mynatt, C., & Sherman, S. J. (1975). Responsibility attribution in groups and individuals : A direct test of the diffusion of responsibility hypothesis. Journal of personality and social psychology, 32(6), 1111–1118. https://doi.org/10.1037/0022-3514.32.6.1111

Naeem, S. B., Bhatti, R., & Khan, A. (2021). An exploration of how fake news is taking over social media and putting public health at risk. Health Information & Libraries Journal, 38(2), 143–149. https://doi.org/https://doi.org/10.1111/hir.12320

Nagl, M. (1974). SF, Occult Sciences, and Nazi Myths. Science Fiction Studies, 1(3), 185–197.

Nandagopal, N., & Elowitz, M. B. (2011). Synthetic Biology : Integrated Gene Circuits. Science, 333(6047), 1244–1248.

National Intelligence Council. (2000). National intelligence estimate : the global infectious disease threat and its implications for the United States. Environ Change Secur Proj Rep(6), 33–65.

Nazi Germany. (1933). Blut und Boden (Blood and Soil). In (pp. Nationalist slogan expressing Nazi Germany's ideal of a racially defined national body).

Nelson, W. M. (2010). Making men : Enlightenment ideas of racial engineering. Am Hist Rev, 115(5), 1364–1394. https://doi.org/10.1086/ahr.115.5.1364

Netanyahu, B. (1995). The Origins of the Inquisition in Fifteenth Century Spain (First edition. ed.). New York, New York : Random House.

Nettesheim, H. C. A. (1993). Three Books of Occult Philosophy (J. Freake, Trans.; D. Tyson, Ed.). Llewellyn.

______. (2009). The Fourth Book of Occult Philosophy : The Companion to Three Books of Occult Philosophy (R. Turner, Trans.; D. Tyson, Ed.). Llewellyn Publications.

Neumann, F. L. (1944). Behemoth : The Structure and Practice of National Socialism, 1933-1944. Toronto; New York [etc.], Oxford University Press.

Newall, V. (2004). "The Jew as a Witch-Figure.". In V. Newall (Ed.), The Witch Figure : Folklore Essays by a Group of Scholars in England Honouring the 75th Birthday of Katharine M. Briggs. Routledge.

Newby, G. A., & Liu, D. R. (2021). In vivo somatic cell base editing and prime editing.

Mol Ther, 29(11), 3107–3124. https://doi.org/10.1016/j.ymthe.2021.09.002
Newman, W. R., & Grafton, A. (2001). Secrets of Nature : Astrology and Alchemy in Early Modern Europe. MIT Press.
Nguyen, C. T., Saputra, Y. M., Huynh, N. V., Nguyen, N. T., Khoa, T. V., Tuan, B. M., Nguyen, D. N., Hoang, D. T., Vu, T. X., Dutkiewicz, E., Chatzinotas, S., & Ottersten, B. (2020a). A Comprehensive Survey of Enabling and Emerging Technologies for Social Distancing—Part I : Fundamentals and Enabling Technologies. IEEE Access, 8, 153479–153507. https://doi.org/10.1109/ACCESS.2020.3018140
______. (2020b). A Comprehensive Survey of Enabling and Emerging Technologies for Social Distancing—Part II : Emerging Technologies and Open Issues. IEEE Access, 8, 154209–154236. https://doi.org/10.1109/ACCESS.2020.3018124
Nicholas, L. H. (1994). The Rape of Europa : The Fate of Europe's Treasures in the Third Reich and the Second World War (First edition. ed.). New York, Knopf.
Nicolle, C. (1995). "Investigations on Typhus." Nobel Lectures : Physiology Or Medicine, 1922–1941. Amsterdam, Elsevier Publishing Co.
Niekisch, E. (1953). Das Reich der niederen Dämonen. Rowohlt.
Nietzsche, F. W. ([1908] 1992). Ecce Homo : How One Becomes What One is (R. J. Translation: Hollingdale & M. Introduction: Tanner, Trans.). Penguin Publishing Group.
______. (1911). The Twilight of the Idols, Or, How to Philosophise with the Hammer. Foulis.
______. (1917). Thus Spoke Zarathustra (T. Common, Trans.). Modern Library.
______. (1967). Beyond Good and Evil : Prelude to a Philosophy of the Future (H. Zimmern, Trans.; [Fourth edition, second impression]. ed.). London, Allen & Unwin.
Nikiforuk, A. (1993). The Fourth Horseman : A Short History of Epidemics, Plagues, Famine and Other Scourges. London, UK, Phoenix.
Nimpf, S., & Keays, D. A. (2017). Is magnetogenetics the new optogenetics? The EMBO Journal, 36(12), 1643–1646. https://doi.org/https://doi.org/10.15252/embj.201797177
Nirenberg, D. (2002). Mass Conversion and Genealogical Mentalities : Jews and Christians in Fifteenth-Century Spain. Past & Present(174), 3–41.
______. (2003). The Birth of the Pariah : Jews, Christian Dualism, and Social Science. Social Research, 70(1), 201–236.
______. (2008). Race and the Middle Ages. In M. R. Greer, et al. (Eds.), Rereading the Black Legend (pp. 71–87). University of Chicago Press. https://doi.org/10.7208/chicago/9780226307244.003.0004
______. (2014). Was There Race before Modernity? The Example of "Jewish" Blood in Late Medieval Spain. In D. Nirenberg (Ed.), Neighboring Faiths : Christianity, Islam, and Judaism in the Middle Ages and Today (pp. 169–190). University of Chicago Press.
Noah, D. L., Huebner, K. D., Darling, R. G., & Waeckerle, J. F. (2002). The history and threat of biological warfare and terrorism. Emerg Med Clin North Am, 20(2), 255–271. https://doi.org/10.1016/s0733-8627(01)00002-5
Noakes, J. (1984a). Nazism and Eugenics : The Background to the Nazi Sterilization Law of 1914 July 1933. In R. Bullen, et al. (Eds.), Ideas into politics: aspects of European history 1880-1950 (pp. 75–94). Croom Helm.

_____. (1984b). Nazism : 1919-1945 : A History in Documents and Eyewitness Accounts, 1919-1945 (Vol. 1). Schocken Books.

Nordenskiöld, E. (1936). The History of Biology : A Survey. New York, New York : Tudor Pub. Co.

North, J. (1991). The Quadrivium. In H. d. Ridder-Symoens (Ed.), A History of the University in Europe: Volume 1: Universities in the Middle Ages (Vol. 1, pp. 337–359). Cambridge University Press. https://doi.org/DOI: 10.1017/CBO9780511599507.013

Nurenberg Military Tribunals. (1954). Trials of War Criminals before the Nurenberg Military Tribunals under Control Council Law No. 10 "Green Series" Vol. IV (0017-811X). Retrieved from the Library of Congress, https://www.loc.gov/item/2011525364_NT_war-criminals_Vol-IV/.

O'Flaherty, W. D. (1980). Dionysus and Siva : Parallel Patterns in Two Pairs of Myths. History of Religions, 20(1/2), 81–111.

Oakesmith, J. (1919). Race & nationality, an inquiry into the origin and growth of patriotism, by John Oakesmith. England, William Heinemann.

Obeng, E. M., Fianu, I., & Danquah, M. K. (2022). Multivalent ACE2 engineering—A promising pathway for advanced coronavirus nanomedicine development. Nano Today, 46. https://doi.org/10.1016/j.nantod.2022.101580

Ogilvie, J., & Annandale, C. (1883). 'Aryan'. In The Imperial Dictionary of the English Language : A Complete Encyclopedic Lexicon, Literary, Scientific, and Technological. https://books.google.com/books?id=NNGD5ENlUPgC

Ohana, D. (2022). Jacob L. Talmon : Mission and Testimony : Political Essays. Liverpool, UK, Liverpool University Press.

Opotow, S. (1990a). Moral Exclusion and Injustice : An Introduction. Journal of Social Issues, 46(1), 1–20. https://doi.org/https://doi.org/10.1111/j.1540-4560.1990.tb00268.x

_____. (1990b). Deterring Moral Exclusion. Journal of Social Issues, 46(1), 173–182. https://doi.org/https://doi.org/10.1111/j.1540-4560.1990.tb00280.x

Ornstein, R. E. (1977). The Psychology of Consciousness (Second ed.). New York, Harcourt Brace Jovanovich.

Ortega, F., & Orsini, M. (2020). Governing COVID-19 without government in Brazil : Ignorance, neoliberal authoritarianism, and the collapse of public health leadership. Global Public Health, 15(9), 1257–1277. https://doi.org/10.1080/17441692.2020.1795223

Otis, L. (1999). Membranes : Metaphors of Invasion in Nineteenth-Century Literature, Science, and Politics. Baltimore, Md, Johns Hopkins University Press.

Outram, D. (2013). Enlightenment. Cambridge, UNKNOWN, Cambridge University Press.

Owen, A. (2004). The Place of Enchantment : British Occultism and The Culture of the Modern (First ed.). Chicago, Ill, University of Chicago Press.

Owen, D. S., Strong, T. B., & Livingstone, R. (Eds.). (2004). The Vocation Lectures. Hackett Pub.

Pagel, A. (1992). Jugendstil and Racism : An Unexpected Alliance. RACAR: revue d'art canadienne / Canadian Art Review, 19(1/2), 97–111.

Pagels, H. R. (1982). The Cosmic Code : Quantum Physics as the Language of Nature. Dover Publications.

Pais, A. (1991). Niels Bohr's Times : In Physics, Philosophy, and Polity. Clarendon Press.

Palló, G. (2000). Rescue and cordon sanitaire : The Rockefeller Foundation in Hungarian public health. Studies in History and Philosophy of Science Part C: Studies in History and Philosophy of Biological and Biomedical Sciences, 31(3), 433–445. https://doi.org/https://doi.org/10.1016/S1369-8486(00)00013-3

Palvi, M. (1941). Economic Foundations of the German Totalitarian State. American Journal of Sociology, 46(4), 469–486.

Parhiz, H., Atochina-Vasserman, E. N., & Weissman, D. (2024). mRNA-based therapeutics : looking beyond COVID-19 vaccines. The Lancet, 403(10432), 1192–1204. https://doi.org/10.1016/s0140-6736(23)02444-3

Park, J. H., Jiang, Y., Zhou, J., Gong, H., Mohapatra, A., Heo, J., Gao, W., Fang, R. H., & Zhang, L. (2021). Genetically engineered cell membrane-coated nanoparticles for targeted delivery of dexamethasone to inflamed lungs. Sci Adv, 7(25). https://doi.org/10.1126/sciadv.abf7820

Paterlini, M. (2021). Covid-19 : Sweden, Norway, and Finland suspend use of Moderna vaccine in young people "as a precaution". BMJ. https://doi.org/10.1136/bmj.n2477

Patone, M., Handunnetthi, L., Saatci, D., Pan, J., Katikireddi, S. V., Razvi, S., Hunt, D., Mei, X. W., Dixon, S., Zaccardi, F., Khunti, K., Watkinson, P., Coupland, C. A. C., Doidge, J., Harrison, D. A., Ravanan, R., Sheikh, A., Robertson, C., & Hippisley-Cox, J. (2021). Neurological complications after first dose of COVID-19 vaccines and SARS-CoV-2 infection. Nature Medicine, 27(12), 2144–2153. https://doi.org/10.1038/s41591-021-01556-7

Pearson, K. (1897). Socialism and Natural Selection. In The Chances of Death, and Other Studies in Evolution (Vol. 1, pp. 103–139). E. Arnold.

______. (1905). National Life from the Standpoint of Science : An Address delivered at Newcastle, November 19, 1900. London, Adam and Charles Black.

______. (1922). Francis Galton, 1822-1922 : A Centenary Appreciation. Cambridge University Press.

Peck, M. S. (1983). People of the Lie : The Hope for Healing Human Evil. Simon and Schuster.

Peleg, K., Bodas, M., Hertelendy, A. J., & Kirsch, T. D. (2021). The COVID-19 pandemic challenge to the All-Hazards Approach for disaster planning. International Journal of Disaster Risk Reduction, 55, 102103. https://doi.org/https://doi.org/10.1016/j.ijdrr.2021.102103

Pellegrino, E. D. (2005). Foreword. In Murderous Medicine : Nazi Doctors, Human Experimentation, and Typhus (pp. xiii–xviii). Praeger Publishers.

Pendas, D. O., Roseman, M., & Wetzell, R. F. (Eds.). (2017). Beyond The Racial State : Rethinking Nazi Germany. Cambridge University Press.

Pennick, N. (1981). Hitler's Secret Sciences : His Quest for the Hidden Knowledge of the Ancients. Neville Spearman.

Perusse, R. I. (1958). Psychological Warfare Reappraised. In D. W.E. & M. Janowitz (Eds.), A Psychological Warfare Casebook (pp. 25–36). Johns Hopkins Press.

Petersen, M. B., Jørgensen, F., & Lindholt, M. F. (2022). Did the European suspension of the AstraZeneca vaccine decrease vaccine acceptance during the COVID-19 pandemic? Vaccine, 40(4), 558–561. https://doi.org/10.1016/j.vaccine.2021.12.026

Petersen-Withorn, C. (Apr 30, 2021). How Much Money America's Billionaires Have Made During The Covid-19 Pandemic https://www.forbes.com/sites/

chasewithorn/2021/04/30/american-billionaires-have-gotten-12-trillion-richer-during-the-pandemic/
Petrie, W. M. F. (1898). Religion and Conscience in Ancient Egypt : Lectures Delivered at University College, London. Methuen.
Peukert, D. (1987). Inside Nazi Germany : Conformity, Opposition, and Racism in Everyday Life (R. Deveson, Trans.). New Haven, New Haven : Yale University Press.
Pfahler, G. (1938). Warum Erziehung trotz Vererbung? (Third ed.). Vieweg+Teubner Verlag.
_____. (1943). System der Typenlehren (Fourth ed.). J. A. Barth.
Phelps, R. H. (1954). Die Hitler-Bibliothek. Deutsche Rundschau, 80, 923–931.
Phelps, R. H. (1961). Theodor Fritsch und der Antisemitismus. Deutsche Rundschau, 87(5), 442–449.
_____. (1963). "Before Hitler Came" : Thule Society and Germanen Orden. The Journal of Modern History, 35(3), 245–261.
Phillips, D. (2015). Francis Bacon and the Germans : Stories from when 'science' meant 'Wissenschaft'. History of Science, 53(4), 378–394. https://doi.org/10.1177/0073275315597609
Phillips, P. T. (2007). One World, One Faith : The Quest for Unity in Julian Huxley's Religion of Evolutionary Humanism. Journal of the History of Ideas, 68(4), 613–633.
Piccaluga, P. P., Di Guardo, A., Lagni, A., Lotti, V., Diani, E., Navari, M., & Gibellini, D. (2022). COVID-19 Vaccine : Between Myth and Truth. Vaccines (Basel), 10(3), 349. https://doi.org/10.3390/vaccines10030349
Picker, H. (1968). Hitlers Tischgesprache Im Führerhauptquartier 1941-1942. Deutscher taschenbuch Verlag.
_____. (2003). Hitlers Tischgespräche im Führerhauptquartier. Propyläen.
Pilz, S., Theiler-Schwetz, V., Trummer, C., Krause, R., & Ioannidis, J. P. A. (2022). SARS-CoV-2 reinfections : Overview of efficacy and duration of natural and hybrid immunity. Environmental Research, 209, 112911. https://doi.org/https://doi.org/10.1016/j.envres.2022.112911
Piper, E. (2005). Alfred Rosenberg : Hitlers Chefideologe. Pantheon Verlag.
Pipes, D. (1997). Conspiracy : How the Paranoid Style Flourishes and where it comes from. In. Free Press.
Plato. (1804). Introduction to The Critias (T. Taylor, Trans.). In The Timaeus and Critias. G. and W. B. Whittaker.
_____. (1955). The Republic of Plato (F. M. Cornford, Trans.). Oxford, Oxford University Press.
_____. (2000). Timaeus (D. J. Zeyl, Trans.). Hackett Publishing Company, Incorporated.
Ploetz, A. (1927). Rasse und Menschheit. Süddeutsche Monatshefte, 24(10), 248–253.
Ploetz, A. J. (1895). Die Tüchtigkeit unsrer Rasse und der Schutz der Schwachen : bein Versuch über *Rassenhygiene* und ihr Verhältniss zu den humen Idealen, besonders zum Socialismus. S. Fischer.
Ploetz, A. J., & Nordenholz, A. (1904). Archiv für Rassen- u. Gesellschafts-Biologie : einschliesslich Rassen- u. Gesellschafts-Hygiene (Vol. I). B. G. Teubner.
Poewe, K. O. (2006). New Religions and the Nazis. London, Routledge.
Pois, R. A. (1986). National socialism and the religion of nature. New York, St. Martin's Press.

Pokorny, L. K. (2025). “The Way, the Truth, and the Light” : Franz Hartmann’s Daodejing. Vienna Journal of East Asian Studies, 16(1), 25–49. https://doi.org/ https://doi.org/10.30965/25217038-12345007

Poley, J. (2013). Rudolf Steiner and the theosophy of greed. In J. M. Cho, et al. (Eds.), Transcultural Encounters Between Germany and India : Kindred Spirits in the 19th and 20th Centuries (pp. 55–67). Taylor & Francis Group.

Poliakov, L. (1955). Histoire de l’antisémitisme : Du Christ aux juifs de cour. Calmann-Lévy.

Poliakov, L. (1965). The History of Anti-Semitism (R. Howard, Trans.). Schocken Books.

_____. (1974). The Aryan Myth : A History of Racist and Nationalist Ideas in Europe (E. Howard, Trans.). London, New York: Basic Books.

Pollefeyt, D. (1999). The Kafkaesque World of the Holocaust: Paradigmatic Shifts in the Ethical Interpretation of the Nazi Genocide. In J. K. Roth (Ed.), Ethics after the Holocaust : Perspectives, Critiques, and Responses. Paragon House.

Popkin, R. H. (1974a). The Philosophical Basis of Modern Racism. In Philosophy and the Civilizing Arts : Essays Presented to Herbert W. Schneider (pp. 126–165). Athens : Ohio University Press.

_____. (1974b). The Philosophical Basis of Eighteenth-Century Racism. Studies in eighteenth-century culture, 3(1), 245–262. https://doi.org/10.1353/sec.1974.0016 (Original work published See also Pagliaro, H. E. (Ed.). (1973). American Society for Eighteenth-Century Studies : Racism in the Eighteenth Century (Vol. 3). Press of Case Western Reserve University. https://books.google.com/books?id=axlmAAAAMAAJ.)

Popper, K. R. (1963). The Open Society and its Enemies (Fourth, Revised ed., Vol. 1: The spell of Plato). Princetow, N. J, Princeton University Press.

Porto, E. M., Komor, A. C., Slaymaker, I. M., & Yeo, G. W. (2020). Base editing : advances and therapeutic opportunities. Nature Reviews Drug Discovery, 19(12), 839–859. https://doi.org/10.1038/s41573-020-0084-6

Pottegård, A., Lund, L. C., Karlstad, Ø., Dahl, J., Andersen, M., Hallas, J., Lidegaard, Ø., Tapia, G., Gulseth, H. L., Ruiz, P. L.-D., Watle, S. V., Mikkelsen, A. P., Pedersen, L., Sørensen, H. T., Thomsen, R. W., & Hviid, A. (2021). Arterial events, venous thromboembolism, thrombocytopenia, and bleeding after vaccination with Oxford-AstraZeneca ChAdOx1-S in Denmark and Norway : population based cohort study. BMJ. https://doi.org/10.1136/bmj.n1114

Prakash, V., Moore, M., & Yáñez-Muñoz, R. J. (2016). Current Progress in Therapeutic Gene Editing for Monogenic Diseases. Molecular Therapy, 24(3), 465–474. https://doi.org/10.1038/mt.2016.5

Pratkanis, A. R., Pratkanis, A., & Aronson, E. (2001). Age of Propaganda : The Everyday Use and Abuse of Persuasion. Henry Holt and Company.

Pringle, H. A. (2006). The Master Plan : Himmler’s Scholars and the Holocaust (First ed.). New York, New York : Hyperion.

Prior, M. E. (1954). Bacon’s Man of Science. Journal of the History of Ideas, 15(3), 348–370. https://doi.org/10.2307/2707759

Proctor, R. (1982). Pawns or Pioneers? The Role of Doctors in the Origins of Nazi Racial Science

_____. (2000). The Nazi War on Cancer (Third Printing ed.). Princeton University Press.

Proctor, R. N. (1988). Racial Hygiene : Medicine Under The Nazis. Harvard University Press.

Proctor, R. N. (1990). Nazi Biomedical Technologies. In T. Casey & L. E. Embree (Eds.), Lifeworld and Technology (Vol. Current Continental Research 009, pp. 17–48). Center for Advanced Research in Phenomenology & University Press of America.

______. (1992). Nazi Doctors, Racial Medicine, and Human Experimentation. In G. J. Annas & M. A. Grodin (Eds.), The Nazi Doctors and the Nuremberg Code : Human Rights in Human Experimentation (pp. 17–31). Oxford University Press.

Provine, W. B. (1971). The Origins of Theoretical Population Genetics. University of Chicago Press.

Pugh, J., Wilkinson, D., Kerridge, I., & Savulescu, J. (2022). Vaccine suspension, risk, and precaution in a pandemic. Journal of Law and the Biosciences, 9(1). https://doi.org/10.1093/jlb/lsab036

Pulzer, P. (1988). The Rise of Political Anti-semitism in Germany & Austria (Revised ed.). London, UK, Peter Halban.

Puskás, I. (1990). Magasthenes and the "Indian Gods" Herakles and Dionysos. Mediterranean Studies, 2, 39–47.

Puzzo, D. A. (1964). Racism and the Western Tradition. Journal of the History of Ideas, 25(4), 579–586. https://doi.org/10.2307/2708188

Qiu, M., Glass, Z., Chen, J., Haas, M., Jin, X., Zhao, X., Rui, X., Ye, Z., Li, Y., Zhang, F., & Xu, Q. (2021). Lipid nanoparticle-mediated codelivery of Cas9 mRNA and single-guide RNA achieves liver-specific *in vivo* genome editing ofAngptl3. Proceedings of the National Academy of Sciences, 118(10). https://doi.org/10.1073/pnas.2020401118

Quinn, M. (1994). The Swastika : Constructing the Symbol (First ed.). London, Routledge.

Raethke, M., van Hunsel, F., Thurin, N. H., Dureau-Pournin, C., Mentzer, D., Kovačić, B., Mirošević Skvrce, N., De Clercq, E., Sabbe, M., Trifirò, G., Luxi, N., Giovanazzi, A., Shakir, S., Klungel, O. H., Schmikli, S., & Sturkenboom, M. (2023). Cohort Event Monitoring of Adverse Reactions to COVID-19 Vaccines in Seven European Countries: Pooled Results on First Dose. Drug Safety, 46(4), 391–404. https://doi.org/10.1007/s40264-023-01281-9

Ramesh, M., Janani, R., Deepa, C., & Rajeshkumar, L. (2022). Nanotechnology-Enabled Biosensors : A Review of Fundamentals, Design Principles, Materials, and Applications. Biosensors (Basel), 13(1). https://doi.org/10.3390/bios13010040

Ramm, R. (2019). Medical Jurisprudence and Rules of the Medical Profession (M. W. Cooper, Trans.). Springer International Publishing.

Rankin, C. H., Abrams, T., Barry, R. J., Bhatnagar, S., Clayton, D. F., Colombo, J., Coppola, G., Geyer, M. A., Glanzman, D. L., Marsland, S., McSweeney, F. K., Wilson, D. A., Wu, C.-F., & Thompson, R. F. (2009). Habituation revisited : An updated and revised description of the behavioral characteristics of habituation. Neurobiology of Learning and Memory, 92(2), 135–138. https://doi.org/10.1016/j.nlm.2008.09.012

Rao, S., Chen, R., LaRocca, A. A., Christiansen, M. G., Senko, A. W., Shi, C. H., Chiang, P. H., Varnavides, G., Xue, J., Zhou, Y., Park, S., Ding, R., Moon, J., Feng, G., & Anikeeva, P. (2019). Remotely controlled chemomagnetic modulation of targeted neural circuits. Nat Nanotechnol, 14(10), 967–973. https://doi.org/10.1038/s41565-019-0521-z

Rauschning, H. (1939). Die Revolution Des Nihilismus. William Heinemann.

_____. (1940). The Voice of Destruction. G. P. Putnam's sons.
_____. (2005). Gespräche mit Hitler : mit dem Schlußkapitel "Hitler privat". Europa-Verlag.
Rawlinson, G., Rawlinson, H., & Wilkinson, J. G. (1880). History of Herodotus : A New English Version (Vol. II). Murray.
Read, J. (1995). From Alchemy to Chemistry. New York, Dover Publications.
Reden, S. v. (2007). Money in Ptolemaic Egypt : from the Macedonian Conquest to the end of the third century BC. Cambridge, UK ;, Cambridge University Press.
Redles, D. (2005). Hitler's Millennial Reich : Apocalyptic Belief and the Search for Salvation (First ed.). New York, New York University Press.
_____. (2012). The Turning Point : The Protocols of the Elders Zion and the Eschatological War between Aryans and the Jews. In R. Landes & S. T. Katz (Eds.), The Paranoid Apocalypse : A Hundred-Year Retrospective on the Protocols of the Elders of Zion (pp. 112–131). New York : New York University Press.
Regalado, A. (2015). Engineering the Perfect Baby : Scientists are developing ways to edit the DNA of tomorrow's children. Should they stop before it's too late? MIT Technology Review, 1–10.
Reich, W. (1970). The Mass Psychology of Fascism (V. R. Carfagno, Trans.). Farrar, Straus & Giroux.
Reichmann, E. G. (1951). Hostages of Civilisation : The Social Sources of National Socialist Anti-Semitism. Boston, Boston : Beacon Press.
Reinach, S. (1892). L'origine des Aryens : histoire d'une controverse (E. Leroux, Ed.). E. Leroux.
Reis, G., Moreira Silva, E. A. d. S., Medeiros Silva, D. C., Thabane, L., Singh, G., Park, J. J. H., Forrest, J. I., Harari, O., Quirino dos Santos, C. V., Guimarães de Almeida, A. P. F., Figueiredo Neto, A. D. d., Savassi, L. C. M., Milagres, A. C., Teixeira, M. M., Simplicio, M. I. C., Ribeiro, L. B., Oliveira, R., Mills, E. J., & Investigators, T. (2021). Effect of Early Treatment With Hydroxychloroquine or Lopinavir and Ritonavir on Risk of Hospitalisation Among Patients With COVID-19: The TOGETHER Randomized Clinical Trial. JAMA Network Open, 4(4), e216468–e216468. https://doi.org/10.1001/jamanetworkopen.2021.6468
Rensmann, L. (2017). Guilt, Resentment, and Post-Holocaust Democracy : The Frankfurt School's Analysis of "Secondary Antisemitism" in the Group Experiment and Beyond [Article]. Antisemitism Studies, 1, 4+.
_____. (2020). The Contemporary Globalization of Political Antisemitism : Three Political Spaces and the Global Mainstreaming of the "Jewish Question" in the Twenty-First Century. Journal of Contemporary Antisemitism, 3(1), 83–108. https://doi.org/10.26613/jca/3.1.46
Resnick, I. M. (2000). Medieval Roots of the Myth of Jewish Male Menses. The Harvard Theological Review, 93(3), 241–263. (Original work published On the Roots of the Myth of Jewish Male Menses in Jacques de Vitry, History of Jerusalem (International Rennert Guest Lecture Series 3; Ramat Gan: Bar-Ilan University, 1998) 1-27.)
Reynaud-Paligot, C. (2014). Construction and Circulation of the Notion of "Race" in the Nineteenth Century. In N. Bancel, et al. (Eds.), The Invention of Race : Scientific and Popular Representations (pp. 87–99). Taylor & Francis Group.
Reznikov, A. (2001). George Orwell's Theory of Language. NE, USA, iUniverse.

Richardson, M. (1990). Philanthropy and the internationality of Learning : The Rockefeller Foundation and National Socialist Germany. Minerva, 28(1), 21–58. https://doi.org/10.1007/BF01096325

Rockefeller Foundation. (1933). The Rockefeller Foundation : Annual Report : 1933. Rockefeller Foundation.

Rodrı, x, & guez-Ocaña, E. (2000). Foreign expertise, political pragmatism and professional elite : The Rockefeller Foundation in Spain, 1919–39. Studies in History and Philosophy of Science Part C: Studies in History and Philosophy of Biological and Biomedical Sciences, 31(3), 447–461. https://doi.org/https://doi.org/10.1016/S1369-8486(00)00014-5

Rohden, A. v. (1934). Verstößt das Gesetz zur Verhütung erbkranken Nachwuchses gegen das Gebot der Nächstenliebe? Neues Volk, 2.

Roland, C. G. (1992). Courage Under Siege : Starvation, Disease, and Death in the Warsaw Ghetto. New York, Oxford University Press.

Rose, P. L. (1982). The Noble Anti-Semitism of Richard Wagner [Wagner. A Biography, Curt von Westernhagen, Mary Whittall; The Diary of Richard Wagner 1865-1882. The Brown Book, Joachim Bergfeld, George Bird; Cosima Wagner's Diaries, Martin Gregor-Dellin, Dietrich Mack, Geoffrey Skelton]. The Historical Journal, 25(3), 751–763.

Rosen, G. (1958). A History of Public Health. New York, MD Publications.

Rosenbaum, R. (2014). Explaining Hitler : the search for the origins of his evil (Updated edition, First Da Capo Press edition. ed.). Boston, MA, Da Capo Press.

Rosenbaum, R. A. (2010). Waking to Danger : Americans and Nazi Germany, 1933-1941. Santa Barbara, CA, USA, ABC-CLIO.

Rosenberg, A. (1934). Der deutsche Ordensstaat : ein neuer Abschnitt in der Entwicklung des nationalsozialistischen Staatsgedankens. Zentralverlag der N.S.D.A.P., F. Eher Nachf.

Ross, E. A. (1920). Foundations of Sociology (R. T. Ely, Ed.). Macmillan.

Ross, E. D. (2021). Differential Hemispheric Lateralization of Emotions and Related Display Behaviors : Emotion-Type Hypothesis. Brain Sciences, 11(8), 1034.

Rossi, P. (1968). Francis Bacon : From Magic to Science (S. Rabinovitch, Trans.). The University of Chicago Press.

Rothgangl, T., Dennis, M. K., Lin, P. J. C., Oka, R., Witzigmann, D., Villiger, L., Qi, W., Hruzova, M., Kissling, L., Lenggenhager, D., Borrelli, C., Egli, S., Frey, N., Bakker, N., Walker, J. A., Kadina, A. P., Victorov, D. V., Pacesa, M., Kreutzer, S.,…Schwank, G. (2021). *In vivo* adenine base editing of PCSK9 in macaques reduces LDL cholesterol levels. Nature Biotechnology, 39(8), 949–957. https://doi.org/10.1038/s41587-021-00933-4

Rotzoll, M., Richter, P., Fuchs, P., Hinz-wessels, A., Topp, S., & Hohendorf, G. (2006). The First National Socialist Extermination Crime : The T4 Program and Its Victims. International Journal of Mental Health, 35(3), 17–29.

Rowbotham, A. H. (1939). Gobineau and the Aryan Terror. The Sewanee Review, 47(2), 152–165.

Rubenstein, R. L. (1975). The Cunning of History : Mass death and the American future (First ed.). New York, Harper & Row.

Rudolph, K. (1987). Gnosis : The Nature and History of Gnosticism. HarperCollins.

Rupnow, D., Veronika Lipphardt, Jens Thiel, & Christina Wessely (Eds.). (2008). Pseudowissenschaft : Konzeptionen von Nichtwissenschaftlichkeit in der

Wissenschaftsgeschichte (Originalausg., 1. Aufl ed.). Suhrkamp.
Rushton, J. P., & Ankney, C. D. (2009). Whole Brain Size and General Mental Ability : A Review. International Journal of Neuroscience, 119(5), 692–732. https://doi.org/10.1080/00207450802325843
Ruttke, F. (1937). Rasse, Recht und Volk : Beiträge zur rassengesetzlichen Rechtslehre. J.F. Lehmann.
Ryback, T. W. (208). Hitler's Private Library : The Books That Shaped His Life. New York, Knopf Doubleday Publishing Group.
Sæbø, M. (1998). On the Way to Canon : Creative Tradition History in the Old Testament. Sheffield, UK, Sheffield Academic Press.
Salmi, H. (2020). Imagined Germany : Richard Wagner's National Utopia (Second Edition. ed.). New York, Peter Lang Publishing.
Salter, W. M. (1915). Nietzsche's Superman. The Journal of Philosophy, Psychology and Scientific Methods, 12(16), 421–438. https://doi.org/10.2307/2013575
Samuel, R. H. (1963). The Origin and Development of the Ideology of National Socialism. Australian Journal of Politics & History, 9(1), 59–77. https://doi.org/https://doi.org/10.1111/j.1467-8497.1963.tb01051.x
Sandford, S. (2018). Kant, race, and natural history. Philosophy & Social Criticism, 44(9), 950–977. https://doi.org/10.1177/0191453718768358
Sandys, J. E. (2009). A History of Classical Scholarship : From the End of the Sixth Century B. C. to the End of the Middle Ages. Cambridge University Press.
Sanford, N. (1973). Authoritarian Personality in Contemporary Perspective. In J. N. Knutson (Ed.), Handbook of Political Psychology ([First edition]. ed., pp. 139–170). Jossey-Bass Publishers.
Santin, A. D., Scheim, D. E., McCullough, P. A., Yagisawa, M., & Borody, T. J. (2021). Ivermectin : a multifaceted drug of Nobel prize-honoured distinction with indicated efficacy against a new global scourge, COVID-19. New Microbes and New Infections, 43, 100924–100924. https://doi.org/10.1016/j.nmni.2021.100924
Santucci, J. A. (2008). The Notion of Race in Theosophy. Nova religio, 11(3), 37–63. https://doi.org/10.1525/nr.2008.11.3.37
Sathyamoorthy, A. J., Patel, U., Paul, M., Savle, Y., Manocha, D., & Ab Wahab, M. N. (2021). COVID surveillance robot : Monitoring social distancing constraints in indoor scenarios. PloS one, 16(12), e0259713. https://doi.org/10.1371/journal.pone.0259713
Saunders, G. R. B., Wang, X., Chen, F., Jang, S. K., Liu, M., Wang, C., Gao, S., Jiang, Y., Khunsriraksakul, C., Otto, J. M., Addison, C., Akiyama, M., Albert, C. M., Aliev, F., Alonso, A., Arnett, D. K., Ashley-Koch, A. E., Ashrani, A. A., Barnes, K. C.,… Vrieze, S. (2022). Genetic diversity fuels gene discovery for tobacco and alcohol use. Nature, 612(7941), 720–724. https://doi.org/10.1038/s41586-022-05477-4
Savage, R. (2007). "Disease Incarnate" : Biopolitical Discourse and Genocidal Dehumanisation in the Age of Modernity. Journal of Historical Sociology, 20(3), 404–440. https://doi.org/10.1111/j.1467-6443.2007.00315.x
Savulescu, J. (2021). Good reasons to vaccinate : mandatory or payment for risk? Journal of Medical Ethics, 47(2), 78–85. https://doi.org/10.1136/medethics-2020-106821
Sayce, A. H. (1888). Address to the Anthropological Section of the British Association at Manchester. The Journal of the Anthropological Institute of Great Britain and Ireland, 17, 166–181. https://doi.org/10.2307/2841598
Scammell, G. V. (1991). Essay and Reflection : On the Discovery of the Americas and

the Spread of Intolerance, Absolutism, and Racism in Early Modern Europe. The International History Review, 13(3), 502–521.

Schäfer, P. (1998). Judeophobia : Attitudes Toward the Jews in the Ancient World. Harvard University Press.

Schafft, G. E. (2004). From Racism to Genocide : Anthropology in the Third Reich. Urbana, Urbana : University of Illinois Press.

Schallmayer, W. (1895). Die drohende physische Entartung der Culturvölker. Heuser.

______. (1903). Vererbung und Auslese im Lebenslauf der Völker, eine staatswissenschaftliche Studie auf Grund der neueren Biologie, von Dr. Wilhelm Schallmayer. G. Fischer.

Schallmayer, W. (1909). Generative Ethik. ARGB, vol. 6, 199–231.

Schallmayer, W. (1918). Vererbung und Auslese : Grundriss der Gesellschaftsbiologie und der Lehre vom Rassedienst, für Rassehygieniker, Bevölkerungspolitiker. Jena Fischer.

Schechter, S. (1915). Higher Criticism—Higher Anti-Semitism. In Seminary addresses & other papers (pp. 35—39). Ark Publishling Co.

Schellinger, U., Anton, A., & Schetsche, M. T. (2015). Pragmatic Occultism in the Military History of the Third Reich. In M. Black & E. Kurlander (Eds.), Revisiting the 'Nazi Occult' : Histories, Realities, Legacies (pp. 157–180). Camden House.

Schertel, E. (2009). Magic : History, Theory, Practice. COTUM - Catalog of the Universal Mind.

Schlake, T., Thess, A., Fotin-Mleczek, M., & Kallen, K.-J. (2014). Developing mRNA-vaccine technologies. RNA Biology, 9(11), 1319–1330. https://doi.org/10.4161/rna.22269

Schleunes, K. A. (1970). The Twisted Road to Auschwitz : Nazi Policy Toward German Jews, 1933-1939. Urbana, Urbana : University of Illinois Press.

Schliemann, H. (1875). Troy and Its Remains : A Narrative of Researches and Discoveries Made on the Site of Ilium, and in the Trojan Plain (L. D. Schmitz, Trans.; P. Smith, Ed.). New York, John Murray.

Schmuhl, H. W. (2005). Grenzüberschreitungen : Das Kaiser-Wilhelm-Institut für Anthropologie, menschliche Erblehre und Eugenik, 1927-1945. Göttingen: Wallstein; Springer.

______. (2008). The Kaiser Wilhelm Institute for Anthropology, Human Heredity and Eugenics, 1927-1945 : Crossing Boundaries (S. Richter, Trans.). Springer Netherlands.

______. (2014). Eugenics and Racial Hygiene : Applied Research Strategies before, during, and after National Socialism. In S. Rubenfeld & S. Benedict (Eds.), Human Subjects Research after the Holocaust (pp. 31–38). Springer International Publishing.

Schneider, W. H. (1996). The history of research on blood group genetics : initial discovery and diffusion. Hist Philos Life Sci, 18(3), 277–303.

Schulle, D. (2001). Das Reichssippenamt : Eine Institution nationalsozialistischer Rassenpolitik. Logos.

Schuman, F. L. (1936). The Nazi Dictatorship : A Study in Social Pathology and the Politics of Fascism. London, A. A. Knopf.

Schwab, K., & Malleret, T. (2020). COVID-19 : The Great Reset. Forum Publishing.

Schwartz, L., Schwartz, M., & Chase, A. (1982). [The Legacy of Malthus : The Social Costs of the New Scientific Racism, Allan Chase]. Contemporary Sociology,

11(5), 501–504. https://doi.org/10.2307/2068386
Schwarzwäller, W. (1989). The Unknown Hitler : His Private Life & Fortune. Stoddart.
Schweikhart, C. (1954). Verkündigung des Pol-Reiches. Mensch und Schicksal, 8(7), 3–6.
Scott, A. (2006). 'Like editing bits of ourselves' : geneticisation and human fate. New Genetics and Society, 25(1), 109–124. https://doi.org/10.1080/14636770600603535
Searle, G. R. (1971). The Quest for National Efficiency : A Study in British Politics and Political Thought, 1899-1914. Oxford, Blackwell.
Sebottendorf, R. v. (1925). Der Talisman des Rosenkreuzers : Roman. Pfullingen.
Sebottendorff, R. v. (1933). Bevor Hitler kam : Urkundliches aus der Frühzeit der nationalsozialistischen Bewegung.
Sedyaningsih, E. R., Isfandari, S., Soendoro, T., & Supari, S. F. (2008). Towards mutual trust, transparency and equity in virus sharing mechanism : the avian influenza case of Indonesia. Annals of the Academy of Medicine, Singapore, 37(6), 482–488. https://doi.org/10.47102/annals-acadmedsg.v37n6p482
Segev, T. (1987). Soldiers of Evil : The Commandants of the Nazi Concentration Camps. New York, McGraw-Hill.
Seidelman, W. E. (1989). Medical selection : Auschwitz antecedents and effluent. Holocaust Genocide Stud, 4(4), 1989. https://doi.org/10.1093/hgs/4.4.435
_____. (1992). Racial Hygiene : Medicine Under the Nazis Robert N. Proctor. Canadian Bulletin of Medical History, 9(1), 148–151. https://doi.org/10.3138/cbmh.9.1.148
Sellon, E. B., & Weber, R. (1992). Theosophy and The Theosophical Society. In A. Faivre & J. Needleman (Eds.), Modern Esoteric Spirituality (Vol. 21 of World Spirituality : An Encyclopedic History of the Religious Quest). Crossroad.
Semmel, B. (1958). Karl Pearson : Socialist and Darwinist. The British Journal of Sociology, 9(2), 111–125. https://doi.org/10.2307/587909
Serjeantson, R. (2018). Francis Bacon's New Atlantis (B. Price, Ed.). Manchester University Press.
Sgro, A., & Blancafort, P. (2020). Epigenome engineering : new technologies for precision medicine. Nucleic Acids Res, 48(22), 12453–12482. https://doi.org/10.1093/nar/gkaa1000
Sharpley, C. F., Koehn, C., & Tier, B. (2022). Aristotle's Concept of Heredity. Journal of the Washington Academy of Sciences, 108(1), 11–30.
Shavit, Y., & Eran, M. (2007). The Hebrew Bible Reborn : From Holy Scripture to the Book of Books : a History of Biblical Culture and the Battles Over the Bible in Modern Judaism (C. Naor, Trans.; Vol. 38, Studia Judaica). Berlin, Germany, Walter de Gruyter.
She, J., Hou, D., Chen, C., Bi, J., & Song, Y. (2022). Challenges of vaccination and herd immunity in COVID-19 and management strategies. The Clinical Respiratory Journal, 16(11), 708–716. https://doi.org/https://doi.org/10.1111/crj.13543
Shen, L., Wang, P., & Ke, Y. (2021). DNA Nanotechnology-Based Biosensors and Therapeutics. Advanced Healthcare Materials, 10(15). https://doi.org/10.1002/adhm.202002205
Shi, P., & Wu, X. (2024). Programmable RNA targeting with CRISPR-Cas13. RNA Biology, 21(1), 575–583. https://doi.org/10.1080/15476286.2024.2351657
Shimabukuro, T., & Nair, N. (2021). Allergic Reactions Including Anaphylaxis After Receipt of the First Dose of Pfizer-BioNTech COVID-19 Vaccine. JAMA, 325(8).

https://doi.org/10.1001/jama.2021.0600
Shipman, M. (2015). Magnetic Nanoparticle Chains Offer New Technique for Controlling Soft Robots. Retrieved Nov. 24 from https://news.ncsu.edu/2015/12/tracy-chains-2015/
Shirer, W. L. (1960). The Rise and Fall of the Third Reich : A History of Nazi Germany. New York, Simon and Schuster.
Sickinger, R. L. (2000). Hitler and the Occult : The Magical Thinking of Adolf Hitler. The Journal of Popular Culture, 34(2), 107–125. https://doi.org/https://doi.org/10.1111/j.0022-3840.2000.3402_107.x
Siegel, A. W. (2014). Some doubts about *in vitro* eugenics as a human enhancement technology. Journal of Medical Ethics, 40(11), 732. https://doi.org/https://doi.org/10.1136/medethics-2013-101511
Siepe, D. (2008). Die Rolle der Wewelsburg in der phantastischen Literatur, in Esoterik und Rechtsextremismus nach 1945. In J. Erik Schulte & J. E. Schulte (Eds.), Die SS, Himmler und die Wewelsburg (Vol. 7, pp. 488–510). Brill | Schöningh. https://doi.org/https://doi.org/10.30965/9783657763740_027
Silverman, W. Z. (1992). Anti-Semitism and Occultism in *fin-de-siècle* France. In B. T. Cooper & M. Donaldson-Evans (Eds.), Modernity and Revolution in Late Nineteenth-century France (pp. 155–163). University of Delaware Press.
Simmel, G. (1908). Soziologie : Untersuchungen Über Die Formen Der Vergesellschaftung. Duncker & Humblot.
Singer, P. W. (2002). AIDS and International Security. Survival, 44(1), 145–158. https://doi.org/10.1080/00396338.2002.9688546
Sisco, H. K. F., & Brummette, J. (2024). mRNA Vaccine Hesitancy : Spreading Misinformation Through Online Narratives. Journal of health communication, 29(8), 538–547. https://doi.org/10.1080/10810730.2024.2379954
Sizer, N. (1888). Forty Years in Phrenology : Embracing Recollections of History, Anecdote, and Experience. Fowler & Wells.
Sizer, P. N. (1909). The Facial Angle. In Cranium, A., The Phrenological Journal and Science of Health: Fowler & Wells Company.
Slight, D. (1937). Disorganization in the Individual and in Society. American Journal of Sociology, 42(6), 840–847.
Smith, B. F., & Peterson, A. F. (Eds.). (1974). Heinrich Himmler : Geheimreden 1933 bis 1945 ; und andere Ansprachen. Propyläen.
Smith, J., Chatterjee, A., & Rodger, E. J. (2025). Clinical epigenetics : recent advances and opportunities. Epigenetics Communications, 5(1), 7. https://doi.org/10.1186/s43682-025-00038-y
Smith, S. (2008). The New Atlantis : Francis Bacon's Theological-Political Utopia? Harvard Theological Review, 101(1), 97–125. https://doi.org/10.1017/s0017816008001740
Smith, W. (1842). Currus. In A Dictionary of Greek and Roman Antiquities.
Smith, W., & Anthon, C. (1851). Dionysus. In A New Classical Dictionary of Greek and Roman Biography, Mythology, and Geography: Partly Based Upon the Dictionary of Greek and Roman Biography and Mythology.
Snyder, L. H. (1926). Human Blood Groups : Their Inheritance and Racial Significance. American Journal of Physical Anthropology, 9(2), 233–263. https://doi.org/10.1002/ajpa.1330090216
Snyder, L. L. (1962). The Idea of Racialism : Its Meaning and History. Princeton

: Van Nostrand.
Society of Antiquaries of Newcastle upon Tyne. (1884). Archaeologia Aeliana, Or, Miscellaneous Tracts Relating to Antiquities (Vol. XLVIII). Sarah Hodgson.
Sofsky, W. (1997). The Order of Terror : The Concentration Camp. Princeton University Press.
Sommer, A. (2009). From Astronomy to Transcendental Darwinism : Carl du Prel (1839–1899). Journal of Scientific Exploration, 23, 59–68.
Sontag, F. (1958). [Religion Without Revelation, Julian Huxley]. Science, 127(3294), 343–343.
Sontag, S. (1989). Illness as Metaphor and AIDS and Its Metaphors. Farrar, Straus & Giroux.
_____. (2001). Illness as Metaphor and AIDS and Its Metaphors. Picador.
Sorokin, P. A. (1928). Contemporary Sociological Theories. New York, Harper & Brothers.
Spanuth, J. (1956). Atlantis : The Mystery Unravelled, by Jurgen Spanuth (First American Edition ed.). Arco publ.
_____. (1979). Atlantis of the North. Sidgwick & Jackson.
Sparke, M., & Levy, O. (2022). Competing Responses to Global Inequalities in Access to COVID Vaccines: Vaccine Diplomacy and Vaccine Charity Versus Vaccine Liberty. Clinical Infectious Diseases, 75(Supplement_1), S86–S92. https://doi.org/10.1093/cid/ciac361
Sparrow, R. (2014). *In vitro* eugenics. J Med Ethics, 40(11), 725–731. https://doi.org/10.1136/medethics-2012-101200
Spelter, J. (1937). Der deutsche Erzieher als Lehrer der Rassenkunde. Verlag Pfeiffer.
Spence, L. (1920). Encyclopaedia of Occultism. London, George Routledge & Sons, Ltd.
_____. (1940). The Occult Causes of the Present War. London, Rider & Co.
Spielvogel, J., & Redles, D. (2011). Hitler's Racial Ideology : Content and Occult Sources. In M. R. Marrus (Ed.), The Nazi Holocaust (Vol. 2 The Origins of the Holocaust), Part One: Racism, thre Occult, and Eugenics, pp. 79–98). De Gruyter.
Stachura, P. D. (1980). The Political Strategy of the Nazi Party, 1919-1933. German Studies Review, 3(2), 261–288. https://doi.org/10.2307/1429723
Stafford, E. (2012). Herakles (First ed.). Abingdon, Oxon ;, Routledge.
Stähle, E. (1934). Blut und Rasse, neue Forschungsergebnisse (Blood and Race : New Research Results). In Die Volksgesundheitswacht, Gilbhart (pp. 4–6).
Stanislas, K. d. R. (1992). Alchemy : The Secret Art. Thames and Hudson.
Stark, W. (1961). Natural and Social Selection. In M. Banton & J. Bronowski (Eds.), Darwinism and the Study of Society : A Centenary Symposium. Tavistock Publications.
Staub, E. (1989). The Roots of Evil : The Origins of Genocide and Other Group Violence. Cambridge [England] ;, Cambridge University Press.
_____. (1990). Moral Exclusion, Personal Goal Theory, and Extreme Destructiveness. Journal of Social Issues, 46(1), 47–64. https://doi.org/https://doi.org/10.1111/j.1540-4560.1990.tb00271.x
_____. (1995). Torture : Psychological and cultural origins. In R. D. Crelinsten & A. P. Schmid (Eds.), The Politics of Pain : Torturers and Their Masters (pp. 99–111). Westview Press.
_____. (1999). The Roots of Evil : Social Conditions, Culture, Personality, and Basic Human Needs. Personality and Social Psychology Review, 3(3), 179–192. https://

doi.org/10.1207/s15327957pspr0303_2

Staudenmaier, P. (2008). Race and Redemption : Racial and Ethnic Evolution in Rudolf Steiner's Anthroposophy. Nova Religio: The Journal of Alternative and Emergent Religions, 11(3), 4–36. https://doi.org/10.1525/nr.2008.11.3.4

______. (2009). Occultism, Race and Politics in German-speaking Europe, 1880—1940 : A Survey of the Historical Literature. European History Quarterly, 39(1), 47–70. https://doi.org/10.1177/0265691408097366

______. (2010). Anthroposophie in Deutschland : Theosophische Weltanschauung und gesellschaftliche Praxis 1884–1945. Aries (Leiden, Netherlands), 10(1), 107–116. https://doi.org/10.1163/156798910X12584583444998

______. (2013). Nazi Perceptions of Esotericism : The Occult as Fascination and Menace. In A. Manthripragada, et al. (Eds.), The Threat and Allure of the Magical : Selected Papers from the 17th Annual Interdisciplinary German Studies Conference, University of California, Berkeley (pp. 25–58). Cambridge Scholars Publishing.

______. (2014). Between Occultism and Nazism : Anthroposophy and the Politics of Race in the Fascist Era. Brill.

______. (2015). Esoteric Alternatives in Imperial Germany : Science, Spirit, and the Modern Occult Revival. In M. Black & E. Kurlander (Eds.), Revisiting the 'Nazi Occult' : Histories, Realities, Legacies (pp. 23–41). Camden House.

Stavish, M. (2007). freemasonry : Rituals, Symbols & History of the Secret Society. Woodbury, Minn, USA, Llewellyn.

Steigmann-Gall, R. (2003). The Holy Reich : Nazi conceptions of Christianity, 1919-1945. Cambridge, UK ;, Cambridge University Press.

Stein, G. J. (1988). Biological Science and the Roots of Nazism. American Scientist, 76(1), 50–58.

Stein, L. (1950). The Racial Thinking of Richard Wagner. New York, Philosophical Library.

Stein, R. (June, 2021). He Inherited A Devastating Disease. A CRISPR Gene-Editing Breakthrough Stopped It.

Steinberg, H.-J. (1972). Sozialismus und deutsche Sozialdemokratie : zur Ideologie der Partei vor dem 1. Weltkrieg. Verlag Neue Gesellschaft.

Steiner, R. (1966). Theosophy of the Rosicrucian : A Course of Fourteen Lectures: Munich, 22nd May-6th June, 1907. Rudolf Steiner Press.

______. (1981). Menschheitsentwicklung und Christus-Erkenntnis : Theosophie und Rosenkreuzertum. Dornach/Schweiz, Rudolf-Steiner-Verlag.

______. (1985a). Die Welträtsel und die Anthroposophie : zweiundzwanzig öffentliche Vorträge, gehalten zwischen dem 5. Oktober 1905 und 3. Mai 1906 im Architektenhaus zu Berlin. Rudolf Steiner Verlag.

______. (1985b). The Apocalypse of St. John : Lectures on the Book of Revelation (CW 104). SteinerBooks.

______. (1990). The Universal Human : The Evolution of Individuality : Four Lectures Given Between 1909 and 1916 in Munich and Bern (C. Bamford & S. H. Seiler, Eds.). Anthroposophic Press.

Steinmeier, F.-W. (2020). Address by President of Germany to the Fifth World Holocaust Forum "Remembering the Holocaust : Fighting Antisemitism. Yad Vashem. Retrieved Nov 12, 2025 from https://www.bundespraesident.de/SharedDocs/Reden/EN/Frank-Walter-Steinmeier/Reden/2020/01/200123-World-Holocaust-

Forum-Yad-Vashem.html
Steinweis, A. E. (2006). Studying the Jew : Scholarly Antisemitism in Nazi Germany. Cambridge, BRAZIL, Harvard University Press.
Stenger, V. J. (1995). The Unconscious Quantum : Metaphysics in Modern Physics and Cosmology. Amherst, N.Y, Prometheus Books.
Stern, F. (1974). The Politics of Cultural Despair : A Study in the Rise of the Germanic Ideology. Berkeley, Berkeley : University of California Press.
_____. (1987). Dreams and Delusions : The Drama of German History (First edition. ed.). New York, Knopf.
Sternberg, R. J. (2003). A Duplex Theory of Hate : Development and Application to Terrorism, Massacres, and Genocide. Review of General Psychology, 7(3), 299–328. https://doi.org/10.1037/1089-2680.7.3.299
Stoller, J. K. (2020). Reflections on leadership in the time of COVID-19. BMJ Leader, 4(2), 77. https://doi.org/10.1136/leader-2020-000244
Strabo. (1854). The Geography of Strabo (H. C. Hamilton & W. Falconer, Trans.; Vol. II). H.G. Bohn.
Strack, H. L. (1909). The Jew and Human Sacrifice : Human Blood and Jewish Ritual, An Historical and Sociological Inquiry (H. Blanchamp, Trans.). England, Bloch publishing co.
Straiton, J. (2019). Genetically Modified Humans : The X-Men of Scientific Research. BioTechniques, 66(6), 249–252. https://doi.org/10.2144/btn-2019-0056
Strassler, R. B. (Ed.). (2007). The Landmark Herodotus : The Histories (A. L. Purvis, Trans.). Pantheon Books.
Strube, J. (2017). Between Occultism and Nazism : Anthroposophy and the Politics of Race in the Fascist Era, written by Peter Staudenmaier. Aries, 17(2), 254–259. https://doi.org/https://doi.org/10.1163/15700593-01702005
Stuckrad, K. v. (2005). Western esotericism : Towards an integrative model of interpretation. Religion, 35(2), 78–97. https://doi.org/https://doi.org/10.1016/j.religion.2005.07.002
Stuurman, S. (2000). Francois Bernier and the invention of racial classification. Hist Workshop J(50), 1–21. https://doi.org/10.1093/hwj/2000.50.1
_____. (2017). The Invention of Humanity : Equality and Cultural Difference in World History. Cambridge, Massachusetts, Harvard University Press.
Sulaiman et al. (1990). Ayat-ayat setan Yahudi. Pustakakarya Grafikatama.
Sullenger, B. A. (2003). Targeted genetic repair : an emerging approach to genetic therapy. J Clin Invest, 112(3), 310–311. https://doi.org/10.1172/JCI19419
Sutherland, A. (2006). I Was A Stranger : A Christian Theology of Hospitality. Nashville, TN, USA, Abingdon Press.
Swindells, M. (March, 2017). Emergency Preparedness, Resilience and Response (EPRR) (Paper: PB.30.03.2017/10). https://www.england.nhs.uk/wp-content/uploads/2017/03/board-paper-300317-item-10.pdf
Taguieff, P.-A. (2004). Rising From the Muck : The New Anti-Semitism in Europe (P. Carmiller, Trans.). Chicago, Ivan R. Dee.
Tahir, M. J., Sawal, I., Essar, M. Y., Jabbar, A., Ullah, I., & Ahmed, A. (2022). Disease X : A hidden but inevitable creeping danger. Infect Control Hosp Epidemiol, 43(11), 1758–1759. https://doi.org/10.1017/ice.2021.342
Tal, U. (1975). Christians and Jews in Germany : Religion, Politics, and Ideology in the Second Reich, 1870-1914 (N. J. Jacobs, Trans.). Ithaca, Ithaca : Cornell

University Press.
Tamar, S. (2021). Blind-sided by privacy? Digital contact tracing, the Apple/Google API and big tech's newfound role as global health policy makers. Ethics and Information Technology, suppl. 1, 23, 45–57. https://doi.org/https://doi.org/10.1007/s10676-020-09547-x
Tambiah, S. J. (1990). Magic, Science and Religion and the Scope of Rationality. Cambridge University Press.
Tang, Y. S., Wang, D., Zhou, C., Ma, W., Zhang, Y. Q., Liu, B., & Zhang, S. (2012). Bacterial magnetic particles as a novel and efficient gene vaccine delivery system. Gene Ther, 19(12), 1187–1195. https://doi.org/10.1038/gt.2011.197
Tatz, C. M., Arnold, P., & Tatz, S. (2006). Genocide Perspectives III : Essays on the Holocaust and Other Genocides. Brandl & Schlesinger.
Taylor, F. S. (1953). The Alchemists : Founders of Modern Chemistry. London, W. Heinemann.
Taylor, H. S., Apelian, S., & Adashi, E. Y. (2025). In-vitro gametogenesis : redefining reproduction. The Lancet Obstetrics, Gynaecology, & Women's Health, 1(2), e90–e91. https://doi.org/10.1016/j.lanogw.2025.100052
Taylor, I. (1906). The Origin of the Aryans : An Account of the Prehistoric Ethnology and Civilisation of Europe. Scribner & Welford.
Tenenbaum, J. (1956). Race and Reich : The Story of an Epoch. New York, New York : Twayne Publishers.
Thapar, R. (1965). A History of India (Vol. 1). Penguin Publishing Group.
_____. (1996). The Theory of Aryan Race and India : History and Politics. Social Scientist, 24(1/3), 3–29. https://doi.org/10.2307/3520116
Thomas, J. M. (2010). The racial formation of medieval Jews : a challenge to the field. Ethnic and Racial Studies, 33(10), 1737–1755. https://doi.org/10.1080/01419871003706766
Thomson, S., & Ip, E. C. (2020). COVID-19 emergency measures and the impending authoritarian pandemic. J Law Biosci, 7(1), lsaa064. https://doi.org/10.1093/jlb/lsaa064
Thorndike, L. (1923). A History of Magic and Experimental Science : The first thirteen centuries of our era (Vol. II.). Macmillan.
Toczek, N. (2015). Haters, Baiters and Would-Be Dictators : Anti-Semitism and the UK Far Right. Taylor & Francis.
Todorov, T. (1993). On Human Diversity : Nationalism, Racism, and Exoticism in French Thought. Cambridge, Mass., Cambridge, Mass. : Harvard University Press.
Tolischus, O. D. (1940). They Wanted War (Fourth ed.). New York, Reynal & Hitchcock.
Tomes, N. (1998). The Gospel of Germs : Men, Women, and the Microbe in American Life. Cambridge, Mass, Harvard University Press.
Tourlamain, G. (2014). "Völkisch" Writers and National Socialism. Oxford, UK, Peter Lang Verlag.
Trachtenberg, J. (1983). The Devil and the Jews : The Medieval Conception of the Jew and its Relation to Modern Antisemitism. Jewish Publication Society.
Tran, D. P., Pham, T. T. T., Wolfrum, B., Offenhausser, A., & Thierry, B. (2018). CMOS-Compatible Silicon Nanowire Field-Effect Transistor Biosensor : Technology Development toward Commercialization. Materials (Basel), 11(5), 785. https://doi.org/10.3390/ma11050785

Tregenza, M. (1977). Belzec Death Camp. Wiener Library Bulletin, 30, 8–25.
Treitel, C. (2004). A Science for the Soul : Occultism and the Genesis of the German Modern. Johns Hopkins University Press.
Trevor-Roper, H. R., Ed. & Trans. (1953). Hitler's Secret Conversations, 1941-1944. Farrar, Straus and Young.
Trotsky, L. (1925). Literature and Revolution (R. Strunsky, Trans.). London, George Allen &Unwin Ltd.
Trunk, I. (1972). Judenrat : The Jewish Councils in Eastern Europe Under Nazi Occupation. New York, Macmillan.
_____. (1989). Epidemics and Mortality in the Warsaw Ghetto, 1939-1942. In Volume 1 (pp. 3–43). De Gruyter Brill, K. G. Saur. https://doi.org/doi:10.1515/9783110968736.3 (Original work published Trunk, I. (1953). Epidemics and Mortality in the Warsaw Ghetto 1939–1942. YIVO Annual 8, 82–122.)
Tucker, P. (2020). A Military-Funded Biosensor Could Be the Future of Pandemic Detection. Retrieved Nov. 25 from https://www.defenseone.com/technology/2020/03/military-funded-biosensor-could-be-future-pandemic-detection/163497/
Turney, J. (1998). Frankenstein's Footsteps : Science, Genetics and Popular Culture. New Haven, Yale University Press.
Unkelbach, C. (2007). Reversing the truth effect : Learning the interpretation of processing fluency in judgments of truth. Journal of Experimental Psychology: Learning, Memory, and Cognition, 33(1), 219–230. https://doi.org/10.1037/0278-7393.33.1.219
Unkelbach, C., & Greifeneder, R. (2018). Experiential fluency and declarative advice jointly inform judgments of truth. Journal of Experimental Social Psychology, 79, 78–86. https://doi.org/https://doi.org/10.1016/j.jesp.2018.06.010
Unkelbach, C., Koch, A., Silva, R. R., & Garcia-Marques, T. (2019). Truth by Repetition : Explanations and Implications. Current Directions in Psychological Science, 28(3), 247–253. https://doi.org/10.1177/0963721419827854
Unkelbach, C., & Rom, S. C. (2017). A referential theory of the repetition-induced truth effect. Cognition, 160, 110–126. https://doi.org/10.1016/j.cognition.2016.12.016
Unkelbach, C., & Stahl, C. (2009). A multinomial modeling approach to dissociate different components of the truth effect. Consciousness and Cognition, 18(1), 22–38. https://doi.org/10.1016/j.concog.2008.09.006
Varshizky, A. (2019). The Metaphysics of Race : Revisiting Nazism and Religion. Central European History, 52(2), 252–288. https://doi.org/10.1017/S0008938919000189
Vartanian, A. (2015). Diderot and Descartes : A Study of Scientific Naturalism in the Enlightenment. Princeton University Press.
Vartija, D. (2020). Revisiting Enlightenment racial classification : time and the question of human diversity. Intellectual History Review, 31(4), 603–625. https://doi.org/10.1080/17496977.2020.1794161
Vecoli, R. J. (1960). Sterilization : A Progressive Measure? The Wisconsin Magazine of History, 43(3), 190–202.
Verschuer, O. F. v. (1944). Erbanlage als Schicksal und Aufgabe. de Gruyter.
Voegelin, E. (1940). The Growth of the Race Idea. The Review of Politics, 2(3), 283–317.
Vogel, G., & Kupferschmidt, K. (2021). New problems erode confidence in AstraZeneca's vaccine. Science, 371(6536), 1294–1295. https://doi.org/10.1126/

science.371.6536.1294

Volkov, S. (1978). Antisemitism as a Cultural Code : Reflections on the History and Historiography of Antisemitism in Imperial Germany. The Leo Baeck Institute Yearbook, 23(1), 25–46. https://doi.org/10.1093/leobaeck/23.1.25

Vraga, E. K., & Bode, L. (2017). I do not believe you : how providing a source corrects health misperceptions across social media platforms. Information, Communication & Society, 21(10), 1337–1353. https://doi.org/10.1080/1369118x.2017.1313883

Wachsmuth, G. (1955). The Life and Work of Rudolf Steiner from the Turn of the Century to His Death. Whittier Books.

Waite, A. E. (1888). The Real History of the Rosicrucians Founded on Their Own Manifestoes, and on Facts and Documents Collected from the Writings of Initiated Brethren. J.W. Bouton.

Waite, R. G. L. (1978). The Psychopathic god : Adolf Hitler. New York, New American Library.

Walbaum, J. (1941). Kampf den Seuchen : Deutscher Aerzte-Einsatz im Osten. Die Aufbauarbeit im Gesundheitswesen des Generalgouvernements. Deutscher Osten.

Walker, D. P. (1975). Spiritual and Demonic Magic : From Ficino to Campanella. Notre Dame [Ind, University of Notre Dame Press.

Walker, N. (1939). Real History of the Swastika. London, Lutterworth Press.

Wallace, A. F. C. (2003). The Social Context of Innovation : Bureaucrats, Families, and Heroes in the Early Industrial Revolution, As Foreseen in Bacon's New Atlantis. University of Nebraska Press.

Wallach, M. A., Kogan, N., & Bem, D. J. (1962). Group influence on individual risk taking. The Journal of Abnormal and Social Psychology, 65(2), 75–86. https://doi.org/10.1037/h0044376

Waller, J. (2007). Becoming Evil : How Ordinary People Commit Genocide and Mass Killing (Second ed.). Oxford; New York, Oxford University Press.

Walsh, C. (1958). [Religion without Revelation, Julian Huxley]. The Journal of Religion, 38(3), 210–211.

Wang, B. (2014). Ido Bachelet Announces 2015 Human Trial of DNA Nanobots to Fight Cancer and soon to Repair Spinal Cords. http://nextbigfuture.com/2014/12/ido-bachelet-announces-2015-human-trial.html

Warren, H. C. (1933). Social nudism and the body taboo. Psychological review, 40(2), 160–183. https://doi.org/10.1037/h0073466

Wartburg, W. P. v., & Liew, J. (1999). Gene Technology and Social Acceptance. University Press of America.

Waxman, Z. V. (2010). Towards an Integrated History of the Holocaust : Masculinity, Femininity, and Genocide. In Years of Persecution, Years of Extermination : Saul Friedlander and the Future of Holocaust Studies (pp. 311–321). Bloomsbury Publishing Plc.

Weale, A. (2010). The SS : A New History. Little, Brown Book Group.

Webb, J. (1669). An Historical Essay : Endeavoring a Probability that the Language of the Empire of China is the Primitive Language. Nath. Brook.

Webb, J. (1974). The Occult Underground. Open Court.

_____. (1976). The Occult Establishment. Open Court Publishing Company.

Wedgwood, C. V. (1941). The Roots of National Socialism : A Review [The Roots of

National Socialism, 1783-1933, Rohan D'O. Butler]. International Affairs Review Supplement, 19(6), 311–313. https://doi.org/10.2307/3026379
Wegner, G. (2002). Anti-Semitism and Schooling under the Third Reich. New York; UK, Taylor & Francis Group.
Weikart, R. (1993). The Origins of Social Darwinism in Germany, 1859-1895. Journal of the History of Ideas, 54(3), 469–488. https://doi.org/10.2307/2710024
_____. (2003). Progress through Racial Extermination : Social Darwinism, Eugenics, and Pacifism in Germany, 1860-1918. German Studies Review, 26(2), 273–294. https://doi.org/10.2307/1433326
_____. (2004). From Darwin to Hitler : Evolutionary Ethics, Eugenics, and Racism in Germany (First edition ed.). New York, Palgrave Macmillan.
_____. (2006). The Impact of Social Darwinism on Anti-Semitic Ideology in Germany and Austria, 1860-1945. In Jewish Tradition and the Challenge of Darwinism (pp. 93–115). University of Chicago Press.
_____. (2013). The Role of Darwinism in Nazi Racial Thought. German Studies Review, 36(3), 537–556.
_____. (2016). Hitler's religion : The Twisted Beliefs That Drove The Third Reich Regnery History. Washington, DC, Regnery History.
Weinberg, R. (2012). The Blood Libel in Eastern Europe. Jewish History, 26(3/4), 275–285.
Weinberger, J. (1976). Science and Rule in Bacon's Utopia : An Introduction to the Reading of the New Atlantis. The American Political Science Review, 70(3), 865–885. https://doi.org/10.2307/1959872
_____. (1985). Science, faith, and Politics : Francis Bacon and the Utopian roots of the Modern Age : A Commentary on Bacon's Advancement of Learning. Ithaca, Cornell University Press.
Weindling, P. J. (1985). Weimar Eugenics : The Kaiser Wilhelm Institute for Anthropology, Human Heredity and Eugenics in Social Context [Article]. Annals of Science, 42(3), 303. https://doi.org/10.1080/00033798500200221
_____. (1988). The Rockefeller Foundation and German Biomedical Sciences, 1920–40 : from Educational Philanthropy to International Science Policy. In N. A. Rupke (Ed.), Science, Politics and the Public Good: Essays in Honour of Margaret Gowing (pp. 119–140). Palgrave Macmillan UK. https://doi.org/10.1007/978-1-349-09514-8_6
_____. (1993). Health, Race and German Politics Between National Unification and Nazism, 1870-1945. Cambridge University Press.
_____. (1997). Philanthropy and World Health : The Rockefeller Foundation and the League of Nations Health Organisation. Minerva, 35(3), 269–281.
_____. (2000). Epidemics and Genocide in Eastern Europe, 1890–1945. Oxford University Press.
_____. (2000). An overloaded ark? The Rockefeller Foundation and refugee medical scientists, 1933–45. Studies in History and Philosophy of Science Part C: Studies in History and Philosophy of Biological and Biomedical Sciences, 31(3), 477–489. https://doi.org/https://doi.org/10.1016/S1369-8486(00)00015-7
_____. (2014). Delousing and Resistance during the Holocaust. In M. A. Grodin (Ed.), Jewish Medical Resistance in the Holocaust. Berghahn Books, Incorporated.
_____. (2017). Intrroduction : a new historiography of the Nazi medical experiments and coerced research. In P. Weindling (Ed.), From Clinic to Concentration Camp

: Reassessing Nazi Medical and Racial Research, 1933-1945 (First ed., pp. 3–32). Taylor & Francis. https://doi.org/https://doi.org/10.4324/9781315583310
Weiner, B. (1980). A cognitive (attribution)-emotion-action model of motivated behavior : An analysis of judgments of help-giving. Journal of personality and social psychology, 39(2), 186–200. https://doi.org/10.1037/0022-3514.39.2.186
_____. (1985). An Attributional Theory of Achievement Motivation and Emotion. Psychological review, 92(4), 548–573. https://doi.org/10.1037/0033-295X.92.4.548
_____. (2006). Social Motivation, Justice, and the Moral Emotions : An Attributional Approach (First ed.). Oxford, Psychology Press.
Weinert, H. (1934). Unsere Eiszeit-Ahnen. Stenger.
Weinreich, M. (1999). Hitler's Professors : The Part of Scholarship in Germany's Crimes Against the Jewish People (Second ed.). Yale University Press.
Weismann, A. (1891). Essays Upon Heredity and Kindred Biological Problems (E. B. Poulton, et al., Eds. Vol. 1). Clarendon Press.
_____. (1893). The Germ-plasm : A Theory of Heredity (W. N. Parker & H. Rōnníeldt, Trans.). C. Scribner's sons.
Weiss, S. F. (1987a). The Race Hygiene Movement in Germany. Osiris, 3, 193–236. https://doi.org/10.1086/368666
_____. (1987b). Race Hygiene and National Efficiency : The Eugenics of Wilhelm Schallmayer. Berkeley, Berkeley : University of California Press.
_____. (1990). The Race Hygiene Movement in Germany, 1904–1945. In The Wellborn Science : Eugenics in Germany, France, Brazil, and Russia. Oxford University Press, Incorporated.
_____. (2010). The Nazi Symbiosis : Human Genetics and Politics in the Third Reich. Chicago, UNITED STATES, University of Chicago Press.
_____. (2012). The Loyal Genetic Doctor, Otmar Freiherr von Verschuer, and the Institut für Erbbiologie und *Rassenhygiene*: Origins, Controversy, and Racial Political Practice. Central European History, 45(4), 631–668.
Welch, D. (1993). Manufacturing a Consensus : Nazi Propaganda and the Building of a 'National Community' (Volksgemeinschaft). Contemporary European History, 2(1), 1–15.
_____. (2004). Nazi Propaganda and the Volksgemeinschaft : Constructing a People's Community. Journal of Contemporary History, 39(2), 213–238.
Wenglinsky, M. (1975). [Obedience to Authority: An Experimental View., Stanley Milgram]. Contemporary Sociology, 4(6), 613–617. https://doi.org/10.2307/2064024
Werner, U. (1999). Anthroposophen in der Zeit des Nationalsozialismus (1933-1945). Oldenburg.
Wertham, F. (1966). A Sign for Cain : An Exploration of Human Violence. New York, New York : Macmillan.
Wevers, R. (2018). Unmasking Biometrics' Biases : Facing Gender, Race, Class and Ability in Biometric Data Collection. TMG Journal for Media History. https://doi.org/10.18146/2213-7653.2018.368
Weyers, W. (1998). Death of medicine in Nazi Germany : Dermatology and dermatopathology under the swastika (A. B. Ackerman, Ed.). Philadelphia, Ardor Scribendi; Lanham : Madison Books.
Wheeler, M. A., Smith, C. J., Ottolini, M., Barker, B. S., Purohit, A. M., Grippo, R.

M., Gaykema, R. P., Spano, A. J., Beenhakker, M. P., Kucenas, S., Patel, M. K., Deppmann, C. D., & Güler, A. D. (2016). Genetically targeted magnetic control of the nervous system. Nat Neurosci, 19(5), 756–761. https://doi.org/10.1038/nn.4265

White, H. (1956). [Psychological Warfare., Paul M. A. Linebarger]. The Journal of Politics, 18(1), 130–132. https://doi.org/10.2307/2126690

White, J. (2015). Emergency Europe. Political Studies, 63(2), 300–318. https://doi.org/10.1111/1467-9248.12072

White, R. (1999). The Rosicrucian Enlightenment Revisited. Lindisfarne Books.

WHO. (2007). The world health report 2007 : a safer future: global public health security in the 21st century. World Health Organization.

______. (2013). Global vaccine action plan 2011-2020 (9241504986).

______. (2021). World health statistics 2021 : monitoring health for the SDGs, sustainable development goals. World Health Organization.

______. (2023). Defining collaborative surveillance : a core concept for strengthening the global architecture for health emergency preparedness, response, and resilience (HEPR). Geneva, World Health Organization.

______. (2024). Pandemic influenza preparedness partnership contribution high-level implementation plan iii 2024-2030. World Health Organization.

Wiese, C. (2005). Challenging Colonial Discourse : Jewish Studies And Protestant Theology In Wilhelmine Germany (B. Harshav & C. Wiese, Trans.; Vol. 10, Studies in Jewish History and Culture Series). Leiden, The Netherlands, Brill.

Wilkinson, L. (2000). Burgeoning visions of global public health: The Rockefeller Foundation, The London School of Hygiene and Tropical Medicine, and the 'Hookworm Connection'. Studies in History and Philosophy of Science Part C : Studies in History and Philosophy of Biological and Biomedical Sciences, 31(3), 397–407. https://doi.org/https://doi.org/10.1016/S1369-8486(00)00018-2

Wilkinson, M. A. (2021). Authoritarian Liberalism and the Transformation of Modern Europe. Oxford University Press.

______. (2022). Authoritarian Liberalism and the Transformation of Modern Europe : Rejoinder. European Law Open, 1(1), 191–208. https://doi.org/10.1017/elo.2021.7

Wilkinson, S. (2010). On the Distinction Between Positive and Negative Eugenics. In M. Häyry, et al. (Eds.), Arguments and Analysis in Bioethics (pp. 115–128). Brill. https://doi.org/https://doi.org/10.1163/9789042028036_011

Williams, C. A. (1916). [Christianopolis. An Ideal State of the Seventeenth Century, Johann Valentin Andreae, Felix E. Held]. The School Review, 24(9), 710–710.

Williams, J. A. (2007). Turning to nature in Germany : hiking, nudism, and conservation, 1900-1940. Stanford, Calif, Stanford University Press.

Williams, R. (1989). Hegemony and Selective Tradition. In Language, Authority, and Criticism: Readings on the School Textbook (pp. 56–). Falmer Press.

Williamson, G. S. (2004). The Longing for Myth in Germany : Religion and Aesthetic Culture from Romanticism to Nietzsche. University of Chicago Press.

Wilson, T. (1896). The Swastika, the Earliest Known Symbol, and Its Migration With Observations on the Migration of Certain Industries in Prehistoric Times. U.S. National Museum.

Wirth, A. (1914). Rasse und Volk. M. Niemeyer.

Wodak, R. (2007). Pragmatics and Critical Discourse Analysis. Pragmatics & Cognition,

15(1), 203–225. https://doi.org/10.1075/pc.15.1.13wod
______. (2018). The Radical Right and Antisemitism. In J. Rydgren (Ed.), The Oxford Handbook of the Radical Right (pp. 61–85). Oxford University Press.
Woiak, J. (2007). Designing a brave new world : eugenics, politics, and fiction. Public Hist, 29(3), 105–129. https://doi.org/10.1525/tph.2007.29.3.105
Wokler, R. (1995). Anthropology and Conjectural History in the Enlightenment. In C. Fox, et al. (Eds.), Inventing Human Science : Eighteenth-Century Domains (pp. 31–52). Berkeley : University of California Press.
Wolff, L., & Cipolloni, M. (Eds.). (2007). The Anthropology of the Enlightenment. Stanford, Calif. : Stanford University Press.
Wolffram, H. (2009a). Crime, Clairvoyance and the Weimar Police. Journal of Contemporary History, 44(4), 581–601. https://doi.org/10.1177/0022009409339436
______. (2009b). The Stepchildren of Science : Psychical Research and Parapsychology in Germany, C. 1870-1939. Boston, UNITED STATES, BRILL.
Wolitski, R. J., Janssen, R. S., Holtgrave, D. R., & Peterson, J. L. (2004). Chapter 40 - The Public Health Response to the HIV Epidemic in the U.S. In G. P. Wormser (Ed.), AIDS and Other Manifestations of HIV Infection (Fourth Edition) (pp. 997–1012). Academic Press. https://doi.org/https://doi.org/10.1016/B978-012764051-8/50042-1
Woltmann, L. (1905). Die Germanen und die Renaissance in Italien. Thüringische Verlagsansatalt.
Wong, C. (2023). UK first to approve CRISPR treatment for diseases : what you need to know. Nature (London), 623(7988), 676–677. https://doi.org/10.1038/d41586-023-03590-6
Woolf, J. R. (2012). The Devil's Hoofs : The Medieval Roots of The Protocols of the Elders of Zion. In R. Landes & S. T. Katz (Eds.), The Paranoid Apocalypse : A Hundred-Year Retrospective on the Protocols of the Elders of Zion (pp. 49–55). New York : New York University Press.
Wu, K., Saha, R., Su, D., Krishna, V. D., Liu, J., Cheeran, M. C., & Wang, J. P. (2020). Magnetic-Nanosensor-Based Virus and Pathogen Detection Strategies before and during COVID-19. ACS Appl Nano Mater, 3(10), 9560–9580. https://doi.org/10.1021/acsanm.0c02048
Wu, P., & Sun, Z. (2024). Culture and political regimes : How have they influenced the effectiveness of COVID-19 response policy? Heliyon, 10(14), e34271. https://doi.org/10.1016/j.heliyon.2024.e34271
Wüst, W. (1942). Indogermanisches Bekenntnis. Ahnenerbe-Stiftung Verlag.
Wüst, W. (1942). Japan und Wir : Ansprache. Berlin, Ahnenerbe-Stiftung Verlag.
Xie, R., Adam, D. C., Hu, S., Cowling, B. J., Gascuel, O., Zhukova, A., & Dhanasekaran, V. (2024). Integrating Contact Tracing Data to Enhance Outbreak Phylodynamic Inference : A Deep Learning Approach. Molecular Biology and Evolution, 41(11). https://doi.org/10.1093/molbev/msae232
Xu, S., Sy, L. S., Hong, V., Farrington, P., Glenn, S. C., Ryan, D. S., Shirley, A. M., Lewin, B. J., Tseng, H.-F., Vazquez-Benitez, G., Glanz, J. M., Fireman, B., McClure, D. L., Hurley, L. P., Yu, O., Wernecke, M., Smith, N., Weintraub, E. S., & Qian, L. (2024). Mortality risk after COVID-19 vaccination : A self-controlled case series study. Vaccine, 42(7), 1731–1737. https://doi.org/10.1016/j.vaccine.2024.02.032
Xu, W., Zhang, S., Qin, H., & Yao, K. (2024). From bench to bedside: cutting-edge

applications of base editing and prime editing in precision medicine. Journal of Translational Medicine, 22(1), 1133. https://doi.org/10.1186/s12967-024-05957-3

Yao, J., Yan, H., Das, S., Klemic, J. F., Ellenbogen, J. C., & Lieber, C. M. (2014). Nanowire nanocomputer as a finite-state machine. Proc Natl Acad Sci U S A, 111(7), 2431–2435. https://doi.org/10.1073/pnas.1323818111

Yates, F. A. (1964). Giordano Bruno and the Hermetic Tradition. University of Chicago Press.

Yates, F. A. (1972). The Rosicrucian Enlightenment. London, UK, Routledge & Kegan Paul.

_____. (1999). The Occult Philosophy in the Elizabethan Age (Vol. VII of The Selected Works of Frances Yates). London ;, Routledge & K. Paul.

Yellen, J. A. (2013). The Spectre of Revolution : Reconsidering Japan's Decision to Surrender. The International History Review, 35(1), 205–226.

Yenne, B. (2010). Hitler's Master of the Dark Arts : Himmler's Black Knights and the Occult Origins of the SS. Quarto Publishing Group USA.

Yeowell, J. (1993). Odinism and Christianity Under the Third Reich. Odinic Rite.

Yewdell, J. W. (2021). Antigenic drift : Understanding COVID-19. Immunity, 54(12), 2681–2687. https://doi.org/10.1016/j.immuni.2021.11.016

Yourish, K., Lai, K. K. R., Ivory, D., & Smith, M. (2020). One-Third of All U.S. Coronavirus Deaths Are Nursing Home Residents or Workers.

Yourman, J. (1939). Propaganda Techniques Within Nazi Germany. Journal of Educational Sociology, 13(3), 148–163. https://doi.org/10.2307/2262307

Yu, M., Nie, S., Qiao, Y., & Ma, Y. (2023). Guillain-Barre syndrome following COVID-19 vaccines : A review of literature. Frontiers in Immunology, 14. https://doi.org/10.3389/fimmu.2023.1078197

Yuan, L., Xiong, Y., Zhang, Y., Gu, S., & Lei, Y. (2025). Epigenome editing based treatment : Progresses and challenges. Molecular Therapy. https://doi.org/10.1016/j.ymthe.2025.08.047

Zander, H. (2011). Rudolf Steiner : die Biografie. Piper.

Zeidman, L. A. (2020). Brain Science under the Swastika. Oxford University Press.

Zewail, A. H. (2010). Science in Diplomacy. Cell, 141(2), 204–207. https://doi.org/https://doi.org/10.1016/j.cell.2010.04.002

Zhang, H.-X., Zhang, Y., & Yin, H. (2019). Genome Editing with mRNA Encoding ZFN, TALEN, and Cas9. Molecular Therapy, 27(4), 735–746. https://doi.org/10.1016/j.ymthe.2019.01.014

Zheng, Y., Li, Y., Zhou, K., Li, T., VanDusen, N. J., & Hua, Y. (2024). Precise genome-editing in human diseases : mechanisms, strategies and applications. Signal Transduction and Targeted Therapy, 9(1), 47. https://doi.org/10.1038/s41392-024-01750-2

Zhirov, N. (1970). Atlantis : Atlantology : Basic Problems (D. Skvirsky, Trans.). Moscow, Progress Publishers.

Zillmann, P. (1898). Die Wald-Loge und Akademie für okkulte Wissenschaften. Neue Metaphysische Rundschau, 1, 226–228.

_____. (1912). Die Wald-Loge. Groß-Lichterfelde,

Zilsel, E. (1945). The Genesis of the Concept of Scientific Progress. Journal of the History of Ideas, 6(3), 325–349. https://doi.org/10.2307/2707296

Zirkle, C. (1946). The Early History of the Idea of the Inheritance of Acquired Characters and of Pangenesis. Transactions of the American Philosophical

Society, 35(2), 91–151. https://doi.org/10.2307/1005592

Zmarzlik, H.-G. (1972). Social Darwinism in Germany, Seen as a Historical Problem. In H. Holborn & R. Manheim (Eds.), Republic to Reich : The Making of the Nazi Revolution; Ten Essays (pp. 435–474). Pantheon Books.

_____. (2011). Social Darwinism in Germany, Seen as a Historical Problem. In M. R. Marrus (Ed.), The Nazi Holocaust (Vol. 2 The Origins of the Holocaust), Part One: Racism, the Occult, and Eugenics, pp. 3–42). De Gruyter.

www.ingramcontent.com/pod-product-compliance
Lightning Source LLC
LaVergne TN
LVHW081936200926
848484LV00009B/2444